Handbook of Natural Computing

Main Editor
Grzegorz Rozenberg

Editors
Thomas Bäck
Joost N. Kok

Handbook of
Natural Computing

Volume 4

With 734 Figures and 75 Tables

 Springer

Editors
Grzegorz Rozenberg
LIACS
Leiden University
Leiden, The Netherlands
and
Computer Science Department
University of Colorado
Boulder, USA

Joost N. Kok
LIACS
Leiden University
Leiden, The Netherlands

Thomas Bäck
LIACS
Leiden University
Leiden, The Netherlands

ISBN 978-3-540-92909-3   ISBN 978-3-540-92910-9 (eBook)
ISBN 978-3-540-92911-6 (print and electronic bundle)
DOI 10.1007/978-3-540-92910-9
Springer Heidelberg Dordrecht London New York

Library of Congress Control Number: 2010933716

Printed on acid-free paper

Springer is part of Springer Science+Business Media (www.springer.com)

Preface

Natural Computing is the field of research that investigates human-designed computing inspired by nature as well as computing taking place in nature, that is, it investigates models and computational techniques inspired by nature, and also it investigates, in terms of information processing, phenomena taking place in nature.

Examples of the first strand of research include neural computation inspired by the functioning of the brain; evolutionary computation inspired by Darwinian evolution of species; cellular automata inspired by intercellular communication; swarm intelligence inspired by the behavior of groups of organisms; artificial immune systems inspired by the natural immune system; artificial life systems inspired by the properties of natural life in general; membrane computing inspired by the compartmentalized ways in which cells process information; and amorphous computing inspired by morphogenesis. Other examples of natural-computing paradigms are quantum computing and molecular computing, where the goal is to replace traditional electronic hardware, by, for example, bioware in molecular computing. In quantum computing, one uses systems small enough to exploit quantum-mechanical phenomena to perform computations and to perform secure communications more efficiently than classical physics, and, hence, traditional hardware allows. In molecular computing, data are encoded as biomolecules and then tools of molecular biology are used to transform the data, thus performing computations.

The second strand of research, computation taking place in nature, is represented by investigations into, among others, the computational nature of self-assembly, which lies at the core of the nanosciences; the computational nature of developmental processes; the computational nature of biochemical reactions; the computational nature of bacterial communication; the computational nature of brain processes; and the systems biology approach to bionetworks where cellular processes are treated in terms of communication and interaction, and, hence, in terms of computation.

Research in natural computing is genuinely interdisciplinary and forms a bridge between the natural sciences and computer science. This bridge connects the two, both at the level of information technology and at the level of fundamental research. Because of its interdisciplinary character, research in natural computing covers a whole spectrum of research methodologies ranging from pure theoretical research, algorithms, and software applications to experimental laboratory research in biology, chemistry, and physics.

Computer Science and Natural Computing

A preponderance of research in natural computing is centered in computer science. The spectacular progress in Information and Communication Technology (ICT) is highly supported by the evolution of computer science, which designs and develops the instruments needed for this progress: computers, computer networks, software methodologies, etc. As ICT has such a tremendous impact on our everyday lives, so does computer science.

However, there is much more to computer science than ICT: it is the science of information processing and, as such, a fundamental science for other disciplines. On one hand, the only common denominator for research done in such diverse areas of computer science is investigating various aspects of information processing. On the other hand, the adoption of Information and Information Processing as central notions and thinking habit has been an important development in many disciplines, biology and physics being prime examples. For these scientific disciplines, computer science provides not only instruments but also a way of thinking.

We are now witnessing exciting interactions between computer science and the natural sciences. While the natural sciences are rapidly absorbing notions, techniques, and methodologies intrinsic to information processing, computer science is adapting and extending its traditional notion of computation, and computational techniques, to account for computation taking place in nature around us. Natural Computing is an important catalyst for this two-way interaction, and this handbook constitutes a significant record of this development.

The Structure of the Handbook

Natural Computing is both a well-established research field with a number of classical areas, and a very dynamic field with many more recent, novel research areas. The field is vast, and so it is quite usual that a researcher in a specific area does not have sufficient insight into other areas of Natural Computing. Also, because of its dynamic development and popularity, the field constantly attracts more and more scientists who either join the active research or actively follow research developments.

Therefore, the goal of this handbook is two-fold:

(i) to provide an authoritative reference for a significant and representative part of the research in Natural Computing, and
(ii) to provide a convenient gateway to Natural Computing for motivated newcomers to this field.

The implementation of this goal was a challenge because this field and its literature are vast — almost all of its research areas have an extensive scientific literature, including specialized journals, book series, and even handbooks. This implies that the coverage of the whole field in reasonable detail and within a reasonable number of pages/volumes is practically impossible.

Thus, we decided to divide the presented material into six areas. These areas are by no means disjoint, but this division is convenient for the purpose of providing a representative picture of the field — representative with respect to the covered research topics and with respect to a good balance between classical and emerging research trends.

Each area consists of individual chapters, each of which covers a specific research theme. They provide necessary technical details of the described research, however they are self-contained and of an expository character, which makes them accessible for a broader audience. They also provide a general perspective, which, together with given references, makes the chapters valuable entries into given research themes.

This handbook is a result of the joint effort of the handbook editors, area editors, chapter authors, and the Advisory Board. The choice of the six areas by the handbook editors in consultation with the Advisory Board, the expertise of the area editors in their respective areas, the choice by the area editors of well-known researchers as chapter writers, and the peer-review for individual chapters were all important factors in producing a representative and reliable picture of the field. Moreover, the facts that the Advisory Board consists of 68 eminent scientists from 20 countries and that there are 105 contributing authors from 21 countries provide genuine assurance for the reader that this handbook is an authoritative and up-to-date reference, with a high level of significance and accuracy.

Handbook Areas

The material presented in the handbook is organized into six areas: Cellular Automata, Neural Computation, Evolutionary Computation, Molecular Computation, Quantum Computation, and Broader Perspective.

Cellular Automata

Cellular automata are among the oldest models of computation, dating back over half a century. The first cellular automata studies by John von Neumann in the late 1940s were biologically motivated, related to self-replication in universal systems. Since then, cellular automata gained popularity in physics as discrete models of physical systems, in computer science as models of massively parallel computation, and in mathematics as discrete-time dynamical systems. Cellular automata are a natural choice to model real-world phenomena since they possess several fundamental properties of the physical world: they are massively parallel, homogeneous, and all interactions are local. Other important physical constraints such as reversibility and conservation laws can be added as needed, by properly choosing the local update rule. Computational universality is common in cellular automata, and even starkly simple automata are capable of performing arbitrary computation tasks. Because cellular automata have the advantage of parallelism while obeying natural constraints such as locality and uniformity, they provide a framework for investigating realistic computation in massively parallel systems. Computational power and the limitations of such systems are most naturally investigated by time- and space-constrained computations in cellular automata. In mathematics — in terms of symbolic dynamics — cellular automata are viewed as endomorphisms of the full shift, that is, transformations that are translation invariant and continuous in the product topology. Interesting questions on chaotic dynamics have been studied in this context.

Neural Computation

Artificial neural networks are computer programs, loosely modeled after the functioning of the human nervous system. There are neural networks that aim to gain understanding of biological neural systems, and those that solve problems in artificial intelligence without necessarily creating a model of a real biological system. The more biologically oriented neural networks model the real nervous system in increasing detail at all relevant levels of information processing: from synapses to neurons to interactions between modules of interconnected neurons. One of the major challenges is to build artificial brains. By reverse-engineering the mammalian brain in silicon, the aim is to better understand the functioning of the (human) brain through detailed simulations. Neural networks that are more application-oriented tend to drift further apart from real biological systems. They come in many different flavors, solving problems in regression analysis and time-series forecasting, classification, and pattern recognition, as well as clustering and compression. Good old multilayered perceptrons and self-organizing maps are still pertinent, but attention in research is shifting toward more recent developments, such as kernel methods (including support vector machines) and Bayesian techniques. Both approaches aim to incorporate domain knowledge in the learning process in order to improve prediction performance, e.g., through the construction of a proper kernel function or distance measure or the choice of an appropriate prior distribution over the parameters of the neural network. Considerable effort is devoted to making neural networks efficient so that large models can be learned from huge databases in a reasonable amount of time. Application areas include, among many others, system dynamics and control, finance, bioinformatics, and image analysis.

Evolutionary Computation

The field of evolutionary computation deals with algorithms gleaned from models of organic evolution. The general aim of evolutionary computation is to use the principles of nature's processes of natural selection and genotypic variation to derive computer algorithms for solving hard search and optimization tasks. A wide variety of instances of evolutionary algorithms have been derived during the past fifty years based on the initial algorithms, and we are now witnessing astounding successes in the application of these algorithms: their fundamental understanding in terms of theoretical results; understanding algorithmic principles of their construction; combination with other techniques; and understanding their working principles in terms of organic evolution. The key algorithmic variations (such as genetic algorithms, evolution strategies, evolutionary programming, and genetic programming) have undergone significant developments over recent decades, and have also resulted in very powerful variations and recombinations of these algorithms. Today, there is a sound understanding of how all of these algorithms are instances of the generic concept of an evolutionary search approach. Hence the generic term "evolutionary algorithm" is nowadays being used to describe the generic algorithm, and the term "evolutionary computation" is used for the field as a whole. Thus, we have observed over the past fifty years how the field has integrated the various independently developed initial algorithms into one common principle. Moreover, modern evolutionary algorithms benefit from their ability to adapt and self-adapt their strategy parameters (such as mutation rates, step sizes, and search distributions) to the needs of the task at hand. In this way, they are robust and flexible metaheuristics for problem-solving even without requiring too much special expertise from their users. The feature of self-adaptation illustrates the ability of evolutionary principles to work on different levels at the same time, and therefore provides a nice demonstration of the universality of the evolutionary principle for search and optimization tasks. The widespread use of evolutionary computation reflects these capabilities.

Molecular Computation

Molecular computing is an emergent interdisciplinary field concerned with programming molecules so that they perform a desired computation, or fabricate a desired object, or control the functioning of a specific molecular system. The central idea behind molecular computing is that data can be encoded as (bio)molecules, e.g., DNA strands, and tools of molecular science can be used to transform these data. In a nutshell, a molecular program is just a collection of molecules which, when placed in a suitable substrate, will perform a specific function (execute the program that this collection represents). The birth of molecular computing is often associated with the 1994 breakthrough experiment by Leonard Adleman, who solved a small instance of a hard computational problem solely by manipulating DNA strands in test tubes. Although initially the main effort of the area was focused on trying to obtain a breakthrough in the complexity of solving hard computational problems, this field has evolved enormously since then. Among the most significant achievements of molecular computing have been contributions to understanding some of the fundamental issues of the nanosciences. One notable example among them is the contribution to the understanding of self-assembly, a central concept of the nanosciences. The techniques of molecular programming were successfully applied in experimentally constructing all kinds of molecular-scale objects or devices with prescribed functionalities. Well-known examples here are self-assembly of Sierpinski triangles, cubes, octahedra, DNA-based logic circuits, DNA "walkers" that move along a track, and autonomous molecular motors. A complementary approach to understanding bioinformation and computation is through studying the information-processing capabilities of cellular organisms. Indeed, cells and nature "compute" by "reading" and "rewriting" DNA through processes that modify DNA (or RNA) sequences. Research into the computational abilities of cellular organisms has the potential to uncover the laws governing biological information, and to enable us to harness the computational power of cells.

Quantum Computation

Quantum computing has been discussed for almost thirty years. The theory of quantum computing and quantum information processing is simply the theory of information processing with a classical notion of information replaced by its quantum counterpart. Research in quantum computing is concerned with understanding the fundamentals of information processing on the level of physical systems that realize/implement the information. In fact, quantum computing can be seen as a quest to understand the fundamental limits of information processing set by nature itself. The mathematical description of quantum information is more complicated than that of classical information — it involves the structure of Hilbert spaces. When describing the structure behind known quantum algorithms, this reduces to linear algebra over complex numbers. The history of quantum algorithms spans the last fifteen years, and some of these algorithms are extremely interesting, and even groundbreaking — the most remarkable are Shor's factorization in polynomial time and Grover's search algorithm. The nature of quantum information has also led to the invention of novel cryptosystems, whose security is not based on the complexity of computing functions, but rather on the physical properties of quantum information. Quantum computing is now a well-established discipline, however implementation of a large-scale quantum computer continues to be extremely challenging, even though quantum information processing primitives, including those allowing secure cryptography, have been demonstrated to be practically realizable.

Broader Perspective

In contrast to the first five areas focusing on more-established themes of natural computing, this area encompasses a perspective that is broader in several ways. First, the reader will find here treatments of certain well-established and specific techniques inspired by nature (e.g., simulated annealing) not covered in the other five areas. Second, the reader will also find application-centered chapters (such as natural computing in finance), each covering, in one chapter, a collection of natural computing methods, thus capturing the impact of natural computing as a whole in various fields of science or industry. Third, some chapters are full treatments of several established research fields (such as artificial life, computational systems biology, evolvable hardware, and artificial immune systems), presenting alternative perspectives and cutting across some of the other areas of the handbook, while introducing much new material. Other elements of this area are fresh, emerging, and novel techniques or perspectives (such as collision-based computing, nonclassical computation), representing the leading edge of theories and technologies that are shaping possible futures for both natural computing and computing in general. The contents of this area naturally cluster into two kinds (sections), determined by the essential nature of the techniques involved. These are "Nature-Inspired Algorithms" and "Alternative Models of Computation". In the first section, "Nature-Inspired Algorithms", the focus is on algorithms inspired by natural processes realized either through software or hardware or both, as additions to the armory of existing tools we have for dealing with well-known practical problems. In this section, we therefore find application-centered chapters, as well as chapters focusing on particular techniques, not otherwise dealt with in other areas of the handbook, which have clear and proven applicability. In the second section, "Alternative Models of Computation", the emphasis changes, moving away from specific applications or application areas, toward more far-reaching ideas. These range from developing computational approaches and "computational thinking" as fundamental tools for the new science of systems biology to ideas that take inspiration from nature as a platform for suggesting entirely novel possibilities of computing.

Handbook Chapters

In the remainder of this preface we will briefly describe the contents of the individual chapters. These chapter descriptions are grouped according to the handbook areas where they belong and given in the order that they appear in the handbook. This section provides the reader with a better insight into the contents, allowing one to design a personal roadmap for using this handbook.

Cellular Automata

This area is covered by nine chapters.

The first chapter, "Basic Concepts of Cellular Automata", by Jarkko J. Kari, reviews some classical results from the theory of cellular automata, relations between various concepts of injectivity and surjectivity, and some basic dynamical system concepts related to chaos in cellular automata. The classical results discussed include the celebrated Garden-of-Eden and Curtis–Hedlund–Lyndon theorems, as well as the balance property of surjective cellular automata. All these theorems date back to the 1960s. The results are provided together with examples that illustrate proof ideas. Different variants of sensitivity to initial conditions and mixing properties are introduced and related to each other. Also undecidability results concerning cellular automata are briefly discussed.

A popular mathematical approach is to view cellular automata as dynamical systems in the context of symbolic dynamics. Several interesting results in this area were reported as early as 1969 in the seminal paper by G.A. Hedlund, and still today this research direction is among the most fruitful sources of theoretical problems and new results. The chapter "Cellular Automata Dynamical Systems", by Alberto Dennunzio, Enrico Formenti, and Petr Kůrka, reviews some recent developments in this field. Recent research directions considered here include subshifts attractors and signal subshifts, particle weight functions, and the slicing construction. The first two concern one-dimensional cellular automata and give precise descriptions of the limit behavior of large classes of automata. The third one allows one to view two-dimensional cellular automata as one-dimensional systems. In this way combinatorial complexity is decreased and new results can be proved.

Programming cellular automata for particular tasks requires special techniques. The chapter "Algorithmic Tools on Cellular Automata", by Marianne Delorme and Jacques Mazoyer, covers classical algorithmic tools based on signals. Linear signals as well as signals of nonlinear slope are discussed, and basic transformations of signals are addressed. The chapter provides results on using signals to construct functions in cellular automata and to implement arithmetic operations on segments. The methods of folding the space–time, freezing, and clipping are also introduced.

The time-complexity advantage gained from parallelism under the locality and uniformity constraints of cellular automata can be precisely analyzed in terms of language recognition. The chapter "Language Recognition by Cellular Automata", by Véronique Terrier, presents results and questions about cellular automata complexity classes and their relationships to other models of computations. Attention is mainly directed to real-time and linear-time complexity classes, because significant benefits over sequential computation may be obtained at these low time complexities. Both parallel and sequential input modes are considered. Separate complexity classes are given also for cellular automata with one-way communications and two-way communications.

The chapter "Computations on Cellular Automata", by Jacques Mazoyer and Jean-Baptiste Yunès, continues with the topic of algorithmic techniques in cellular automata. This chapter uses the basic tools, such as signals and grids, to build natural implementations of common algorithms in cellular automata. Examples of implementations include real-time multiplication of integers and the prime number sieve. Both parallel and sequential input and output modes are discussed, as well as composition of functions and recursion.

The chapter "Universalities in Cellular Automata", by Nicolas Ollinger, is concerned with computational universalities. Concepts of universality include Turing universality (the ability to compute any recursive function) and intrinsic universality (the ability to simulate any other cellular automaton). Simulations of Boolean circuits in the two-dimensional case are explained in detail in order to achieve both kinds of universality. The more difficult one-dimensional case is also discussed, and seminal universal cellular automata and encoding techniques are presented in both dimensions. A detailed chronology of important papers on universalities in cellular automata is also provided.

A cellular automaton is reversible if every configuration has only one previous configuration, and hence its evolution process can be traced backward uniquely. This naturally corresponds to the fundamental time-reversibility of the microscopic laws of physics. The chapter "Reversible Cellular Automata", by Kenichi Morita, discusses how reversible cellular automata are defined, as well as their properties, how they are designed, and their computing abilities. After providing the definitions, the chapter surveys basic properties of injectivity and surjectivity. Three design methods of reversible cellular automata are provided: block rules, partitioned, and second-order cellular automata. Then the computational power of reversible cellular automata is discussed. In particular, simulation methods of irreversible cellular automata, reversible Turing machines, and some other universal systems are given to clarify universality of reversible cellular automata. In spite of the strong constraint of reversibility, it is shown that reversible cellular automata possess rich information processing capabilities, and even very simple ones are computationally universal.

A conservation law in a cellular automaton is a statement of the invariance of a local and additive energy-like quantity. The chapter "Conservation Laws in Cellular Automata", by Siamak Taati, reviews the basic theory of conservation laws. A general mathematical framework for formulating conservation laws in cellular automata is presented and several characterizations are summarized. Computational problems regarding conservation laws (verification and existence problems) are discussed. Microscopic explanations of the dynamics of the conserved quantities in terms of flows and particle flows are explored. The related concept of dissipating energy-like quantities is also discussed.

The chapter "Cellular Automata and Lattice Boltzmann Modeling of Physical Systems", by Bastien Chopard, considers the use of cellular automata and related lattice Boltzmann methods as a natural modeling framework to describe and study many physical systems composed of interacting components. The theoretical basis of the approach is introduced and its potential is illustrated for several applications in physics, biophysics, environmental science, traffic models, and multiscale modeling. The success of the technique can be explained by the close relationship between these methods and a mesoscopic abstraction of many natural phenomena.

Neural Computation

This area is covered by ten chapters.

Spiking neural networks are inspired by recent advances in neuroscience. In contrast to classical neural network models, they take into account not just the neuron's firing rate, but also the time moment of spike firing. The chapter "Computing with Spiking Neuron Networks", by Hélène Paugam-Moisy and Sander Bohte, gives an overview of existing approaches to modeling spiking neural neurons and synaptic plasticity, and discusses their computational power and the challenge of deriving efficient learning procedures.

Image quality assessment aims to provide computational models to predict the perceptual quality of images. The chapter "Image Quality Assessment — A Multiscale Geometric Analysis-Based Framework and Examples", by Xinbo Gao, Wen Lu, Dacheng Tao, and Xuelong Li, introduces the fundamentals and describes the state of the art in image quality assessment. It further proposes a new model, which mimics the human visual system by incorporating concepts such as multiscale analysis, contrast sensitivity, and just-noticeable differences. Empirical results clearly demonstrate that this model resembles subjective perception values and reflects the visual quality of images.

Neurofuzzy networks have the important advantage that they are easy to interpret. When applied to control problems, insight about the process characteristics at different operating regions can be easily obtained. Furthermore, nonlinear model predictive controllers can be developed as a nonlinear combination of several local linear model predictive controllers that have analytical solutions. Through several applications, the chapter "Nonlinear Process Modelling and Control Using Neurofuzzy Networks", by Jie Zhang, demonstrates that neurofuzzy networks are very effective in the modeling and control of nonlinear processes.

Similar to principal component and factor analysis, independent component analysis is a computational method for separating a multivariate signal into additive subcomponents. Independent component analysis is more powerful: the latent variables corresponding to the subcomponents need not be Gaussian and the basis vectors are typically nonorthogonal. The chapter "Independent Component Analysis", by Seungjin Choi, explains the theoretical foundations and describes various algorithms based on those principles.

Neural networks has become an important method for modeling and forecasting time series. The chapter "Neural Networks for Time-Series Forecasting", by G. Peter Zhang, reviews some recent developments (including seasonal time-series modeling, multiperiod forecasting, and ensemble methods), explains when and why they are to be preferred over traditional forecasting models, and also discusses several practical data and modeling issues.

Support vector machines have been extensively studied and applied in many domains within the last decade. Through the so-called kernel trick, support vector machines can efficiently learn nonlinear functions. By maximizing the margin, they implement the principle of structural risk minimization, which typically leads to high generalization performance. The chapter "SVM Tutorial — Classification, Regression and Ranking", by Hwanjo Yu and Sungchul Kim, describes these underlying principles and discusses support vector machines for different learning tasks: classification, regression, and ranking.

It is well known that single-hidden-layer feedforward networks can approximate any continuous target function. This still holds when the hidden nodes are automatically and randomly generated, independent of the training data. This observation opened up many possibilities for easy construction of a broad class of single-hidden-layer neural networks. The chapter "Fast Construction of Single-Hidden-Layer Feedforward Networks", by Kang Li, Guang-Bin Huang, and Shuzhi Sam Ge, discusses new ideas that yield a more compact network architecture and reduce the overall computational complexity.

Many recent experimental studies demonstrate the remarkable efficiency of biological neural systems to encode, process, and learn from information. To better understand the experimentally observed phenomena, theoreticians are developing new mathematical approaches and tools to model biological neural networks. The chapter "Modeling Biological Neural Networks", by Joaquin J. Torres and Pablo Varona, reviews some of the most popular models of neurons and neural networks. These not only help to understand how living systems perform information processing, but may also lead to novel bioinspired paradigms of artificial intelligence and robotics.

The size and complexity of biological data, such as DNA/RNA sequences and protein sequences and structures, makes them suitable for advanced computational tools, such as neural networks. Computational analysis of such databases aims at exposing hidden information that provides insights that help in understanding the underlying biological principles. The chapter "Neural Networks in Bioinformatics", by Ke Chen and Lukasz A. Kurgan, focuses on proteins. In particular it discusses prediction of protein secondary structure, solvent accessibility, and binding residues.

Self-organizing maps is a prime example of an artificial neural network model that both relates to the actual (topological) organization within the mammalian brain and at the same time has many practical applications. Self-organizing maps go back to the seminal work of Teuvo Kohonen. The chapter "Self-organizing Maps", by Marc M. Van Hulle, describes the state of the art with a special emphasis on learning algorithms that aim to optimize a predefined criterion.

Evolutionary Computation

This area is covered by thirteen chapters.

The first chapter, "Generalized Evolutionary Algorithms", by Kenneth De Jong, describes the general concept of evolutionary algorithms. As a generic introduction to the field, this chapter facilitates an understanding of specific instances of evolutionary algorithms as instantiations of a generic evolutionary algorithm. For the instantiations, certain choices need to be made, such as representation, variation operators, and the selection operator, which then yield particular instances of evolutionary algorithms, such as genetic algorithms and evolution strategies, to name just a few.

The chapter "Genetic Algorithms — A Survey of Models and Methods", by Darrell Whitley and Andrew M. Sutton, introduces and discusses (including criticism) the standard genetic algorithm based on the classical binary representation of solution candidates and a theoretical interpretation based on the so-called schema theorem. Variations of genetic algorithms with respect to solution representations, mutation operators, recombination operators, and selection mechanisms are also explained and discussed, as well as theoretical models of genetic algorithms based on infinite and finite population size assumptions and Markov chain theory concepts. The authors also critically investigate genetic algorithms from the perspective of identifying their limitations and the differences between theory and practice when working with genetic algorithms. To illustrate this further, the authors also give a practical example of the application of genetic algorithms to resource scheduling problems.

The chapter "Evolutionary Strategies", by Günter Rudolph, describes a class of evolutionary algorithms which have often been associated with numerical function optimization and continuous variables, but can also be applied to binary and integer domains. Variations of evolutionary strategies, such as the $(\mu+\lambda)$-strategy and the (μ,λ)-strategy, are introduced and discussed within a common algorithmic framework. The fundamental idea of self-adaptation of strategy parameters (variances and covariances of the multivariate normal distribution used for mutation) is introduced and explained in detail, since this is a key differentiating property of evolutionary strategies.

The chapter "Evolutionary Programming", by Gary B. Fogel, discusses a historical branch of evolutionary computation. It gives a historical perspective on evolutionary programming by describing some of the original experiments using evolutionary programming to evolve finite state machines to serve as sequence predictors. Starting from this canonical evolutionary programming approach, the chapter also presents extensions of evolutionary programming into continuous domains, where an attempt towards self-adaptation of mutation step sizes has been introduced which is similar to the one considered in evolutionary strategies. Finally, an overview of some recent applications of evolutionary programming is given.

The chapter "Genetic Programming — Introduction, Applications, Theory and Open Issues", by Leonardo Vanneschi and Riccardo Poli, describes a branch of evolutionary algorithms derived by extending genetic algorithms to allow exploration of the space of computer programs. To make evolutionary search in the domain of computer programs possible, genetic programming is based on LISP S-expression represented by syntax trees, so that genetic programming extends evolutionary algorithms to tree-based representations. The chapter gives an overview of the corresponding representation, search operators, and technical details of genetic programming, as well as existing applications to real-world problems. In addition, it discusses theoretical approaches toward analyzing genetic programming, some of the open issues, as well as research trends in the field.

The subsequent three chapters are related to the theoretical analysis of evolutionary algorithms, giving a broad overview of the state of the art in our theoretical understanding. These chapters demonstrate that there is a sound theoretical understanding of capabilities and limitations of evolutionary algorithms. The approaches can be roughly split into convergence velocity or progress analysis, computational complexity investigations, and global convergence results.

The convergence velocity viewpoint is represented in the chapter "The Dynamical Systems Approach — Progress Measures and Convergence Properties", by Silja Meyer-Nieberg and Hans-Georg Beyer. It demonstrates how the dynamical systems approach can be used to analyze the behavior of evolutionary algorithms quantitatively with respect to their progress rate. It also provides a complete overview of results in the continuous domain, i.e., for all types of evolution strategies on certain objective functions (such as sphere, ridge, etc.). The chapter presents results for undisturbed as well as for noisy variants of these objective functions, and extends the approach to dynamical objective functions where the goal turns into optimum tracking. All results are presented by means of comparative tables, so the reader gets a complete overview of the key findings at a glance.

The chapter "Computational Complexity of Evolutionary Algorithms", by Thomas Jansen, deals with the question of optimization time (i.e., the first point in time during the run of an evolutionary algorithm when the global optimum is sampled) and an investigation of upper bounds, lower bounds, and the average time needed to hit the optimum. This chapter presents specific results for certain classes of objective functions, most of them defined over binary search spaces, as well as fundamental limitations of evolutionary search and related results on the "no free lunch" theorem and black box complexity. The chapter also discusses the corresponding techniques for analyses, such as drift analysis and the expected multiplicative distance decrease.

Concluding the set of theoretical chapters, the chapter "Stochastic Convergence", by Günter Rudolph, addresses theoretical results about the properties of evolutionary algorithms concerned with finding a globally optimal solution in the asymptotic limit. Such results exist for certain variants of evolutionary algorithms and under certain assumptions, and this chapter summarizes the existing results and integrates them into a common framework. This type of analysis is essential in qualifying evolutionary algorithms as global search algorithms and for understanding the algorithmic conditions for global convergence.

The remaining chapters in the area of evolutionary computation report some of the major current trends.

To start with, the chapter "Evolutionary Multiobjective Optimization", by Eckart Zitzler, focuses on the application of evolutionary algorithms to tasks that are characterized by multiple, conflicting objective functions. In this case, decision-making becomes a task of identifying good compromises between the conflicting criteria. This chapter introduces the concept and a variety of state-of-the-art algorithmic concepts to use evolutionary algorithms for approximating the so-called Pareto front of solutions which cannot be improved in one objective without compromising another. This contribution presents all of the required formal concepts, examples, and the algorithmic variations introduced into evolutionary computation to handle such types of problems and to generate good approximations of the Pareto front.

The term "memetic algorithms" is used to characterize hybridizations between evolutionary algorithms and more classical, local search methods (and agent-based systems). This is a general concept of broad scope, and in order to illustrate and characterize all possible instantiations, the chapter "Memetic Algorithms", by Natalio Krasnogor, presents an algorithmic engineering approach which allows one to describe these algorithms as instances of generic patterns. In addition to explaining some of the application areas, the chapter presents some theoretical remarks, various different ways to define memetic algorithms, and also an outlook into the future.

The chapter "Genetics-Based Machine Learning", by Tim Kovacs, extends the idea of evolutionary optimization to algorithmic concepts in machine learning and data mining, involving applications such as learning classifier systems, evolving neural networks, and genetic fuzzy systems, to mention just a few. Here, the application task is typically a data classification, data prediction, or nonlinear regression task — and the quality of solution candidates is evaluated by means of some model quality measure. The chapter covers a wide range of techniques for applying evolutionary computation to machine learning tasks, by interpreting them as optimization problems.

The chapter "Coevolutionary Principles", by Elena Popovici, Anthony Bucci, R. Paul Wiegand, and Edwin D. de Jong, deals with a concept modeled after biological evolution in which an explicit fitness function is not available, but solutions are evaluated by running them against each other. A solution is evaluated in the context of the other solutions, in the actual population or in another. Therefore, these algorithms develop their own dynamics, because the point of comparison is not stable, but coevolving with the actual population. The chapter provides a fundamental understanding of coevolutionary principles and highlights theoretical concepts, algorithms, and applications.

Finally, the chapter "Niching in Evolutionary Algorithms", by Ofer M. Shir, describes the biological principle of niching in nature as a concept for using a single population to find, occupy, and keep multiple local minima in a population. The motivation for this approach is to find alternative solutions within a single population and run of evolutionary algorithms, and this chapter discusses approaches for niching, and the application in the context of genetic algorithms as well as evolutionary strategies.

Molecular Computation

This area is covered by eight chapters.

The chapter "DNA Computing — Foundations and Implications", by Lila Kari, Shinnosuke Seki, and Petr Sosík, has a dual purpose. The first part outlines basic molecular biology notions necessary for understanding DNA computing, recounts the first experimental demonstration of DNA computing by Leonard Adleman in 1994, and recaps the 2001 milestone wet laboratory experiment that solved a 20-variable instance of 3-SAT and thus first demonstrated the potential of DNA computing to outperform the computational ability of an unaided human. The second part describes how the properties of DNA-based information, and in particular the Watson–Crick complementarity of DNA single strands, have influenced areas of theoretical computer science such as formal language theory, coding theory, automata theory, and combinatorics on words. More precisely, it explores several notions and results in formal language theory and coding theory that arose from the problem of the design of optimal encodings for DNA computing experiments (hairpin-free languages, bond-free languages), and more generally from the way information is encoded on DNA strands (sticker systems, Watson–Crick automata). Lastly, it describes the influence that properties of DNA-based information have had on research in combinatorics on words, by presenting several natural generalizations of classical concepts (pseudopalindromes, pseudoperiodicity, Watson–Crick conjugate and commutative words, involutively bordered words, pseudoknot bordered words), and outlining natural extensions in this context of two of the most fundamental results in combinatorics of words, namely the Fine and Wilf theorem and the Lyndon–Schützenberger result.

The chapter "Molecular Computing Machineries — Computing Models and Wet Implementations", by Masami Hagiya, Satoshi Kobayashi, Ken Komiya, Fumiaki Tanaka, and Takashi Yokomori, explores novel computing devices inspired by the biochemical properties of biomolecules. The theoretical results section describes a variety of molecular computing models for finite automata, as well as molecular computing models for Turing machines based on formal grammars, equality sets, Post systems, and logical formulae. It then presents molecular computing models that use structured molecules such as hairpins and tree structures. The section on wet implementations of molecular computing models, related issues, and applications includes: an enzyme-based DNA automaton and its applications to drug delivery, logic gates and circuits using DNAzymes and DNA tiles, reaction graphs for representing various dynamics of DNA assembly pathways, DNA whiplash machines implementing finite automata, and a hairpin-based implementation of a SAT engine for solving the 3-SAT problem.

The chapter "DNA Computing by Splicing and by Insertion–Deletion", by Gheorghe Păun, is devoted to two of the most developed computing models inspired by DNA biochemistry: computing by splicing, and computing by insertion and deletion. DNA computing by splicing was defined by Tom Head already in 1987 and is based on the so-called splicing operation. The splicing operation models the recombination of DNA molecules that results from cutting them with restriction enzymes and then pasting DNA molecules with compatible ends by ligase enzymes. This chapter explores the computational power of the splicing operation showing that, for example, extended splicing systems starting from a finite language and using finitely many splicing rules can generate only the family of regular languages, while extended splicing systems starting from a finite language and using a regular set of rules can generate all recursively enumerable languages. Ways in which to avoid the impractical notion of a regular infinite set of rules, while maintaining the maximum computational power, are presented. They include using multisets and adding restrictions on the use of rules such as permitting contexts, forbidding contexts, programmed splicing systems, target languages, and double splicing. The second model presented, the insertion–deletion system, is based on a finite set of axioms and a finite set of contextual insertion rules and contextual deletion rules. Computational power results described here include the fact that insertion–deletion systems with context-free insertion rules of words of length at most one and context-free deletion rules of words of unbounded length can generate only regular languages. In contrast, for example, the family of insertion–deletion systems where the insertion contexts, deletion contexts, and the words to be inserted/deleted are all of length at most one, equals the family of recursively enumerable languages.

The chapter "Bacterial Computing and Molecular Communication", by Yasubumi Sakakibara and Satoshi Hiyama, investigates attempts to create autonomous cell-based Turing machines, as well as novel communication paradigms that use molecules as communication media. The first part reports experimental research on constructing *in vivo* logic circuits as well as efforts towards building *in vitro* and *in vivo* automata in the framework of DNA computing. Also, a novel framework is presented to develop a programmable and autonomous *in vivo* computer in a bacterium. The first experiment in this direction uses DNA circular strands (plasmids) together with the cell's protein-synthesis mechanism to execute a finite state automaton in *E. coli*. Molecular communication is a new communication paradigm that proposes the use of molecules as the information medium, instead of the traditional electromagnetic waves. Other distinctive features of molecular communication include its stochastic nature, its low energy consumption, the use of an aqueous transmission medium, and its high compatibility with biological systems. A molecular communication system starts with a sender (e.g., a genetically modified or an artificial cell) that generates molecules, encodes information onto the molecules (called information molecules), and emits the information molecules into a propagation environment (e.g., aqueous solution within and between cells). A molecular propagation system (e.g., lipid bilayer vesicles encapsulating the information molecules) actively transports the information molecules to an appropriate receiver. A receiver (e.g., a genetically modified or an artificial cell) selectively receives the transported information molecules, and biochemically reacts to the received information molecules, thus "decoding" the information. The chapter describes detailed examples of molecular communication system designs, experimental results, and research trends.

The chapter "Computational Nature of Gene Assembly in Ciliates", by Robert Brijder, Mark Daley, Tero Harju, Nataša Jonoska, Ion Petre, and Grzegorz Rozenberg, reviews several approaches and results in the computational study of gene assembly in ciliates. Ciliated protozoa contain two functionally different types of nuclei, the macronucleus and the micronucleus. The macronucleus contains the functional genes, while the genes of the micronucleus are not functional due to the presence of many interspersing noncoding DNA segments. In addition, in some ciliates, the coding segments of the genes are present in a permuted order compared to their order in the functional macronuclear genes. During the sexual process of conjugation, when two ciliates exchange genetic micronuclear information and form two new micronuclei, each of the ciliates has to "decrypt" the information contained in its new micronucleus to form its new functional macronucleus. This process is called gene assembly and involves deleting the noncoding DNA segments, as well as rearranging the coding segments in the correct order. The chapter describes two models of gene assembly, the intermolecular model based on the operations of circular insertion and deletion, and the intramolecular model based on the three operations of "loop, direct-repeat excision", "hairpin, inverted-repeat excision", and "double-loop, alternating repeat excision". A discussion follows of the mathematical properties of these models, such as the Turing machine computational power of contextual circular insertions and deletions, and properties of the gene assembly process called invariants, which hold independently of the molecular model and assembling strategy. Finally, the template-based recombination model is described, offering a plausible hypothesis (supported already by some experimental data) about the "bioware" that implements the gene assembly.

The chapter "DNA Memory", by Masanori Arita, Masami Hagiya, Masahiro Takinoue, and Fumiaki Tanaka, summarizes the efforts that have been made towards realizing Eric Baum's dream of building a DNA memory with a storage capacity vastly larger than the brain. The chapter first describes the research into strategies for DNA sequence design, i.e., for finding DNA sequences that satisfy DNA computing constraints such as uniform melting temperature, avoidance of undesirable Watson–Crick bonding between sequences, preventing secondary structures, avoidance of base repeats, and absence of forbidden sequences. Various implementations of memory operations, such as access, read, and write, are described. For example, the "access" to a memory word in Baum's associative memory model, where a memory word consists of a single-stranded portion representing the address and a double-stranded portion representing the data, can be implemented by using the Watson–Crick complement of the address fixed to a solid support. In the Nested Primer Molecular Memory, where the double-stranded data is flanked on both sides by address sequences, the data can be retrieved by Polymerase Chain Reaction (PCR) using the addresses as primer pairs. In the multiple hairpins DNA memory, the address is a catenation of hairpins and the data can be accessed only if the hairpins are opened in the correct order by a process called DNA branch migration. After describing implementations of writable and erasable hairpin memories either in solution or immobilized on surfaces, the topic of *in vivo* DNA memory is explored. As an example, the chapter describes how representing the digit 0 by regular codons, and the digit 1 by wobbled codons, was used to encode a word into an essential gene of *Bacillus subtilis*.

The chapter "Engineering Natural Computation by Autonomous DNA-Based Biomolecular Devices", by John H. Reif and Thomas H. LaBean, overviews DNA-based biomolecular devices that are autonomous (execute steps with no external control after starting) and programmable (the tasks executed can be modified without entirely redesigning the DNA nanostructures). Special attention is given to DNA tiles, roughly square-shaped DNA nanostructures that have four "sticky-ends" (DNA single strands) that can specifically bind them to other tiles via Watson–Crick complementarity, and thus lead to the self-assembly of larger and more complex structures. Such tiles have been used to execute sequential Boolean computation via linear DNA self-assembly or to obtain patterned 2D DNA lattices and Sierpinski triangles. Issues such as error correction and self-repair of DNA tiling are also addressed. Other described methods include the implementation of a DNA-based finite automaton via disassembly of a double-stranded DNA nanostructure effected by an enzyme, and the technique of whiplash PCR. Whiplash PCR is a method that can achieve state transitions by encoding both transitions and the current state of the computation on the same DNA single strand: The free end of the strand (encoding the current state) sticks to the appropriate transition rule on the strand forming a hairpin, is then extended by PCR to a new state, and finally is detached from the strand, this time with the new state encoded at its end. The technique of DNA origami is also described, whereby a scaffold strand (a long single DNA strand, such as from the sequence of a virus) together with many specially designed staple strands (short single DNA strands) self-assemble by folding the scaffold strand — with the aid of the staples — in a raster pattern that can create given arbitrary planar DNA nanostructures. DNA-based molecular machines are then described such as autonomous DNA walkers and programmable DNA nanobots (programmable autonomous DNA walker devices). A restriction-enzyme-based DNA walker consists of a DNA helix with two sticky-ends ("feet") that moves stepwise along a "road" (a DNA nanostructure with protruding "steps", i.e., single DNA strands).

The chapter "Membrane Computing", by Gheorghe Păun, describes theoretical results and applications of membrane computing, a branch of natural computing inspired by the architecture and functioning of living cells, as well as from the organization of cells in tissues, organs, or other higher-order structures. The cell is a hierarchical structure of compartments, defined by membranes, that selectively communicate with each other. The computing model that abstracts this structure is a membrane system (or P system, from the name of its inventor, Gheorghe Păun) whose main components are: the membrane structure, the multisets of objects placed in the compartments enveloped by the membranes, and the rules for processing the objects and the membranes. The rules are used to modify the objects in the compartments, to transport objects from one compartment to another, to dissolve membranes, and to create new membranes. The rules in each region of a P system are used in a maximally parallel manner, nondeterministically choosing the applicable rules and the objects to which they apply. A computation consists in repeatedly applying rules to an initial configuration of the P system, until no rule can be applied anymore, in which case the objects in a priori specified regions are considered the output of the computation. Several variants of P systems are described, including P systems with symport/antiport rules, P systems with active membranes, splicing P systems, P systems with objects on membranes, tissue-like P systems, and spiking neural P systems. Many classes of P systems are able to simulate Turing machines, hence they are computationally universal. For example, P systems with symport/antiport rules using only three objects and three membranes are computationally universal. In addition, several types of P systems have been used to solve NP-complete problems in polynomial time, by a space–time trade-off. Applications of P systems include, among others, modeling in biology, computer graphics, linguistics, economics, and cryptography.

Quantum Computation

This area is covered by six chapters.

The chapter "Mathematics for Quantum Information Processing", by Mika Hirvensalo, contains the standard Hilbert space formulation of finite-level quantum systems. This is the language and notational system allowing us to speak, describe, and make predictions about the objects of quantum physics. The chapter introduces the notion of quantum states as unit-trace, self-adjoint, positive mappings, and the vector state formalism is presented as a special case. The physical observables are introduced as complete collections of mutually orthogonal projections, and then it is discussed how this leads to the traditional representation of observables as self-adjoint mappings. The minimal interpretation, which is the postulate connecting the mathematical objects to the physical world, is presented. The treatment of compound quantum systems is based mostly on operative grounds. To provide enough tools for considering the dynamics needed in quantum computing, the formalism of treating state transformations as completely positive mappings is also presented. The chapter concludes by explaining how quantum versions of finite automata, Turing machines, and Boolean circuits fit into the Hilbert space formalism.

The chapter "Bell's Inequalities — Foundations and Quantum Communication", by Časlav Brukner and Marek Żukowski, is concerned with the nature of quantum mechanics. It presents the evidence that excludes two types of hypothetical deterministic theories: neither a nonlocal nor a noncontextual theory can explain quantum mechanics. This helps to build a true picture of quantum mechanics, and is therefore essential from the philosophical point of view. The Bell inequalities show that nonlocal deterministic theories cannot explain the quantum mechanism, and the Kochen–Specker theorem shows that noncontextual theories are not possible as underlying theories either. The traditional Bell theorem and its variants, GHZ and CHSH among them, are presented, and the Kochen–Specker theorem is discussed. In this chapter, the communication complexity is also treated by showing how the violations of classical locality and noncontextuality can be used as a resource for communication protocols. Stronger-than quantum violations of the CHSH inequality are also discussed. They are interesting, since it has been shown that if the violation of CHSH inequality is strong enough, then the communication complexity collapses into one bit (hence the communication complexity of the true physical world seems to settle somewhere between classical and stronger-than quantum).

The chapter "Algorithms for Quantum Computers", by Jamie Smith and Michele Mosca, introduces the most remarkable known methods that utilize the special features of quantum physics in order to gain advantage over classical computing. The importance of these methods is that they form the core of designing discrete quantum algorithms. The methods presented and discussed here are the quantum Fourier transform, amplitude amplification, and quantum walks. Then, as specific examples, Shor's factoring algorithm (quantum Fourier transform), Grover search (amplitude amplification), and element distinctness algorithms (quantum random walks) are presented. The chapter not only involves traditional methods, but it also contains discussion of continuous-time quantum random walks and, more importantly, an extensive presentation of an important recent development in quantum algorithms, viz., tensor network evaluation algorithms. Then, as an example, the approximate evaluation of Tutte polynomials is presented.

The chapter "Physical Implementation of Large-Scale Quantum Computation", by Kalle-Antti Suominen, discusses the potential ways of physically implementing quantum computers. First, the DiVincenzo criteria (requirements for building a successful quantum computer) are presented, and then quantum error correction is discussed. The history, physical properties, potentials, and obstacles of various possible physical implementations of quantum computers are covered. They involve: cavity QED, trapped ions, neutral atoms and single electrons, liquid-form molecular spin, nuclear and electron spins in silicon, nitrogen vacancies in diamond, solid-state qubits with quantum dots, superconducting charge, flux and phase quantum bits, and optical quantum computing.

The chapter "Quantum Cryptography", by Takeshi Koshiba, is concerned with quantum cryptography, which will most likely play an important role in future when quantum computers make the current public-key cryptosystems unreliable. It gives an overview of classical cryptosystems, discusses classical cryptographic protocols, and then introduces the quantum key distribution protocols BB84, B92, and BBM92. Also protocol OTU00, not known to be vulnerable under Shor's algorithm, is presented. In future, when quantum computers are available, cryptography will most probably be based on quantum protocols. The chapter presents candidates for such quantum protocols: KKNY05 and GC01 (for digital signatures). It concludes with a discussion of quantum commitment, oblivious transfer, and quantum zero-knowledge proofs.

The complexity class BQP is the quantum counterpart of the classical class BPP. Intuitively, BQP can be described as the class of problems solvable in "reasonable" time, and, hence, from the application-oriented point of view, it will likely become the most important complexity class in future, when quantum computers are available. The chapter "BQP-Complete Problems", by Shengyu Zhang, introduces the computational problems that capture the full hardness of BQP. In the very fundamental sense, no BQP-complete problems are known, but the promise problems (the probability distribution of outputs is restricted by promise) bring us as close as possible to the "hardest" problems in BQP, known as BQP-complete promise problems. The chapter discusses known BQP-complete promise problems. In particular, it is shown how to establish the BQP-completeness of the Local Hamiltonian Eigenvalue

Sampling problem and the Local Unitary Phase Sampling problem. The chapter concludes with an extensive study showing that the Jones Polynomial Approximation problem is a BQP-complete promise problem.

Broader Perspective

This area consists of two sections, "Nature-Inspired Algorithms" and "Alternative Models of Computation".

Nature-Inspired Algorithms

This section is covered by six chapters.

The chapter "An Introduction to Artificial Immune Systems", by Mark Read, Paul S. Andrews, and Jon Timmis, provides a general introduction to the field. It discusses the major research issues relating to the field of Artificial Immune Systems (AIS), exploring the underlying immunology that has led to the development of immune-inspired algorithms, and focuses on the four main algorithms that have been developed in recent years: clonal selection, immune network, negative selection, and dendritic cell algorithms; their use in terms of applications is highlighted. The chapter also covers evaluation of current AIS technology, and details some new frameworks and methodologies that are being developed towards more principled AIS research. As a counterpoint to the focus on applications, the chapter also gives a brief outline of how AIS research is being employed to help further the understanding of immunology.

The chapter on "Swarm Intelligence", by David W. Corne, Alan Reynolds, and Eric Bonabeau, attempts to demystify the term Swarm Intelligence (SI), outlining the particular collections of natural phenomena that SI most often refers to and the specific classes of computational algorithms that come under its definition. The early parts of the chapter focus on the natural inspiration side, with discussion of social insects and stigmergy, foraging behavior, and natural flocking behavior. Then the chapter moves on to outline the most successful of the computational algorithms that have emerged from these natural inspirations, namely ant colony optimization methods and particle swarm optimization, with also some discussion of different and emerging such algorithms. The chapter concludes with a brief account of current research trends in the field.

The chapter "Simulated Annealing", by Kathryn A. Dowsland and Jonathan M. Thompson, provides an overview of Simulated Annealing (SA), emphasizing its practical use. The chapter explains its inspiration from the field of statistical thermodynamics, and then overviews the theory, with an emphasis again on those aspects that are important for practical applications. The chapter then covers some of the main ways in which the basic SA algorithm has been modified by various researchers, leading to improved performance for a variety of problems. The chapter briefly surveys application areas, and ends with several useful pointers to associated resources, including freely available code.

The chapter "Evolvable Hardware", by Lukáš Sekanina, surveys this growing field. Starting with a brief overview of the reconfigurable devices used in this field, the elementary principles and open problems are introduced, and then the chapter considers, in turn, three main areas: extrinsic evolution (evolving hardware using simulators), intrinsic evolution (where the evolution is conducted within FPGAs, FPTAs, and so forth), and adaptive hardware (in which real-world adaptive hardware systems are presented). The chapter finishes with an overview of major achievements in the field.

The first of two application-centered chapters, "Natural Computing in Finance — A Review", by Anthony Brabazon, Jing Dang, Ian Dempsey, Michael O'Neill, and David Edelman, provides a rather comprehensive account of natural computing applications in what is, at the time of writing (and undoubtedly beyond), one of the hottest topics of the day. This chapter introduces us to the wide range of different financial problems to which natural computing methods have been applied, including forecasting, trading, arbitrage, portfolio management, asset allocation, credit risk assessment, and more. The natural computing areas that feature in this chapter are largely evolutionary computing, neural computing, and also agent-based modeling, swarm intelligence, and immune-inspired methods. The chapter ends with a discussion of promising future directions.

Finally, the chapter "Selected Aspects of Natural Computing", by David W. Corne, Kalyanmoy Deb, Joshua Knowles, and Xin Yao, provides detailed accounts of a collection of example natural computing applications, each of which is remarkable or particularly interesting in some way. The thrust of this chapter is to provide, via such examples, an idea of both the significant impact that natural computing has already had, as well as its continuing significant promise for future applications in all areas of science and industry. While presenting this eclectic collection of marvels, the chapter also aims at clarity and demystification, providing much detail that helps see how the natural computing methods in question were applied to achieve the stated results. Applications covered include Blondie24 (the evolutionary neural network application that achieves master-level skill at the game of checkers), the design of novel antennas using evolutionary computation in conjunction with developmental computing, and the classic application of learning classifier systems that led to novel fighter-plane maneuvers for the USAF.

Alternative Models of Computation

This section is covered by seven chapters.

The chapter "Artificial Life", by Wolfgang Banzhaf and Barry McMullin, traces the roots, raises key questions, discusses the major methodological tools, and reviews the main applications of this exciting and maturing area of computing. The chapter starts with a historical overview, and presents the fundamental questions and issues that Artificial Life is concerned with. Thus the chapter surveys discussions and viewpoints about the very nature of the differences between living and nonliving systems, and goes on to consider issues such as hierarchical design, self-construction, and self-maintenance, and the emergence of complexity. This part of the chapter ends with a discussion of "Coreworld" experiments, in which a number of systems have been studied that allow spontaneous evolution of computer programs. The chapter moves on to survey the main theory and formalisms used in Artificial Life, including cellular automata and rewriting systems. The chapter concludes with a review and restatement of the main objectives of Artificial Life research, categorizing them respectively into questions about the origin and nature of life, the potential and limitations of living systems, and the relationships between life and intelligence, culture, and other human constructs.

The chapter "Algorithmic Systems Biology — Computer Science Propels Systems Biology", by Corrado Priami, takes the standpoint of computing as providing a philosophical foundation for systems biology, with at least the same importance as mathematics, chemistry, or physics. The chapter highlights the value of algorithmic approaches in modeling, simulation, and analysis of biological systems. It starts with a high-level view of how models and experiments can be tightly integrated within an algorithmic systems biology vision, and then deals in turn with modeling languages, simulations of models, and finally the postprocessing of results from biological models and how these lead to new hypotheses that can then re-enter the modeling/simulation cycle.

The chapter "Process Calculi, Systems Biology and Artificial Chemistry", by Pierpaolo Degano and Andrea Bracciali, concentrates on the use of process calculi and related techniques for systems-level modeling of biological phenomena. This chapter echoes the broad viewpoint of the previous chapter, but its focus takes us towards a much deeper understanding of the potential mappings between formal systems in computer science and systems interpretation of biological processes. It starts by surveying the basics of process calculi, setting out their obvious credentials for modeling concurrent, distributed systems of interacting parts, and mapping these onto a "cells as computers" view. After a process calculi treatment of systems biology, the chapter goes on to examine process calculi as a route towards artificial chemistry. After considering the formal properties of the models discussed, the chapter ends with notes on some case studies showing the value of process calculi in modeling biological phenomena; these include investigating the concept of a "minimal gene set" prokaryote, modeling the nitric oxide-cGMP pathway (central to many signal transduction mechanisms), and modeling the calyx of Held (a large synapse structure in the mammalian auditory central nervous system).

The chapter on "Reaction–Diffusion Computing", by Andrew Adamatzky and Benjamin De Lacy Costello, introduces the reader to the concept of a reaction–diffusion computer. This is a spatially extended chemical system, which processes information via transforming an input profile of ingredients (in terms of different concentrations of constituent ingredients) into an output profile of ingredients. The chapter takes us through the elements of this field via case studies, and it shows how selected tasks in computational geometry, robotics, and logic can be addressed by chemical implementations of reaction–diffusion computers. After introducing the field and providing a treatment of its origins and main achievements, a classical view of reaction–diffusion computers is then described. The chapter moves on to discuss varieties of reaction–diffusion processors and their chemical constituents, covering applications to the aforementioned tasks. The chapter ends with the authors' thoughts on future developments in this field.

The chapter "Rough–Fuzzy Computing", by Andrzej Skowron, shifts our context towards addressing a persistent area of immense difficulty for classical computing, which is the fact that real-world reasoning is usually done in the face of inaccurate, incomplete, and often inconsistent evidence. In essence, concepts in the real world are vague, and computation needs ways to address this. We are hence treated, in this chapter, to an overarching view of rough set theory, fuzzy set theory, their hybridization, and applications. Rough and fuzzy computing are broadly complementary approaches to handling vagueness, focusing respectively on capturing the level of distinction between separate objects and the level of membership of an object in a set. After presenting the basic concepts of rough computing and fuzzy computing in turn, in each case going into some detail on the main theoretical results and practical considerations, the chapter goes on to discuss how they can be, and have been, fruitfully combined. The chapter ends with an overview of the emerging field of "Wisdom Technology" (Wistech) as a paradigm for developing modern intelligent systems.

The chapter "Collision-Based Computing", by Andrew Adamatzky and Jérôme Durand-Lose, presents and discusses the computations performed as a result of spatial localizations in systems that exhibit dynamic spatial patterns over time. For example, a collision may be between two gliders in a cellular automaton, or two separate wave fragments within an excitable chemical system. This chapter introduces us to the basics of collision-based computing and overviews collision-based computing schemes in 1D and 2D cellular automata as well as continuous excitable media. Then, after some theoretical foundations relating to 1D cellular automata, the chapter presents a collision-based implementation for a 1D Turing machine and for cyclic tag systems. The chapter ends with discussion and presentation of "Abstract Geometrical Computation", which can be seen as collision-based computation in a medium that is the continuous counterpart of cellular automata.

The chapter "Nonclassical Computation — A Dynamical Systems Perspective", by Susan Stepney, takes a uniform view of computation, in which inspiration from a dynamical systems perspective provides a convenient way to consider, in one framework, both classical discrete systems and systems performing nonclassical computation. In particular, this viewpoint presents a way towards computational interpretation of physical embodied systems that exploit their natural dynamics. The chapter starts by discussing "closed" dynamical systems, those whose dynamics involve no inputs from an external environment, examining their computational abilities from a dynamical systems perspective. Then it discusses continuous dynamical systems and shows how these too can be interpreted computationally, indicating how material embodiment can give computation "for free", without the need to explicitly implement the dynamics. The outlook then broadens to consider open systems, where the dynamics are affected by external inputs. The chapter ends by looking at constructive, or developmental, dynamical systems, whose state spaces change during computation. These latter discussions approach the arena of biological and other natural systems, casting them as computational, open, developmental, dynamical systems.

Acknowledgements

This handbook resulted from a highly collaborative effort. The handbook and area editors are grateful to the chapter writers for their efforts in writing chapters and delivering them on time, and for their participation in the refereeing process.

We are indebted to the members of the Advisory Board for their valuable advice and fruitful interactions. Additionally, we want to acknowledge David Fogel, Pekka Lahti, Robert LaRue, Jason Lohn, Michael Main, David Prescott, Arto Salomaa, Kai Salomaa, Shinnosuke Seki, and Rob Smith, for their help and advice in various stages of production of this handbook. Last, but not least, we are thankful to Springer, especially to Ronan Nugent, for intense and constructive cooperation in bringing this project from its inception to its successful conclusion.

Leiden; Edinburgh; Nijmegen;   Grzegorz Rozenberg (Main Handbook Editor)
Turku; London, Ontario   Thomas Bäck (Handbook Editor and Area Editor)
October 2010   Joost N. Kok (Handbook Editor and Area Editor)
David W. Corne (Area Editor)
Tom Heskes (Area Editor)
Mika Hirvensalo (Area Editor)
Jarkko J. Kari (Area Editor)
Lila Kari (Area Editor)

Editor Biographies

Prof. Dr. Grzegorz Rozenberg

Prof. Rozenberg was awarded his Ph.D. in mathematics from the Polish Academy of Sciences, Warsaw, and he has since held full-time positions at Utrecht University, the State University of New York at Buffalo, and the University of Antwerp. Since 1979 he has been a professor at the Department of Computer Science of Leiden University, and an adjunct professor at the Department of Computer Science of the University of Colorado at Boulder, USA. He is the founding director of the Leiden Center for Natural Computing.

Among key editorial responsibilities over the last 30 years, he is the founding Editor of the book series Texts in Theoretical Computer Science (Springer) and Monographs in Theoretical Computer Science (Springer), founding Editor of the book series Natural Computing (Springer), founding Editor-in-Chief of the journal Natural Computing (Springer), and founding Editor of Part C (Theory of Natural Computing) of the journal Theoretical Computer Science (Elsevier). Altogether he's on the Editorial Board of around 20 scientific journals.

He has authored more than 500 papers and 6 books, and coedited more than 90 books. He coedited the "Handbook of Formal Languages" (Springer), he was Managing Editor of the "Handbook of Graph Grammars and Computing by Graph Transformation" (World Scientific), he coedited "Current Trends in Theoretical Computer Science" (World Scientific), and he coedited "The Oxford Handbook of Membrane Computing" (Oxford University Press).

He is Past President of the European Association for Theoretical Computer Science (EATCS), and he received the Distinguished Achievements Award of the EATCS "in recognition of his outstanding scientific contributions to theoretical computer science". Also he is a cofounder and Past President of the International Society for Nanoscale Science, Computation, and Engineering (ISNSCE).

He has served as a program committee member for most major conferences on theoretical computer science in Europe, and among the events he has founded or helped to establish are the International Conference on Developments in Language Theory (DLT), the International

Conference on Graph Transformation (ICGT), the International Conference on Unconventional Computation (UC), the International Conference on Application and Theory of Petri Nets and Concurrency (ICATPN), and the DNA Computing and Molecular Programming Conference.

In recent years his research has focused on natural computing, including molecular computing, computation in living cells, self-assembly, and the theory of biochemical reactions. His other research areas include the theory of concurrent systems, the theory of graph transformations, formal languages and automata theory, and mathematical structures in computer science.

Prof. Rozenberg is a Foreign Member of the Finnish Academy of Sciences and Letters, a member of the Academia Europaea, and an honorary member of the World Innovation Foundation. He has been awarded honorary doctorates by the University of Turku, the Technical University of Berlin, and the University of Bologna. He is an ISI Highly Cited Researcher.

He is a performing magician, and a devoted student of and expert on the paintings of Hieronymus Bosch.

Prof. Dr. Thomas Bäck

Prof. Bäck was awarded his Ph.D. in Computer Science from Dortmund University in 1994, for which he received the Best Dissertation Award from the Gesellschaft für Informatik (GI). He has been at Leiden University since 1996, where he is currently full Professor for Natural Computing and the head of the Natural Computing Research Group at the Leiden Institute of Advanced Computer Science (LIACS).

He has authored more than 150 publications on natural computing technologies. He wrote a book on evolutionary algorithms, "Evolutionary Algorithms in Theory and Practice" (Oxford University Press), and he coedited the "Handbook of Evolutionary Computation" (IOP/Oxford University Press).

Prof. Bäck is an Editor of the book series Natural Computing (Springer), an Associate Editor of the journal Natural Computing (Springer), an Editor of the journal Theoretical Computer Science (Sect. C, Theory of Natural Computing; Elsevier), and an Advisory Board member of the journal Evolutionary Computation (MIT Press). He has served as program chair for all major conferences in evolutionary computation, and is an Elected Fellow of the International Society for Genetic and Evolutionary Computation for his contributions to the field.

His main research interests are the theory of evolutionary algorithms, cellular automata and data-driven modelling, and applications of these methods in medicinal chemistry, pharmacology and engineering.

Prof. Dr. Joost N. Kok

Prof. Kok was awarded his Ph.D. in Computer Science from the Free University in Amsterdam in 1989, and he has worked at the Centre for Mathematics and Computer Science in Amsterdam, at Utrecht University, and at the Åbo Akademi University in Finland. Since 1995 he has been a professor in computer science, and since 2005 also a professor in medicine at Leiden University. He is the Scientific Director of the Leiden Institute of Advanced Computer Science, and leads the research clusters Algorithms and Foundations of Software Technology.

He serves as a chair, member of the management team, member of the board, or member of the scientific committee of the Faculty of Mathematics and Natural Sciences of Leiden University, the ICT and Education Committee of Leiden University, the Dutch Theoretical Computer Science Association, the Netherlands Bioinformatics Centre, the Centre for Mathematics and Computer Science Amsterdam, the Netherlands Organisation for Scientific Research, the Research Foundation Flanders (Belgium), the European Educational Forum, and the International Federation for Information Processing (IFIP) Technical Committee 12 (Artificial Intelligence).

Prof. Kok is on the steering, scientific or advisory committees of the following events: the Mining and Learning with Graphs Conference, the Intelligent Data Analysis Conference, the Institute for Programming and Algorithms Research School, the Biotechnological Sciences Delft–Leiden Research School, and the European Conference on Machine Learning and Principles and Practice of Knowledge Discovery in Databases. And he has been a program committee member for more than 100 international conferences, workshops or summer schools on data mining, data analysis, and knowledge discovery; neural networks; artificial intelligence; machine learning; computational life science; evolutionary computing, natural computing and genetic algorithms; Web intelligence and intelligent agents; and software engineering.

He is an Editor of the book series Natural Computing (Springer), an Associate Editor of the journal Natural Computing (Springer), an Editor of the journal Theoretical Computer Science (Sect. C, Theory of Natural Computing; Elsevier), an Editor of the Journal of Universal

Computer Science, an Associate Editor of the journal Computational Intelligence (Wiley), and a Series Editor of the book series Frontiers in Artificial Intelligence and Applications (IOS Press).

His academic research is concentrated around the themes of scientific data management, data mining, bioinformatics, and algorithms, and he has collaborated with more than 20 industrial partners.

Advisory Board

Area Editors

Table of Contents

Volume 1

Volume 3

Volume 4

List of Contributors

Andrew Adamatzky
Department of Computer Science
University of the West of England
Bristol
UK
andrew.adamatzky@uwe.ac.uk

Paul S. Andrews
Department of Computer Science
University of York
UK
psa@cs.york.ac.uk

Masanori Arita
Department of Computational Biology,
Graduate School of Frontier Sciences
The University of Tokyo
Kashiwa
Japan
arita@k.u-tokyo.ac.jp

Wolfgang Banzhaf
Department of Computer Science
Memorial University of Newfoundland
St. John's, NL
Canada
banzhaf@cs.mun.ca

Hans-Georg Beyer
Department of Computer Science
Fachhochschule Vorarlberg
Dornbirn
Austria
hans-georg.beyer@fhv.at

Sander Bohte
Research Group Life Sciences
CWI
Amsterdam
The Netherlands
s.m.bohte@cwi.nl

Eric Bonabeau
Icosystem Corporation
Cambridge, MA
USA
eric@icosystem.com

Anthony Brabazon
Natural Computing Research and
Applications Group
University College Dublin
Ireland
anthony.brabazon@ucd.ie

Andrea Bracciali
Department of Computing Science and
Mathematics
University of Stirling
UK
braccia@cs.stir.ac.uk

Robert Brijder
Leiden Institute of Advanced Computer
Science
Universiteit Leiden
The Netherlands
robert.brijder@uhasselt.be

Časlav Brukner
Faculty of Physics
University of Vienna
Vienna
Austria
caslav.brukner@univie.ac.at

Anthony Bucci
Icosystem Corporation
Cambridge, MA
USA
anthony@icosystem.com

Ke Chen
Department of Electrical and Computer
Engineering
University of Alberta
Edmonton, AB
Canada
kchen1@ece.ualberta.ca

Seungjin Choi
Pohang University of Science and
Technology
Pohang
South Korea
seungjin@postech.ac.kr

Bastien Chopard
Scientific and Parallel Computing Group
University of Geneva
Switzerland
bastien.chopard@unige.ch

David W. Corne
School of Mathematical and Computer
Sciences
Heriot-Watt University
Edinburgh
UK
dwcorne@macs.hw.ac.uk

Mark Daley
Departments of Computer Science and
Biology
University of Western Ontario
London, Ontario
Canada
daley@csd.uwo.ca

Jing Dang
Natural Computing Research and
Applications Group
University College Dublin
Ireland
jing.dang@ucd.ie

Edwin D. de Jong
Institute of Information and Computing
Sciences
Utrecht University
The Netherlands
dejong@cs.uu.nl

Kenneth De Jong
Department of Computer Science
George Mason University
Fairfax, VA
USA
kdejong@gmu.edu

Benjamin De Lacy Costello
Centre for Research in Analytical, Material
and Sensor Sciences, Faculty of Applied
Sciences
University of the West of England
Bristol
UK
ben.delacycostello@uwe.ac.uk

Kalyanmoy Deb
Department of Mechanical Engineering
Indian Institute of Technology
Kanpur
India
deb@iitk.ac.in

Pierpaolo Degano
Dipartimento di Informatica
Università di Pisa
Italy
degano@di.unipi.it

Marianne Delorme
Laboratoire d'Informatique Fondamentale
de Marseille (LIF)
Aix-Marseille Université and CNRS
Marseille
France
delorme.marianne@orange.fr

Ian Dempsey
Pipeline Financial Group, Inc.
New York, NY
USA
ian.dempsey@pipelinefinancial.com

Alberto Dennunzio
Dipartimento di Informatica
Sistemistica e Comunicazione, Università
degli Studi di Milano-Bicocca
Italy
dennunzio@disco.unimib.it

Kathryn A. Dowsland
Gower Optimal Algorithms, Ltd.
Swansea
UK
k.a.dowsland@btconnect.com

Jérôme Durand-Lose
LIFO
Université d'Orléans
France
jerome.durand-lose@univ-orleans.fr

David Edelman
School of Business
UCD Michael Smurfit Graduate Business
School
Dublin
Ireland
david.edelman@ucd.ie

Gary B. Fogel
Natural Selection, Inc.
San Diego, CA
USA
gfogel@natural-selection.com

Enrico Formenti
Département d'Informatique
Université de Nice-Sophia Antipolis
France
enrico.formenti@unice.fr

Xinbo Gao
Video and Image Processing System Lab,
School of Electronic Engineering
Xidian University
China
xbgao@ieee.org

Shuzhi Sam Ge
Social Robotics Lab
Interactive Digital Media Institute, The
National University of Singapore
Singapore
elegesz@nus.edu.sg

Masami Hagiya
Department of Computer Science
Graduate School of Information Science
and Technology
The University of Tokyo
Tokyo
Japan
hagiya@is.s.u-tokyo.ac.jp

Tero Harju
Department of Mathematics
University of Turku
Finland
harju@utu.fi

Mika Hirvensalo
Department of Mathematics
University of Turku
Finland
mikhirve@utu.fi

Satoshi Hiyama
Research Laboratories
NTT DOCOMO, Inc.
Yokosuka
Japan
hiyama@nttdocomo.co.jp

Guang-Bin Huang
School of Electrical and Electronic
Engineering
Nanyang Technological University
Singapore
egbhuang@ntu.edu.sg

Thomas Jansen
Department of Computer Science
University College Cork
Ireland
t.jansen@cs.ucc.ie

Nataša Jonoska
Department of Mathematics
University of South Florida
Tampa, FL
USA
jonoska@math.usf.edu

Jarkko J. Kari
Department of Mathematics
University of Turku
Turku
Finland
jkari@utu.fi

Lila Kari
Department of Computer Science
University of Western Ontario
London
Canada
lila@csd.uwo.ca

Sungchul Kim
Data Mining Lab, Department of Computer
Science and Engineering
Pohang University of Science and
Technology
Pohang
South Korea
subright@postech.ac.kr

Joshua Knowles
School of Computer Science and
Manchester Interdisciplinary
Biocentre (MIB)
University of Manchester
UK
j.knowles@manchester.ac.uk

Satoshi Kobayashi
Department of Computer Science
University of Electro-Communications
Tokyo
Japan
satoshi@cs.uec.ac.jp

Ken Komiya
Interdisciplinary Graduate School of
Science and Engineering
Tokyo Institute of Technology
Yokohama
Japan
komiya@dis.titech.ac.jp

Takeshi Koshiba
Graduate School of Science and
Engineering
Saitama University
Japan
koshiba@mail.saitama-u.ac.jp

Tim Kovacs
Department of Computer Science
University of Bristol
UK
kovacs@cs.bris.ac.uk

Natalio Krasnogor
Interdisciplinary Optimisation Laboratory,
The Automated Scheduling, Optimisation
and Planning Research Group, School of
Computer Science
University of Nottingham
UK
natalio.krasnogor@nottingham.ac.uk

Lukasz A. Kurgan
Department of Electrical and Computer
Engineering
University of Alberta
Edmonton, AB
Canada
lkurgan@ece.ualberta.ca

Petr Kůrka
Center for Theoretical Studies
Academy of Sciences and Charles
University in Prague
Czechia
kurka@cts.cuni.cz

Thomas H. LaBean
Department of Computer Science and
Department of Chemistry and Department
of Biomedical Engineering
Duke University
Durham, NC
USA
thomas.labean@duke.edu

Kang Li
School of Electronics, Electrical Engineering
and Computer Science
Queen's University
Belfast
UK
k.li@ee.qub.ac.uk

Xuelong Li
Center for OPTical IMagery Analysis and
Learning (OPTIMAL), State Key Laboratory
of Transient Optics and Photonics
Xi'an Institute of Optics and Precision
Mechanics, Chinese Academy of Sciences
Xi'an, Shaanxi
China
xuelong_li@opt.ac.cn

Wen Lu
Video and Image Processing System Lab,
School of Electronic Engineering
Xidian University
China
luwen@mail.xidian.edu.cn

Jacques Mazoyer
Laboratoire d'Informatique Fondamentale
de Marseille (LIF)
Aix-Marseille Université and CNRS
Marseille
France
mazoyerj2@orange.fr

Barry McMullin
Artificial Life Lab, School of Electronic
Engineering
Dublin City University
Ireland
barry.mcmullin@dcu.ie

Silja Meyer-Nieberg
Fakultät für Informatik
Universität der Bundeswehr München
Neubiberg
Germany
silja.meyer-nieberg@unibw.de

Kenichi Morita
Department of Information Engineering,
Graduate School of Engineering
Hiroshima University
Japan
morita@iec.hiroshima-u.ac.jp

Michele Mosca
Institute for Quantum Computing and
Department of Combinatorics &
Optimization
University of Waterloo and St. Jerome's
University and Perimeter Institute for
Theoretical Physics
Waterloo
Canada
mmosca@iqc.ca

Nicolas Ollinger
Laboratoire d'informatique fondamentale
de Marseille (LIF)
Aix-Marseille Université, CNRS
Marseille
France
nicolas.ollinger@lif.univ-mrs.fr

Michael O'Neill
Natural Computing Research and
Applications Group
University College Dublin
Ireland
m.oneill@ucd.ie

Hélène Paugam-Moisy
Laboratoire LIRIS – CNRS
Université Lumière Lyon 2
Lyon
France
and
INRIA Saclay – Ile-de-France
Université Paris-Sud
Orsay
France
helene.paugam-moisy@univ-lyon2.fr
hpaugam@lri.fr

Gheorghe Păun
Institute of Mathematics of the Romanian
Academy
Bucharest
Romania
and
Department of Computer Science and
Artificial Intelligence
University of Seville
Spain
gpaun@us.es
george.paun@imar.ro

Ion Petre
Department of Information Technologies
Åbo Akademi University
Turku
Finland
ipetre@abo.fi

Riccardo Poli
Department of Computing and Electronic
Systems
University of Essex
Colchester
UK
rpoli@essex.ac.uk

Elena Popovici
Icosystem Corporation
Cambridge, MA
USA
elena@icosystem.com

Corrado Priami
Microsoft Research
University of Trento Centre for
Computational and Systems Biology
(CoSBi)
Trento
Italy
and
DISI
University of Trento
Trento
Italy
priami@cosbi.eu

Mark Read
Department of Computer Science
University of York
UK
markread@cs.york.ac.uk

John H. Reif
Department of Computer Science
Duke University
Durham, NC
USA
reif@cs.duke.edu

Alan Reynolds
School of Mathematical and Computer
Sciences
Heriot-Watt University
Edinburgh
UK
a.reynolds@hw.ac.uk

Grzegorz Rozenberg
Leiden Institute of Advanced Computer
Science
Universiteit Leiden
The Netherlands
and
Department of Computer Science
University of Colorado
Boulder, CO
USA
rozenber@liacs.nl

Günter Rudolph
Department of Computer Science
TU Dortmund
Dortmund
Germany
guenter.rudolph@tu-dortmund.de

Yasubumi Sakakibara
Department of Biosciences and Informatics
Keio University
Yokohama
Japan
yasu@bio.keio.ac.jp

Lukáš Sekanina
Faculty of Information Technology
Brno University of Technology
Brno
Czech Republic
sekanina@fit.vutbr.cz

Shinnosuke Seki
Department of Computer Science
University of Western Ontario
London
Canada
sseki@csd.uwo.ca

Ofer M. Shir
Department of Chemistry
Princeton University
NJ
USA
oshir@princeton.edu

Andrzej Skowron
Institute of Mathematics
Warsaw University
Poland
skowron@mimuw.edu.pl

Jamie Smith
Institute for Quantum Computing and
Department of Combinatorics &
Optimization
University of Waterloo
Canada
ja5smith@iqc.ca

Petr Sosík
Institute of Computer Science
Silesian University in Opava
Czech Republic
and
Departamento de Inteligencia Artificial
Universidad Politécnica de Madrid
Spain
petr.sosik@fpf.slu.cz

Susan Stepney
Department of Computer Science
University of York
UK
susan.stepney@cs.york.ac.uk

Kalle-Antti Suominen
Department of Physics and Astronomy
University of Turku
Finland
kalle-antti.suominen@utu.fi

Andrew M. Sutton
Department of Computer Science
Colorado State University
Fort Collins, CO
USA
sutton@cs.colostate.edu

Siamak Taati
Department of Mathematics
University of Turku
Finland
siamak.taati@gmail.com

Masahiro Takinoue
Department of Physics
Kyoto University
Kyoto
Japan
takinoue@chem.scphys.kyoto-u.ac.jp

Fumiaki Tanaka
Department of Computer Science
Graduate School of Information Science
and Technology
The University of Tokyo
Tokyo
Japan
fumi95@is.s.u-tokyo.ac.jp

Dacheng Tao
School of Computer Engineering
Nanyang Technological University
Singapore
dacheng.tao@gmail.com

Véronique Terrier
GREYC, UMR CNRS 6072
Université de Caen
France
veroniqu@info.unicaen.fr

Jonathan M. Thompson
School of Mathematics
Cardiff University
UK
thompsonjm1@cardiff.ac.uk

Jon Timmis
Department of Computer Science and
Department of Electronics
University of York
UK
jtimmis@cs.york.ac.uk

Joaquin J. Torres
Institute "Carlos I" for Theoretical and
Computational Physics and Department of
Electromagnetism and Matter Physics,
Facultad de Ciencias
Universidad de Granada
Spain
jtorres@ugr.es

Marc M. Van Hulle
Laboratorium voor Neurofysiologie
K.U. Leuven
Leuven
Belgium
marc@neuro.kuleuven.be

Leonardo Vanneschi
Department of Informatics, Systems and
Communication
University of Milano-Bicocca
Italy
vanneschi@disco.unimib.it

Pablo Varona
Departamento de Ingeniería Informática
Universidad Autónoma de Madrid
Spain
pablo.varona@uam.es

Darrell Whitley
Department of Computer Science
Colorado State University
Fort Collins, CO
USA
whitley@cs.colostate.edu

R. Paul Wiegand
Institute for Simulation and Training
University of Central Florida
Orlando, FL
USA
wiegand@ist.ucf.edu

Xin Yao
Natural Computation Group, School of
Computer Science
University of Birmingham
UK
x.yao@cs.bham.ac.uk

Takashi Yokomori
Department of Mathematics, Faculty of
Education and Integrated Arts and
Sciences
Waseda University
Tokyo
Japan
yokomori@waseda.jp

Hwanjo Yu
Data Mining Lab, Department of Computer
Science and Engineering
Pohang University of Science and
Technology
Pohang
South Korea
hwanjoyu@postech.ac.kr

Jean-Baptiste Yunès
Laboratoire LIAFA
Université Paris 7 (Diderot)
France
jean-baptiste.yunes@liafa.jussieu.fr

G. Peter Zhang
Department of Managerial Sciences
Georgia State University
Atlanta, GA
USA
gpzhang@gsu.edu

Jie Zhang
School of Chemical Engineering and
Advanced Materials
Newcastle University
Newcastle upon Tyne
UK
jie.zhang@newcastle.ac.uk

Shengyu Zhang
Department of Computer Science and
Engineering
The Chinese University of Hong Kong
Hong Kong S.A.R.
China
syzhang@cse.cuhk.edu.hk

Eckart Zitzler
PHBern – University of Teacher Education,
Institute for Continuing Professional
Education
Bern
Switzerland
eckart.zitzler@phbern.ch
eckart.zitzler@tik.ee.ethz.ch

Marek Żukowski
Institute of Theoretical Physics and
Astrophysics
University of Gdansk
Poland
marek.zukowski@univie.ac.at
fizmz@univ.gda.pl

Broader Perspective – Nature-Inspired Algorithms

David W. Corne

47 An Introduction to Artificial Immune Systems

Mark Read[1] · *Paul S. Andrews*[2] · *Jon Timmis*[3]
[1]Department of Computer Science, University of York, UK
markread@cs.york.ac.uk
[2]Department of Computer Science, University of York, UK
psa@cs.york.ac.uk
[3]Department of Computer Science and Department of Electronics,
University of York, UK
jtimmis@cs.york.ac.uk

G. Rozenberg et al. (eds.), *Handbook of Natural Computing*, DOI 10.1007/978-3-540-92910-9_47,
© Springer-Verlag Berlin Heidelberg 2012

Abstract

The field of artificial immune systems (AIS) comprises two threads of research: the employment of mathematical and computational techniques in the modeling of immunology, and the incorporation of immune system metaphors in the development of engineering solutions. The former permits the integration of immunological data and sub-models into a coherent whole, which can be of value to immunologists in the facilitation of immunological understanding, hypothesis testing, and the direction of future research. The latter attempts to harness the perceived properties of the immune system in the solving of engineering problems. This chapter concentrates on the latter: the development and application of immune inspiration to engineering solutions.

1   Introduction

Artificial Immune Systems (AIS) is a branch of biologically inspired computation focusing on many aspects of immune systems. AIS development can be seen as having two target domains: the provision of solutions to engineering problems through the adoption of immune system inspired concepts; and the provision of models and simulations with which to study immune system theories.

The motivation for building immune-inspired solutions to engineering problems arises from the identification of properties within the immune system that are attractive from an engineering perspective. These include (de Castro and Timmis 2002a): the self-organization of huge numbers of immune cells; the distributed operation of the immune system throughout the body; pattern recognition and anomaly detection to enable the immune system to recognize pathogens; and optimization and memory to improve and remember immune responses. AIS take inspiration from these properties and associated immune processes, and have been defined as:

▶   "adaptive systems, inspired by theoretical immunology and observed immune function, principles and models, which are applied to problem solving." (de Castro and Timmis 2002a)

The field of AIS also encompasses modeling and simulation techniques to understand the immune system in general (see Timmis et al. (2008a) for a review), however, this chapter focuses on immune-inspired systems for engineering problems.

This chapter is not intended as an extensive review chapter, but its purpose is to present a general introduction to the area and provide discussion on the major research issues relating to the field of AIS. Therefore, in this chapter we briefly explore the underlying immunology that has served as an inspiration for the development of immune-inspired algorithms. We have chosen not to focus on the modeling aspect of AIS, but rather on the main algorithms that have been developed over recent years. This is undertaken in ❷ Sect. 3 where we discuss four main immune-inspired algorithms that dominate the literature, namely, clonal selection, immune networks, negative selection, and dendritic cell algorithms, and highlight their usage in terms of applications. ❷ Section 4 follows with a discussion on AIS and how researchers have begun to evaluate current AIS and describes new frameworks and methodologies that aim to help develop AIS in a more principled manner. We also briefly discuss the application of AIS to a variety of different domains and the types of applications that AIS might be better suited to, and finally we provide a very brief outline of the modeling approaches that can be found in the literature that are employed to help further our understanding of immunology. ❷ Section 5 provides a chapter summary.

2 The Immune System

Immunology concerns the study of the immune system and the effects of its operation on the body. The immune system is normally defined in relation to its perceived function: a defense system that has evolved to protect its host from pathogens (harmful microorganisms such as bacteria, viruses, and parasites) (Goldsby et al. 2003). It comprises a variety of specialized cells that circulate and monitor the body, various extracellular molecules, and immune organs that provide an environment within which immune cells interact, mature, and respond. The collective action of immune cells and molecules forms a complex network leading to the detection and recognition of pathogens within the body. This is followed by a specific effector response aimed at eliminating the pathogen. This recognition and response process is very complicated with many details not yet properly understood.

In mammals, the immune system can be classified into two components based on functionality: a less specific component termed *innate* immunity and a more specific component termed *adaptive* (or acquired) immunity. The mechanisms of innate immunity are generic defense mechanisms that are nonspecific to particular examples of pathogen, but act against general classes of pathogen. They are encoded within the genes of the species, and do not adapt during the lifetime of the individual. Examples include the inflammatory response, phagocytic immune cells (those that can ingest and kill pathogens), anatomic barriers such as skin, and physiologic barriers such as temperature.

By contrast, the mechanisms of adaptive immunity enable the immune system to adapt to previously unseen pathogens based upon exposure to them (Goldsby et al. 2003). This is achieved through a learning mechanism that operates during the lifetime of the individual. Additionally, once exposed to a pathogen, memory mechanisms exist to allow the immune system to remember the *shape* of the pathogen. This enables a faster and more effective secondary response that can be elicited against the pathogen if it is encountered again. The adaptive and innate arms of the immune system interact to provide the body with a comprehensive defense mechanism against pathogens.

All immune cells, and the majority of other cells of the body, possess protein molecules on their surface that act as receptors to other extracellular molecules. When a sufficiently strong chemical bond occurs between a receptor and another molecule (a ligand), a cascade of intracellular signals is initiated, the outcome of which depends on the initiating receptors. This process provides the immune system with a mechanism for recognition at the molecular level. Two types of immune cell receptors exist: innate receptors that have evolved to recognize specific molecules; and the unique receptors of lymphocytes that are generated during the life time of the individual to recognize previously unseen molecules. The latter of these molecules are generically known as *antigens*, a term given to any molecular structure that can chemically bind to the unique receptors of adaptive immune cells, known as T and B-cells. The antigen receptors of the B-cell are called *antibodies*, and those of the T-cell are called T cell receptors (TCR). They are both generated via a stochastic process, and are vital to the body's adaptive immune response. Communication between immune cells involves a number of immune molecules. They include cytokines, immune cell receptors, antibodies, enzymes, plasma proteins, and adhesion molecules. The cytokines, for example, are signaling molecules secreted by both immune and other bodily cells, which are then detected via specific cellular receptors. Many different types of cytokine exist and their effects include the activation, differentiation, growth, movement, and death of many types of cells (Cohen 2000).

2.1 Motivation for Immune Inspired Engineering Solutions

Why is it that engineers are attracted to the immune system for inspiration? The immune system exhibits several properties that engineers recognize as being desirable in their systems. Timmis and Andrews (2007), Timmis et al. (2008a), de Castro and Timmis (2002a) have identified these as the following.

Distribution and self-organization. The behavior of the immune system is deployed through the actions of billions of agents (cells and molecules) distributed throughout the body. Their collective effects can be highly complex with no central controller. An organized response emerges as a system-wide property derived from the low-level agent behaviors. These immune agents act concurrently making immune processes naturally parallelized.

Learning, adaption, and memory. The immune system is capable of recognizing previously unseen pathogens, thus exhibiting the ability to learn. Learning implies the presence of memory, and the immune system is able to "remember" previously encountered pathogens, as demonstrated by the phenomenon of primary and secondary immune responses. The first time a pathogen is encountered, an immune response (the primary response) is elicited; the next time that pathogen is encountered, a faster and often more aggressive response is mounted (the secondary response).

Pattern recognition. Through its various receptors and molecules, the immune system is capable of recognizing a diverse range of patterns. This is accomplished through receptors that perceive antigenic materials in differing contexts (processed molecules, whole molecules, additional signals, etc.). Receptors of the innate immune system vary little, whilst receptors of the adaptive immune system, such as antibodies and T-cell receptors, are subject to huge diversity.

Classification. The immune system is very effective at distinguishing harmful substances (typically viewed as *nonself*) from the body's own tissues (typically viewed as *self*), and directing its actions accordingly. From a computational perspective, it does this with access to only a single class of data, self-molecules (Stibor et al. 2005). The creation of a system that effectively classifies data into two classes, having been trained on examples from only one, is a challenging task.

3 Engineering Artificial Immune Systems

de Castro and Timmis (2002a) have proposed a flexible and generic layered approach to the development of immune-inspired engineering solutions, shown in ❷ *Fig. 1*. This framework identifies the main design decisions that need to be addressed in the deployment of an immune-inspired engineering solution: representations, affinity measures, and immune algorithms.

◘ Fig. 1
The layered framework approach to constructing AIS solutions. (Taken from de Castro and Timmis (2002a).)

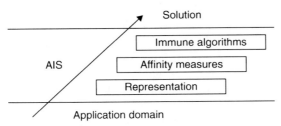

Given a particular application domain, an appropriate representation of the data must be chosen. In AIS, this typically follows the notion of *shape-space* (Perelson and Oster 1979). Here, molecules m (such as receptors and antigen) exist as points in a shape space S, and can be represented as a string of attributes $m = <m_1, m_2, \ldots, m_L>$ in an L dimensional space, $m \in S^L$. The attributes m_i will represent aspects of a problem domain: patterns to be recognized, functional values to be optimized, combinations of input and proposed actions, etc. de Castro and Timmis (2002a) suggest four data types of which these attributes may belong: real valued; integer valued; hamming valued, finite length strings composed of digits with a finite alphabet; and categorically valued, where values include items such as "name," or "color." The affinity measures are functions or criteria through which interactions of the AIS elements are quantified. They are highly dependent on the representation chosen, for example, continuous variables typically employ the Euclidean distance measure, whereas bit string representations may use the hamming distance. Work in McEwan et al. (2008) provides a convincing critique of the shape-space paradigm and discusses the limitations of such an approach. Their paper is discussed in more detail in ❯ Sect. 4.1.

The highest layer of the framework details the selection of an immune-inspired algorithm to operate over the immune elements of the system. Various types of immune-inspired algorithms exist, which can operate independently of the choice of representation and affinity measure, adding dynamics to the algorithm populations based on measurements that the affinity functions provide. Despite this, immune algorithms should be chosen with care based on the problem's data (Freitas and Timmis 2007).

In what is considered to be one of the first papers in AIS, Farmer et al. (1986) examined the immune system in the context of classifier systems, essentially highlighting the parallels of the immune network theory (Jerne 1974) and artificial intelligence. AIS has since been applied to a large range of domains that can broadly be classified as learning, anomaly detection, and optimization problems (Hart and Timmis 2008). Four main classes of AIS algorithm have been applied to these problems and each is outlined below.

3.1 Clonal Selection Theory-Inspired AIS

Clonal selection-based algorithms attempt to capture mechanisms of the antigen-driven proliferation of B-cells that results in their improved binding abilities. Using a process known as affinity maturation, the receptors of B-cell are mutated and subsequent B-cell selection results in a population of B-cells with better overall affinity for the antigen.

Clonal selection algorithms capture the properties of learning, memory, adaption, and pattern recognition (Timmis et al. 2008a).

A generic clonal selection inspired algorithm, based on CLONALG (de Castro and Von Zuben 2002, 2000), is presented in ❷ *Algorithm 1*. A set of patterns (antigens) is input to the algorithm, and output is a set of memory B-cells capable of recognizing unseen patterns. A randomly initialized set of B-cells are preferentially selected based on their affinity for the antigen. The higher affinity cells are cloned proportionally to their affinity, and mutated at a rate inversely proportional to affinity. The higher affinity clones will replace the lower affinity cells of the previous generation. Very high affinity clones compete for a place in the set of memory cells. This algorithm can be tailored toward optimization problems by removing the antigen set *S*, and directly representing the function or domain to be optimized as the affinity function. As clonal selection algorithms employ mutation and selection of a population of candidate solutions, they tend to be similar to other evolutionary algorithms (Newborough and Stepney 2005).

In ❷ *Algorithm 1*, a *generic* clonal selection algorithm is outlined, however, there are many variants in the literature that have been augmented and altered to fit specific application areas. For example, work in Watkins et al. (2004) developed a reinforcement learning approach known as AIRS (artificial immune recognition system), based on the ideas of clonal selection for the classification of unseen data items. In effect AIRS is an instance creation algorithm which acts as a preprocessor to the k-nearest neighbor approach that has been found to perform well on certain types of classification problems (Secker and Freitas 2007). In the context of dynamic learning, work by Kim and Bentley (2002a, b, c) developed a network intrusion detection system based on a dynamic variant of the clonal selection paradigm that was capable of identifying potential attacks to computer networks in an online manner and then be able to, in a limited manner, adapt to new types of attacks. As a final example, work by Kelsey and Timmis (2003), and Cutello et al. (2004a, b, 2005) have developed particularly effective optimization algorithms based on variants of clonal selection

---

Algorithm 1   A generic clonal selection algorithm, based on CLONALG (de Castro and Von Zuben 2000, 2002)

input: S = a set of antigens, representing data elements to be recognized.
output: M = set of memory B-cells capable of classifying unseen data elements.
begin
   Generate set of random specificity B-cells B.
   for all antigens $ag \in S$ **do**
      Calculate affinity of all B-cells $b \in B$ with ag.
      Select highest affinity B-cells, perform affinity proportional cloning, place clones in C.
      for all B-cell clones $c \in C$ **do**
         Mutate c at rate inversely proportional to affinity.
         Determine affinity of c with ag.
      end for
      Copy all $c \in C$ into B.
      Copy the highest affinity clones $c \in C$ into memory set M.
      Replace lowest affinity B-cells $b \in B$ with randomly generated alternatives.
   end for
end

by making use of novel selection and mutation mechanisms tailored specifically for certain types of optimization problems.

3.2 Immune Network Theory AIS

The immune network theory as proposed by Jerne (1974) views the immune system as a regulated network of molecules and cells that recognize each other which acts in a self-organizing manner to produce memory, even in the absence of antigen. B-cells interact via receptors to stimulate and suppress each other. This forms a regulatory network that represents an *internal image* of the antigenic patterns that the immune system observes (Farmer et al. 1986).

As with clonal selection, the immune network theory has provided inspiration for many algorithms ranging from optimization to machine learning (de Castro and Timmis 2002b; Honorio et al. 2007; Timmis and Neal 2000; de Castro and Von Zuben 2001; Bezerra et al. 2004). From a machine learning perspective, many of the systems are unsupervised and produce an instant reduction of the data space. They present clusters of this reduced data as networks of connected B-cells, where a B-cell may be considered a point m in the shape space S^L discussed above. The motivation for such algorithms is that the resulting networks highlight structures inherent in the data set and reduce the dimensionality and complexity of the data (Neal 2003). A generic immune network algorithm, based on aiNet (de Castro and Von Zuben 2001), is presented in ❷ *Algorithm 2*. It is a modified version of CLONALG that incorporates a mechanism of suppression amongst B-cells.

In aiNet, data items are represented as antigen which B-cells (detectors) recognize. Like clonal section algorithms (❷ *Algorithm 1*), affinity maturation produces B-cells with differing specificities, and competition removes the worst of these cells from the population. A suppressive mechanism then prunes cells of similar specificities from the population. The resulting network of B-cells is then representative of clusters within the data.

Despite possessing suppressive mechanisms, early immune network algorithms suffered from an excess of B-cells, which hindered run time efficiency and rendered the resulting networks overly complex (Timmis and Neal 2000). To address this, work by Timmis and Neal (2000) incorporated the notion of an artificial recognition ball (ARB), a bounded area surrounding a point in antigenic space. All B-cells exhibiting specificities within an ARB's area are represented by that ARB, thus removing the requirement to explicitly represent each of them. To further regulate the network's population size, ARBs lie in competition with one another for a share of finite system-wide resource; ARBs that are unable to claim sufficient resource are removed from the network. Resource is allocated on the basis of ARB stimulation, derived from antigen affinity, and from low affinity to the other ABRs with which they are linked. Hence, the pressures of the algorithm are to derive clusters of linked but well spread out ARBs that represent structure in the data.

A similar, but modified, immune network algorithm was published by Neal (2003). Both cloning and hypermutation are absent in this algorithm; new ARBs are created from antigen that fall outside the range of existing ARBs in the network. The algorithm does not incorporate any stopping criteria, and can be used to create cluster-based representations of dynamically changing data. This algorithm removed the requirement for central control over the allocation of resources; ARBs are responsible for determining their own stimulation and acting accordingly. The nature of the stimulation calculation prevents ARB population explosion and

---

Algorithm 2   A generic immune network algorithm, based on aiNet (de Castro and Von Zuben 2001) (Taken from Timmis et al. (2008a).)

input: S = a set of antigens, representing data elements to be clustered, nt network affinity
   threshold, ct clonal pool threshold, h number of highest affinity clones, a number of new
   antibodies to introduce.
output: N = set of memory detectors capable of classifying unseen patterns.
begin
   Generate set of random specificity B-cells N.
   repeat
      for all antigens $ag \in S$ **do**
         Calculate affinity of all B-cells $b \in N$ with ag.
         Select highest affinity B-cells, perform affinity proportional cloning, place clones in C.
         for all B- cell clones $c \in C$ **do**
            Mutate c at rate inversely proportional to affinity.
            Determine affinity of c with ag.
         end for
         Select h highest affinity clones $c \in C$ and place in D.
         Remove all elements of D whose affinity with ag is less than ct.
         Remove elements of D whose affinity with other elements in D is less than ct.
         Insert remaining elements of D into N.
      end for
      Determine affinity between each pair of B-cells in N.
      Systemically remove all B cells whose affinity to another B cell is less than nt.
      Introduce a new, randomly generated, B-cells into N.
   until a stopping condition has been satisfied
end

---

renders the algorithm robust regarding exact parameter values. The algorithm captures well the properties of self-organization and population regulation as exhibited by the immune system. Galeano et al. (2005) provides a good review of many other immune networks that appear in the literature.

3.3   Negative Selection AIS

Inspired by the observation that the immune system protects the host body from invading pathogens, early AIS mapped these qualities to the invasion of computers and computer networks by viruses, worms, and intruders. The concept of self–nonself discrimination provided the basis for the development of various security-based AIS algorithms. Specifically, a process called negative selection was used, as inspiration, to derive a set of detectors capable of recognizing only nonself. An example of an algorithm based on Forrest et al. (1994) is shown in ❷ *Algorithm 3*.

This algorithm was applied to protecting a computer from unauthorized changes, such as infection with a virus. There are two main stages to the algorithm: the generation of detectors; and the online monitoring of data and programs for changes. In the detector-generation stage, the collection of self strings S represents data and programs stored on the computer. The randomly generated detectors D are matched against elements in S, and those $d \in D$ that match (based on an affinity function) are removed. In Forrest et al. (1994) (this work was the

---

Algorithm 3   Generic negative selection algorithm (Based on Forrest et al. (1994).)

input: S = set of self strings characterizing benign, normal data.
output: A = Stream of nonself strings detected.
begin
    Create empty set of detector strings D ▷ Generation of detector strings
    Generate random strings C.
    for all random strings $c \in C$ **do**
        for all self strings $s \in S$ **do**
            if c matches s **then**
                Discard c
            else
                Place c in D
            end if
        end for
    end for
    while There exist protected strings p to check **do**   ▷ Detection stage
        Retrieve protected string p
        for all detector strings $d \in D$ **do**
            if p matches d **then**
                Place p in A and output.   ▷ Nonself string detected
            end if
        end for
    end while
end

---

first instance of negative selection being employed in the context of computer security) an affinity function that checked for the similarity of r consecutive characters at any point in the detector and self strings, called the r-contiguous matching rule, was used. The randomly generated detectors that are not removed from the detector collection are used to check for alterations to the system.

Negative selection algorithms have not been constrained to detection of viruses; they have also found application as intrusion detection systems. In this context the self strings S could be a concatenation of source IP, destination IP, and port addresses (Forrest and Beauchemin 2007). The detector-generation stage would be executed during a time when the network was known to be secure. Consequently, a match during the monitoring phase could indicate an anomalous connection, an intrusion. A large amount of work has been dedicated to the development of negative selection algorithms in a variety of application areas and from a theoretical perspective (Balthrop et al. 2002; Gonzalez and Dasgupta 2003; Esponda et al. 2004).

Despite a considerable amount of examples in the literature, it has been argued that negative selection suffers several drawbacks. Defining self can prove problematic; in the case of a network the total variety of safe packets can be enormous, the logistics of capturing this self set can prove difficult. In deriving the set D a huge quantity of randomly generated detectors that match self will have been deleted, thus it can become very inefficient (Freitas and Timmis 2007). Furthermore, algorithms of this variety have been seen to suffer certain scaling problems: as the universe in which self and nonself elements are defined grows (reflecting the complexity of the detection problem), the number of detectors required to effectively cover the nonself space becomes difficult to generate (Stibor et al. 2005; Timmis et al. 2008b).

3.4 Danger Theory AIS

It has been suggested that the integration of mechanisms derived from *danger theory* (Matzinger 1994) could provide for more effective intrusion detection algorithms than traditional negative selection approaches (Aickelin and Cayzer 2002). Rather than monitor for the explicit presence of the intruder, danger-theory-inspired systems could be alerted by the anomalous intruder behavior. The shift in emphasis is subtle, but significant. Such an intrusion detection system would monitor for signs of "danger," such as abnormalities in memory usage or disk activity, unexpected or unwarranted frequencies of file changes (Aickelin and Cayzer 2002; Aickelin et al. 2003).

An interesting consequence of danger-inspired AIS lies in the interpretation of the *danger zone*. In vivo, this is the spatial neighborhood from where the danger signals originate. In the artificial domain, this concept need not be spatial, Secker et al. (2003) place the danger zone in the temporal domain. The concept of danger signals provides danger-theory-inspired engineering solutions with several advantages over self- and nonself-inspired approaches. Danger signals restrict the domain of nonself to a manageable size, remove the requirement to observe all self, and instill adaptability regarding scenarios where self and nonself boundaries are dynamic (Aickelin and Cayzer 2002).

The main danger-theory-inspired algorithm that has been developed is the dendritic cell algorithm (DCA) (Greensmith et al. 2005, 2006a). The DCA is a signal-processing algorithm, inspired by the behavior of dendritic cells. These reside in the body tissues and collect antigen and other (danger) signals that provide a picture of the current state of the tissues. This picture determines whether the antigen has been collected in a safe or dangerous context, and causes dendritic cells to change into a *semi-mature* or *mature* state. The task of the DCA is to classify data items (antigens) as being either benign or malignant in nature. Antigen are associated with concentrations of pathogen associated molecular pattern (PAMP) signal, danger signal, safe signal, and pro-inflammatory signals. These signals are derived from real biological signals and are mapped onto attributes associated with the data items as follows (Greensmith et al. 2006a):

- *PAMP.* A known signature of abnormal behavior. This attribute of the data item is highly indicative of an anomaly.
- *Danger signal.* A moderate degree of confidence that this attribute of the data item is associated with abnormal behavior.
- *Safe signal.* Indicative of normal system operation.
- *Pro-inflammatory signal.* A general sign of system distress.

The main challenge in implementing the DCA is defining how these signals map onto the data items derived from the problem domain (Greensmith et al. 2006a).

The DCA, shown in ❷ *Algorithm 4*, operates by maintaining a pool of dendritic cells (DCs). From this pool, dendritic cells are randomly selected to sample data items (and related signals) that are presented to the algorithm in a sequential manner. Based on the signals received, dendritic cells produce *semi-mature* and *mature* cytokines (immune signaling molecules). At the end of antigen processing, DCs are assigned semi-mature or mature status according to the levels of the cytokines produced. Every data item is then classified as being benign or malignant on the basis of a majority vote amongst the DCs that sampled it, each voting in accordance to its level of maturity.

Through its focus on behavioral consequences (derived from the signals described above) as opposed to physical presence (in the case of negative selection algorithms), the

---

Algorithm 4   The Dendritic Cell Algorithm (DCA) (Greensmith et al. 2005)

input: $S =$ a set of antigens, representing data elements classified as safe or dangerous.
output: $K =$ set of antigens classified as safe.
       $L =$ set of antigens classified as dangerous.
begin
  Create *DC* pool of 100 dendritic cells.
  for all antigen *ag* $\in S$ **do**   ▷ Perform signal processing on *ag*
    for 10 randomly selected dendritic cells *dc* $\in DC$ **do**
      Sample *ag*.
      Update *dc.danger*, *dc. PAMP*, and *dc.safe* signals based on *ag*.
      Calculate and update concentration of *dc.semimatureCytokine* output cytokine.
      Calculate and update concentration of *dc.matureCytokine* output cytokine.
      Calculate and update concentration of *dc.coStimulatory* output molecules.
      if concentration of *dc.coStimulatory* > threshold **then**
        Remove *dc* from *DC* and place in *M*.
        Insert new *dc* into *DC*.
      end if
    end for
  end for

  for all dendritic cells *dc* $\in$ *M* **do**   ▷ Differentiation of dendritic cells.
    if concentration of *dc.semimatureCytokine* > *dc.matureCytokine* **then**
      dc.class = semi/mature.
    else
      dc.class = mature.
    end if
  end for

  for all antigen *ag* $\in S$ **do**   ▷ Classification of antigens
    for all dendritic cells *dc* $\in$ *M* that sampled *ag* **do**
      Calculate if *ag* presented in mature or semimature context by *dc*.
    end for
    if *ag* presented as semimature majority of time **then**
      Place *ag* in *K*.   ▷ *ag* is benign
    else
      Place *ag* in *L*.   ▷ *ag* is malignant
    end if
  end for
end

DCA is able to operate in the presence of dynamically changing environments. However, in its current state (Greensmith et al. 2005, 2006a, b), the DCA is not able to operate in a true online fashion; data must be collected *a priori* and classification is performed as a final batch operation. Hence, anomalies cannot be detected as they occur. A second potential problem for the DCA is that misclassification can occur around the boundaries where data items switch between *safe* and *dangerous* contexts. This is due to multiple sampling of antigen by each DC. The consequence is that the DCA will exhibit significant misclassification when applied to problems where context switches in the data items are frequent (Greensmith et al. 2005). In order to overcome the limitation of operating in an off-line manner, Lay and Bate (2007) have developed a real-time, online DCA that is capable of altering schedule overruns in real-time operating systems. The DCA has also been used for behavior classification on a robotics platform (de Castro et al. 2007a).

4 Reflections and Projections

Artificial immune systems has matured into a well recognized field that tackles a broad range of problem domains. This is best illustrated from the proceedings of the International Conference of Artificial Immune Systems ICARIS (Timmis et al. 2003; Nicosia et al. 2004; Jacob et al. 2005; Bersini and Carneiro 2006; de Castro et al. 2007b; Bentley et al. 2008). The field is now at a stage where a number of researchers are reflecting upon its contributions to the wider academic and engineering communities. A number of these reflections and proposed future directions for AIS are assessed here.

4.1 Evaluation of Current AIS

Hart and Timmis (2008) analyze a large collection of AIS engineering applications and categorize these into three classes of problem: anomaly detection, optimization, and clustering and classification. Considering key works from each class in turn, they attempt to assess and evaluate whether the application of AIS brings any benefits that could not be derived from applying alternative, existing techniques to the problem. Their criteria asserts that it is not sufficient to simply outperform other algorithms on benchmark tests; to be truly successful, the AIS must contain features that are not present in alternative paradigms.

Anomaly detection AIS are assessed by Hart and Timmis (2008) as having had limited success, but the authors make note of recent advances that danger-theory-inspired algorithms have provided, and state that significant breakthroughs are still possible. For optimization problems, it is concluded that although optimization-based AIS can and will provide comparable performance to existing methods, they will not offer any distinguishing features that cannot be found elsewhere. For classification and clustering applications, the authors conclude that the naturally distributed nature of some AIS algorithms allows for natural parallelization and distribution across several processors, offering something potentially distinctive. Regarding operation over dynamic data sets, the authors state that by definition, AIS algorithms incorporate some notion of memory, and could therefore outperform alternative learning systems which are purely reactive in nature.

Though their assessment of AIS accomplishments concludes that many are not truly successful, the authors note that this is partly due to several shortcomings that have characterized AIS design and application to date (Hart and Timmis 2008). These include: the methodology through which AIS algorithms capture their inspiring immunology; the attention paid to the effects that certain design decisions impose when engineering AIS systems; the theoretical understanding of AIS algorithms; and the nature of the problems to which AIS have been applied.

In a similar vein to Hart and Timmis (2008), Garrett (2005) studies various AIS to attempt to answer the question of whether AIS research has delivered anything *useful* to date. A useful algorithm in this context is defined by being *distinct* and *effective*. An algorithm's distinctiveness is assessed through criteria covering the algorithm's internal representation of the problem and potential solutions, and its computational components. Effectiveness is assessed on the algorithm's performance, including the path through which solutions are obtained, the quality of results obtained through its application to benchmark problems, and the speed at which results can be obtained. In combining the two sets of criteria, an algorithm is said to be useful if it is both effective and distinctive.

The fact that work reflecting on the state of AIS is being conducted is encouraging, and is healthy for the discipline. However, it should be noted that the method and criteria employed by Garrett (2005) in arriving at its conclusions has been criticized for being more of an exercise in classification than in detailed evaluation, and for being highly subjective in nature (Timmis et al. 2008a). Additionally, the criteria focuses on performance in relation to benchmark problems. It has been suggested that a downfall of AIS research to date has been its repeated application to benchmark problems, and to areas for which many quality solutions already exist (Hart and Timmis 2008; Timmis et al. 2008a). The effectiveness criteria do not reflect the need for AIS to carve its own niche (Hart and Timmis 2008; Timmis et al. 2008a), and provide quality solutions in a problem domain that no other technique can match.

McEwan et al. (2008) question the appropriateness of the shape space representation for AIS with respect to machine-learning problems. Typical machine-learning problems entail data sets of very high dimensionality. In such a scenario, the adoption of the shape space representation can lead to the "curse of dimensionality": as the dimension of the space increases linearly, its volume increases exponentially, and the quality of locality that affinity measures attempt to discern becomes meaningless as all points approach in equidistance to one another. It has been noted by Stibor et al. (2005) that the task of generating, maintaining, and exploiting an effective set of detectors within such a high-dimensional space is computationally intractable.

As an alternative, McEwan et al. (2008) propose marrying the machine-learning technique of *boosting* with immune inspiration. Boosting proposes a strong learning strategy that is derived as a compound decision between multiple (slightly better than random) weak learners. The authors draw analogy to the cooperative nature in which many varieties of immune cells with differing specificities and recognition targets are able to cooperatively mount an effective immune response that hones on a specific target (Cohen's *correspondence* (Cohen 2000)).

4.2 Inspiration, Frameworks, and Methodologies

In recent years, there has been a gradual shift in some AIS toward paying more attention to the underlying biological system that serves as inspiration. For example, the development of the DCA (see ❷ Sect. 3.4) involved the input from real biological experimentation as inspiration. However, there was no reported sophisticated biological modeling to understand the underlying biology as is suggested by Stepney et al. (2005) and Timmis et al. (2006). Other examples of this shift back to the underlying biology include Wilson and Garrett (2004) and Jacob et al. (2004), who have used modeling techniques to build AIS in order to understand underlying immune properties.

The majority of AIS such as those detailed in ❷ Sects. 3.1–3.3 have taken their inspiration from well-established immunological perspectives. In contrast, Andrews and Timmis (2005, 2007) advocate exploiting conflicting immune theories as a rich source of potential ideas for the engineer. This is an approach that has been successfully carried out by Aickelin and

Cayzer (2002), Secker et al. (2003), and Greensmith et al. (2005) in exploiting danger theory and the development of the DCA. AIS can draw significantly more inspiration from the immune system, and the immunological debate surrounding its higher functions, than the relatively simplistic subset of concepts that have served thus far.

Taking this approach, the AIS engineer enjoys the freedom of adopting various immunological models and concepts that best suit the application domain. It is, however, essential to ensure that the concepts employed are correctly abstracted and reasoned about to accurately capture the emergent phenomena from which they are inspired. It has been argued that various immune-inspired algorithms have been hampered with a lack of biological accuracy (Timmis 2007). Typically, immune-inspired algorithms have fallen prey to "reasoning by metaphor," wherein their operation and structures bear only a weak resemblance to the biological phenomenon that inspired them (Stepney et al. 2005), and, thus, consequentially fail to unlock their full potential (Hart and Timmis 2008).

To combat the problems associated with this apparent weakness in biological metaphors, a conceptual framework approach to the development of bio-inspired algorithms has been proposed (Stepney et al. 2005), shown in ❷ *Fig. 2*. This provides a structure and methodology for biological investigation, abstraction, modeling, and ultimately the construction of algorithms. The process should be interdisciplinary, involving at the very least biologists, mathematicians, and computer scientists. The framework aims to facilitate a better understanding of the targeted underlying biological concepts and to ultimately build more powerful bio-inspired algorithms whilst simultaneously gaining a better understanding of which application domains these algorithms are best suited to.

The first step in the conceptual framework approach is to probe the biological system through observation and experimentation. These probes are biased toward extracting information concerning the particular biological phenomena of interest. From the information gained, careful abstraction and mathematical modeling will highlight the central processes responsible for the observed biological phenomena. Analytical computational models may be constructed, which allow for the execution and animation of any underlying model, and can

◻ Fig. 2
The conceptual framework approach to deriving biologically inspired algorithms (Stepney et al. 2005).

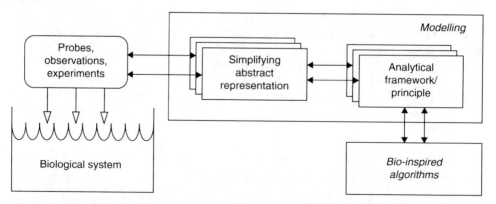

provide a deeper insight into its workings. The observations and mechanisms perceived at this stage will be free from any particular application bias. Finally, these insights can serve as design principles for bio-inspired algorithms, which may be applied to non-biological problems (Stepney et al. 2005; Hart and Timmis 2008).

A number of AIS works have been inspired by the conceptual framework principles. These include: a computational model of degenerate T-cell receptors (Andrews and Timmis 2006) and adaptable degenerate immune cell receptors (Andrews and Timmis 2008); and an instantiation of an artificial cytokine network (Hone and van den Berg 2007), which examined the behavior of the network to elicit any useful properties that could be applied to solving engineering problems (Read et al. 2008). Newborough and Stepney (2005) also apply many of the conceptual framework ideas to produce a generic framework for population-based bio-inspired algorithms including genetic algorithms, negative selection, clonal selection, particle swarm optimization, and ant colony optimization.

The conceptual framework of Stepney et al. (2005) also influenced Twycross and Aickelin (2005) who present a general meta-framework for models incorporating innate immunity. A table of six general properties of the innate immune system is presented and it is claimed that AIS will need to incorporate properties such as these to realize functions of the immune system. Similarly, Guzella et al. (2007) highlight a class of T cell, T regulatory cells, as inspiration for AIS. They suggest that incorporating these cells might lead to more biologically plausible models and algorithms that achieve better results in real-life problems.

While the conceptual framework offers a structured methodology for the development of immune- (and other biologically) inspired algorithms, the deployment of these AIS in a particular engineering context also requires careful consideration. Through their examination of AIS application to classification problems Freitas and Timmis (2007) note several considerations, frequently overlooked, which can significantly affect an algorithm's suitability and performance. They state that the implementor of an AIS algorithm should note the nature of the problem's data, and chose a representation that intuitively maps the data's characteristics. Altering the data to suit a particular representation, in particular, discarding data that is of a different type (e.g., disposing of categorical data to fit a continuous valued representation), is bad practice. Rather, the immune-inspired algorithm's representation should be tailored to suit the problem's data.

Freitas and Timmis (2007) also advise careful consideration of the choice of affinity measure for the chosen representation. An affinity measure can be associated with an inductive bias: some basis through which one hypothesis will be favored over another. An inductive bias is not an undesirable trait, it forms the basis of learning. Yet, care must be taken to ensure that the inductive bias incurred is appropriate for the problem at hand. For example, certain affinity measures have a positional bias, whereby the order of data within the representation can affect the outcome of the affinity measure. If the order of the data is irrelevant to the problem being tackled, then an affinity measure yielding a positional bias might be an inappropriate choice. This work is supported by empirical investigations into the effects of different affinity measures by Hart and Ross (2004) and Hart (2005).

4.3 Application Domains

It has been suggested by Hart and Timmis (2008) that there will be little benefit from applying AIS algorithms to problems of a static nature, over existing and established paradigms. The authors conjecture that the distinctive "killer application" niche for AIS will require algorithms to exhibit the following properties (quoted verbatim):

- They will be *embodied*.
- They will exhibit *homeostasis*.
- They will benefit from interactions between *innate* and *adaptive* immune models.
- They will consist of *multiple, heterogeneous interacting, communicating components*.
- Components can be easily and naturally *distributed*.
- They will be required to perform *life-long learning*.

Recent applications of AIS in novel problem domains have started to show indications of satisfying these properties, which are reviewed here.

A central function of the immune system is its cooperation with the endocrine and neural systems in the provision of homeostasis to the host (Hart and Timmis 2008). Homeostasis is "the tendency of a system, esp. the physiological system of higher animals, to maintain internal stability, owing to the coordinated response of its parts to any situation or stimulus tending to disturb its normal condition or function" American Psychological Association (APA). Hence, since the domain in which in vivo immune systems operate is inherently dynamic; it is not unreasonable to surmise that immune-inspired algorithms might be particularly well suited to operation in dynamic environments. The immune system's potential as inspiration for homeostasis in robotics is investigated by Owens et al. (2007). Here, homeostasis requires: the system to perceive the environment from multiple perspectives to overcome the inherent problems of sensory malfunction; a repertoire of innate responses that can affect change in the environment or the system directly; the cognition that facilitates the selection of an appropriate effector action in response to perceived input state; and the ability to adaptively correlate sensory information and effector mechanisms, such that its actions can dynamically evolve with a changing environment. Similarly, Neal et al. (2006) outline an endocrine-immune-inspired homeostatic control system. The artificial immune system allows for low-level faults (e.g., an overheating motor) to be corrected locally (e.g., by turning on a local fan), while integration with an artificial endocrine system allows for chronic faults to propagate inflammation throughout the robot's systems. System-wide inflammation influences the higher level function of the robot in a global attempt to rectify the fault, for example, the decision by the robot to stop moving, thus allowing the motor to cool down.

The potential for AIS application in the domain of real-time systems was demonstrated by Lay and Bate (2007), who employed the dendritic cell algorithm in the detection of process deadline over runs in an embedded system. The analysis of process executions, and the insurance that all deadlines are met is typically performed statically during the development process. By incorporating adaptive AIS techniques, it is hoped that the system is rendered robust, while simultaneously reducing development time and costs.

Embodiment in bio-inspired engineering has been investigated by Stepney (2007), who examines the intimate coupled nature of a system and its environment. This includes their perceptions and consequent reactions in perturbing one another through complex high bandwidth feedback networks. The environment is open, with a quantitatively large and rich variety of information flowing through it, while the system exhibits highly nonlinear dynamics; small input perturbations need not equate to small behavioral modifications. A consequence of embodiment is the coevolution of the environment with the system. In the context of the danger theory of the immune system (Matzinger 2002), immune cells (system) have learnt to perceive danger signals just as the body (environment) has learnt to provide them. Pathogens experience evolutionary pressure to evade detection, thus contributing to the environment's dynamics (Stepney 2007). Thus, the two are intimately bound. This concurs with the argument for the complex systems view of immunology presented by Cohen (2000). Thus, for engineers to truly capture the complexity of the biology from which they derive their inspiration, they must embody the artificial system within its artificial environment, rather than deliberately engineer the interfaces, sensors, and actuators through which the system interacts with its environment.

Though no AIS currently satisfies the conceptual features of embodiment as outlined in Stepney (2007), Bentley et al. (2005) go some way in addressing similar issues by suggesting that a layer is missing from AIS design. They outline the concept of an artificial tissue layer acting as an interface between a problem space and an AIS. The tissue layer performs some data preprocessing before presenting it to the AIS, allowing for the incorporation of domain-specific knowledge and the integration of several data sources, and is the medium through which the AIS responds. An analogy is drawn by Bentley et al. (2005) between the artificial tissue providing an innate response and the AIS providing the adaptive response. As a preliminary investigation, two tissue algorithms are presented by Bentley et al. (2005). In a similar work, Twycross and Aickelin (2006) present a framework that facilitates the complete encapsulation of an AIS, providing: artificial anatomical compartments within which the AISs immune elements may operate; and generic receptors for other immune cells, antigens, and cytokines contained within the compartment. The framework was partially motivated through the possibility to evaluate the performance of several alternative AIS algorithms on the same problem, but the manner in which it interfaces the AIS with the environment, and performs preprocessing is interesting from the perspective of embodiment.

4.4 Modeling and Simulating the Immune System

In recent years, building models and simulations of the immune system have become an important aspect of AIS, both as stand-alone pieces of work, and as steps toward producing engineering applications using methodologies such as the conceptual framework approach of Stepney et al. (2005). Forrest and Beauchemin (2007) note that there is a vast range of modeling approaches applicable to modeling the immune system, each with their own advantages and disadvantages operating at different levels of abstraction.

An overview of many mathematical techniques used for modeling the immune system is provided by Perelson and Weisbuch (1997). A large number of these approaches involve the use of differential equations, although other techniques can be applied, such as Boolean networks (Weisbuch and Atlan 1988) and the work of Kelsey et al. (2008) which present and analyze a Markov chain model of a cytokine network. Recently, process calculi have been applied to models of the immune system such as Owens et al. (2008). Process calculi are formal languages from computer science that are used to specify concurrent systems. As biological systems are inherently concurrent, these types of languages seem well suited to biological modeling.

Forrest and Beauchemin (2007) provide a review of many of the modeling approaches in immunology, with a focus on agent based modeling (ABM). In ABM, components such as cells (and sometimes molecules) are represented individually as *agents*, rather than as homogenous populations such as in differential equation techniques. Different agent types typically represent different immune cell types. These agent types are encoded with simple rules extracted from the real biology that govern how they behave and interact. ABM techniques typically employ an explicit notion of space, such as that used in cellular automata-like models (Kleinstein and Seiden 2000). The advantage of ABM is that it allows the observation of agent population dynamics as they emerge from the interactions of individual agents. An example of ABM is Beauchemin et al. (2006) who investigate the dynamics of in vitro infection with a strain of influenza. ABM has also been used to study more computational aspects of applied AIS such as a series of work by Hart and Ross (2004), Hart (2005, 2006), and Hart et al. (2006). These works use a simulation of an idiotypic network to investigate how different models of shape-space and affinity affect the dynamics of the network, such as memory capacity and the structures formed, emphasizing the need for careful choice of parameters in the engineered systems.

Diagramatic tools have also been used to model the immune system, the most widely used being the unified modeling language (UML) (Fowler 2000), which consists of a set of 13 different types of diagram that can model different aspects of structure and behavior. The advantage of the UML is its non-domain-specific nature and subsequent ability to capture abstractions. The UML (and related diagrams such as statecharts) have started to become a powerful tool in modeling aspects of biological systems. By far the most advanced use of the UML and statecharts in immunology is that of Efroni et al. (2003), who have built a sophisticated and predictive model of T cell maturation in the thymus using a tool called reactive animation, which combines the use of statecharts and other UML diagrams. In addition to the UML, there are other techniques used in software engineering, which Bersini (2006) suggests can facilitate the development and *communication* of immune modeling. These include object-oriented technologies such as object-oriented programming and design patterns (Gamma et al. 1995). The perceived benefit is the clarification of immune objects and their relationships. To support this, Bersini (2006) provides an example of how clonal selection can be modeled with a simple state diagram.

5   Summary

This chapter is intended as an overview of the area of artificial immune systems (AIS), predominantly from an engineering solutions perspective. It is not meant to be an exhaustive bibliography, but serves to illustrate that AIS is an area of great diversity, actively reflecting upon itself, and expanding into new areas and meeting new challenges. The spectrum of AIS research ranges from the modeling of immune systems in aid of immunological study, to the development of algorithms for specific engineering applications. While the predominant focus of this chapter has been on the algorithmic aspect of AIS, the other aspects of AIS research are no less significant. The principled development of immune-inspired algorithms that capture, in a more than superficial manner, the properties and characteristics of the immune system are equally valuable to the discipline's continuing success. This is highlighted to different degrees in Stepney et al. (2005) and Timmis et al. (2006) who advocate the careful consideration of the underlying biological system, the use of modeling to help understand that system, and the principled abstraction of algorithms and general frameworks from those models.

It is worth noting that there are many different aspects of the immune system that have served as inspiration for AIS. Only the main strands have been reviewed in this chapter, namely: clonal selection, immune networks, negative selection, and danger theory. However, there remain many untapped possibilities that are worthy of consideration and study within principled frameworks, such as the conceptual framework of Stepney et al. (2005). In addition, the area of AIS should seek out challenging application areas that exploit the immune metaphor further than it has to date.

As pointed out by Timmis et al. (2008a), a recent paper by Cohen (2007) identifies three types of AIS researcher. The first type he calls the literal school, they build artificial systems that attempt to perform analogous tasks to the actual immune system (e.g., build computer security systems that discriminate between self and nonself); the second type are those of the metaphorical school who take inspiration from the immune system and build artificial systems based on analogies (so the application may be far from analogous to what the immune system does); and a third type of researcher aims to understand immunity through the development of computational and mathematical models. This goes to illustrate the diversity of research that lies within the discipline of AIS, and renders it a promising avenue for truly interdisciplinary research.

Acknowledgments

Mark Read is sponsored by the Department of Computer Science, University of York, and Paul Andrews is supported by EPSRC grant number EP/E053505/1.

References

Aickelin U, Cayzer S (2002) The danger theory and its application to artificial immune systems. In: Timmis J, Bentley PJ (eds) ICARIS 2002: Proceedings of the 1st international conference on artificial immune systems, University of Kent Printing unit, Canterbury, UK, September 2002, pp 141–148

Aickelin U, Bentley PJ, Cayzer S, Kim J, McLeod J (2003) Danger theory: The link between AIS and IDS? In: Bentley PJ, Hart E (eds) ICARIS 2003: 2nd international conference on artificial immune systems, Edinburgh, Scotland, September 2003. Lecture notes in computer science, vol 2787. Springer, New York, pp 147–155

American Psychological Association (APA): Homeostasis. (n.d.). Dictionary.com Unabridged (v 1.1). Retrieved June 25, 2008, from Dictionary.com Web site: http://dictionary.reference.com/browse/homeostasis

Andrews PS, Timmis J (2005) Inspiration for the next generation of artificial immune systems. In: Jacob C, Pilat ML, Bentley PJ, Timmis J (eds) ICARIS 2005: 4th international conference on artificial immune systems, Banff, Canada, April 2005. Lecture notes in computer science, vol 3627. Springer, Heidelberg, pp 126–138

Andrews PS, Timmis J (2006) A computational model of degeneracy in a lymph node. In: Bersini H, Carneiro J (eds) ICARIS 2006: 5th international conference on artificial immune systems, Oeiras, Portugal, September 2006. Lecture notes in computer science, vol 4163. Springer, Berlin, pp 164–177

Andrews PS, Timmis J (2007) Alternative inspiration for artificial immune systems: exploiting Cohen's cognitive immune model. In: Flower D, Timmis J (eds) In silico-immunology. Springer, New York, Chap 7 (2007)

Andrews PS, Timmis J (2008) Adaptable lymphocytes for artificial immune systems. In: Bentley PJ, Lee D, Jung S (eds) ICARIS 2008: 7th international conference on artificial immune systems, Phuket, Thailand, August 2008. Lecture notes in computer science, vol 4163. Springer, Berlin, pp 376–386

Balthrop J, Esponda F, Forrest S, Glickman M (2002) Coverage and generalisation in an artificial immune system. In: Genetic and evolutionary computation. Morgan Kaufmann Publishers Inc., San Francisco, CA, USA, July 2002, pp 3–10

Beauchemin C, Forrest S, Koster FT (2006) Modeling influenza viral dynamics in tissue. In: Bersini H, Carneiro J (eds) ICARIS 2006: 5th international conference on artificial immune systems, Oeiras, Portugal, September 2006. Lecture notes in computer science, vol 4163. Springer, Berlin, pp 23–36

Bentley PJ, Greensmith J, Ujjin S (2005) Two ways to grow tissue for artificial immune systems. In: Jacob C, Pilat ML, Bentley PJ, Timmis J (eds) ICARIS 2005: 4th international conference on artificial immune systems, Banff, Canada, April 2005. Lecture notes in computer science, vol 3627. Springer, Heidelberg, pp 139–152

Bentley PJ, Lee D, Jung S (eds) (2008) ICARIS 2008: 7th international conference on artificial immune systems, Phuket, Thailand, August 2008. Lecture notes in computer science, vol 5132. Springer, New York http://www.artificial-immune-systems.org/icaris.shtml

Bersini H (2006) Immune system modeling: the OO way. In: Bersini H, Carneiro J (eds) ICARIS 2006: 5th international conference on artificial immune systems, Oeiras, Portugal, September 2006. Lecture notes in computer science, vol 4163. Springer, Berlin, pp 150–163

Bersini H, Carneiro J (eds) (2006) ICARIS 2006: 5th international conference on artificial immune systems, Oeiras, Portugal, September 2006. Lecture notes in computer science, vol 4163. Springer, Berlin http://www.artificial-immune-systems.org/icaris.shtml

Bezerra GB, de Castro LN, Zuben FJV (2004) A hierachical immune network applied to gene expression data. In: ICARIS 2004: Proceedings of the 3rd international conference on artificial immune systems. Catania, Springer, Berlin/Heidelberg, September 2004, pp 14–27

Cohen IR (2000) Tending Adam's garden: evolving the cognitive immune self. Elsevier Academic Press, London, UK

Cohen IR (2007) Real and artificial immune systems: computing the state of the body. Nat Rev Immunol 7:569–574

Cutello V, Nicosia G, Pavone M (2004a) Exploring the capability of immune algorithms: a characterization of hypermutation operators. In: Nicosia G, Cutello V, Bentley PJ, Timmis J (eds) ICARIS 2004: 3rd international conference on artificial immune systems? Catania, Italy, September 2004. Lecture notes in computer science, vol 3239. Springer, Berlin, pp 263–276

Cutello V, Nicosia G, Pavone M (2004b) An immune algorithm with hyper-macromutations for the Dill's 2D hydrophobic-hydrophilic model. IEEE congress on evolutionary computation, CEC 2004, Portland, Oregon, USA, June 19–23, 2004. IEEE Press, 1:1074–1080

Cutello V, Narzisi G, Nicosia G, Pavone M (2005) Clonal selection algorithms: A comparative case study using effective mutation potentials. In: Jacob C, Pilat ML, Bentley PJ, Timmis J (eds) ICARIS 2005: 4th international conference on artificial immune systems, Banff, Canada, April 2005. Lecture notes in computer science, vol 3627. Springer, Heidelberg, pp 263–276

de Castro LN, Timmis J (2002a) Artificial immune systems: a new computational approach. Springer-Verlag, London

de Castro L, Timmis J (2002b) An artificial immune network for multi modal optimisation. In: WCCI: Proceedings of the world congress on computational intelligence, Honolulu, HI, May 2002. IEEE, New York, NY, USA, pp 699–704

de Castro LN, Von Zuben FJ (2000) The clonal selection algorithm with engineering applications. In: Proceedings of GECCO'00, workshop on artificial immune systems and their applications. Las Vegas, NV

de Castro LN, Von Zuben FJ (2001) aiNet: an artificial immune network for data analysis. Idea Group Publishing, Hershey, PA, pp 231–259

de Castro LN, Von Zuben FJ (2002) Learning and optimization using the clonal selection principle. IEEE Trans Evol Comput 6(2):239–251

de Castro LN, Von Zuben FJ, Knidel H (eds) (2007a) The application of a dendritic cell algorithm to a robotic classifier. In: ICARIS 2007: Proceedings of 6th international conference on artificial immune systems, Santos, Brazil, August 2007. Lecture notes in computer science, vol 4628. Springer, Berlin

de Castro LN, Von Zuben FJ, Knidel H (eds) (2007b) ICARIS 2007: Proceedings of 6th international conference on artificial immune systems, Santos, Brazil, August 2007. Lecture notes in computer science, vol 4628. Springer, Berlin http://www.artificial-immune-systems.org/icaris.shtml

Efroni S, Harel D, Cohen IR (2003) Towards rigorous comprehension of biological complexity: modeling, execution, and visualization of thymic t-cell maturation. Gen Res 13:2485–2497

Esponda F, Forrest S, Helman P (2004) A formal framework for positive and negative detection schemes. IEEE Trans Syst Man Cybern B Cybern 34(1):357–373

Farmer JD, Packard NH, Perelson AS (1986) The immune system, adaptation, and machine learning. Phys D 2(1–3):187–204

Flower D, Timmis J (eds) (2007) In silico immunology. Springer, New York

Forrest S, Beauchemin C (2007) Computer immunology. Immunol Rev 216(1):176–197

Forrest S, Perelson AS, Allen L, Cherukuri R (1994) Self-nonself discrimination in a computer. In: SP '94: Proceedings of the 1994 IEEE symposium on security and privacy, Oakland, CA, May 1994. IEEE Computer Society, Washington DC, pp 202–212

Fowler M (2000) UML distilled: a brief guide to the standard object modeling language, 2nd edn. Addison-Wesley, Reading, MA

Freitas A, Timmis J (2007) Revisiting the foundations of artificial immune systems for data mining. IEEE Trans Evol Comput 11(4):521–540

Galeano JC, Veloza-Suan A, González FA (2005) A comparative analysis of artificial immune network models. In: GECCO 2005: Proceedings of the genetic and evolutionary computation conference, Washington, DC, June 2005. Springer, Berlin

Gamma E, Helm R, Johnson R, Vlissides J (1995) Design patterns: elements of reusable object-oriented software. Addison-Wesley, Reading, MA

Garrett S (2005) How do we evaluate artificial immune systems? Evol Comput 13(2):145–177

Goldsby RA, Kindt TJ, Osborne BA, Kuby J (2003) Immunology, 5th edn. W. H. Freeman and Company, New York

Gonzalez FA, Dasgupta D (2003) Anomaly detection using real-valued negative selection. Genet Programming Evolvable Mach 4(4):383–403

Greensmith J, Aickelin U, Cayzer S (2005) Introducing dendritic cells as a novel immune-inspired algorithm for anomaly detection. In: Jacob C, Pilat ML, Bentley PJ, Timmis J (eds) ICARIS 2005: 4th international conference an artificial immune systems, Banff, Canada, April 2005. Lecture notes in computer science, vol 3627. Springer, Heidelberg, pp 153–167

Greensmith J, Aickelin U, Twycross J (2006a) Articulation and clarification of the dendritic cell algorithm. In: Bersini H, Carneiro J (eds) ICARIS 2006: 5th international conference on artificial immune systems, Oeiras, Portugal, September 2006. Lecture notes in computer science, vol 4163. Springer, Berlin, pp 404–417

Greensmith J, Twycross J, Aickelin U (2006b) Dendritic cells for anomaly detection. In: CEC 2006: IEEE congress on evolutionary computation, Vancouver, Canada, July 2006, pp 664–671

Guzella TS, Mota-Santos TA, Caminhas WM (2007) Regulatory t cells: inspiration for artificial immune systems. In: de Castro LN, Von Zuben FJ, Knidel H (eds) ICARIS 2007: of 6th international conference on artificial immune systems, Santos, Brazil, August 2007. Lecture notes in computer science, vol 4628. Springer, Berlin, pp 312–323

Hart E (2005) Not all balls are round: an investigation of alternative recognition-region shapes. In: Jacob C, Pilat ML, Bentley PJ, Timmis J (eds) ICARIS 2005: 4th international conference on artificial immune systems, Banff, Canada, April 2005. Lecture notes in computer science, vol 3627. Springer, Heidelberg, pp 29–42

Hart E (2006) Analysis of a growth model for idiotypic networks. In: Bersini H, Carneiro J (eds) ICARIS 2006: 5th international conference on artificial immune systems, Oeiras, Portugal, September 2006. Lecture notes in computer science, vol 4163. Springer, Berlin, pp 66–80

Hart E, Ross P (2004) Studies on the implications of shape-space models for idiotypic networks.

In: Nicosia G, Cutello V, Bentley PJ, Timmis J (eds) ICARIS 2004: 3rd international conference on artificial immune systems, Catania, Italy, September 2004. Lecture notes in computer science, vol 3239. Springer, Berlin, pp 413–426

Hart E, Timmis J (2008) Application areas of AIS: the past, the present and the future. J Appl Soft Comput 8(1):191–201

Hart E, Bersini H, Santos F (2006) Tolerance vs intolerance: How affinity defines topology in an idiotypic network. In: Bersini H, Carneiro J (eds) ICARIS 2006: 5th international conference on artificial immune systems, Oeiras, Portugal, September 2006. Lecture notes in computer science, vol 4163. Springer, Berlin, pp 109–121

Hone A, van den Berg H (2007) Modelling a cytokine network (special session: Foundations of artificial immune systems). In: Foundations of computational intelligence, Honolulu, HI, April 2007. IEEE, New York, pp 389–393

Honorio L, Leite da Silva A, Barbosa D (2007) A gradient-based artificial immune system applied to optimal power flow problems. In: de Castro LN, Von Zuben FJ, Kneidel H (eds) ICARIS 2007: 6th international conference on artificial immune systems, Santos, Brazil, August 2007. Lecture notes in computer science, vol 4628. Springer, Berlin, pp 1–12

Jacob C, Litorco J, Lee L (2004) Immunity through swarms: Agent-based simulations of the human immune system. In: Nicosia G, Cutello V, Bentley PJ, Timmis J (eds) ICARIS 2004: 3rd international conference on artificial immune systems, Calania, Italy, September 2004. Lecture notes in computer science, vol 3239. Springer, Berlin, pp 400–412

Jacob C, Pilat ML, Bentley PJ, Timmis J (eds) (2005) ICARIS 2005: 4th international conference on Artificial immune systems, Banff, Canada, April 2005. Lecture notes in computer science, vol 3627. Springer, Heidelberg http://www.artificial-immune-systems.org/icaris.shtml

Jerne NK (1974) Towards a network theory of the immune system. Ann Immunol (Inst Pasteur) 125C:373–389

Kelsey J, Timmis J (2003) Immune inspired somatic contiguous hypermutation for function optimisation. In: GECCO 2003: Genetic and evolutionary computation conference, Chicago, IL, July 2003. Springer, New York, pp 207–218

Kelsey J, Henderson B, Seymour R, Hone A (2008) A stochastic model of the interleukin (IL)-1β network. In: Bentley PJ, Lee D, Jung S (eds) ICARIS 2008: 7th international conference on artificial immune systems, Phuket, Thailand, August 2008. Lecture notes in computer science, vol 5132. Springer, New York, pp 1–11

Kim J, Bentley PJ (2002a) A model of gene library evolution in the dynamic clonal selection algorithm. In: Timmis J, Bentley PJ (eds) ICARIS 2002: Proceedings of the 1st international conference on artificial immune systems. University of Kent Printing Unit, Canterbury, UK, September 2002, pp 182–189

Kim J, Bentley P (2002b) Immune memory in the dynamic clonal selection algorithm. In: Timmis J, Bentley PJ (eds) ICARIS 2002: Proceedings of the 1st international conference on artificial immune systems. University of Kent Printing Unit, Canterbury, UK, September 2002, pp 59–67

Kim J, Bentley PJ (2002c) Towards an artificial immune system for network intrusion detection: an investigation of dynamic clonal selection. In: CEC2002: Proceedings of the 2002 congress on evolutionary computation. Honolulu, HI, May 2002

Kleinstein SH, Seiden PE (2000) Simulating the immune system. Comput Sci Eng 2(4):69–77

Lay N, Bate I (2007) Applying artificial immune systems to real-time embedded systems. In: IEEE congress on evolutionary computation 2007, Singapore, September 2007, pp 3743–3750

Matzinger P (1994) Tolerance, danger, and the extended family. Annu Rev Immunol 12:991–1045

Matzinger P (2002) The danger model: a renewed sense of self. Science 296(5566):301–305

McEwan C, Hart E, Paechter B (2008) Boosting the immune system. In: Bentley PJ, Lee D, Jung S (eds) ICARIS 2008: 7th international conference on artificial immune systems, Phuket, Thailand, August 2008. Lecture notes in computer science, vol 5132. Springer, New York, pp 316–327

Neal M (2003) Meta-stable memory in an artificial immune network. In: Timmis J, Bentley PJ, Hart E (eds) ICARIS 2003: 2nd international conference on artificial immune systems, Edinburgh, Scotland, September 2003. Lecture notes in computer science, vol 2787. Springer, New York, pp 168–180

Neal M, Feyereisl J, Rascunà R, Wang X (2006) Don't touch me, I'm fine: robot autonomy using an artificial innate immune system. In: Bersini H, Carneiro J (eds) ICARIS 2006: 5th international conference on artificial immune systems, Oeiras, Portugal, September 2006. Lecture notes in computer science, vol 4163. Springer, Berlin, pp 349–361

Newborough J, Stepney S (2005) A generic framework for population-based algorithms, implemented on multiple FPGAs. In: Jacob C, Pilat ML, Bentley PJ, Timmis J (eds) ICARIS 2005: 4th international conference on artificial immune systems, Banff, Canada, April 2005. Lecture notes in computer science, vol 3627. Springer, Heidelberg, pp 43–55

Nicosia G, Cutello V, Bentley PJ, Timmis J (eds) (2004) ICARIS 2004: 3rd international conference on artificial immune systems, Catania, Italy, September 2004. Lecture notes in computer science, vol 3239. Springer, Berlin http://www.artificial-immune-systems.org/icaris.shtml

Owens ND, Timmis J, Greensted AJ, Tyrell AM (2007) On immune inspired homeostasis for electronic systems. In: de Castro CN, Von Zuben FJ, Knidel H (eds) ICARIS 2007: 6th international conference on artificial immune systems, Santos, Brazil, August 2007. Lecture notes in computer science, vol 4628. Springer, Berlin, pp 216–227

Owens NDL, Timmis J, Greensted A, Tyrrell A (2008) Modelling the tunability of early t cell signalling events. In: Bentley PJ, Lee D, Jung S (eds) ICARIS 2008: 7th international conference on artificial immune systems, Phuket, Thailand, August 2008. Lecture notes in computer science, vol 5132. Springer, New York, pp 12–23

Perelson AS, Oster GF (1979) Theoretical studies of clonal selection: Minimal antibody repertoire size and reliability of self-non-self discrimination. J Theor Biol 81(4):645–670

Perelson AS, Weisbuch G (1997) Immunology for physicists. Rev Mod Phys 69(4):1219–1267

Read M, Timmis J, Andrews PS (2008) Empirical investigation of an artificial cytokine network. In: Bentley PJ, Lee D, Jung S (eds) ICARIS 2008: 7th international conference on artificial immune systems, Phuket, Thailand, August 2008. Lecture notes in computer science, vol 5132. Springer, New York, pp 340–351

Secker A, Freitas A (2007) WAIRS: Improving classification accuracy by weighting attributes in the AIRS classifier. In: Proceedings of the congress on evolutionary computation, Singapore, September 2007. IEEE Press, Singapore, pp 3759–3765

Secker A, Freitas A, Timmis J (2003) A danger theory inspired approach to web mining. In: Timmis J, Bentley PJ, Hart E (eds) ICARIS 2003: 2nd international conference on artificial immune systems, Edinburgh, Scotland, September 2003. Lecture notes in computer science, vol 2787. Springer, New York, pp 156–167

Stepney S (2007) Embodiment. In: Flower D, Timmis J (eds) In silico immunology. Springer, New York, Chap 12

Stepney S, Smith RE, Timmis J, Tyrrell AM, Neal MJ, Hone ANW (2005) Conceptual frameworks for artificial immune systems. Int J Unconventional Comput 1(3):315–338

Stibor T, Mohr P, Timmis J, Eckert C (2005) Is negative selection appropriate for anomaly detection? In: GECCO '05: Proceedings of the 2005 conference on genetic and evolutionary computation, Washington, DC, June 2005. ACM, New York, pp 321–328. doi: http://doi.acm.org/10.1145/1068009.1068061

Timmis J (2007) Artificial immune systems – today and tomorrow. Nat Comput 6(1):1–18

Timmis J, Andrews PS (2007) A beginners guide to artificial immune systems. In: Flower D, Timmis J (eds) In silico immunology. Springer, New York, Chap 3 (2007)

Timmis J, Bentley PJ (eds) (2002) ICARIS 2002: Proceedings of the 1st international conference on artificial immune systems, University of Kent Printing Unit, Canterbury, UK, September 2002

Timmis J, Neal MJ (2000) A resource limited artificial immune system for data analysis. In: Proceedings of ES2000 – Research and development in intelligent systems XVII, Cambridge, UK, December 2000. URL http://www.cs.kent.ac.uk/pubs/2000/1121, pp 19–32

Timmis J, Bentley P, Hart E (eds) (2003) ICARIS 2003: 2nd international conference on artificial immune systems, Edinburgh, Scotland, September 2003. Lecture notes in computer science, vol 2787. Springer, New York http://www.artificial-immune-systems. org/icaris.shtml

Timmis J, Amos M, Banzhaf W, Tyrrell A (2006) "Going back to our roots": second generation biocomputing. Int J Unconventional Comput 2(4):349–382

Timmis J, Andrews P, Owens N, Clark E (2008a) An interdisciplinary perspectives on artificial immune systems. Evol Intell 1(1):5–26

Timmis J, Hone A, Stibor T, Clark E (2008b) Theoretical advances in artificial immune systems. J Theor Comput Sci. doi: 10.1016/j.tcs.2008.02.011

Twycross J, Aickelin U (2005) Towards a conceptual framework for innate immunity. In: Jacob C, Pilat ML, Bentley PJ, Timmis J (eds) ICARIS 2005: 4th international conference on artificial immune systems, Banff, Canada, April 2005. Lecture notes in computer science, vol 3627. Springer, Heidelberg, pp 112–125

Twycross J, Aickelin U (2006) Libtissue – implementing innate immunity. In: IEEE congress on evolutionary computation, Vancouver, Canada, July 2006. pp 499–506

Watkins A, Timmis J, Boggess L (2004) Artificial immune recognition system (AIRS): an immune-inspired supervised machine learning algorithm. Genet Programming Evolvable Mach 5(3):291–317. URL citeseer.ist.psu.edu/watkins04artificial.html

Weisbuch G, Atlan H (1988) Control of the immune response. J Phys A Math Gen 21(3):189–192

Wilson WO, Garrett SM (2004) Modelling immune memory for prediction and computation. In: Nicosia G, Cutello V, Bentley PJ, Timmis J (eds) ICARIS 2004: 3rd international conference on artificial immune systems, Catania, Italy, September 2004. Lecture notes in computer science, vol 3239. Springer, Berlin, pp 386–399

48 Swarm Intelligence

David W. Corne[1] · *Alan Reynolds*[2] · *Eric Bonabeau*[3]
[1]School of Mathematical and Computer Sciences, Heriot-Watt University, Edinburgh, UK
dwcorne@macs.hw.ac.uk
[2]School of Mathematical and Computer Sciences, Heriot-Watt University, Edinburgh, UK
a.reynolds@hw.ac.uk
[3]Icosystem Corporation, Cambridge, MA, USA
eric@icosystem.com

G. Rozenberg et al. (eds.), *Handbook of Natural Computing*, DOI 10.1007/978-3-540-92910-9_48,
© Springer-Verlag Berlin Heidelberg 2012

Abstract

Increasing numbers of books, websites, and articles are devoted to the concept of "swarm intelligence." Meanwhile, a perhaps confusing variety of computational techniques are seen to be associated with this term, such as "agents," "emergence," "boids," "ant colony optimization," and so forth. In this chapter, we attempt to clarify the concept of swarm intelligence and its associations, and to provide a perspective on its inspirations, history, and current state. We focus on the most popular and successful algorithms that are associated with swarm intelligence, namely, ant colony optimization, particle swarm optimization, and (more recently) foraging algorithms, and we cover the sources of natural inspiration with these foci in mind. We then round off the chapter with a brief review of current trends.

1 Introduction

Nature provides inspiration to computer scientists in many ways. One source of such inspiration is the way in which natural organisms behave when they are in groups. Consider a swarm of ants, a swarm of bees, a colony of bacteria, or a flock of starlings. In these cases and in many more, biologists are of the opinion (and as we have often seen for ourselves) that the group of individuals itself exhibits a behavior that the individual members do not, or cannot. In other words, if the group itself is considered as an individual – the *swarm* – in some ways, at least, it seems to be more intelligent than any of the individuals within it.

This observation is the seed for a cloud of concepts and algorithms, some of which have become associated with swarm intelligence. Indeed, it turns out that swarm intelligence is only closely associated with a small portion of this cloud. If one searches nature for scenarios in which a collection of agents exhibit behavior that the individuals do not (or cannot), it is easy to find entire and vast subareas of science, especially in the biosciences. For example, any biological organism seems to exemplify this concept, when the individual organism is considered as the "swarm" and its cellular components as the agents.

One might consider brains, and nervous systems, in general, as a supreme exemplar of this concept, when individual neurons are considered as the agents. Or one might zoom in on certain inhomogeneous sets of biomolecules as our "agents," and herald gene transcription, say, as an example of swarm behavior. Fortunately, for the sake of this chapter's brevity and depth, it turns out that the swarm intelligence literature has come to refer to a small and rather specific set of observations and associated algorithms. This is not to say that computer scientists are uninspired by the totality of nature's wonders that exhibit such "more than the sum of the parts" behavior – much of this volume makes it clear that this is not so at all. However, if the focus is on the specific concept of swarm intelligence and the attempt to define it intentionally, the result might be a useful behavior that emerges from the cooperative efforts of a group of individual agents in which

1. The individual agents are largely homogeneous
2. The individual agents act asynchronously in parallel
3. There is little or no centralized control
4. Communication between agents is largely effected by some form of stigmergy
5. The "useful behavior" is relatively simple (finding a good place for food, or building a nest – not writing a symphony, or surviving for many years in a dynamic environment)

So, swarm intelligence is not about how collections of cells yield brains (which falls foul of at least items 1, 4, and 5), and it is not about how individuals form civilizations (violating mainly items 4 and 5), and it is not about such things as the life cycle of the slime mould (item 5). However, it is about individuals cooperating (knowingly or not) to achieve a definite goal, such as, ants finding the shortest path between their nest and a good source of food, or bees finding the best sources of nectar within the range of their hive. These and similar natural processes have led directly to families of algorithms that have proved to be very substantial contributions to the sciences of computational optimization and machine learning.

So, originally inspired, respectively, by certain natural behaviors of swarms of ants, and flocks of birds, the backbone of swarm intelligence research is built mainly upon two families of algorithms: ant colony optimization, and particle swarm optimization. Seminal works on ant colony optimization were by Dorigo et al. (1991) and Colorni et al. (1992a, b), and particle swarm optimization harks back to Kennedy and Eberhart (1995). More recently, alternative inspirations have led to new algorithms that are becoming accepted under the swarm intelligence umbrella; among these are search strategies inspired by bee swarm behavior, bacterial foraging, and the way in which ants manage to cluster and sort items. Notably, this latter behavior is explored algorithmically in a subfield known as swarm robotics. Meanwhile, the way in which insect colonies collectively build complex and functional constructions is a very intriguing study that continues to be carried out in the swarm intelligence arena. Finally, another field that is often considered in the swarm intelligence community is the synchronized movement of swarms, in particular, the problem of defining simple rules for individual behavior that led to realistic and natural behavior in a simulated swarm. "Reynolds' rules" provided a general solution to this problem in 1987, and this can be considered an early triumph for swarm intelligence, which has been exploited much in the film and entertainment industries.

In the remainder of this chapter we expand on each of these matters. ❷ Section 2 gives an account of the natural behaviors that have inspired the main swarm intelligence algorithms. ❷ Section 3 then discusses the more prominent algorithms that have been inspired by the techniques in ❷ Sect. 2, and ❷ Sect. 4 notes some current trends and developments and offers some concluding remarks.

2 Inspiration from Nature

2.1 Social Insects and Stigmergy

Ants, termites, and bees, among many other insect species, are known to have a complex social structure. Swarm behavior is one of several emergent properties of colonies of such so-called *social insects*. A ubiquitous characteristic that is seen again and again in such scenarios is *stigmergy*. Stigmergy is the name for the *indirect* communication that seems to underpin cooperation among social insects (as well as between cells, or between arbitrary entities, so long as the communication is *indirect*).

The term was introduced by Pierre-Paul Grassé in the late 1950s (Grassé 1959). Quite simply, stigmergy means communication via signs or cues placed in the environment by one entity, which affect the behavior of other entities who encounter them. Stigmergy was originally defined by Grassé in his research on the construction of termite nests. ❷ *Figure 1* shows a simplified schematic of a termite nest. We will say more about termite nests in

■ **Fig. 1**
A highly simplified schematic of a termite nest.

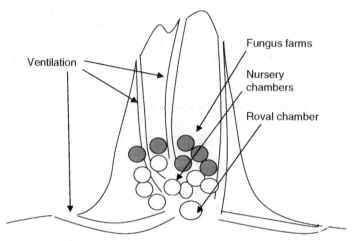

❯ Sect. 2.1.3, but for now it suffices to point out that these can be huge structures, several meters high, constructed largely from mud and from the saliva of termite workers. Naturally, the complexity and functionality of the structure is quite astounding, given what we understand to be the cognitive capabilities of a single termite.

Following several field trips to Africa in the late 1930s and 1940s studying termites and their nests, among other things, Grassé showed that the regulation and the coordination of the nest-building activity did not depend on the termite workers themselves, but was instead achieved by the nest itself. That is, some kind of stimulating configuration of materials triggers a response in a termite worker, where that response transforms the configuration into another configuration that may, in turn, trigger yet another, possibly different, action performed by the same termite or by any other termite worker in the colony. This concept of stigmergy was attractive and stimulating, but at the time, and often today, it was and is often overlooked by students of social insects, because it leaves open the important operational issue of how the specific trigger–response configurations and stimuli must be organized in time and space to allow appropriate coordination. But despite the general vagueness of Grassé's formulation, stigmergy is recognized as a very profound concept, the consequences of which are still to be fully explored. Stigmergy is not only of potential importance for our understanding of the evolution and maintenance of social behavior in animals, from communally breeding species to highly social insects, but it is also turning out to be a crucial concept in other fields, such as artificial intelligence, robotics, or the social, political, and economic sciences. Meanwhile, in the arena of natural computing, stigmergy is the fundamental concept behind one of the main swarm intelligence algorithms, as well as several others.

Apart from termite nests, another exemplary case of stigmergy in nature is that of pheromone deposition. Ants deposit pheromone along their paths as they travel; an ant striking out on its own will detect pheromone trails, and prefer to follow such trails already travelled. In general, the concept of stigmergy captures underlying commonalities in (usually) insect behaviors that are underpinned by indirect communication. This covers more emergent behaviors than trail following, and (the original inspiration for the term) the construction of structures such as termite mounds and bee hives. Stigmergy also seems key to behaviors such as brood sorting and cemetery clustering – some ant species are known to spatially cluster their young into age-groups within the nest, and they keep their nests tidy by removing dead nest mates and piling them into clusters outside.

The phenomenon of stigmergy has early evolutionary roots; it is now used to explain the morphologies of multicellular organisms, seashells, and so forth. Essentially, individual cells position themselves in a way influenced by deposits left behind by their colleagues or precursors. A useful way to think of it is that stigmergic communication involves a "stigmergy structure," which is like a notepad, or an actual structure, built from cues left by individuals.

The structure itself may be a spatially distributed accumulation of pheromone, or a partially built hive, or a partially constructed extracellular matrix. The structure itself influences the behaviors of the individuals that "read" it, and these individuals usually also add to the structure. Army ants find their directions of travel influenced by pheromone trails, and they add to the trails themselves. Termites are triggered by particular patterns that they see locally in the partially built mound, and act in simple and specific ways, resulting in additions to the structure itself. An authoritative overview of stigmergy-associated behaviors in nature is by Bonabeau et al. (1997), while Theraulaz (1994) provides a comprehensive survey of self-organization processes in insect colonies. As hinted above, when the stigmergic processes often observed in nature are considered, the most prominent sources of inspiration from the swarm intelligence viewpoint are those of navigation to food sources, sorting/clustering, and the collective building of structures. Each of these processes are briefly considered next.

2.1.1 Natural Navigation

Navigation to food sources seems to depend on the deposition of pheromone by individual ants. In the natural environment, the initial behavior of a colony of ants in seeking a new food source is for individual ants to wander randomly. When an ant happens to find a suitable food source it will return to its colony; throughout, the individual ants have been laying pheromone trails. Subsequent ants setting out to seek food will sense the pheromone laid down by their precursors, and this will influence the path they take up. Over time, of course, pheromone trails evaporate. However, consider what happens in the case of a particularly close food source (or, alternatively, a faster or safer route to a food source). The first ant to find this source will return relatively quickly. Other ants that take this route will also return relatively quickly, so that the best routes will enjoy a greater frequency of pheromone laying over time, becoming strongly fancied by other ants. The overall collective behavior amounts to finding the best path to a nearby food source, and there is enough stochasticity in the process to avoid convergence to poor local optima – trail evaporation ensures that suboptimal paths discovered earlier are not converged upon too quickly, while individual ants maintain stochasticity in their choices, being influenced by but not enslaved by the strongest pheromone trail they sense. ❯ *Figure 2* shows a simple illustration, indicating how ants will converge via stigmergy toward a safer and faster way to cross a flow of water between their nest and a food source.

❯ *Figure 2* shows three contrived snapshots of a simple scenario over time. On the left, ants need to cross a narrow stream of flowing water toward a tempting food source, and three crossings – fallen twigs – present themselves. Initially, ants are equally likely to try each one. Each of these ants lays its trail of pheromone as it makes its journey toward the right. Over time, however, the path toward the middle twig becomes less laid with pheromone, simply

□ **Fig. 2**
Convergence to a safer crossing over time.

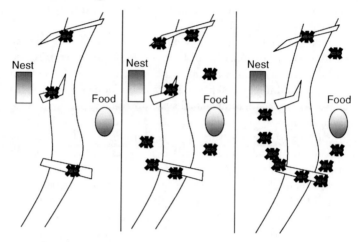

because, unlike the situation with the upper and lower twigs, there are no ants laying pheromone *on their way back* from that particular path. Eventually, on the right, we see several ants following the path defined by the lower twig, both attracted by, and further multiplying, the steady buildup of pheromone on this path, which is faster than that on the path defined by the upper twig. Although the upper twig provides a fairly short journey, it is more perilous, since it is quite narrow, and several ants fall off before being able to strengthen the path.

Actually, this is one of the simplest and most straightforward activities in social insects related to pheromones. The term "pheromone" itself was promoted for this context in 1959 (the late 1950s was clearly a fruitful period for swarm intelligence vocabulary) (Karlson and Luscher 1959), to encompass the broad range of biologically active chemicals used by insects for varieties of communicative purposes. The context in which we describe it above is known in biology as "recruitment," referring broadly to tasks in which individuals discover an opportunity (usually a food source) and need to recruit others to help exploit it. However, there are many other behaviors that are associated with pheromones, such as indicating alarms or warnings, interactions between queens and workers, and mating. An excellent source of further information is Vander Meer et al. (1998b), and in particular Vander Meer et al. (1998a).

2.1.2 Natural Clustering

Turning now to a different style of swarm behavior, it is well known that ant species (as well as other insect species) exhibit emergent sorting and clustering of objects. Two of the most well-known examples are the clustering of corpses of dead ants into cemeteries (achieved by the species *Lasius niger*, *Pheidole pallidula*, and *Messor sancta* (Depickere et al. 2004), and the arrangement of larvae into similar age groups (so-called brood-sorting, achieved by *Leptothorax unifasciatus* (Franks and Sendova-Franks 1992)).

For example, in *L. unifasciatus* ant colonies, the ants' young are organized into concentric rings called annuli. Each ring comprises young of a different type. The youngest (eggs, and micro-larvae) are clustered together in the central cluster. As one moves outward from the center, progressively older groups of larvae are encountered. Also, the spaces between these rings increase as one moves outward.

In cemetery formation, certain ant species are known to cluster their dead into piles, with individual piles maintained at least a minimal distance from each other. In this way, corpses are removed from the living surroundings, and cease to be a hindrance to the colony. One aspect of this behavior in particular is that it is arguably not exemplary of swarm behavior; that is, it is perhaps not *collective* intelligence. The explanatory model that seems intuitively correct and confirmed by observation (Theraulaz et al. 2002) is one in which an individual operates according to quite simple rules while otherwise generally wandering around randomly:

1. If not carrying a corpse, and a single corpse (or quite small cluster of corpses) is encountered, pick it up.
2. If carrying a corpse and a relatively large cluster of corpses is encountered, put it down.

A single ant could achieve the clustering observed in nature that seems to operate according to these rules (Theraulaz et al. 2002). However, the emergent clusters are produced faster when a collection of ants are involved. Gaubert et al. (2007) is a useful reference for discussion of mathematical and other models of these behaviors. Meanwhile, a steady line of recent research is investigating computational clustering methods that are directly inspired by these natural phenomena (Handl and Meyer 2007). The first such inspired clustering algorithm was proposed by Lumer and Faieta (1994), closely based on the Deneubourg et al. (1991) model of the natural process. In recent work (Handl et al. 2006), an ant-based clustering method called ATTA is tested, and the case is made convincingly that ant-based clustering algorithms certainly have a niche in data mining, performing particularly well on problems where the number of clusters are not known in advance, and where the clusters themselves are highly variable in size and shape. In this article, our focus stays with optimization and we will say a little more about clustering; the reader is again referred to a review by Handl and Meyer (2007) for further study on this topic.

2.1.3 Natural Construction

We now consider the extraordinary collective behavior that leads to the construction of achievements such as wasp nests, termite mounds, and bee hives. Brood-sorting, considered above, exemplifies a simple structure that arises from collective behavior. However, the more visible and impressive structures such as termite mounds have always impressed observers, and often confounded one when one tries to imagine how such simple minds can lead to such creations. As will be clear from context, stigmergy seems to be the key to understanding these buildings; patterns inherent in partial elements of structures are thought to trigger simple rule-based behavior in the insect, which in turn changes the perceived patterns, and so on, until a complete hive or nest is built. Much computation-based study has been made of this by Bonabeau, Theraulaz, and coworkers.

In nature, the sizes of such social insect structures can reach an astounding 30 m in diameter (Grassé 1984). An impressive example of the complexity of these structures comes from the African termites *Macrotermitinae*, "the fungus growers." In a mature nest of this species, there are typically seven distinct elements of structure (this description is adapted from Bonabeau et al. 1998):

1. The protective and ribbed cone-shaped outer walls (also featuring ventilation ducts).
2. Brood chambers within the central "hive" area. They have a laminar structure and contain the nurseries where the young termites are raised.
3. The hive consists of thin horizontal lamellae supported by pillars.
4. A flat floor structure, sometimes exhibiting cooling vents in a spiral formation.
5. The royal chamber: a thick walled enclosure for the queen, with small holes in the walls to allow workers in and out. This is usually well protected underneath the hive structure, and is where the queen lays her eggs.
6. Garden areas dedicated to cultivating fungi. These are arranged around the hive, and have a comb-like structure, arranged between the central hive and the outer walls.
7. Tunnels and galleries constructed both above and below ground which provide pathways from the termite mound to the colony's known foraging sites.

So, how does a collection of termites make such a structure? Perhaps this is the most astonishing example of natural swarm intelligence at work. The observations and descriptions of such structures from the biology literature have tended to focus on *description*, elucidating further and finer details from a variety of species, but have done little to clarify the mechanism. However, computational simulation work, such as Theraulaz and Bonabeau (1995), has indicated how such behavior can emerge from collections of "micro-rules," where patterns of the growing structure perceived by an individual termite (or ant, wasp, etc.) act as a stigmergic trigger, perhaps in tandem with other environmental influences, leading to a specific response that adds a little new structure. Theraulaz and Bonabeau (1995) and Bonabeau et al. (2000) have shown how specific collections of such micro-rules can lead to, in simulation, a variety of emergent structures, each of which seems convincingly similar to wasp nests from specific wasp species. ❷ *Figure 3* shows examples of three artificial nests, constructed in this fashion, that closely resemble the nest structures of three specific wasp species; several more are presented in Theraulaz and Bonabeau (1995) and in Bonabeau et al. (2000).

❷ *Figure 4* shows some examples of micro-rules of the kind that can lead to the types of structures shown in ❷ *Fig. 3*. A micro-rule simply describes a three-dimensional pattern of "bricks"; in the case of the experiments that led to ❷ *Fig. 3*, a brick has a hexagonal horizontal cross section, and there are two types of bricks (it was observed that at least two different brick types seem necessary for interesting results). The three-dimensional pattern describes the immediate neighborhood of a single central brick, including the seven hexagonal cells above it (the upper patches of hexagons in ❷ *Fig. 4*), the six that surround it, and the seven below it. In the figure, a "white" hexagonal cell is empty – meaning no brick here; otherwise there are two kinds of bricks, distinguished by different shadings in the figure. A micro-rule expresses the following building instruction: "if the substructure defined by this pattern is found, with the central cell empty, then add the indicated type of brick in the central position."

Intuition suggests how the construction by a collection of agents such as wasps of artifacts such as those in ❷ *Fig. 3*, or even more complex artifacts, may be facilitated by specific collections of micro-rules; however, that does not make it easy to design a set of micro-rules for a specific target construction. Sets of micro-rules achieving the illustrated results were obtained by using a carefully designed evolutionary algorithm. Interested readers should

◘ Fig. 3
Figure 2 from Bonabeau et al. (2000), reproduced with permission. These show results from artificial colonies of ten wasps, operating under the influence of stimulus–response micro-rules based on patterns in a 3D hexagonal brick lattice. (a) A nest-like architecture resembling the nests of *Vespa* wasps, obtained after 20,000 building steps; (b) an architecture resembling the nests of *Parachartergus* wasps, obtained after 20,000 building steps; (c) resembling *Chatergus* nests, obtained after 100,000 building steps; (d, e) showing internal structure of (c).

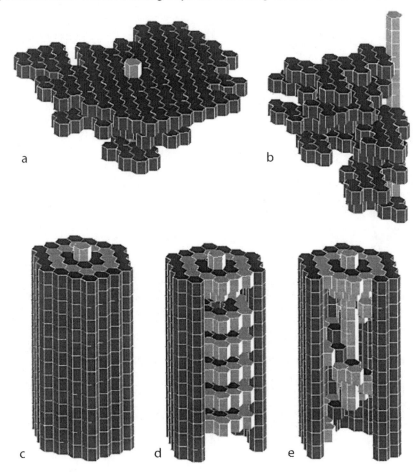

consult Bonabeau et al. (2000) for further details, including analyses of the operation of the emerging rule sets, revealing the requirement for various types of coordination implicitly built into the micro-rule collection.

Meanwhile, it is a long way from wasp nests to termite mounds, especially mounds of the complexity hinted at above. However, by considering and extending partial models for elements of termite mounds, in Bonabeau et al. (1998) some basis is provided for the suspicion that such complexity may be explained by the interaction of stimulus–response-based ("micro-rule") processes and pheromone-based triggers that modify the stimulus–response behavior, unfolding over time as a controlled morphogenetic process.

◻ Fig. 4
An illustration of two "micro-rules" from the space of such rules that can lead to structures such as those in ❷ *Fig. 3*. A single micro-rule defines a building instruction based on matching a pattern of existing bricks. A single column of three groups of seven hexagonal cells is a micro-rule, by describing a structure around the neighborhood of the central cell (which is empty in the matching pattern). The building instruction is to fill the central cell with the indicated type of "brick." In this figure, two examples of micro-rules are shown, each further illustrated by "before" and "after" building patterns on the right.

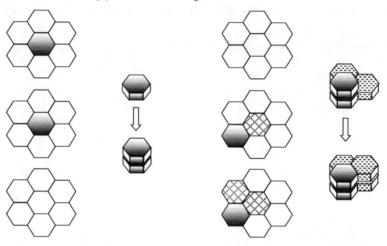

Finally, we note that much of what was discussed in this subsection forms part of the basis for the new field of "swarm robotics." This area of research (Sahin 2005; Mondada et al. 2005) focuses on what may be achievable by collections of small, simple robots furnished with relatively non-sophisticated ways to communicate. Chief among the motivating elements of such research are the qualities of robustness, flexibility, and scalability that swarm robots could bring to a number of tasks, ranging through agriculture, construction, exploration, and other fields. Imagine that one wishes to build a factory on Mars, for example. One might imagine this could be performed, in some possible future, by a relatively small collection of very sophisticated and intelligent robots. However, in the harsh conditions of Mars, one can expect that loss or destruction of one or more of these is quite likely. Swarms of simple robots, instead, are far more robust to loss and damage, and may present an altogether more manageable, shorter-term, and more adaptable approach than the "super-robot" style.

As yet, swarm robotics has not risen to such heights, although there is published discussion along these lines, drawing from the sources of natural inspiration already discussed (Cicirello and Smith 2001). Meanwhile, interesting and useful behaviors have been demonstrated in swarm robotics projects, after, in almost all cases, considerable work in design, engineering, and construction of the individual "bots." Supporting these studies, a large part of swarm robotics research is into how to design the individual robots' behaviors in such a way that the swarm (or team) achieves an overall goal or behavior. Unsurprisingly to many, it turns out that the judicious use of evolutionary computation proves effective for this difficult design problem (Waibel et al. 2009). However, hand-design or alternative principled methods for designing behaviors remains a backbone of this research, especially when the desired overall behavior is complex, involving many tasks (e.g., Ducatelle et al. 2009), and in general there are several emerging issues in swarm robotics that have sparked active current lines of research, such as the problem of "interference" – swarm robots, whether cooperating on the same task or not, often physically interfere with each others' operation (Pini et al. 2009) – and the problem of achieving tasks with minimal energy requirements (Roberts et al. 2008).

Relatively impressive behaviors from swarm robotics research has so far included cooperative transport of one or more objects, and cooperation toward moving up a vertical step (as large as the bots involved); readers may visit the European project "Swarmbots" and "Swarmanoids" websites for explanation and many other resources, at http://www.swarm-bots.org/, and at http://www.swarmanoid.org/ as well as similar resources such as http://www.pherobot.com/. Hardware and related technology issues remain a bottleneck that still inhibit a full exploration of social insect swarm intelligence in robotics; however, this work continues toward that end and will be observed with great interest.

2.2 Foraging

There are broadly two types of natural processes that go by the term "foraging," and in turn provide sources of inspiration for optimization (or resource allocation) methods. In both cases, the overall emergent behavior is that the swarm finds and exploits good food sources, adaptively moves to good new sources as current ones become depleted, and does all of this with efficient expenditure of energy (as opposed to, e.g., brute force search of their environment). The means by which this behavior is achieved is rather different in these two sources of inspiration.

In one case, that of "bacterial foraging" (Passino 2002), individual bacteria are (essentially) directed toward rich areas via *chemotaxis*; that is, they exist in an environment in which their food source diffuses, so they can detect and respond to its presence. In particular, chemotaxis refers to movement along a chemical gradient. An individual *Escherichia coli* bacterium has helical appendages called *flagellae* which spin either clockwise or anticlockwise (one can think of them as analogous to propellers). When they spin in one direction, the bacterium will "tumble"; this is an operation which ends up moving the bacterium a short distance, and leaving it with an essentially random new orientation. When the flagellae spin in the other direction, the bacterium's movement will be a "run" – this is a straight-line movement in the direction the bacterium was facing at the beginning, and continues as long as the flagellae continue to spin in the same direction.

In a nutrient-free and toxin-free environment, an individual bacterium will alternate between clockwise and anticlockwise movement of its flagellae. So, it tumbles, runs, tumbles, runs, and so forth. The effect of this behavior is a random search for nutrients. However, when the bacterium encounters an increasing nutrient gradient, that is, a higher concentration of nutrient in its direction of movement, its internal chemistry operates in a way that causes the runs to be of longer duration. It still alternates between tumbles and runs, but maintains longer run lengths so long as the gradient continues to increase. The effect of this is to explore and exploit the food source, moving upward along the nutrient gradient, while maintaining an element of stochastic exploration.

In addition, under certain conditions we know that bacteria secrete chemicals that attract each other. There is speculation that this can happen in response to nutrient-rich environments, so that additional bacteria are recruited to exploit the food source, where the attractive secretions build further on the attraction provided by the chemical gradient. Also, there is evidence that bacteria release such an attractant under stressful conditions too, which in turn may be a protective response; as they swarm into a sizeable cluster many individuals are protected from the stressful agent. These self-attractant and chemotactic behaviors are known to lead to pattern formations under certain conditions (Budrene and Berg 1991). These and many other details have been elucidated for *E. coli* and similar bacteria via careful experimentation, for example: Berg and Brown (1972), Segall et al. (1986), and DeRosier (1998).

The other broad style of efficient collective foraging behavior is that exhibited by the honeybee (among other insects). When a bee discovers a food source some distance from the hive, it returns to the hive and *communicates the direction, distance, and quality* of the food source to other bees. The details of this communication, achieved by specialized "dances," are quite remarkable, and have emerged from a series of ingenious experiments and observations, largely by Karl von Frisch (1967). The essential details are these: in context, the dance is performed in alignment with a particular aspect of the hive structure, which provides an absolute reference against which the bee audience can perceive specific angles. The main dance is the "waggle" dance, which consists of a straight-line movement, during which the bee waggles from side to side along the way. This straight-line movement is done upward at a particular angle from the vertical. At the end of the straight-line part, the bee loops round to the starting point and repeats the dance (actually, it alternates the direction of this loop, drawing a figure "8"). The angle of this dance from vertical indicates to the bee audience the direction they need to take with respect to the current position of the sun. Among various extraordinary aspects of this, it is known that the bee automatically corrects for movement of the sun during the day, and communicates the correct direction. Also, at times, the bee will pause its dance and allow watching bees to sample the nectar it is carrying, giving an indication of the quality of the food source.

More interesting from the algorithmic viewpoint, however, is that the abundance of the food source is communicated by the duration of the dance (essentially, the number of times the figure "8" is repeated). An individual enjoying this performance may or may not decide to follow these directions and be "recruited" to this particular source. Such an individual may also be exposed to rival performances. However, the longer the duration, the more bees will see this dance, and the more will be recruited to this dance rather than others. In this way, the bee colony sports the emergent behavior of smart resource allocation, with more bees assigned to better sources, and adaptation over time as returning bees gradually provide shorter and shorter dances as the source becomes depleted.

As we will see later, both bee and bacterial foraging have been taken as the inspiration for general optimization methods, as well as for approaches to the specific problem domain of optimal resource allocation.

2.3 Flocking

Perhaps the most visible phenomenon that brings to mind swarm intelligence is the travelling behavior of groups (flocks, swarms, herds, etc.) of individuals that all are familiar with. The mesmerizing behavior of large flocks of starlings is a common morning sight over river estuaries. Swarms of billions of monarch butterflies, herds of wildebeest, schools of tuna fish, swarms of bees, all share common emergent behaviors, chiefly being:

1. The individuals stay close to each other, but not too close, and there seem to be no collisions.
2. Swarms change direction smoothly, as if the swarm was a single organism.
3. *Unlike* a single organism, yet still smoothly and cleanly, swarms sometimes pass directly through narrow obstacles (in the way that a stream of water passes around a vertical stick placed centrally in the stream's path).

In some ways, such swarm behavior is arguably less mysterious than other emergent behaviors; it seems clear that we might be able to explain this behavior via a built-in predisposition for individuals to stay with their colleagues, and we can readily imagine how evolution will have favored such behavior: there is safety in numbers. However, the devil is in the detail, and it took seminal work by Reynolds (1987) to outline and demonstrate convincing mechanisms that can explain these behaviors. Reynolds' work was within the computer graphics community, and has had a volcanic impact there. Now known as "Reynolds' rules," the recipe that achieves realistic swarm behavior (with some, but not obtrusively much, parameter investigation needed depending on the species simulated) is this triplet of steering behaviors to be followed by each individual in a swarm:

> **Separation**: steer to avoid coming too close to others.
> **Alignment**: steer toward the mean heading of others.
> **Cohesion**: steer toward the mean *position* of others.

A basic illustration of each rule is given in ❷ *Fig. 5*. In the figure, the common terminology of "boid" is taken to refer to an individual in a flock. The figure shows examples of the adjustments that might be made under the guidance of the rules. To understand how realistic swarm simulation works, it is important to note that each boid has its own perceptual field – that is, it could only "see" a certain distance, and had a specific field of view (e.g., boids cannot see behind them). The adjustments it makes to its velocity at any time are therefore a function of the positions and velocities of the boids in its perceptual field, rather than a function of the flock as a whole.

The rules are key ingredients to a realistic appearance in simulated flocks, but there are several other details, particularly regarding obstacle avoidance and goal-seeking behavior. Interested readers may consult Reynolds (1987) and the many works that cite it. It is important to note that these rules are not strictly nature inspired, in the sense that Reynolds was not attempting to explain natural swarming behavior, he was simply attempting to emulate it. However, the resulting behavior was found to agree well with observations of natural flocking behavior (e.g., Partidge (1982) and Potts (1984)), and Reynolds (1987) reported that "many people who view these animated flocks immediately recognize them as a representation of a natural flock, and find them similarly delightful to watch." These techniques are now common in the film industry; among the earliest uses were in the film Batman Returns (1992, director Tim Burton), in which Reynolds' rules lay behind the simulated bat swarms and flocks of penguins.

Meanwhile, natural flocking behavior also turns out to be one of the sources of inspiration for the highly popular and successful particle swarm optimization algorithm, which appears in the next section as one of the prominent flagships for swarm intelligence. It is not obvious why flocking behavior might lead to an optimization algorithm; however, it soon becomes clear when the dynamics of flocking and the tendency of optimization landscapes to be locally smooth are considered. In the case of bacterial foraging, the dynamics of the natural behavior are such that individuals will tend to congregate around good areas. With the bacterial

◘ **Fig. 5**

Illustrating Reynolds' rules, which lead to natural-looking behavior in simulated swarms. *Upper*: Separation: each boid makes an adjustment to velocity which prevents it from coming too close to the flockmates in its perceptual field. *Middle*: Alignment: a boid adjusts its heading toward the average of those in its perceptual field. *Lower*: a boid makes an adjustment to velocity that moves it toward the mean position of the flockmates in its perceptual field.

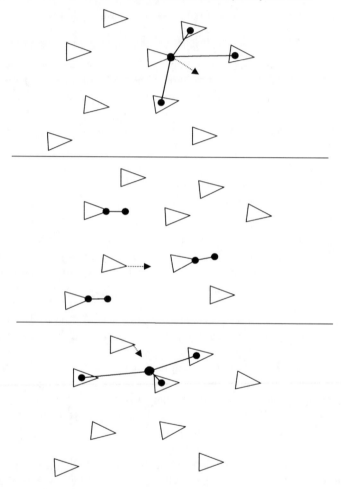

example, nature provides mechanisms for suggesting appropriate directions of movement, while there is a clear goal for the bacterial colony to achieve – find nutrient-rich (and toxin free) areas. For the current context in particular, the secretion of attractant chemicals is a mechanism that promotes bacteria swarming together, while an individual's position in its environment directly provides it with a level of "fitness" that it can sense in terms of nutrient concentration.

When the flocking behavior in birds is considered, however, Reynolds' work provides clues about appropriate ways to move together as a swarm, but there is no clear mirror of a "fitness" in the environment. Often, birds will migrate from A to B, knowing where they are going, rather than seeking new environments. However, if flocks *did* have a goal to move toward "fitter" positions in the landscape they travel, then it becomes intuitively reasonable to consider the cohesive swarm behavior as a sensible way to achieve local exploration around fit areas, perhaps enabling the sensing of even fitter areas that may then sway the overall movement of the flock. In this way, flocking behavior combines with a little algorithm engineering to achieve a very successful optimization mechanism.

3 Two Main Concepts for Swarm Intelligence Algorithms

When we consider the impact of swarm intelligence so far on computer science, two families of algorithms clearly stand out in terms of the amount of work published, degree of current activity, and the overall impact on industry. One such family is inspired directly by the pheromone-trail following behavior of ant species, and this field is known as ant colony optimization (ACO). The other such family is inspired by flocking and swarming behavior, and the main exemplar algorithm family is known as particle swarm optimization (PSO). Also in this family are algorithms based on bacterial foraging, and some of the algorithms that are based on bee foraging; these share with PSO the broad way in which the natural phenomenon is mapped onto the concept of search within a landscape. In this section, these two main families will be discussed in turn.

3.1 Ant Colony Optimization

Ant colony optimization (ACO) was introduced in 1996 via an algorithm called "Ant System" (AS) (Dorigo et al. 1996). The basic approach used in AS remains highly characteristic of most ACO methods in current use, as we will describe next.

Recall that, in the natural case, an ant finds a path from its nest to a food source by following the influences of pheromone trials laid down by ants that have previously sought food (and usually returned). AS, and ACO algorithms in general, mirror aspects of this behavior quite faithfully. In short, an artificial ant builds a solution to the optimization problem at hand, and lays down simple "artificial pheromone" along the route it took toward that solution. Following artificial ants then build solutions of their own, but are influenced by the pheromone trails left behind by their precursors. This is the essential idea, and starts to indicate the mapping from the natural to the artificial case. However, there are various further issues necessary to consider making this an effective optimization algorithm. This is discussed further in the next section, with a focus on the mapping from the basic ideas of ACO to applications in optimization.

3.1.1 Applying ACO to Optimization Problems

In order to apply AS to an optimization problem, the problem needs to be represented in such a way that any candidate solution to it is a specific path along a network. This network can be conceived as having a single start node, from which (usually) every ant starts, and a single finish node, reaching which indicates that the path taken encodes a complete solution. A clear example might be the network of roads in a city, where each junction of roads is a node in the network, and each road is an arc or edge between nodes. Consider the problem of finding the shortest path between a specific junction A and another specific junction B. In this case, A and B are clearly the start and finish nodes, and one can imagine an ACO approach which maps very closely indeed to the natural case. However, with a little thought, it is clear that each individual ant's path construction should be constrained so that it does not return to a junction it has already visited (unless this is a valid move for the problem under consideration). Also, though one might choose to simulate the preferential recruitment of new ants to shorter paths by closely following the natural case, it seems more sensible and straightforward to make the pheromone trail strength directly a function of the solution quality. That is, when an ant has completed its path, one evaluates the quality of its solution, and renders things such that better solutions lead to stronger pheromone deposits along its arcs. These pheromone deposits will decay over time; however, just as in the natural case, we can see that this will prevent premature convergence to poor solutions that happen to be popular in the early stages.

Finally, since such information is often available, and would seem useful in cases where ants have large numbers of choices, one might bias the paths available at each junction with the aid of a simple heuristic evaluation of the potential of that arc. For a shortest path problem, for example, this could be based on how much closer to B each arc would leave the ant.

With such considerations in mind, one can envisage artificial ants travelling the road network from A to B via distinct but sensible routes. At each junction, the ant senses the pheromone levels that await it at each of the arcs it can feasibly take. These levels are made from many components; arcs that are highly attractive will probably enjoy the remnants of trails from prior ants that have reported good solutions, and/or may have a good heuristic component. Arcs with low pheromone levels will probably be losers in the heuristic stakes, and have seen little activity that has led to good solutions; however, the ant may still choose such an arc, since our algorithm is stochastic.

Finally, as an individual artificial ant arrives at B, it retrospectively lays pheromone on the path it took, where the strength of that pheromone trail will reflect the quality of its solution. The next artificial ant starts from A and sees a slightly updated pheromone trail (stigmergic) pattern, and so it continues.

To apply this method to other problems, one simply needs (implicitly, at least) a network-based representation of the problem as described. If one is solving the travelling salesperson problem (TSP), for example, the network is the complete graph of the cities, each arc between cities indicates the distance or cost of that arc, and in this case an individual ant can start anywhere. As one follows an individual ant's route, one sensibly constrains its potential next-hops to avoid cycles, and along the way one may bias its choices simply by using the distance of the next arc, and it retrospectively lays pheromone once it has completed a tour of all cities. In general, an optimization problem can always be approached in this way, by suitable choices of semantics for nodes and arcs, and well-designed routines for generating and constraining an ant's available choices at each junction.

This explication can now be finished by clarifying the AS algorithm, which in fact has already been covered verbosely in the above. Once the transformation of the problem has been designed, so that an ant's path through a network provides a candidate solution, the algorithm cycles through the following two steps until a suitable termination condition is reached:

Solution Construction: A number of ants individually construct solutions based on the current pheromone trail strengths (initially, pheromone is randomly distributed). Each ant steps through the network choosing among feasible paths. At each choice point, the ant chooses among available arcs according to a function of the pheromone strength on each arc, and of the heuristic values of each arc. In the original, and still commonly used version of this function (see Dorigo et al. (1996) and many more), the pheromone and heuristic components for each arc are exponentiated to parameters α and β respectively, allowing for tuning of the level to which the algorithm relies on exploration and heuristic support. Also, the overall attractiveness value of each arc is scaled so that the ant can treat these values as a probability distribution over its available choices.

Pheromone Update: When the ants' paths are complete, for each ant, the corresponding solution is evaluated, and pheromone is laid on each arc travelled in proportion to the overall solution quality. Also, the component of pheromone strength that arises from earlier ants is reduced. Quite simply, for any particular arc, its updated pheromone strength p_{new} is $(1 - \rho)$ $p_{old} + \rho f$, where ρ controls the speed of pheromone decay, and f accumulates the overall quality of solutions found which involved that arc in the current iteration.

ACO has now been applied to very many problems and (clearly, or we would probably not have devoted so much time to it) has been very successful, especially when hybridized with local search or some other metaheuristic (in such hybridized algorithms, an ant will typically use the additional heuristic for a short time to find an improvement locally to its solution). Initially demonstrated for the TSP (Dorigo et al. 1996), there are an enormous number of applications of ACO now published. We mention vehicle routing (Gambardella et al. 1999), rule discovery (Parpinelli et al. 2002), and protein/ligand docking (Korb et al. 2007) just to give some initial idea of the range of applications. To discover more, recent surveys include Blum (2005a) and Gutjahr (2007), the latter concentrating on theoretical analyses. Meanwhile, Socha and Dorigo (2008) show how to apply ACO to continuous domains (essentially, ants select parameters via a probability density function, rather than a discrete distribution over a fixed set of arcs).

3.1.2 Ant-Based Routing in Telecommunications

The basic ACO idea of exploiting pheromone-trail-based recruitment is also the inspiration for a healthy subarea of research in communications networks; therein, ant-inspired algorithms are designed to assist with network routing and other network tasks, leading to systems that combine high performance with a high level of robustness, able to adapt with current network traffic and robust to network damage. Early and prominent studies in this line were by Schoonderwoerd et al. (1996, 1997), which were soon built upon by di Caro and Dorigo's AntNet (1998). To explain this application area, and the way that ACO ideas are applied therein, it will be helpful to first explore the problem and the associated solution that was studied in Schoonderwoerd et al.'s seminal work.

Schoonderwoerd et al. were concerned with load balancing in telecommunication networks. The task of a telecommunication network is to connect calls that can arise at any node, and which may need to be routed to any other node. The networks themselves, as a function of the capacities of the constituent equipment at each node, cannot guarantee successful call connections in all cases, but they aim to maintain overall acceptable performance under standard conditions. At very busy times, and/or if a particular node is overwhelmingly flooded with calls, then typically many calls will be "dropped." It is worth noting that there are problems with central control of such systems (landline telecommunication networks, mobile networks, traffic networks, and so forth). To achieve centralized control in a way that manages load-balancing or any other such target, several disadvantages are apparent. The controller usually needs current knowledge about many aspects of the entire system, which in turn necessitates using communication links from every part of the system to the controller itself. Centrally managed control mechanisms therefore scale badly as a result of the rapidly increasing overheads of this communication, interfering with the performance of the system itself. Also, failure in the controller will often lead to complete failure of the complete system.

Schoonderwoerd et al.'s ant-inspired approach, which remains a central part of the majority of more recent ant-based approaches, was to replace the routing tables in such networks with so-called "pheromone tables." Networks of the type of interest invariably have a routing table at each node, specifying which "next-hop" neighboring node to pass an incoming call to, given the ultimate destination of that call. In Schoonderwoerd et al.'s "Ant-Based Control" (ABC) method, the routing table at a network node instead provided *n* probability distributions over its neighboring nodes, one for each of the *n* possible destinations in the network. When a routing decision is to be made, it is made stochastically according to these probabilities – that is, it is most likely that the next-hop with the highest probability will be taken, but there is a chance that the next-hop with the lowest probability will be taken instead. The entries in the pheromone table were considered analogous to pheromone trail strengths, and changed adaptively during the operation of the network. Updating of these trails in ABC is very simple – whenever a call is routed from node A to B, the entries for A in node B's routing table are all increased, with corresponding reductions to other entries. However, this simple idea has obvious intuitive benefits; first, by testing various routing decisions over time (rather than deterministic decisions), the process effectively monitors the current health of a wide variety of different routing strategies; when a link is over-used, or down, this naturally leads to diminution in its probability of use, since the associated entries in routing tables will not be updated, and hence will naturally reduce as alternatives are updated. Also, as it turns out, the pheromone levels can adapt quite quickly to changes in call patterns and loads. The ABC strategy turned out to be surprisingly effective, despite its simplicity, when Schoonderwoerd et al. compared it with a contemporary agent-based strategy developed by Appleby and Steward (1994), and found it superior over a wide range of different situations.

Research in ant-based approaches for decentralized management is increasingly very active (e.g., Hussein and Saadawi 2003; Rosati et al. 2008; Di Caro et al. 2008). The essential idea, to replace static built-in routing strategies with stochastic "pheromone tables" or similar, is applicable in almost all modern communication scenarios, ranging through ad hoc computer networks, mobile telephone networks, and various layers of the Internet. Ongoing research continues to explore alternative strategies for making the routing decisions, controlling the updates to pheromone trails, and so forth, while investigating various distinct application domains, and continuing to find competitive or better performance than alternative state-of-the-art methods used in network engineering.

3.2 Particle Swarm Optimization and Foraging

PSO was established in 1995 with Kennedy and Eberhart's paper in the international joint conference on neural networks (IJCNN) (Kennedy and Eberhart 1995). The paper described a rather simple algorithm (and time has seen no need to alter its straightforward fundamentals), citing Craig Reynolds' work as inspiration (Reynolds 1987), along with slightly later work in the modeling of bird flocks (Heppner and Grenander 1990). The basic idea is to unite the following two notions: (a) the behavior of a flock of birds moving in 3D space toward some goal; (b) a swarm of solutions to an optimization problem, moving through the multidimensional search space toward good solutions.

Thus, a "particle" is equated with a candidate solution to an optimization problem. Such a particle has both a *position* and a *velocity*. Its position is, in fact, precisely the candidate solution it currently represents. Its velocity is a displacement vector in the search space, which (it hopes) will contribute toward a fruitful change in its position at the next iteration.

The heart of the classic PSO algorithm is in the step which calculates a new position for the particle based on three influences. The inspiration from Reynolds (1987) is clear, but the details are quite different, and, of course, exploit the fact that the particle is moving in a search space and can measure the "fitness" of any position. The influences – the components that lead to the updated position – are:

Current velocity: the particle's current velocity (obviously).

Personal best: the particle remembers the fittest position it has yet encountered, called the personal best. A component of its updated velocity is the direction from its current position to the personal best.

Global best: every particle in the swarm is aware of the best position that any particle has yet discovered (i.e., the best of the personal bests). The final component of velocity update, shared by all particles, is a vector in the direction from its current position to this globally best-known position.

Following a random initialization of positions and velocities, the evaluation of the fitness of the particles' current positions, and consequent initialization of the personal bests and global best, PSO proceeds in a remarkably straightforward manner. First, each particle updates its velocity by adding a vector in each of the above three component directions. To provide these vectors, in the classic algorithm, the current velocity component is left undisturbed, while the personal and global best components are each scaled by a random scalar drawn uniformly from [0,2]. The resulting vector is used to update the current velocity, and the new velocity vector is used to update the current position. The new position is evaluated, book-keeping is done to update personal and global bests, and then this is repeated.

Kennedy and Eberhart initially reported that PSO appeared to do very well over a wide range of test problems, including its use as an alternative to backpropagation for training an artificial neural network (Kennedy and Eberhart 1995). Perhaps helped by the ease of implementation of this algorithm (remarkably few lines of code are needed for the classic algorithm), an avalanche of papers began to follow, almost invariably adding to the evidence that this algorithm provides a very substantial contribution to optimization practice. Naturally, this field is now rich in variants and extensions to the original design – a number of recent surveys are available (e.g., Reyes-Sierra and Coello 2006; Yang et al. 2007) – while the published applications are as varied as one might expect from such a generally applicable algorithm.

3.2.1 Bacteria and Bees

Newer to the ranks of swarm-intelligence-based optimization, and yet to prove quite as widely successful, are techniques inspired by bacterial and bee foraging. For the most part, these algorithms follow the broad direction of PSO, in that individuals in a swarm represent solutions moving through a landscape, with the fitness of their current solution easily obtained by evaluating their position. Meanwhile, just as with PSO, an individual's movement through this landscape is influenced by the movements and discoveries of other individuals. The fine details of a bacterial foraging algorithm (BFA), however, are quite distinct, and in one of the more popular methods draw quite closely from what is known (and briefly touched upon above) about bacterial swarming in nature. Passino (2002) presents a fine and detailed explication of both the natural case and the BFA. It turns out that BFA-style algorithms are enjoying quite some success in recent application to a range of engineering problems (e.g., Niu et al. 2006; Tripathy and Mishra 2007; Guney and Basbug 2008).

Also inspiring, so far, a small following are algorithms that are inspired by bee foraging behavior. The authoritative sources for this are Quijano and Passino's papers, respectively, outlining the design and theory (Quijano and Passino 2007a), and application (Quijano and Passino 2007b) of a bee foraging algorithm. In Quijano and Passino (2007a) the design of a bee-foraging algorithm is presented in intimate connection with an elaboration of the mechanisms of natural bee foraging (such as briefly described earlier). The algorithm is as much a model of the natural process as it is a routine applicable to certain kinds of problem. Considering individual bees as resources, the concept here is to use bee foraging behavior as a way to ideally distribute those resources in the environment, and maintain an ideal distribution over time as it adapts to changing patterns of supply. Just as natural bees maintain an efficient distribution of individuals among the available sources of nectar, the idea is that this can be mapped to control problems which aim to maintain a distribution of resources (such as power or voltage) in such a way that some goal is maintained (such as even temperature or maximal efficiency). In Quijano and Passino (2007b), we see the algorithm tested successfully on a dynamic voltage-allocation problem, in which the task is to maintain a uniform and maximal temperature across an interconnected grid of temperature zones.

Finally, we note that bee-foraging behavior has also directly inspired techniques for internet search, again based on the notion of maintaining a maximally effective use of server resources, adapting appropriately and effectively to the relative richness of new discoveries (Walker 2000; Nakrani and Tovey 2003).

4 Current Trends and Concluding Notes

We have pointed to a number of survey papers and other works from which the reader can attain a full grasp of the current activity in swarm intelligence algorithms, but in this brief section we attempt a few notes that outline major current trends, and then we wrap up.

A notable trend in recent work on PSO, and indeed on metaheuristics in general, is toward the creation of hybrid algorithms. While themes from evolutionary computation continue to be incorporated in PSO (Shi et al. 2005), others have explored the idea of hybridization with less frequently used techniques such as scatter search and path relinking (Yin et al. 2007), immune system methods (Zhang and Wu 2008), and, indeed, ACO (Holden and Freitas 2007). Meanwhile, the range of problems to which PSO may be applied has been greatly increased with the development of multi-objective forms of PSO (Coello Coello et al. 2004).

Other work has involved the use of multiple swarms. This may allow each swarm to optimize a different section of the solution (van den Bergh and Engelbrecht 2004). Alternatively, each swarm may be configured differently to take advantage of the strengths of different PSO variants (e.g., Jordan et al. 2008), in an attempt to create a more reliable algorithm that can be applied to a wide range of problem domains.

The themes of multi-objective optimization and hybridization equally apply to recent research into ACO. While multi-objective ACO is a more mature field than multi-objective PSO (see, e.g., Mariano and Morales 1999), work continues in categorizing and comparing multi-objective approaches to ACO (e.g., García-Martínez et al. 2007), in creating generic frameworks for multi-objective ACO and in creating new multi-objective variants (e.g., Alaya 2007). Recent applications have seen single-objective ACO hybridized with genetic algorithms (Lee et al. 2008), beam search (Blum 2005b), and immune systems (Yuan et al. 2008) and multi-objective ACO used in combination with dynamic programming (Häckel et al. 2008) and integer linear programming (Doerner et al. 2006).

Other recent work has seen ACO adapted for use in continuous domains (Dreo and Siarry 2006; Socha and Dorigo 2008), while research continues into variations of ACO and new algorithm features, for example, different types of pheromone and the use of dominance rules to warn ants from searching among solutions known to be of low quality (Lin et al. 2008).

Recent work on bacterial foraging algorithms has concentrated on exploiting the effectiveness of the local search ability of the algorithm, while adapting it to improve the global search ability on high-dimensional and multimodal problems. With this aim, bacterial foraging has been hybridized with more effective global optimizers such as genetic algorithms (Chen et al. 2007; Kim et al. 2007) and PSO (Tang et al. 2007; Biswas et al. 2007).

In conclusion, we attempted to demystify the concept of swarm intelligence, and, after touring through the chief sources of natural inspiration, we distilled the essence of its impact and presence in computer science down to two major families of algorithms for optimization. No less intriguing and exciting additional topics in the swarm intelligence arena, that have also been discussed, are stigmergic construction, ant-based clustering, and swarm robotics. It is abundantly clear that the natural inspirations from swarming ants, bees, and birds (among others) have provided new ideas for optimization algorithms that have extended the state of the art in performance on many problems, sometimes with and sometimes without additional tailoring and hybridization. Ant-based clustering also seems to provide a valuable contribution, while swarm robotics, stigmergy-based construction, and a variety of other emerging subtopics have considerable promise, and will doubtless develop in directions rather difficult to foresee.

References

Alaya I (2007) Ant colony optimization for multi-objective optimization problems. In: Proceedings of the 19th IEEE international conference on tools with artificial intelligence. Patras, Greece, pp 450–457

Appleby S, Steward S (1994) Mobile software agents for control in telecommunications networks. BT Technol J 12(2):104–113

Berg H, Brown D (1972) Chemotaxis in Escherichia coli analysed by three-dimensional tracking. Nature 239:500–504

van den Bergh F, Engelbrecht AP (2004) A cooperative approach to particle swarm optimization. IEEE Trans Evol Comput 8(3):225–239

Biswas A, Dasgupta S, Das S, Abraham A (2007) Synergy of PSO and bacterial foraging optimization – A

comparative study on numerical benchmarks. In: Corchado E et al. (eds) Innovations in hybrid intelligent systems. Advances in soft computing, vol 44. Springer, Germany, pp 255–263

Blum C (2005a) Ant colony optimization: Introduction and recent trends. Phys Life Rev 2(4):353–373

Blum C (2005b) Beam-ACO – hybridizing ant colony optimization with beam search: an application to open shop scheduling. Comp Oper Res 32(6):1565–1591

Bonabeau E, Guérin S, Snyers D, Kuntz P, Theraulaz G (2000) Three-dimensional architectures grown by simple 'stigmergic' agents. Biosystems 56:13–32

Bonabeau E, Theraulaz G, Deneubourg J-L, Aron S, Camazine S (1997) Self-organization in social insects. Trends Ecol Evol 12(5):188–193

Bonabeau E, Theraulaz G, Deneubourg J-L, Franks NR, Rafelsberger O, Joly J-L, Blanco S (1998) A model for the emergence of pillars, walls and royal chambers in termite nests. Phil Trans Royal Soc B Biol Sci 353(1375):1561–1576

Budrene E, Berg H (1991) Complex patterns formed by motile cells of Escherichia coli. Nature 349: 630–633

Chen T-C, Tsai P-W, Chu S-C, Pan J-S (2007) A novel optimization approach: bacterial-GA foraging. In: Proceedings of the second international conference on innovative computing, information and control (ICICIC). IEEE Computer Press, Washington, DC, p 391

Cicirello VA, Smith SF (2001) Wasp nests for self-configurable factories. In: Proceedings of fifth international conference on autonomous agents. ACM, New York, pp 473–480

Coello Coello CA, Toscano Pulido G, Salazar Lechuga M (2004) Handling multiple objectives with particle swarm optimization. IEEE Trans Evol Comp 8 (3):256–279

Colorni A, Dorigo M, Maniezzo V (1992a) Distributed optimization by ant colonies. In: Varela F, Bourgine P (eds) Proceedings of the first European conference on artificial life, Elsevier, Paris, France, pp 134–142

Colorni A, Dorigo M, Maniezzo V (1992b) An investigation of some properties of an ant algorithm. In: Männer R, Manderick B (eds) Proceedings of the parallel problem solving from nature conference (PPSN 92), Elsevier, Brussels, Belgium, pp 509–520

Deneubourg J-L, Goss S, Franks N, Sendova-Franks A, Detrain C, Chretien L (1991) The dynamics of collective sorting: Robot-like ants and ant-like robots. In: Arcady-Meyer J, Wilson S (eds) From animals to animats: proceedings of first international conference on simulation of adaptive behavior. MIT Press, Cambridge, pp 356–365

Depickere S, Fresneau D, Deneubourg J-L (2004) Dynamics of aggregation in Lasius niger (Formicidae): Influence of polyethism. Insectes Sociaux 51(1): 81–90

DeRosier D (1998) The turn of the screw: the bacterial flagellar motor. Cell 93:17–20

Di Caro G, Dorigo M (1998) AntNet: distributed stigmergetic control for communications networks. JAIR 9:317–365

Di Caro G, Ducatelle F, Gambardella LM (2008) Theory and practice of ant colony optimization for routing in dynamic telecommunications networks. In: Sala N, Orsucci F (eds) Reflecting interfaces: the complex coevolution of information technology ecosystems. Idea Group, Hershey

Doerner KF, Gutjahr WJ, Hartl RF, Strauss C, Stummer C (2006) Pareto ant colony optimization with ILP

preprocessing in multiobjective project portfolio selection. Eur J Oper Res 171:830–841

Dorigo M, Maniezzo V, Colorni A (1991) The ant system: an autocatalytic optimizing process. Technical Report No. 91-016 Revised. Politecnico di Milano, Italy

Dorigo M, Maniezzo V, Colorni A (1996) Ant system: Optimization by a colony of co-operating agents. IEEE Trans Syst Man Cybernetics – Part B: Cybernetics 26(1):29–41

Dreo J, Siarry P (2006) An ant colony algorithm aimed at dynamic continuous optimization. Appl Math Comput 181:457–467

Ducatelle F, Förster A, Di Caro G, Gambardella LM (2009) New task allocation methods for robotic swarms. In: Ninth IEEE/RAS conference on autonomous robot systems and competitions. Castelo Branco, Portugal, May 2009

Franks NR, Sendova-Franks A (1992) Brood sorting by ants: Distributing the workload over the work-surface. Behav Ecol Sociobiol 30(2):109–123

Gambardella LM, Taillard É, Agazzi G (1999) MACS-VRPTW: a multiple ant colony system for vehicle routing problems with time windows. In: Corne D, Dorigo M, Glover F (eds) New ideas in optimization. McGraw-Hill, London, pp 63–76

García-Martínez C, Cordón O, Herrera F (2007) A taxonomy and an empirical analysis of multiple objective ant colony optimization algorithms for the bi-criteria TSP. Eur J Oper Res 180:116–148

Gaubert L, Redou P, Harrouet F, Tisseau J (2007) A first mathematical model of brood sorting by ants: Functional self organisation without swarm-intelligence. Ecol Complexity 4:234–241

Grassé P-P (1959) La reconstruction du nid et les coordinations inter-individuelles chez Bellicositermes Natalensis et Cubitermes sp. La théorie de la stigmergie: essai d'interprétation du comportement des termites constructeurs. Insectes Sociaux 6:41–84

Grassé P-P (1984) Termitologia, Tome II – Fondation des sociétés construction. Masson, Paris

Guney K, Basbug S (2008) Interference suppression of linear antenna arrays by amplitude-only control using a bacterial foraging algorithm. Prog Electromagnet Res 79:475–497

Gutjahr WJ (2007) Mathematical runtime analysis of ACO algorithms: Survey on an emerging issue. Swarm Intell 1(1):59–79

Häckel S, Fischer M, Zechel D, Teich T (2008) A multi-objective ant colony approach for pareto-optimization using dynamic programming. In: Proceedings of the tenth annual conference on genetic and evolutionary computation (GECCO). ACM, New York, pp 33–40

Handl J, Meyer B (2007) Ant-based and swarm-based clustering. Swarm Intell 1(2):95–113

Handl J, Knowles J, Dorigo M (2006) Ant-based clustering and topographic mapping. Artif Life 12 (1):35–61

Heppner F, Grenander U (1990) A stochastic nonlinear model for coordinated bird flocks. In: Krasner S (ed) The ubiquity of chaos. AAAS, Washington, DC

Holden N, Freitas AA (2007) A hybrid PSO/ACO algorithm for classification. In: Proceedings of the 2007 GECCO conference companion on genetic and evolutionary computation. London, UK, pp 2745–2750

Hussein O, Saadawi T (2003) Ant routing algorithm for mobile ad-hoc networks (ARAMA). In: Proceedings of IEEE conference on performance, computing and communications, Phoenix, Arizona, USA, pp 281–290

Jordan J, Helwig S, Wanka R (2008) Social interaction in particle swarm optimization, the ranked FIPS, and adaptive multi-swarms. In: Proceedings of the genetic and evolutionary computation conference (GECCO). Atlanta, Georgia, USA, pp 49–56

Karlson P, Luscher M (1959) Pheromones: A new term for a class of biologically active substances. Nature 183:155–176

Kennedy J, Eberhart R (1995) Particle swarm optimization. In: Proceedings of IEEE international joint conference on neural networks. IEEE Press, Piscataway, pp 1942–1948

Kim DH, Abraham A, Cho JH (2007) A hybrid genetic algorithm and bacterial foraging approach for global optimization. Inf Sci 177:3918–3937

Korb O, Stützle T, Exner TE (2007) An ant colony optimization approach to flexible protein-ligand docking. Swarm Intell 1(2):115–134

Lee Z-J, Su S-F, Chuang C-C, Liu K-H (2008) Genetic algorithm with ant colony optimization (GA-ACO) for multiple sequence alignment. Appl Soft Comput 8:55–78

Lin BMT, Lu CY, Shyu SJ, Tsai CY (2008) Development of new features of ant colony optimization for flow-shop scheduling. Int J Prod Econ 112:742–755

Lumer E, Faieta B (1994) Diversity and adaptation in populations of clustering ants. In: Cliff D et al. (eds) From animals to animats 3: Proceedings of third international conference on simulation of adaptive behaviour. MIT Press, Cambridge, pp 501–508

Mariano CE, Morales E (1999) MOAQ: An ant-Q algorithm for multiple objective optimization problems. In: Banzhaf W, Daida J, Eiben AE, Garzon MH, Honavar V, Jakiela M, Smith RE (eds) Proceedings of the genetic and evolutionary computation conference (GECCO 99). Orlando, Florida, USA, pp 894–901

Mondada F, Gambardella LM, Floreano D, Nolfi S, Deneubourg J-L, Dorigo M (2005) The cooperation of swarm-bots: physical interactions in collective robotics. IEEE Robot Automat Mag 12(2):21–28

Nakrani S, Tovey C (2003) On honey bees and dynamic allocation in an internet server ecology. In: Proceedings of second international workshop on the mathematics and algorithms of social insects

Niu B, Zhu Y, He X, Zeng X (2006) Optimum design of PID controllers using only a germ of intelligence. In: Proceedings of sixth world congress on intelligent control and automation. IEEE Press, Piscataway, NJ, pp 3584–3588

Parpinelli RS, Lopes HS, Freitas AA (2002) Data mining with an ant colony optimization algorithm. IEEE Trans Evol Comput 6(4):321–332

Partridge BL (1982) The structure and function of fish schools. Scient Am June:114–123

Passino KM (2002) Biomimicry of bacterial foraging for distributed optimization and control. IEEE Cont Syst Mag June:52–68

Pini G, Brutschy A, Birattari M, Dorigo M (2009) Interference reduction through task partitioning in a robotic swarm. In: Sixth international conference on informatics in control, automation and robotics (ICINCO 09). Milan, Italy

Potts WK (1984) The chorus-line hypothesis of manoeuvre coordination in avian flocks. Lett Nat 309:344–345

Quijano N, Passino KM (2007a) Honey bee social foraging algorithms for resource allocation. Part I: Algorithm and theory. In: Proceedings of 2007 American control conference. New York, USA, pp 3383–3388

Quijano N, Passino KM (2007b) Honey bee social foraging algorithms for resource allocation. Part II: Application. In: Proceedings of 2007 American control conference, New York City, New York, USA, pp 3389–3394

Reyes-Sierra M, Coello Coello CA (2006) Multi-objective particle swarm optimizers: A survey of the state-of-the-art. Int J Comput Intell Res 2(3):287–308

Reynolds C (1987) Flocks, herds and schools: A distributed behavioral model. Comput Grap 21 (4):25–34

Roberts J, Zufferey J, Floreano D (2008) Energy management for indoor hovering robots. In: IEEE (eds) IEEE/RSJ international conference on intelligent robots and systems (IROS-2008). Nice, France

Rosati L, Berioli M, Reali G (2008) On ant routing algorithms in ad hoc networks with critical connectivity. Ad Hoc Netw 6(6):827–859

Sahin E (2005) Swarm robotics: From sources of inspiration to domains of application. In: Swarm robotics. LNCS, vol 3342. Springer, Berlin, pp 10–20

Schoonderwoerd R, Holland O, Bruten J, Rothkrantz L (1996) Ant-based load balancing in telecommunications networks. Adap Behav 5(2):169–207

Schoonderwoerd R, Holland O, Bruten J (1997) Ant-like agents for load balancing in telecommunications networks. In: Proceedings of the first international conference on autonomous agents. ACM, New York, pp 209–216

Segall J, Block S, Berg H (1986) Temporal comparisons in bacterial chemotaxis. PNAS 83:8987–8991

Shi XH, Liang YC, Lee HP, Lu C, Wang LM (2005) An improved GA and a novel PSO-GA-based hybrid algorithm. Inf Process Lett 93:255–261

Socha K, Dorigo M (2008) Ant colony optimization for continuous domains. Eur J Oper Res 185:1155–1173

Tang WJ, Wu QH, Saunders JR (2007) A bacterial swarming algorithm for global optimization. In: Proceedings of the 2007 IEEE congress on evolutionary computation (CEC 2007). IEEE Service Center, Piscataway, pp 1207–1212

Theraulaz G (1994) Du super-organisme à l'intelligence en essaim: modèles et représentations du fonctionnement des sociétés d'insectes. In: Bonabeau E, Theraulaz G (eds) Intelligence collective. Hermes, Paris, pp 29–109

Theraulaz G, Bonabeau E (1995) Modelling the collective building of complex architectures in social insects with lattice swarms. J Theor Biol 177(4):381–400

Theraulaz G, Bonabeau E, Nicolis SC, Sole RV, Fourcassie V, Blanco S, Fournier R, Joly J-L, Fernandez P, Grimal A, Dalle P, Deneubourg J-L (2002) Spatial patterns in ant colonies. PNAS 99 (15):9645–9649

Tripathy M, Mishra S (2007) Bacteria foraging-based solution to optimize both real power loss and voltage stability limit. IEEE Trans Power Syst 22 (1):240–248

Vander Meer RK, Alonso LE (1998a) Pheromone directed behaviour in ants. In: Vander Meer RK et al. (eds) Pheromone communication in social insects. Westview, Boulder, CO, pp 159–192

Vander Meer RK, Breed M, Winston M, Espelie KE (eds) (1998b) Pheromone communication in social insects. Westview, Boulder, CO, pp 368

von Frisch K (1967) The dance language and orientation of bees. Harvard University Press, Cambridge, MA

Waibel M, Keller L, Floreano D (2009) Genetic team composition and level of selection in the evolution of multi-agent systems. IEEE Trans Evol Comput 13 (3):648–660

Walker RL (2000) Dynamic load balancing model: Preliminary assessment of a biological model for a pseudo-search engine. In: Parallel and distributed processing. LNCS, vol 1800. Springer, Berlin, pp 620–627

Yang B, Chen Y, Zhao Z (2007) Survey on applications of particle swarm optimization in electric power systems. In: IEEE international conference on control and automation. Guangzhou, China, pp 481–486

Yin P-Y, Glover F, Laguna M, Zhu J-X (2007) Scatter PSO – A more effective form of particle swarm optimization. In: Proceedings of the IEEE congress on evolutionary computation (CEC 2007). IEEE Press, Piscataway, NJ, pp 2289–2296

Yuan H, Li Y, Li W, Zhao K, Wang D, Yi R (2008) Combining immune with ant colony algorithm for geometric constraint solving. In: Proceedings of the 2008 workshop on knowledge discovery and data mining. IEEE Computer Society, Washington, DC, pp 524–527

Zhang R, Wu C (2008) An effective immune particle swarm optimization algorithm for scheduling job shops. In: Proceedings of the third IEEE conference on industrial electronics and applications. Singapore, pp 758–763

49 Simulated Annealing

Kathryn A. Dowsland[1] · *Jonathan M. Thompson*[2]
[1]Gower Optimal Algorithms, Ltd., Swansea, UK
k.a.dowsland@btconnect.com
[2]School of Mathematics, Cardiff University, UK
thompsonjm1@cardiff.ac.uk

G. Rozenberg et al. (eds.), *Handbook of Natural Computing*, DOI 10.1007/978-3-540-92910-9_49,
© Springer-Verlag Berlin Heidelberg 2012

Abstract

Since its introduction as a generic heuristic for discrete optimization in 1983, simulated annealing (SA) has become a popular tool for tackling both discrete and continuous problems across a broad range of application areas. This chapter provides an overview of the technique with the emphasis being on the use of simulated annealing in the solution of practical problems. A detailed statement of the algorithm is given, together with an explanation of its inspiration from the field of statistical thermodynamics. This is followed by a brief overview of the theory with emphasis on those results that are important to the decisions that need to be made for a practical implementation. It then goes on to look at some of the ways in which the basic algorithm has been modified in order to improve its performance in the solution of a variety of problems. It also includes a brief section on application areas and concludes with general observations and pointers to other sources of information such as survey articles and websites offering downloadable simulated annealing code.

1 Introduction

Combinatorial optimization problems occur whenever there is a requirement to select the best option from a finite set of alternatives. They arise in a wide range of application areas including business, medicine, and engineering, and are renowned for being relatively simple to state but are often very difficult to solve. Until the 1980s much of the research effort into the solution of such problems concentrated on exact solution techniques, that is, techniques that are guaranteed to produce optimal solutions. Although such approaches cut down considerably on the computational requirements of complete enumeration, for many classes of problems the number of steps needed to guarantee optimality still grows exponentially with problem size – a factor that has serious repercussions for the scalability of any solution approach. By the end of the 1960s, these ideas had been formalized with the definition of an efficient algorithm as one in which the number of steps required increases polynomially with problem size (Edmonds 1965), and the conjecture that there are some problems for which no efficient solution algorithm exists. Subsequent work in theoretical computer science during the 1970s resulted in the now well-established theory of computational complexity and NP-completeness due to Cook (1971, 1972) and Karp (1972). A full treatment of these developments can be found in Garey and Johnson (1979). From a practical point-of-view, the key finding of this work is that there are many combinatorial optimization problems for which efficient algorithms that can guarantee optimality may never be found. This fact, combined with wider availability and decreasing cost of computer power during the early 1980s, lead to a change of focus from exact techniques. Instead, many researchers concentrated their efforts on the search for generic heuristic approaches that could be applied to any combinatorial optimization problem and produce high quality, if not optimal, results. Simulated annealing is one of a number of heuristics developed at this time that derive their inspiration from the natural world. In the case of simulated annealing, this inspiration comes from the behavior of fluids when they are subjected to controlled cooling such as in the production of large crystals. Since its introduction in 1983 there has been a wealth of literature on both theoretical and practical aspects. The purpose of this chapter is to provide an overview of simulated annealing, with emphasis on its use in the solution of practical problems. Nevertheless, some theoretical results are necessary in order to get the best out of a simulated annealing implementation and so these are covered in ❯ Sect. 4. More detailed treatments can be found in Aarts and Korst (1989), Aarts et al. (2005), and Salamon et al. (2002). It should also be noted that although simulated annealing was first introduced as a heuristic for combinatorial optimization, it is now also used for continuous optimization.

Simulated annealing (SA) can be regarded as an improved version of the older technique of local search. In a local search heuristic, an initial solution is gradually improved by considering small perturbations or changes; for example, changing the value of a single variable, or swapping the values of two variables. The set of solutions that can be produced in this way is known as the neighborhood of the original solution and its elements are referred to as neighbors. Starting from the initial solution, a trial solution is chosen from the set of neighbors. If this is "better" than the current solution, then it replaces it. This process continues until no improving solution exists. The replacement of the current solution by the trial solution is known as a neighborhood move. The problem with this type of approach is that the search terminates at a local, rather than global, optimum. This problem can be partially overcome by repeating the algorithm from different starting solutions, or by using more complex neighborhoods, thus widening the scope of the search. A third option used in heuristics, such as SA and tabu search (Glover 1989; Gendreau and Potvin 2005), is to allow some non-improving moves to be accepted. However, to ensure that the overall trajectory of the search is in the direction of improvement, this must be done in a controlled manner. In the case of SA, the control mechanism is inspired by the way in which thermodynamic energy is reduced when a substance is subjected to controlled cooling. These ideas are formalized in the following section.

2 Combinatorial Optimization and Local Search

A combinatorial optimization problem (S, f) consists of a finite space, S, of all possible solutions and a cost function $f : S \rightarrow \Re$ that assigns a real cost to each $s \in S$. For a minimization problem the objective is to find s^* such that $f(s^*) \leq f(s) \; \forall s \in S$.

In order to apply a local search heuristic, we need to define a neighborhood structure $N(S)$ on the solution space S. For each $s \in S$, $N(s)$ defines a subset of S whose members are close to s in some way. Given the above definitions, the generic local search algorithm can be stated as follows:

Procedure local search
    Begin
        Initialise (s_0)
        $s_c = s_0$
        Repeat
            Select $s \in N(s_c)$
            If $f(s) < f(s_c)$ then $s_c = s$
            Until $f(s) \geq f(s_c) \; \forall s \in N(s_c)$
    End

The above definition requires further detail concerning the selection of s. There are two common strategies; random descent and steepest descent. In random descent, the neighborhood is sampled uniformly and the first improving solution is selected, while in steepest descent, the whole neighborhood is evaluated and the lowest cost neighbor is selected.

Two examples of combinatorial optimization problems with appropriate neighborhood definitions are given below.

2.1.1 Example 1: The p-median Problem (Teitz and Bart 1968)

Given a set I of n supply locations, a set J of m demand locations, a cost matrix C, where c_{ij} is the cost of supplying demand location j from supply location i, and an integer constant p, find a subset of p supply locations such that the total cost of supplying J is minimized. Any feasible solution is a partition of I into a set P, containing p elements of I, and the set $I\backslash P$ containing the remaining elements of I. The set of neighboring solutions is defined as those solutions that can be obtained by swapping an element of P for an element in $I\backslash P$.

2.1.2 Example 2: The Traveling Salesman Problem (TSP)

Given a set of n cities and an $n \times n$ matrix D where d_{ij} defines the distance between city i and city j, find a tour (i.e., a circular route visiting each city exactly once and then returning to the starting point) that minimizes the total distance traveled. The set of feasible solutions is the set of permutations of the cities, and any permutation can be regarded as the set of links between adjacent cities in the permutation (the nth city is regarded as being adjacent to the first city). A suitable neighborhood definition is then the set of solutions that can be obtained by removing two edges and then reconnecting the two resulting paths in such a way as to produce a feasible tour. This is illustrated in ❷ *Fig. 1*. Note that if the removed links are (i_1, i_2) and (j_1, j_2) the new links must be (j_1, i_1) and (j_2, i_2) if the result is to be a feasible tour. Note also that the segment between i_1 and j_2 will be reversed. Thus, if the distances are symmetric, tours in $N(s)$ will differ only by the difference in the four edges involved in the swap. However, in the case of the asymmetric problem, this may not be such a good neighborhood as the cost of the reversed segment will also have changed.

◻ **Fig. 1**
An example of a 2-opt neighborhood move for the TSP.

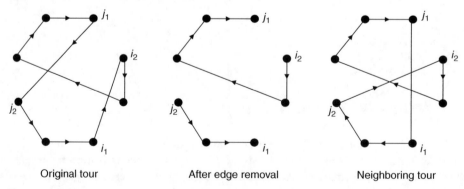

Original tour   After edge removal   Neighboring tour

The advantage of local search is that it is a generic approach that simply requires the definition of a neighborhood structure on the solution space. Thus, it can be applied to almost any combinatorial optimization problem. The disadvantage is that it will converge to a local rather than a global optimum, and thus the quality of the final solution is dependent on the starting point. As a simple illustration, consider the one-dimensional function shown in ❷ *Fig. 2*. Let the set of neighbors of a given point, s, be the two points adjacent to s. Then, if we apply local search from solution P the result will be the local optimum at Q, whereas, if we start from R we may reach Q or S. Both solutions are clearly worse than the global optimum at T. Although this is a very simple example, the same behavior is apparent in multidimensional solution spaces associated with more complex problems.

This difficulty can be ameliorated somewhat by selecting several starting solutions. For realistically sized problems, simply selecting a set of random starting points is rarely sufficient, however, techniques such as greedy randomized adaptive search procedure (GRASP) (Feo et al. 1994) that attempt to use heuristic rules to obtain promising starting solutions and adjust these using feedback from previous iterations have had some success. An alternative approach is to define larger neighborhoods, for example, swapping k locations for the p-median problem, or k edges for the TSP. Such neighborhoods are referred to as k-opt neighborhoods. This can improve results and, in general, as k increases, the number of local optima decreases. Unfortunately, the size of such neighborhoods is exponential in k and so only small values of k are practical and solution quality remains highly dependent on the starting solution. For example, in ❷ *Fig. 2* if we increase the neighborhood to those solutions within two steps of the current solution we will not be able to reach the global optimum from S, or any point to the left of S. More intelligent ways of expanding the neighborhood such as those used in variable neighborhood search or very large neighborhood search (VLNS) have been successful for some problems. The third option for dealing with local optima is to allow some uphill moves in a controlled manner. This is the method employed by simulated annealing.

◼ **Fig. 2**
An example landscape for a one-dimensional function.

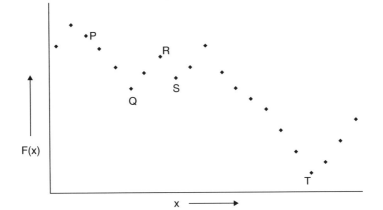

3 Overview of Simulated Annealing

Simulated annealing for combinatorial optimization was introduced by Kirkpatrick et al. (1983) and independently by Černy (1985). The approach is based on an algorithm to simulate the cooling of material in a heat bath, due to Metropolis et al. (1953), a process known as annealing. When solid material is heated past its melting point and then cooled again, the structural properties of the resulting material depend on the residual energy in the material, which in turn depends on the rate of cooling. For example, large crystals represent a state of low energy and result from very slow cooling, whereas fast cooling or quenching results in crystals with high energy, resulting in imperfections. The laws of thermodynamics state that for a substance in a state of equilibrium at an ambient temperature T, the probability of an increase in energy of magnitude δE is given by

$$P(\delta E) = \exp\left(\frac{-\delta E}{kT}\right) \tag{1}$$

where k is a constant known as Boltzmann's constant. Metropolis proposed a simulation model to determine the energy of a system at its ground (or frozen) state, that is, when the temperature is reduced to its limiting value. In the simulation, the material is considered as a system of atoms. For each possible configuration of atoms, the energy of the system can be calculated. At each iteration, one atom in the current configuration is subjected to a small displacement and the increase in energy, δE, is calculated. If $\delta E \leq 0$ a new configuration is accepted automatically. Otherwise it is accepted with probability $P(\delta E)$. This is facilitated by generating a random number, r, in the interval $[0,1)$ and accepting the new configuration if $r < P(\delta E)$. The algorithm can be used to simulate the annealing process by repeatedly reducing the value of T once the system has reached equilibrium at the current temperature, until the system freezes at its ground state. Note that this approach can be regarded as a random descent approach in which some uphill moves are accepted according to ❷ Eq. 1. Kirkpatrick formalized these ideas by mapping the elements of the Metropolis algorithm onto the elements of a combinatorial optimization problem as given in ❷ *Table 1*. Note that as Boltzmann's constant is a physical constant, it is essentially suppressed and the term kT is considered as a single constant, t, which is usually referred to as the temperature.

Although this analogy is an attractive one, in order to be of use in an optimization context it must work effectively for problems where the landscape of the energy function over the states of the system displays similar features to those found in combinatorial optimization problems. This is not the case for most fluids as all the atoms are alike and are therefore

◘ Table 1

Mapping of the Metropolis algorithm onto a combinatorial optimization problem

Thermodynamic simulation	Combinatorial optimization
System states	Feasible solutions
Energy	Cost
Change of state	Neighborhood move
Temperature	Control parameter
Frozen state	Heuristic solution

interchangeable in the solution, resulting in a single ground state. However, there are some classes of fluids that do not obey these rules. Kirkpatrick uses the example of spin-glasses. These are materials that exhibit a feature known as frustration, caused by competing sources of magnetic energy within the system. This has a similar effect on the solution landscape to the interaction between conflicting objectives and constraints in a combinatorial optimization problem and results in many different possible ground states. Thus the analogy between searching the solution space for a combinatorial optimization problem and the annealing of spin-glasses was considered worthy of further experimental investigation. Kirkpatrick's first experiments were based on a partitioning problem in computer design in which the cost function had the same form as the energy function for spin glasses. This was followed by work on problems with more general cost functions including the TSP.

The mapping in ❯ *Table 1* allows one to convert any local search algorithm into an annealing algorithm by accepting non-improving moves according to ❯ Eq. 1. The probability of accepting such a move will be dependent on the magnitude of the increase and the temperature. Thus small increases will be accepted more often than large ones and the probability of accepting any such moves will gradually reduce to 0 as the temperature approaches 0.

A formal statement of the simulated annealing algorithm is given below.

Procedure simulated annealing
Begin
    Initialise (s_0)
    $s_c = s_0$
    Repeat
      For *its* = 1, *nrep* do
      Begin
        Select $s \in N(s_c)$
        $\delta = f(s) -- f(s_c)$
        If $\delta < 0$ then $s_c = s$
        else
          $r = U[0,1)$
          if $exp(-\delta/t) > r$ then $s_c = s$
      End
      $t = \alpha(t)$
    Until stopping condition = 'true'
End

where $U[0,1)$ is a random number in the range $[0,1)$.

In its original form, the Metropolis algorithm returns the current solution when the stopping point is reached. However, this may not be the minimum cost solution visited during the search and it is now common practice to return the best solution found rather than the final one.

The above procedure is a generic statement of the algorithm and a number of decisions must be made in order to implement it for the solution of a particular problem. These can be partitioned into two categories; those relating to the cooling process (the generic decisions) and those relating to the way in which the problem is modeled within a local search framework (the problem-specific decisions). The generic decisions include the starting temperature, the rate of cooling (governed by the function α and the parameter *nrep*) and the stopping condition. The problem-specific decisions involve the definition of the solution space, the cost function, and the choice of neighborhood structure. These combine to define a solution landscape over which the search takes place.

Empirical research has shown that both types of decisions need to be made with care as they influence both the speed of the algorithm and the quality of the solutions obtained, while theoretical results have shown that under the right conditions asymptotic convergence to an optimal solution can be guaranteed. Unfortunately, it has also been shown that in order to guarantee a solution that is arbitrarily close to the optimum, the required number of iterations is exponential in problem size, that is, no better than an exact approach based on complete enumeration. Thus, the theory does not lead directly to a set of generic decisions. Nevertheless, it does provide some pointers as to what factors should be taken into account when making both types of decisions. We therefore review some of the results in the next section before the generic and problem-specific decisions are discussed in more detail.

4   Theory

Much of the theoretical background to simulated annealing is derived from concepts in statistical mechanics and from the theory of Markov chains. This section gives a very brief introduction to these two concepts and cites a number of sources that go into more detail.

For a constant temperature parameter, t, the behavior of simulated annealing can be modeled as a Markov chain. Thus, Markov chain theory underpins much of the theoretical work on SA. In the model, the solutions correspond to the possible states of the system and for any pair of solutions i and j the probability that the system will move from state i to state j is given by the product of the probability of generating solution j as a neighbor of i, denoted by $G_{ij}(t)$ and the probability that j will be accepted, denoted by $A_{ij}(t)$. In the basic model, it is usually assumed that each neighbor of i will be selected with equal probability and that the acceptance probability is given by ❯ Eq. 1. Thus we have

$$
P_{ij}(t) = \begin{cases} G_{ij}(t)A_{ij}(t) & \text{if } i \neq j \\ 1 - \sum_{l \in S, l \neq i} G_{il}(t)A_{il}(t) & \text{if } i = j \end{cases}
$$

where

$$
G_{ij}(t) = \begin{cases} \frac{1}{|N(i)|} & \text{if } j \in N(i) \\ 0 & \text{otherwise} \end{cases} \quad \text{for any } t
$$

and

$$
A_{ij}(t) = \begin{cases} 1 & \text{if } f(j) \leq f(i) \\ \exp\left(\frac{(f(i)-f(j))}{t}\right) & \text{otherwise} \end{cases}
$$

As long as the neighborhoods are strongly connected (i.e., it is possible to reach any solution from any other using a sequence of neighborhood moves) Markov chain theory states that there exists a unique stationary distribution. This distribution defines the probability of reaching solution i after an infinite number of moves at temperature t (denoted $Q_t(i)$), for each solution i. It is given by

$$Q_t(i) = \frac{\exp\left(-\frac{f(i)}{t}\right)}{\sum\limits_{j \in S} \exp\left(-\frac{f(j)}{t}\right)}$$

Moreover, if we define the set of optimal solutions to be S^* then

$$\lim_{t \downarrow 0} Q_t(i) = \begin{cases} \frac{1}{|S^*|} & \text{if } i \in S^* \\ 0 & \text{otherwise} \end{cases}$$

This implies asymptotic convergence to an optimal solution as long as the process reaches the stationary distribution at each temperature. See Aarts and Korst (1989) for a more detailed treatment of the derivation of these results.

A corresponding result can be obtained by modeling the problem as an inhomogeneous Markov chain, in which t is reduced after each move, as long as the sequence of temperatures $t_0, t_1, \ldots, t_k$, obey certain assumptions. In particular, Hajek (1988) showed that the rate of cooling should be related to the shape of the solution landscape and that the sequence of temperatures should satisfy $t_k \geq c/\log(2+k)$, where c is the depth of the deepest local (but not global) minimum. Although bounds on c have been derived for some combinatorial optimization problems (Kern 1986), it has also been shown that calculating c is itself an NP-hard problem. Even if c can be approximated, the number of steps required by Hajek's formula for even moderately sized problems is too large to be practically useful. Other results have suggested that the number of moves required to guarantee a stationary distribution that is arbitrarily close to the optimum will typically require a number of iterations that is exponential in problem size. For example Mitra et al. (1986) show that for the inhomogeneous case, the number of moves required is greater than the size of the solution space, while Aarts and Van Laarhoven (1985) showed that for the homogeneous model the number of steps required at each temperature would be quadratic in the size of the solution space.

The results presented above made quite stringent assumptions concerning the generation and acceptance probabilities that correspond to the basic statement of SA. However, it is possible to prove convergence under more general conditions. In particular, the generation probabilities do not need to be uniform over the neighborhood as long as they are symmetric, that is $G_{ij}(t) = G_{ji}(t)$.

In addition to results based on Markov chain theory, it has also been shown that other concepts from the field of statistical thermodynamics are relevant in the context of optimization. These include the specific heat and entropy. The specific heat of a system is defined as the rate of change in energy with respect to temperature and is given by σ_T^2/kT^2, where σ_T^2 is the variance in energy at the current temperature T. In a physical system, a high level of specific heat indicates a fundamental change in the state of the system, and it has been suggested that the corresponding situation in an optimization context may signal the need for slow cooling. For example, in the solution of the TSP, Kirkpatrick et al. (1983) observed an increase in specific heat once the basic structure of the tour was in place. A related concept is entropy, which is a measure of the disorder of a system. The entropy at equilibrium can be defined as $S_T = -\sum\limits_{i \in S} Q_T(i) \ln(Q_T(i))$. Fleischer and Jacobson (1999) carried out an empirical study based on different implementations for the maximum clique problem and showed that the expected cost of the final solution was inversely related to the entropy. From an optimization viewpoint, it can be shown that as long as equilibrium is reached at each value of t, both the expected cost and entropy decrease monotonically. Furthermore, the probability of finding an optimal solution at each temperature increases monotonically as t decreases. For a full derivation of these results, refer to Aarts and Korst (1989).

Although the above results do not provide any definitive rules for defining polynomial time cooling schedules with any performance guarantees, they still provide some guidance when making problem-specific decisions. These will be discussed in the following sections.

5 Generic Decisions

The generic decisions define the cooling schedule and comprise the starting temperature t_0, the final temperature t_f, and the temperature reduction function α. Each of these is discussed in turn.

5.1 Starting Temperature

If the final solution is to be independent of the starting point then the initial temperature must be hot enough to allow free movement through the solution space. In some cases, this can be estimated from the problem data. For example, it may be easy to calculate the mean or maximum change in cost over all uphill neighborhood moves. This can then be used to estimate the temperature at which such a move would be accepted with a reasonable probability. One commonly used formula is $t_0 = (-U)/\ln(R_{acc})$, where U is the mean increase in cost over all uphill moves and R_{acc} is the required acceptance ratio. Even if it is not possible to estimate U theoretically, it can be estimated by conducting a random walk on the solution space. However, as pointed out by Triki et al. (2005), if the landscape contains some shallow and some very deep local optima, then it may be better to base the starting temperature on an estimate of the maximum, rather than the mean. An alternative is simply to start from a low temperature and to increase it rapidly until the desired acceptance ratio is reached. At this point cooling can begin. In fact, this method is closer to the physical analogy in which a solid is heated past its melting point before being annealed slowly. This approach was taken by Kirkpatrick et al. (1983) in their experiments on a chip-placement problem arising in computer design.

5.2 Reduction Function

The success of any simulated annealing algorithm is highly sensitive to the rate at which the temperature parameter is reduced. This is governed by the function α and the number of steps at each temperature, *nrep*. It is apparent from the theory that t needs to be reduced slowly. Many different schedules have been suggested and two of the most popular are a geometric schedule based on the homogeneous model and a schedule introduced by Lundy and Mees (1986) based on the inhomogeneous model. The geometric schedule takes the form $\alpha(t) = at$, where $a < 1$. Empirical evidence from the literature supports the need for slow cooling and typical values for a are in the range $0.8 - 0.99$ with a bias to higher values, that is, to slower cooling. The value of *nrep* is often related to the size of the neighborhood, and may vary from temperature to temperature. For example, it is important to spend enough time at low temperatures to ensure that the regions around a local optimum have been fully explored.

Thus, it is common to increase *nrep* as the temperature is reduced. *nrep* may also be related more directly to the state of the search by using feedback from the search statistics. This results in a dynamic schedule. Simple dynamic schedules include reducing the temperature with reference to the number of moves accepted, the ratio of accepted moves to trial moves (the acceptance rate), or the ratio of accepted moves to cost increasing moves. As any of these statistics may be very small at low temperatures it is usual to impose a maximum number of moves per temperature as well. The Lundy and Mees schedule is based on the inhomogeneous model and reduces the temperature every iteration according to $\alpha(t) = t/(1+\beta t)$ where β is a suitably small value. Lundy and Mees suggest that β should be significantly smaller than $1/D$, where D is an upper bound on the maximum cost difference between any two points in the solution space. It has been shown that if the parameters for these two schedules are such that t is reduced at a similar rate over a similar range then there is little to choose between them in terms of solution speed or quality. Note that as the gradient of the Lundy and Mees schedule gets shallower as temperature decreases, an equivalent geometric schedule would involve increasing *nrep* with every decrease in t.

In all the above schedules, the function α is constant and predetermined. Suitable parameters are often derived by trial and error and may not be robust, even for different instances of the same problem. As the theory of simulated annealing is based on the concept of equilibrium at each temperature, it has been argued that an adaptive schedule based on this idea may be a better choice. One way of achieving this is to ensure that equilibrium is reached at the initial temperature and then reduce the temperature by a sufficiently small amount so that equilibrium will be reached quickly at the new temperature. This will be true for t_0 if it is high enough to allow free movement between solutions. Thus all that is necessary is to ensure that each temperature value is sufficiently close to its predecessor. A number of reduction functions have been suggested based on estimates of the mean energy at different temperatures. An early example is the schedule of Van Laarhoven and Aarts (1987). The aim of their schedule is for the stationary distributions at consecutive temperatures to be close. This will be true if $Q_{\alpha(t)}(i) < (1+\varepsilon)Q_t(i)\ \forall i, t$, and for a suitably small value of ε. This leads to the derivation of the function

$$\alpha(t) = t\frac{1}{1 + \frac{\ln(1+\varepsilon)}{3\sigma_t}t}$$

where σ_t is the standard deviation of the cost at temperature t. In practice, suitably small values of ε and suitable starting and ending temperatures mean that this cooling schedule is usually too slow. One of the most popular schedules in practice is due to Huang et al. (1986). The reduction in temperature is determined by the difference in the average cost at consecutive temperatures. They use

$$\alpha(t) = t \cdot \exp(-\lambda t/\sigma_t)$$

where λ is an estimate of Δ/σ_t where Δ is the difference in the expected cost at t and $\alpha(t)$. They suggest using a value of 0.7 for λ.

Although both these schedules are designed to cope with different solution landscapes, we note that neither is parameter free. More recently, Triki et al. (2005) suggested a more general representation of this type of schedule using

$$\alpha(t) = t \cdot \left(1 - t\frac{\Delta(t)}{\sigma_t^2}\right)$$

where $\Delta(t)$ is an estimate of Δ as defined above. They point out that an appropriate choice of $\Delta(t)$ will yield both Van Laarhoven and Huang's schedules, as well as other schedules from the literature. They suggest an implementation in which the value of $\Delta(t)$ is initialized to a quantity proportional to σ (estimated from a random walk) and *nrep* is governed by a limit on the number of iterations and the number of acceptances, whichever occurs first. At this point, if the decrease in cost is sufficiently close to the target $\Delta(t)$, equilibrium is assumed and the temperature is reduced. Otherwise, another set of repetitions is carried out at this temperature. If the test for equilibrium is failed four times at the same temperature, the assumption is that the temperature was reduced too rapidly so the temperature is increased and the target function Δ is decreased.

5.3 Freezing Point

In theory, the temperature should be reduced to zero when no further uphill moves are possible and the system will terminate at a local minimum. In practice, a frozen state where no further moves are possible is often reached well before this. Lundy and Mees suggest that if the objective is to produce a solution within ε of the local optimum with probability θ then the termination point should satisfy $t_f \leq \varepsilon / \ln((|S|-1)/\theta)$. Others have suggested stopping after a predefined number of iterations without acceptance, or when the acceptance ratio falls below a certain value. Alternatively, if the average cost is measured over a small number of moves, the system may be considered frozen when the difference in consecutive calculations falls below a predefined parameter.

6 Problem-Specific Decisions

The problem-specific decisions cover the definition of the solution space, neighborhood structure, and cost function. These factors combine to define the solution landscape. Empirical results suggest that this has a significant impact on the success of a simulated annealing implementation. This coincides with the theoretical results such as Hajek's and fits well with the concept of entropy, as at the same temperature sparsely connected regions of the solution space display lower levels of entropy than denser areas, indicating that there is a smaller choice of acceptable moves. As with the generic decisions it is not possible to derive a set of definitive rules that can be applied to a given problem. Nevertheless, it is possible to outline some desirable properties. These are based on common sense, empirical investigations, and theoretical results. For many problems it may not be possible to meet all these guidelines at the same time, and some of them may even be contradictory.

The first consideration is to maintain the conditions for convergence. As outlined in the previous section, the original convergence results relied on a set of stringent conditions, but it has since been shown that it is sufficient that every solution be reachable from every other by a chain of neighborhood moves. This is usually easy to verify for well-structured problems, but may be more difficult for messy problems that occur in practice.

A second consideration stems from the requirement for slow cooling. This means that it will be necessary to execute a large number of iterations within the available computing time. Thus the calculations involved in each iteration need to be as efficient as possible. Two processes that are related to the solution landscape are required at each iteration. The first is the generation of a randomly selected trial solution from the set of neighbors of the current solution. For many problems, a combination of the most obvious definitions of solution space and neighborhood structure results in a fast and simple neighborhood generation routine. For example, for the p-median problem all that is required is to select a member of P and a member of $I\backslash P$. This can be achieved by generating two integer random numbers within the appropriate ranges. However, this may not be the case for complex neighborhood definitions or if the solution space is highly constrained. In such cases, it may be beneficial to relax some of the constraints in the definition of the solution space and to include a penalty term in the cost function to guide the search away from infeasible solutions. This will be illustrated in the graph coloring example in the next section.

The second routine that is required at every iteration is the calculation of the difference in cost between the current solution and the trial solution. One way of doing this is to calculate the cost of the trial solution from scratch, but this is often unnecessary as it is more efficient to calculate the change in cost. For example, in the traveling salesman example, as long as the problem is symmetric, the change in cost is just the difference between the cost of the two removed edges and that of the two replacement edges. The cost calculation for the p-median problem is more complex. In an optimal solution, each demand point will be allocated to the nearest element of P. Thus changing one supply point may change the allocations of any of the demand points. Nevertheless, it is still worthwhile calculating the cost in terms of those demand points that have been reallocated rather than completely evaluating the new solution. The efficiency of such calculations can be improved by using appropriate data structures. It is worth noting that even if suitable data structures require a significant amount of updating every time a move is accepted, the number of accepted moves over a whole run is small compared to the number of trial moves evaluated. Thus their use may still save time.

In order to reach good solutions, it is desirable that the cooling schedule be as close as possible to the ideal. Hajek's result suggests that neighborhood/cost function combinations that give rise to spiky landscapes with deep troughs should be avoided. On the other hand, it is also clear that the cost function should guide the search over the landscape, rather than behave like an unbiased random walk. Thus, flat plateau-like regions resulting from groups of neighboring solutions of equal cost should also be avoided. This point will also be illustrated in the graph coloring example below. A number of theoretical and empirical results also suggest that the rate of cooling should be related to the size of the solution space and/or neighborhoods. This suggests that these should be kept as small as possible. In the case of the solution space, this observation conflicts with the idea of increasing the solution space to allow some infeasible solutions as suggested above. In some cases, it may be possible to reduce the set of solutions by applying a reduction technique to the problem beforehand. For many classical combinatorial optimization problems (e.g., set covering, maximum clique, steiner problem in graphs), such techniques have been well documented. It is however worth remembering that, by definition, these techniques remove points from the solution space and therefore care must be taken that this does not destroy the reachability property. Even if this is not the case the reduction may remove neighbors that are close in cost to the current solution, resulting in a landscape with a spiky topography and lower entropy that will be more difficult to search. In the case of the neighborhoods, it is generally agreed that neighborhoods should be small, so that neighborhood moves are truly local. Nevertheless, it is not always beneficial to use the smallest possible neighborhood as larger neighborhoods may lead to a smoother topography. Conversely, Goldstein and Waterman (1988) suggest that if the landscape is too smooth, it becomes too easy to escape from valleys and thus good optima may be missed. However, at low temperatures, they are likely to need more iterations in order to ensure that they are sampled adequately so that the local minimum in the region is located.

The above discussion suggests that the combination of solution space, neighborhood structure, and cost function suggested for the local search solution for the TSP should also work well for simulated annealing. Neighbors are easy to generate and the change in cost is simple to calculate. The size of the solution space is $(n-1)!/2$ and the size of the neighborhood is $n(n-1)/2$. Thus the neighborhood is small compared to the size of the solution space. It is also easy to see that any solution can be reached from any other by a chain of feasible moves. In the case of the p-median problem, the local search framework is also suitable for SA, although the change in cost function cannot be calculated quite as easily as in the TSP case. The fact that more elements in the cost function are subject to change also suggests that the solution landscape may have more spikes and troughs. The next section looks at one further example – that of graph coloring, for which the problem-specific decisions present a greater challenge.

6.1 Graph Coloring

The graph coloring problem can be stated as follows. Given a graph $G(V, E)$ with vertex set V and edge set E find an allocation of colors to the vertices, such that adjacent vertices are given different colors and the number of colors used is minimized.

Unlike the traveling salesman problem, prior to the 1980s the graph coloring problem was not regarded as a natural candidate for local search. Nevertheless, there has been considerable interest in simulated annealing for graph coloring. The most obvious definition for the solution space is the set of all feasible colorings. These can be represented as partitions of the set of vertices into subsets, such that there are no edges between two vertices in the same subset. All vertices within a subset are considered to be of the same color, so the objective is to minimize the number of subsets. A natural definition for a neighborhood of a given solution is those partitions that can be obtained by moving a single vertex into a new subset. However, there is a problem with this, in that not all moves will result in a feasible solution as edges may be introduced into the subsets involved in the move. One can overcome this by limiting the moves to those subsets that are feasible. This could mean that some vertices will not have any feasible moves and also makes the generation of a neighbor more complex. A more fundamental problem is that it is easy to generate instances where the reachability condition is violated under this neighborhood. Even if the solution landscape remains connected, paths between different regions may be sparse thus making it difficult for the search to move away from the region containing the starting solution.

One way of overcoming this is to use a larger neighborhood. Morgenstern and Shapiro (1989) suggested a neighborhood based on Kempe chains. If we consider that part of the graph relating to just two colors or subsets, then it will consist of one or more components, each of which is completely disconnected from the others. Such components are known as Kempe chains. Swapping the colors (or subsets) of the vertices within one of these components will not introduce any edges between vertices in either subset, and thus the result is also a feasible coloring. This is illustrated in ❷ *Fig. 3*. Morgenstern and Shapiro suggest defining the neighborhood as those solutions that can be reached by swapping the colors of a single Kempe chain. There is a computational overhead in generating such moves, and in calculating their effect on the cost function. Nevertheless, Kempe chain neighborhoods have been shown to outperform simple neighborhoods for some classes of graph. In addition, Thompson and

☐ **Fig. 3**

Example of a Kempe chain neighborhood move for the graph coloring problem.

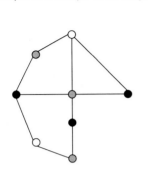 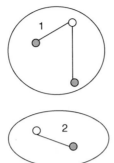 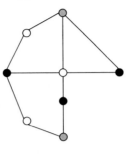

Graph colored in black, white and grey

The 2 components of the grey-white kempe chain

New coloring after swapping component 1

Dowsland (1998), Wright (1996), and Tuga et al. (2007) have all demonstrated their effectiveness in solving graph coloring-type problems arising in scheduling and timetabling.

An alternative to larger neighborhoods is to relax the definition of the solution space to include infeasible solutions and to penalize these with a suitable penalty term in the cost function. Examples of this approach will be given after a more general discussion of the cost function.

The natural definition of the cost function is the number of colors used. This is not suitable for a simulated annealing implementation as it will lead to large plateau-like areas in which the costs of neighboring solutions are identical and there is no information to guide the search toward lower cost solutions. For example, a solution in which there are k subsets each with a similar number of vertices will have the same cost as a solution with k subsets in which one subset has just one vertex. Clearly, the latter is preferable to the former as it requires just one move to reach an improved solution requiring $(k-1)$ subsets. Aarts and Korst (1989) suggest a cost function of the form $\sum_{i=1}^{q} w_i(|V_i| - \lambda|E_i|)$ where V_i, E_i are the sets of vertices and edges in the ith subset, λ is a weighting factor such that $\lambda > 1$ and q is an upper bound on the number of colors needed. $w_1, \ldots, w_q$ is a decreasing sequence of weights based on the number of vertices in subset i, and the subset index, designed to encourage lower indexed sets to be emptied in favor of those with higher indices. The optimal solution with respect to this cost function is guaranteed to be optimal in terms of the original objective, but the weights can be very large and therefore cause computational difficulties. For practical purposes, a relaxation of the condition on the weights is suggested, but this no longer guarantees that the optimal solution will correspond to an optimal coloring. Johnson et al. (1991) suggest a simpler solution based on a term to encourage large subsets and a penalty term for infeasible colorings. The form of their function is:

$$-\sum_{i=1}^{q} |V_i|^2 + \sum_{i=1}^{q} |V_i||E_i|.$$

Although optimal solutions using this function may not correspond to optimal colorings, they are guaranteed to be local optima.

An alternative way of dealing with the cost is to determine an upper bound, q, on the number of colors required, define the solution space as the set of all partitions of the vertices into q subsets, and simply minimize the sum of the $|E_i|$ values. Once this is reduced to zero, q can be reduced and the process repeated. The last value of q for which a zero-valued solution was achieved is taken as the final solution. The problem here is the requirement to find an optimal solution in a series of annealing runs. However, for practical problems, the objective is sometimes simply to find a coloring in q colors for a given q. Johnson et al. (1991) compare this approach with their penalty methods and a Kempe chain-based implementation. They conclude that for problems where the global optimum based on the penalty cost implies several more colors than the optimal coloring, the q partition method can often do better than the Kempe-chain approach. Their main conclusion is that none of the three implementations outperformed the others on all classes of graph, emphasizing the difficulty in finding the right combination of problem-specific definitions for a particular problem type.

7 Enhancements and Modifications

The previous sections have dealt with implementations of simulated annealing that adhere to the analogy with the physical cooling process. Thus the theoretical results on convergence hold for the optimization process. However, as we have seen, run times for cooling schedules that are able to give any performance guarantees tend to be too long to be of practical value. Thus, very few real implementations have the benefit of such theoretical backup. It is also the case that the theory behind many cooling schedules is based on the assumption that the returned solution is the final, and not necessarily the best solution. It is therefore arguable that there is no reason to stick rigidly to the original Metropolis simulation, and that any heuristic that searches the solution space using neighborhood moves and allows uphill moves that are controlled using a probability function based on temperature may be an equally good candidate for solving optimization problems in practice. For some generalizations the convergence proofs will still hold. Even if they do not, the heuristic may still perform well enough. This section examines a number of ways in which successful implementations have bent the rules of the classical Metropolis-based algorithm. In some cases, these modifications improve the performance of SA on problems for which it may not seem ideally suited, for example, those with spiky solution landscapes, or cost functions that are expensive to calculate. In others, they simply improve the efficiency of the search, thus allowing larger problems to be tackled.

7.1 Acceptance Probability

An obvious way of relaxing the conditions of the physical analogy is to move away from the Boltzmann distribution in determining whether or not an uphill move should be accepted. Its use in the original heuristic is entirely due to the laws of thermodynamics, which have no meaning in terms of an optimization problem. Nevertheless, it is still used by the majority of implementations. This is no doubt due to the fact that it does possess some desirable properties: solutions are accepted according to the magnitude of increase in the cost function, δ; the probability of acceptance decreases monotonically as δ increases; and this probability tends to zero as $\delta \to \infty$. On the other hand there is some motivation for using a different distribution. Johnson et al. (1989) showed that in a straightforward implementation of annealing to a graph partitioning problem, the calculation of $\exp(-\delta/t)$ took up approximately one-third of the execution time. They concluded that it may be beneficial to use a simpler function, thereby allowing time for additional iterations. They suggested using a linear function that approximated the exponential over an appropriate range of values but found that better results could be obtained if $\exp(-\delta/t)$ was calculated for all possible δ, t combinations beforehand and stored in a look-up table. This is obviously only possible if the number of different cost function values is reasonably small.

A very simple scheme suggested by Dueck and Sheuer (1990) and by Moscato and Fontanari (1990) is known as threshold accepting. The idea here is that t represents a threshold on the values of δ that will lead to acceptance and the probability is defined as $P(\delta) = 1$ if $\delta \leq t$, $= 0$ otherwise. Threshold accepting has proved successful for some problems and less so for others. More recently, Penna (1994) and Tsallis and Stariolo (1996) suggested a more complex acceptance probability of the form

$$P(\delta) = \begin{cases} 1 & \text{if } \delta \leq 0 \\ (1 - (1-q)\frac{\delta}{t})^{\frac{1}{1-q}} & \text{if } \delta > 0 \text{ and } (1-q)\frac{\delta}{t} \leq 1 \\ 0 & \text{otherwise} \end{cases}$$

where q is an additional parameter such that $q \neq 1$. This allows larger jumps to be accepted occasionally depending on the value of q. If $q \to 1$ then this is equivalent to the standard Metropolis algorithm. Franz and Hoffmann (2003) have since shown that a slight modification of the Tsallis acceptance probability produced by scaling t by a factor of $(2-q)$ not only tends to the Metropolis algorithm as $q \to 1$, but also tends to threshold accepting as $q \to -\infty$. They were able to show that given an ideal cooling schedule results improve as $q \to -\infty$. Given that threshold accepting was introduced as a simplification of SA designed purely for easy implementation this is a somewhat surprising result. However, it is worth noting that performance depends on an appropriate schedule and far more is known about good schedules for the Boltzmann acceptance rules than for threshold accepting.

Many other practical implementations have been shown to benefit from changes to the acceptance rules. For example, Dowsland (1993a) found that the solution landscape for a two-dimensional rectangle packing problem contained large peaks and troughs so that moderate increases in cost were needed to escape local optima. When the temperature was high enough to accept such moves with a reasonable probability, too many small uphill moves were made. This was overcome by adding a positive constant to the change in cost. This has the effect of moving to a flatter part of the exponential curve, thus decreasing the probability of accepting small moves relative to moderate ones, while still ensuring that very large uphill moves are accepted with extremely low probability. There has also been some success with simple functions. For example, Ogbu and Smith (1990) and Vakharia and Chang (1990) use probabilities that are independent of δ on different sequencing problems, while others have used a linear function such as δ/t (Brandimarte et al. 1987).

Wright (2001) suggests an alternative rule for problems in which the objective is made up of a number of different terms, which he refers to as subcosts. This is often the case for problems with hard and soft constraints, where the objective is often to minimize a weighted linear function based on the violated soft constraints. In such cases, the conflict between the constraints may mean that it is necessary to violate several constraints in different groups in order to gain any improvement in another. Therefore, if traditional acceptance probabilities are used, the search may get trapped in regions where certain constraints are never satisfied. In order to overcome this, Wright suggests subcost-guided simulated annealing, which uses the acceptance criterion

$$P(\delta) = \exp\left(\frac{-\delta.\exp(\frac{-\Theta B}{\delta})}{t}\right)$$

where B is the best improvement in any of the subcosts and Θ is a problem-specific parameter. This has the effect of improving the chances of accepting a move that results in a large improvement in one of the subcosts at the expense of several smaller increases in the others.

We have seen that it may be beneficial to include infeasible solutions in the solution space and to penalize them in the cost function. However, as pointed out by Abramson (1991), finding the right weights for such penalties can be difficult. Weights that are high enough to ensure convergence to a feasible solution tend to mean that the search becomes trapped early in a sub-region of the solution space as infeasible solutions tend never to be accepted even at moderate temperatures. Pedamallu and Ozdamar (2008) suggest a novel acceptance criterion for this situation. Instead of considering a combined cost, they monitor the sequence of feasible and infeasible solutions separately and make acceptance decisions independently within each sequence. These two ideas are incorporated into two hybrid simulated annealing methods. They also reduce the probability of accepting an uphill move by only considering acceptance if the last n trial moves were also uphill moves.

7.2   Cooling

Most simulated annealing runs that start with a high enough temperature to allow free movement and cool until the system appears to be frozen do much of their useful work in the middle part of the run. When the temperature is high the search will accept almost any solution, each one of which could have been generated as the initial solution. This part of the search can also be computationally expensive as the high acceptance rate means that there are more solution updates to be carried out. In order to overcome this, some implementations may use rapid cooling at this stage, for example, reducing temperature after just a few moves have been accepted. Others suggest starting at a lower temperature where the rate of acceptance is much lower than that suggested by the theory. Conversely, at low temperatures when few moves are accepted, SA becomes a slower version of random descent in which equal cost moves as well as improving moves are accepted. This can lead to cycling around local optima without actually making any progress. This is sometimes avoided by replacing this phase of the algorithm by a random descent phase and stopping once a local optimum has been reached (e.g., Merlot et al. (2003)).

Many practical implementations set the temperature so that just the middle phase of the cooling schedule is used. However, as long as we return the best solution found during the search, it can be argued that overall cooling may not be necessary. Boese and Kahng (1994) show that under these conditions optimal schedules may involve heating, even at the end of the search. In a practical context, Connolly (1990) suggested using a constant temperature, as long as that temperature is appropriate for the problem at hand. After experimenting with temperatures that resulted in a high measure of specific heat and finding that they did not result in high quality solutions, he suggests starting the process with a fairly quick run using the Lundy and Mees schedule over a wide temperature range, and then using most of the available time using a fixed temperature, defined as that temperature where the minimum cost solution was located. Results on the quadratic assignment problem were better than those obtained with standard cooling.

Connolly reports that a constant temperature allowed a good balance between downhill and uphill moves, thus encouraging the search to get to the bottom of local minima, but also allowing it to escape again. For spiky landscapes this may not be sufficient, and the idea of reheating as a means of escape from deep valleys has been used in a range of different ways. Glover and Greenberg (1989) point out that this is perhaps a more intuitive approach as it corresponds to making a concerted effort to move uphill once it becomes clear that further progress in the current valley is unlikely. It is worth noting that Kirkpatrick et al. did use some reheating in their initial study of the TSP, using a graphical representation of the tours to identify when the search seemed to be stuck and using manual intervention to reheat. Such manual intervention is obviously not practical for most cases and so a number of ways of automating the reheating process have been suggested. We have already mentioned Triki et al. (2005) who reheat slightly when it appears that the latest cooling step may have lowered the temperature too far. Several practical implementations reheat once a frozen state has been reached. For example, Anagnostopoulos et al. (2006) raise the temperature to twice the value where the best solution was found. This is repeated until the number of reheats without an improvement reaches a given limit. Other examples include Azizi and Zolfaghari (2004) who use the cooling function $\alpha(t) = t_0 + \lambda \ln(1+r)$ where r is a count of the number of consecutive uphill moves up to the current iteration and λ is a control parameter. The initial temperature is set to a *low* value, which is equal to the minimum value that t can take. At the start, the low temperature means that the search should move rapidly downhill. Each time an uphill move is selected, the temperature increases, thereby giving it more impetus to climb out of local minima. As soon as a downhill move is accepted, the temperature reverts to its minimum value. Dowsland (1993a) also suggests an undulating schedule. In this case, the decision as to whether to raise or lower temperature is based on whether the last move was accepted or rejected. If a move is accepted, then temperature is reduced and vice versa. Both cooling and reheating are carried out using the Lundy and Mees formula with different values of β. If $\beta_c = k\beta_h$ where β_c is used for cooling and β_h is used for heating, then the ratio of rejected to accepted moves will be approximately k. Dowsland first used this schedule with constant k for a rectangle packing problem. If k is reduced periodically, then this schedule can be used to replace standard cooling. This has been used with some success for examination scheduling (Thompson and Dowsland 1996), and as a way of ensuring that the acceptance rate drops monotonically with temperature in an implementation of SA that involves periodically selecting a new neighborhood (Dowsland et al. 2007).

7.3 Neighborhoods

In the basic algorithm, it is assumed that neighborhoods are well defined and do not change at all during the search. There are many practical examples where nonconstant definitions of the neighborhood have been shown to lead to improvements in solution time or quality. One popular ploy is to adjust the neighborhood as the temperature changes. For example, for problems such as packing or placement problems, where neighborhood moves involve moving an element over a finite distance, it is common to reduce the maximum distance allowed as the temperature reduces. Sechen et al. (1988) applied this idea to a simulated annealing approach to a VLSI design problem in which rectangular blocks are to be placed on a chip in such a way as to minimize a combination of cost factors. Valid moves are a horizontal or vertical transition of a single block. As only small transitions tend to be accepted at low temperatures, the authors argue that time is wasted sampling and rejecting such moves, and so the maximum transition length is decreased as temperature decreases. Similar strategies have been used in SA implementations for packing problems. Such strategies are not limited to problems where moves involve physical distances and are also applied where neighborhood moves involve changing the value of a single variable where an upper limit on the magnitude of the change is reduced gradually.

A reduction in the neighborhood may also save time in situations where not all entities in the solution are contributing to cost. For example, consider a scheduling problem in which the cost function represents a series of penalties for undesirable features in the schedule and neighborhood moves involve moving a single event. Not all events will contribute to the cost and so moving such an event will not lead to an improvement. Many implementations therefore exclude these moves from the neighborhood. For example, in solving the rectangle packing problem Dowsland (1993a) found it beneficial to move only those pieces causing overlap, and in solving an examination scheduling problem. Thompson and Dowsland (1998) move only those exams causing clashes. While this strategy is a good idea for many problems, it should be used with care because it may destroy the reachability property. In view of this, Tovey (1988) suggests that the full neighborhood should be used with a given probability and the reduced neighborhood used the rest of the time.

In the context of other local search algorithms, it has been shown that using different neighborhoods, either with a fixed probability throughout the search, or at different times in the search, can be very effective (Hansen and Mladenovic 2005). Such a strategy is not straightforward in a simulated annealing context, as using different neighborhoods at different temperatures will have a significant effect on the transition probabilities, and therefore the expected cost may no longer decrease monotonically with temperature. Nevertheless, neighborhoods with significant variations have been used in some implementations. For example, Dowsland et al. (2007) use several different neighborhoods in the solution of a shipper rationalization problem and attempts to overcome this problem by using the acceptance rate to set the temperature using the undulating schedule described in the last section.

7.4 Neighborhood Sampling

In addition to the definition of the neighborhood, the transition probabilities are also affected by the probability with which each neighbor is sampled. In ❷ Sect. 4, it was assumed that every neighbor was generated with an equal probability. In practice, sampling may not be uniform. In some cases, the choice of nonuniform sampling is made to simplify the neighborhood generation procedure, while in others it is an attempt to guide the search more effectively.

A simple modification that has been used in a number of implementations is the use of cyclic, as opposed to random sampling. The rationale behind this is twofold. Close to local optima, there may only be one or two moves that lead to an improvement in cost. At low temperatures it is desirable that such moves be made. However, even if the number of trials is greater than the size of the neighborhood such moves may not be sampled and eventually an uphill move may be accepted and the search may move away from the region without the local optimum having been visited. Cyclic sampling ensures that every move is considered before any is considered for a second time. Cyclic sampling also has the advantage of saving time as it does not require the generation of one or more random numbers in order to select a neighbor. Nevertheless, it is important to impose a random ordering on the cycle at the start of the run in order to avoid introducing any form of bias. This can be illustrated by considering an approach taken by Connolly (1990) for the quadratic assignment problem in which a set of n items are to be allocated to n locations so that each location contains a single item. The natural neighborhood for this problem is to swap two items. For any solution we can therefore consider the neighborhood moves as the set of (unordered) pairs of items. If we take the natural ordering then all swaps for item 1 would be considered before those for item 2, etc. This may introduce unwanted bias and so Connolly randomizes the list and then cycles through the list in this new order. This resulted in improved performance over random sampling in terms of both time and solution quality. This example also illustrates a third potential benefit of cyclic sampling. If the list of neighborhood moves can be represented in such a way that it is valid for all solutions, as in this case, then when a move is accepted, instead of starting from the beginning of the list it makes sense to carry on from the current position. In this case, cyclic sampling will help to avoid the search cycling through a small set of solutions as it behaves like a (long) tabu list. For example, if items a and b are swapped, all other swaps will be considered before a and b can be swapped back again. Others who have used cyclic sampling to ensure the search is sufficiently diverse include Osman (1993). It should be noted, however, that other researchers have found that cyclic sampling has an adverse effect on solution quality for their particular implementations.

At low temperatures, time may be wasted in sampling many solutions (often more than once) without an acceptance. Greene and Supowit (1986) suggest the alternative of determining the probability of acceptance for each neighborhood move and then using these to weight the sampling distribution, that is, they combine the two probabilities $A_{ij}(t)$ and $G_{ij}(t)$ into a single acceptance probability. As the transition probabilities remain the same, this will not change the behavior of the algorithm. Greene and Supowit apply their method to a graph partitioning problem and show that the required probabilities are easy to maintain. For other problems this may not be the case, but this approach can be used at a local minimum at low temperatures. In this situation, it may take a long time before an uphill move is accepted. If some form of sampling without replacement is used, as each neighbor is generated, its cost can be recorded. If no move is selected before the neighborhood is exhausted, the appropriate probabilities can be calculated and a move selected according to these values.

Perhaps the most intuitive use of nonuniform sampling is to bias the selection toward low cost neighbors. For example Lin et al. (2008) consider a truck and trailer routing problem and instead of making the accept/reject decision on a single neighbor, they generate several neighbors and then use the best of these as the trial solution. Ropke and Pisinger (2006) apply a very large neighborhood search within a simulated annealing framework for a pick up and delivery problem. VLNS is a local search method that uses neighborhoods that are very large. This obviously smoothes out the solution landscape, but means that it is not practical to evaluate the whole neighborhood or even a large proportion of it. Thus VLNS is only effective when there is some form of optimization algorithm (or good fast heuristic) that can find an optimal (good) member of the neighborhood. If the neighborhood is optimized, then simulated annealing may not work well as the idea behind SA is that the pressure to move downhill should increase slowly throughout the search, and the selection of a neighbor should therefore be sufficiently random to maintain an appropriate level of flexibility. Ropke and Pisinger try to avoid this problem by introducing randomization in two ways. First, one of a number of different heuristics can be used to select the "best" element to remove/insert. This choice is random, but is biased according to previous performance. Second noise is added to the calculated cost before applying the acceptance/rejection criterion. Computational tests show that this latter strategy gave better results than a version without noise. Anagnostopoulos et al. (2006) compare a neighborhood that includes some very large moves with one that includes only simpler moves. They conclude that the former neighborhood gives the best results. In the case of differentiable real-valued functions the concept of large neighborhoods is very attractive as local optima can be found easily using gradient methods. Wales and Scheraga (1999) coined the term "basin hopping" for a simulated annealing solution to a clustering problem, in which each trial solution was a local optimum. One problem with this approach is that the jump between starting points needs to be large enough to converge to a different basin, but close enough to maintain the spirit of a local, rather than random, search.

In the case of real-valued functions, the simplest way of generating a neighbor is to select solutions from a hyper-sphere with a given radius centered at the current solution. Alternatively, the distance is sometimes selected according to the Boltzmann distribution. This may mean that it takes a long time for the search to reach new areas of the solution space. In order to avoid this, it has been suggested that the distance moved is generated from other distributions with fatter tails, thus increasing the probability of generating larger jumps. In variable step sized simulated annealing, the distance is adjusted according to the acceptance ratio. If it is high, the search is considered to be far from the optimal and the size is increased, and vice versa (Kalivas 1992). In fast simulated annealing (Szu and Hartley 1987), the Cauchy distribution is used for this purpose.

In all the above examples, the choice of nonuniform sampling was made to improve the search. For some problems, the most straightforward way(s) of neighbor generation may not correspond to uniform sampling. For example, timetabling problems can be regarded as graph coloring problems in which the feasible solutions are the feasible colorings in q colors, and the objective is to optimize a cost function relating to the quality of the timetable. Suppose we define a neighborhood move as changing the color of a single vertex. As seen in ❷ Sect. 6, if the solution is to remain feasible, not all colors will be feasible for a given vertex. Therefore, if a neighbor is generated by selecting a vertex and then a new color for that vertex, the selection policy will tend to move vertices from "fuller" colors to "emptier" ones. An alternative with different bias might be to select two colors and then select a vertex that is free to move from the first color to the second. Such bias can affect solution results. For example, Thompson and Dowsland (1998) used this approach in the solution for an examination scheduling problem. They found that when the objective was to even out the number of students sitting exams on each day, then the first strategy performed well, but when the objective was to minimize back-to-back exams, when solutions in which alternate full and empty sessions are best, the second strategy was better. Similar results were reported using two different generators for a Kempe chain neighborhood on the same problem. Bias was also found to be a problem in Dowsland (1993b). In this case, the two natural sampling strategies introduced different biases both of which adversely affected solution quality. The final implementation used them both in alternate iterations, which removed the bias and yielded good results.

7.5   Cost Function

The standard simulated annealing algorithm does not assume any restrictions on the cost function, except that it should be well defined and constant. In practice, there are situations in which it may be necessary or advantageous to break these rules. The most obvious scenario is the case of a stochastic cost function or one that is subject to noise. There is a growing body of literature on meta-heuristics, including simulated annealing, for such problems. Gelfand and Mitter (1989) show that if the noise in the objective function is normally distributed, then the convergence properties of SA still hold for noisy functions, while Gutjahr and Pflug (1996) generalize these results to other distributions. Others consider the situation in which the cost of a solution is estimated by sampling either from an appropriate distribution or from a simulation run. Some such as Alrefaei and Andradottir (1999) suggest storing all feasible solutions visited during the search. They then suggest that, when suitably adjusted to take account of the number of neighbors, the number of times a solution is accepted is a good measure of solution quality. Whatever measure of cost or acceptance strategy is employed, storage of all solutions is only feasible if the solution space is small.

When dealing with stochastic costs, the use of statistical tests to aid acceptance/rejection decisions is very common. These may be used simply to determine whether a move is a downhill or an uphill move. In the latter case, mean costs are then used to calculate the change in cost δ. Alternatively a t-test can be used to determine whether additional sampling is required in order to make a decision with confidence. Much of the work on stochastic cost functions is either theoretical or involves empirical analysis of small unrealistic problems and there are relatively few papers showing that these ideas work well in practice. Exceptions include Bulgak and Sanders (1988). A survey of SA for stochastic combinatorial optimization can be found in Bianchi et al. (2008).

A second reason for using an estimated cost function is that it may be computationally expensive to calculate the true cost. In ❷ Sect. 6 we stated that if possible this situation should be avoided, but in some cases it may be worth compromising on this in order to improve other problem-specific factors. Tovey (1988) suggests that in such situations it may be possible to use a simpler approximation function. Rather than using the approximation at every iteration he suggests occasionally taking the time to calculate the true cost value. The difference between the real and estimated values at these points are then used to adjust the acceptance probabilities to compensate for the error. For example, if the true value is calculated 10% of the time and it is found that the estimated value was too high, then 90% of the time this solution will have a lower probability of acceptance. Therefore, on this occasion, the probability is increased to compensate. A drawback of this approach is that if the optimal solution is visited but its cost is overestimated it may not be returned as the best solution. An alternative approach is to use an approximation for the trial solution, but once a solution is accepted, to calculate its true value. In this way there may be some errors in the decision as to whether or not to accept a solution, but the true value of all accepted solutions will be known and the minimum of these can be retained as the final solution.

Even if the cost function is well defined and the minimum cost solutions are true global minima, it may be worthwhile considering an adjustment. This is particularly true when the cost includes one or more penalty terms relating to hard or soft constraints. Although the penalty weights may be set so that the cost is a true reflection of the relative quality of the solutions, these may not be the best weights for the annealing algorithm. For example, Wright (1991) argues that weights should reflect not only the importance of each constraint but also its difficulty. Anagnostopoulos et al. (2006) suggest that the weights need not be constant throughout the search. They adjust the weight of a penalty term up or down whenever a new best solution is accepted depending on whether the solution is infeasible or feasible. Fleischer and Jacobson's (1999) study on the relationship between entropy and the expected cost of the final solution includes experiments with a weighted penalty function for the maximum clique problem. They conclude that results can be significantly improved if the penalty function is weighted in such a way as to smooth the solution landscape, thereby increasing the entropy of the system.

Rather than smoothing the landscape at the start, Hamacher and Wenzel (1999) adjust the landscape through changes to the cost function as the search progresses. Their approach, named stochastic tunneling (STUN), is designed not only to make it easy to descend into local optima, but also to escape from them again. They do this by adjusting the cost function so that the part of the landscape that is higher than the best solution found so far is smoothed, while the troughs in lower areas are exaggerated. This is achieved by making the transformation $f(i) \rightarrow 1 - \exp(-\lambda(f(x) - f(x_{min})))$ where x_{min} is the best solution to date and λ is a scaling factor. However they note that close to the global minimum this function becomes too smooth and insufficient downhill moves are made. For this reason they recommend alternating STUN with low temperature annealing. A later paper (Hamacher 2006) details further improvements.

7.6 With Other Methods

There are a large number of empirical investigations and practical case studies that show that a stand alone simulated annealing heuristic can be a very good solution approach for a wide range of problems. Nevertheless, in some cases, performance may be improved by combining it with other methods. These may be used as preprocessors to find a good starting solution or to transform the problem in some way to make it easier to solve; as post processors to make further improvements to the final solution; or as ingredients of the search process itself. The latter case may simply involve alternating a simulated annealing phase with one or more other search algorithms, or may be a true hybrid in which another method is embedded within the simulated annealing process or vice versa. The volume of papers and different possibilities for combination and hybridization are too great to provide a complete overview, so this section cites a few examples to give a flavor of this type of work.

The most common form of preprocessing is to use a heuristic to find a good starting solution instead of generating a purely random starting point. In this case, it makes sense to start with a lower temperature than normal so that all the benefits of the good solution are not destroyed. This can save time but care must be taken that this does not make the search entirely dependent on the starting solution. This not only involves ensuring that the temperature is not too low, but also that the starting solution should not be too good – for example, by using some form of randomized greedy construction rather than a totally greedy one. Bonomi and Lutton (1984) suggest an interesting way of initializing a solution for the TSP. They cluster the cities into regions and then find a good tour through the centers of the clusters. Within each cluster, the ordering is random but the clusters are ordered according to the tour. A different form of preprocessing is suggested by Chams et al. (1987). They point out that for partitioning problems such as graph coloring where the objective is to minimize the number of partitions required, a greedy algorithm that builds a solution one partition at a time is able to make good decisions early on and it is only in the later part of the construction that the myopic nature of such an approach becomes a problem. They therefore suggest building a partial solution in this way and then using SA to minimize the number of partitions required by the remaining elements. They report promising results using this approach for the graph coloring problem. More recently, Burke et al. (2008) report success with a similar approach for a stock-cutting problem. In this case, the simulated annealing phase is also hybridized with a construction algorithm and further details are given later in this section.

Another possibility is to transform the problem, either into an easier to solve instance of the same problem, or into an equivalent problem. The possibility of reducing the size of the solution space has already been mentioned. Fleischer and Jacobson (1999) suggest that more dramatic transformations may be beneficial and illustrate their ideas by comparing the performance of SA in the solution of a maximum clique problem using a direct model and after transforming the problem into a satisfiability problem. They conclude that the level of entropy differs between the two models and that better solutions tend to result from the model with higher entropy.

The most common form of post-processing is to apply a descent phase starting from the best simulated annealing solution. This ensures that at least a local minimum has been found. In some cases, the neighborhood used for such a search may be different to that used for the annealing phase.

By far the largest body of literature covers other forms of hybridization. An approach that is becoming increasingly common for problems such as cutting and packing is to hybridize simulated annealing with a construction heuristic. The solution space is defined as a permutation of the elements of the solution, and simulated annealing explores this space. Each permutation is transformed into a solution to the original problem by applying the construction heuristic to the pieces in the order given by the permutation. Early implementations of this strategy were often abandoned as the cost function calculation proved too expensive but this is now less of a problem. For example, Burke et al. (2008) use simulated annealing to order the pieces in a two-dimensional packing problem and then pack them using a commonly used heuristic placement policy.

Search heuristics other than simulated annealing also involve acceptance/rejection decisions. There are several examples in which such decisions are made using a simulated annealing acceptance criterion. For example, in some genetic algorithm implementations, a decision as to whether or not a child is to replace its parent is required. While standard practice is to select the fittest of the two, some implementations replace the parent with the child even if the child is worse using the Boltzmann distribution. Examples of this type of approach include Pakhira (2003). Another class of algorithms that need to make acceptance/rejection decisions are hyperheuristics. A hyperheuristic is essentially a high-level heuristic that selects from a number of low-level heuristics to use at different points of the search. This choice may be made in a number of ways but is usually based on some measure of recent performance. Bai et al. (2006) compare a simulated annealing acceptance rule with other options and show that the SA rule works well.

Local search is often used as an ingredient of other search heuristics. For example, in GRASP it is an integral part, while it is now almost always used in an ant colony optimization (ACO) algorithm, and is frequently included in a genetic algorithm, when it is known as a memetic algorithm. In the simplest cases, the local search is a simple downhill search, but some implementations use simulated annealing in this role. Examples include Liu et al. (2000) in a GRASP approach to the frequency assignment problem, Chen et al. (2005) in an ACO heuristic to solve a vehicle routing problem, Guo et al. (2006) in a memetic algorithm for a knapsack problem, and Ge et al. (2007) in a particle swarm optimization algorithm to solve the job shop scheduling problem. However, unless simulated annealing is to become the primary search mechanism the time allowed must be kept relatively low. As success depends on slow cooling, the use of SA in this context has received some criticism.

The above examples involve the use of SA within another algorithm. A further class of hybrids involve the embedding of other methods within SA. Mention has already been made of VLNS and basin hopping in which optimization techniques are used to search large neighborhoods. In the case of discrete problems, examples include the use of dynamic programming, network flow models, and shortest path algorithms, while for continuous differentiable landscapes gradient methods can be applied.

In other cases, simulated annealing implementations may be enhanced by the use of ingredients from other approaches. For example, a tabu list is sometimes applied to ensure that cycling is avoided (Altiparmak and Karaoglan 2008), or a population of simulated annealing runs is used, with the worst performing members being periodically killed off and replaced by solutions found by the better runs (van Hentenryck and Vergados 2007).

Finally, there are many examples of implementations that involve several different heuristics including SA that are run independently at different points in the search.

8 Applications

As illustrated by the examples in the previous sections, simulated annealing has been used across a broad spectrum of application areas. A comprehensive survey is beyond the scope of this chapter. Instead, a few key application areas are highlighted and some examples are cited to illustrate the diversity of problems where simulated annealing has been applied successfully.

One of the earliest practical applications was in electronic engineering where it became a popular approach for placement and routing problems in VLSI design (Sechen and Sangio-vanni-Vincentelli 1988; Chandy and Banerjee 1996; Wong et al. 1998). Since then it has been applied across a wide range of engineering applications. Examples include reactive power planning (Chen and Ke 2004; Jwo et al. 1995), optimizing fuel cells (Outeiro et al. 2008; Wishart et al. 2006), designing building frames (Paya et al. 2008), concrete bridge frame construction (Perea et al. 2008), and the design of steel frame structures (Degertekin 2007).

The operational research community was also quick to embrace simulated annealing and in 1990 Eglese (1990) published a paper entitled "simulated annealing: A general tool for operational research" that was recently selected by the European Association of Operational Research Societies (EURO) as one of the 30 most influential papers published in the *European Journal of Operational Research* between 1975 and 2005. Initial interest was in benchmark instances of classical problems such as the traveling salesman problem, graph coloring, etc. but this soon shifted to real-life problems in a variety of application areas. These include the fields of scheduling (Sekiner and Kurt 2007; Thompson and Dowsland 1998), routing (Tavakkoli-Moghaddam et al. 2007; Van Breedam 1995), location problems (Erdemir et al. 2008), and packing problems (Egeblad and Pisinger 2009; Gomes and Oliveira 2006).

Due to its ability to solve messy problems without any restrictions on the form of the objective function or the structure of the constraints, SA remains a popular choice of optimizer for a variety of scientific applications. A particularly vibrant area of research is in medicine, where it has been used to plan dosages in radiotherapy treatment (Jacob et al. 2008; Kubicky et al. 2008; Morton et al. 2008) and to optimize parameters in medical simulation models, for example, Marsh et al. (2007) and Choi et al. (2004). It is also widely used in the field of earth sciences and agriculture. Applications include land use allocation (Sante-Riveira et al. 2008), the study of gravity and seismic data (Yu et al. 2008), estimating sea surface temperature (Arai and Sakakibara 2006), optimizing water distribution in irrigation canals (Monem and Namdarian 2005) and determining harvest schedules (Crowe and Nelson 2005). Examples of other fields include astronomy (Cornish and Porter 2007), genetics (Tewari et al. 2008), chemistry (De Andrade et al. 2008), and economics (Chen and Yeh 2001).

9 Conclusions

It has been suggested (Nissen 1995) that the cycle of expectation associated with any innovation goes through four phases: disbelief, euphoria, disappointment, and maturity or abandonment. ❷ *Figure 4* illustrates this pattern. This has certainly been true in the case of simulated annealing. When it was first introduced there was much skepticism about its ability to do as well as more traditional methods. Empirical evidence from studies such as those by Johnson et al. (1989, 1991) coupled with the results on theoretical convergence led to a period when SA was hailed as a generic solution approach that was not only easy to apply to almost any combinatorial optimization problem, but was also backed up by a mathematical theory. This led to a great deal of interest among practitioners from a range of different fields, as well as those interested in gaining a deeper understanding of its theoretical behavior. These studies led to two conclusions. First, there are no performance guarantees for implementations that run in a reasonable amount of computer time. Second, a successful implementation usually requires knowledge of the problem structure when making the problem-specific decisions and careful fine tuning of the cooling schedule. This resulted in the caveat that while SA is easy to get working, it is difficult to get it working well, and during this period of disillusionment, many researchers turned their attention to the development of other meta-heuristic techniques. Since then, further work on good, polynomial-time, adaptive cooling schedules has ameliorated some of the difficulties associated with the generic decisions, while a range of

❏ **Fig. 4**
The Nissen model of expectation for innovations.

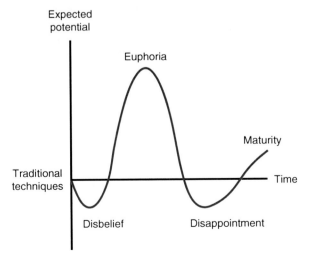

successful implementations on both real and artificial problems have been published. This has led to SA reaching maturity as a useful addition to the optimization toolkit.

As outlined in ❷ Sect. 7, there has been a lot of research into ways of modifying the standard Metropolis algorithm in order to increase the range of problems that can be tackled successfully. Some of these are backed up by theory and/or are based on sound reasoning and have become common in many practical implementations. An obvious example is the use of reheating, which makes sense as the best, rather than the final, solution is returned. Others are ad hoc methods that have been shown to be successful on the problems to which they have been applied but have often not been taken up by others. Although these ideas may work for other problems, before bending the rules of SA too far, it is worth considering whether or not another heuristic, or some form of hybrid, may be more appropriate. Unfortunately there is little guidance on this. There have been a number of articles and larger projects comparing the performances of different heuristics on both benchmark and practical problems. For example, methods based on simulated annealing have performed strongly in two recent international timetabling competitions, with a simulated annealing implementation winning the 2003 competition (Kostuch 2004) and a GRASP-simulated annealing hybrid coming second to a constraint-satisfaction approach in the examination scheduling track in 2008 (Gogos et al. 2008). Other examples where simulated annealing has been shown to outperform other heuristics, or beat previous best-known results on benchmark problems include Egebald and Pisinger (2009), Rodriguez-Tello et al. (2008), and Zolfaghari and Liang (2002). Egebald and Pisinger consider two- and three-dimensional knapsack packing problems and produce competitive results on benchmark datasets, and Rodriguez-Tello et al. improve several best known results for benchmark problems for the minimum linear arrangement problem. Zolfaghari and Liang demonstrate that simulated annealing outperforms both genetic algorithm and tabu search methods on machine-grouping problems. However, the results from comparative studies need to be interpreted with care as the solution quality of any implementation depends as much on the skill of its designer as on the underlying heuristic. This point is clearly illustrated in Tiourine et al. (1995) in which different research groups applied one or more solution approach to the same problem with simulated annealing, tabu search, and genetic algorithms each being implemented by more than one group. The resulting implementations were then scored on a number of different factors. In all three cases, the implementations from different groups resulted in a different set of scores.

In the last 3 decades, there has been a range of competitive new heuristics, many also inspired by nature. In spite of this, an examination of the number of citations of the term "simulated annealing" in Scopus indicates that research is still prospering (see ❷ Fig. 5). The occurrence of simulated annealing-based solutions in the scientific literature increased between 2002 and 2004 and has remained steady for the last few years, suggesting that it is still regarded as a strong candidate for the solution of practical problems. In 2007, there were more than twice as many references for simulated annealing than there were for "tabu search" (including "taboo search") or "ant colony." Thus it remains a vibrant and exciting area of research. This is in part due to the fact that the claim of its early advocates remains true. It is very easy to get it working. For some problems a straightforward implementation using the most obvious definitions of solution space, neighborhood structure, and cost function, combined with a simple geometric or Lundy and Mees cooling schedule, may be all that is required. If this is not the case, then more sophisticated problem-specific decisions or a more complex cooling schedule can be the focus of further development. It is worth noting here that when undertaking such development, graphical information can be invaluable. If solutions are

◘ **Fig. 5**
The number of citations of the term "simulated annealing" in Scopus.

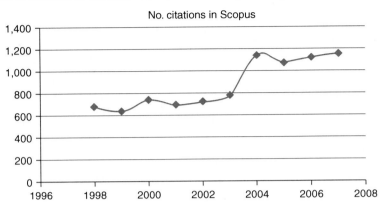

easy to visualize (as with routing or packing problems) many researchers have found it worthwhile to build in routines that show how the current solution changes as the run progresses. Failing this, plots of statistics such as best cost, average cost, acceptance rate, entropy, specific heat, etc. often provide useful clues.

As shown in ❷ *Fig. 5*, simulated annealing continues to be of interest to researchers working in both practical and theoretical areas. This chapter has given a brief overview, with our main focus being the practical solution of combinatorial optimization problems. For a slightly different viewpoint, Aarts et al. (2005), Salamon et al. (2002), and Henderson et al. (2003) are suggested. Freely downloadable simulated annealing software is available from a variety of sources. Examples include a C language downloadable code of adaptive simulated annealing, where the algorithm parameters are automatically adjusted according to the progress of the algorithm (www.ingber.com), code for solving constrained, single-objective problems (http://manip.crhc.uiuc.edu/CSA/), a general purpose simulated annealing package designed for parallel machines (http://wwwcs.uni-paderborn.de/fachbereich/AG/monien/SOFTWARE/PARSA), and a framework for simulated annealing to be implemented within Matlab (http://www.frostconcepts.com/software/index.html).

References

Aarts EHL, Korst JHM (1989) Simulated annealing and Boltzmann machines. Wiley, Chichester

Aarts EHL, Van Laarhoven PJM (1985) Statistical cooling: a general approach to combinatorial optimisation problems. Philips J Res 40:193–226

Aarts EHL, Korst JHM, Michiels W (2005) Simulated annealing. In: Burke EK, Kendall G (eds) Search methodologies. Springer, New York, pp 187–210

Abramson D (1991) Constructing school timetables using simulated annealing: sequential and parallel algorithms. Manag Sci 37:98–113

Alrefaei MH, Andradottir S (1999) A simulated annealing algorithm with constant temperature for discrete stochastic optimisation. Manag Sci 45:748–764

Altiparmak F, Karaoglan I (2008) An adaptive tabu-simulated annealing for concave cost transportation problems. J Operational Res Soc 59:331–341

Anagnostopoulos A, Michel L, Van Hentenryck P, Vergados YA (2006) Simulated annealing approach to the traveling tournament problem. J Scheduling 9:177–193

Arai K, Sakakibara J (2006) Estimation of sea surface temperature, wind speed and water vapour with microwave radiometer data based on simulated annealing. Adv Space Res 37(12):2202–2207

Azizi N, Zolfaghari S (2004) Adaptive temperature control for simulated annealing: a comparative study. Comput Operations Res 31(4):2439–2451

Bai R, Burke EK, Kendall G, McCollum B (2006) A simulated annealing hyper-heuristic for university course timetabling. In: Burke EK, Rudova H (eds) In: Proceedings of PATAT 2006, Brno, Czech Republic, August–September 2006. Lecture notes in computer science, vol 3867. Springer, Heidelberg

Bianci L, Dorigo M, Gambardella LM, Gutjahr WJ (2008) A survey on metaheuristics for stochastic combinatorial optimisation. Nat Comput. DOI 101007, Online September 2008

Boese KD, Kahng AB (1994) Best-so-far vs. where-you-are: implications for optimal finite time annealing. Syst Control Lett 22:71–78

Bonomi E, Lutton JL (1984) The N-city travelling salesman problem: statistical mechanics and the Metropolis algorithm. SIAM Rev 26:551–568

Brandimarte P, Conterno R, Laface P (1987) FMS production scheduling by simulated annealing. In: Micheletti GF (ed) Proceedings of the 3rd international conference on simulation in manufacturing, Turin, Italy, November 1987. Springer, Berlin, pp 235–245

Bulgak AA, Sanders JL (1988) Integrating a modified simulated annealing algorithm with the simulation of a manufacturing system to optimise buffer sizes in automatic assembly systems. In: Abrams M, Haigh P, Comfort J (eds) Proceedings of the 1988 winter simulation conference, San Diego, CA, December 1998. IEEE Press, Piscataway, NJ, pp 684–690

Burke EK, Kendall G, Whitwell G (2008) A simulated annealing enhancement of the best-fit heuristic for the orthogonal stock cutting problem. INFORMS J Comput 21(3):505–516

Černy V (1985) A thermodynamical approach to the travelling salesman problem: an efficient simulation algorithm. J Optimization Theory Appl 45:41–55

Chams M, Hertz A, De Werra D (1987) Some experiments with simulated annealing for colouring graphs. Eur J Operational Res 32:260–266

Chandy JA, Banerjee P (1996) Parallel simulated annealing strategies for VLSI cell placement. In: Proceedings of the 9th conference on VLSI design, Bangalore, India, January 1996. IEEE Computer Society, Washington, pp 37–42

Chen YL, Ke YL (2004) Multi-objective VAr planning for large-scale power systems using projection-based two-layer simulated annealing algorithms. IEE Proc Generation, Transm Distribution 151(4):555–560

Chen S-H, Yeh C-H (2001) Evolving traders and the business school with genetic programming: a new architecture of the agent based artificial stock market. J Econ Dyn Control 25 (3–4):363–393

Chen C-H, Ting C-J, Chang P-C (2005) Applying a hybrid ant colony system to the vehicle routing problem. Lect Notes Comput Sci 3483:417–426

Choi K-S, Sun H, Heng P-A (2004) An efficient and scalable deformable model for virtual reality-based medical applications. Artif Intell Med 32(1):51–69

Connolly DT (1990) An improved annealing scheme for the QAP. Eur J Operational Res 46:93–100

Cook SA (1971) The complexity of theorem procedures. In: Proceedings of 3rd ACM symposium on the theory of computing, Shaker Heights, OH, 1971. ACM, New York, pp 151–158

Cook SA (1972) An overview of computational complexity. Commun ACM 26:400–408

Cornish NJ, Porter EK (2007) The search for massive black hole binaries with LISA. Classical Quantum Gravity 24(23):5729–5755

Crowe KA, Nelson JD (2005) An evaluation of the simulated annealing algorithm for solving the area-restricted harvest-scheduling model against optimal benchmarks. Can J Forest Res 35(10):2500–2509

De Andrade MD, Nascimento MAC, Mundim KC, Sobrinho AMC, Malbouisson LAC (2008) Atomic basis sets optimization using the generalized simulated annealing approach: new basis sets for the first row elements. Int J Quantum Chem 108 (13):2486–2498

Degertekin SO (2007) A comparison of simulated annealing and genetic algorithm for optimum design of nonlinear steel space frames. Struct Multidisciplinary Optimization 34(4):347–359

Dowsland KA (1993a) Some experiments with simulated annealing techniques for packing problems. Eur J Operational Res 68:389–399

Dowsland KA (1993b) Using simulated annealing for efficient allocation of students to practical classes. In: Vidal RVV (ed) Applied simulated annealing. Lecture notes in economics and mathematical systems, vol 396. Springer-Verlag, Berlin

Dowsland KA, Thompson JM (1998) A robust simulated annealing based examination timetabling system. Comput Oper Res 25:637–648

Dowsland KA, Soubeiga E, Burke EK (2007) A simulated annealing based hyperheuristic for determining shipper sizes for storage and transportation. Eur J Oper Res 179(3):759–774

Dueck G, Sheuer T (1990) Threshold accepting: a general purpose optimization algorithm appearing superior to simulated annealing. J Comput Phys 90:161–175

Edmonds J (1965) Paths, trees and flowers. Can J Maths 17:449–467

5

Egeblad J, Pisinger D (2009) Heuristic approaches for the two and three dimensional knapsack packing problem. Comput Oper Res 36(4):1026–1049

Eglese RW (1990) Simulated annealing: a general tool for operational research. Eur J Oper Res 46(3): 271–281

Erdemir ET, Batta R, Spielman S, Rogerson PA, Blatt A, Flanigan M (2008) Location coverage models with demand originating from nodes and paths: application to cellular network design. Eur J Oper Res 190 (3):610–632

Feo TA, Resende MGC, Smith SH (1994) A greedy randomised adaptive search procedure for maximum independent set. Oper Res 42:860–878

Fleischer M, Jacobson SH (1999) Information theory and the finite time behavior of the simulated annealing algorithm: experimental results. INFORMS J Comput 11:35–43

Franz A, Hoffmann KH (2003) Threshold accepting as limit case for a modified Tsallis statistics. Appl Math Lett 16:27–31

Garey MR, Johnson DS (1979) Computers and intractability. WH Freeman, San Francisco, CA

Ge H, Du W, Qian F (2007) A hybrid algorithm based on particle swarm optimisation and simulated annealing for job shop scheduling. In: Proceedings of ICNC 2007. Third International Conference on Natural Computation, vol 3, Haikou, China, August 2007. IEEE Computer Society, Washington, pp 715–719

Gelfand SB, Mitter SK (1989) Simulated annealing with noisy or imprecise measurements. J Opt Theory Appl 69:49–62

Gendreau M, Potvin JY (2005) Tabu search. In: EK Burke, G Kendall (eds) Introductory tutorials in optimisation, decision support and search methodology. Springer, New York, pp 165–186

Glover F (1989) Tabu search part 1. ORSA J Comput 1:190–206

Glover F, Greenberg HJ (1989) New approaches for heuristic search: a bilateral link with artificial intelligence. Eur J Oper Res 39:119–130

Gogos C, Alefragis P, Housos E (2008) A multi-staged algorithmic process for the solution of the examination timetabling problem. In: Burke EK, Gendreau M (eds) The 7th international conference on the practice and theory of automated timetabling, Montreal, Canada, August 2008

Goldstein L, Waterman MS (1988) Neighbourhood size in the simulated annealing algorithm. Am J Math Manag Sci 8:409–423

Gomes AM, Oliveira JF (2006) Solving irregular strip packing problems by hybridising simulated annealing and linear programming. Eur J Oper Res 171(3):811–829

Greene JW, Supowit KJ (1986) Simulated annealing without rejected moves. IEEE Trans Comput Aided Des CAD-5:221–228

Guo XP, Yang GK, Zhiming W, Huang ZH (2006) A hybrid fine-tuned multi-objective memetic algorithm. IEICE Trans Fundam Electron Commun Comput Sci E89A(3):790–797

Gutjahr WJ, Pflug GCh (1996) Simulated annealing for noisy cost functions. J Global Optimisation 8(1):1–13

Hajek B (1988) Cooling schedules for optimal annealing. Math Oper Res 13:311–329

Hamacher K (2006) Adaptation in stochastic tunnelling global optimisation of complex potential energy landscapes. Europhys Lett 74:944–950

Hamacher K, Wenzel W (1999) Scaling behaviour of stochastic minimisation algorithms in a perfect funnel landscape. Phys Rev E 59:938–941

Hansen P, Mladenovic N (2005) Variable neighbourhood search. In: Burke EK, Kendall G (eds) Search methodologies. Springer, New York, pp 211–238

Henderson D, Jacobson SH, Johnson AW (2003) The theory and practice of simulated annealing. In: Glover F, Kochenberger GA (eds) The handbook of metaheuristics, International series in operations research and management science, vol 57. Springer, New York

Huang MD, Romeo F, Sangiovanni-Vincentelli AL (1986) An efficient general cooling schedule for simulated annealing. In: Proceedings of IEEE international conference on computer aided design, Santa Clara, CA, November 1986. IEEE Computer Society, Washington, pp 381–384

Jacob D, Raben A, Sarkar A, Grimm J, Simpson L (2008) Anatomy-based inverse planning simulated annealing optimization in high-dose-rate prostate brachytherapy significant dosimetric advantage over other optimization techniques. Int J Radiat Oncol Biol Phys 72(3):820–827

Johnson DS, Aragon CR, McGeoch LA, Schevon C (1989) Optimization by simulated annealing: an experimental evaluation; part I, graph partitioning. Oper Res 37:865–892

Johnson DS, Aragon CR, McGeoch LA, Schevon C (1991) Optimization by simulated annealing: an experimental evaluation; part II, graph coloring and number partitioning. Oper Res 39:378–406

Jwo W-S, Liu C-W, Liu C-C, Hsiao Y-Y (1995) Hybrid expert system and simulated annealing approach to optimal reactive power planning. IEE Proc Generation, Transm Distribution 142(4):381–385

Kalivas JH (1992) Optimization using variations of simulated annealing. Chemometrics Intell Lab Syst 15(1):1–12

Karp RM (1972) Reducibility amongst combinatorial problems. In: Miller RE, Thatcher JW (eds)

Complexity of computer computations. Plenum Press, New York, pp 85–103

Kern W (1986) On the depth of combinatorial optimisation problems. University of Koln Technical Report 8633

Kirkpatrick CD, Gellat CD, Vecchi MP (1983) Optimisation by simulated annealing. Science 220:671–680

Kubicky CD, Yeh BM, Lessard E, Joe BN, Speight JL, Pouliot J, Hsu I-C (2008) Inverse planning simulated annealing for magnetic resonance imaging-based intracavitary high dose-rate brachytherapy for cervical cancer. Bracytherapy 7(3):242–247

Kostuch PA (2004) The university course timetabling problem with a 3-phase method. In: Burke EK, Trick M (eds) The practice and theory of automated timetabling V. Lecture notes in computer science, vol 3616. Springer-Verlag, Berlin, pp 109–125

Lin S, Yu VF, Chou S-Y (2008) Solving the truck and trailer problem based on a simulated annealing heuristics. Comput Oper Res, Available online 17-4-2008 (corrected proof)

Liu X, Pardalos PM, Rajasekaran S, Resende MGC (2000) A GRASP for frequency assignment in mobile radio networks. In: Badrinath BR, Hsu F, Pardalos PM, Rajasejaran S (eds) Mobile networks and computing. DIMACS series on discrete mathematics and theoretical computer science, vol 52. American Mathematical Society, Providence, RI, pp 195–201

Lundy M, Mees A (1986) Convergence of an annealing algorithm. Math Programming 34:111–124

Marsh RE, Riauka TA, McQuarrie SA (2007) Use of a simulated annealing algorithm to fit compartmental models with an application to fractal pharmacokinetics. J Pharm Pharm Sci 10(2):167–178

Merlot LTG, Boland N, Hughes BD, Stuckey PJ (2003) A hybrid algorithm for the examination timetabling problem. Lect Notes Comput Sci 2740:207–231

Metropolis N, Rosenbluth AW, Rosenbluth MN, Teller AH, Teller E (1953) Equation of state calculation by fast computing machines. J Chem Phys 21:1087–1091

Mitra D, Romeo F, Sangiovanni-Vincentelli AL (1986) Convergence and finite time behaviour of simulated annealing. Adv Appl Probability 18:747–771

Monem MJ, Namdarian R (2005) Application of simulated annealing (SA) techniques for optimal water distribution in irrigation canals. Irrigation Drainage 54(4):365–373

Morgenstern C, Shapiro H (1989) Chromatic number approximation using simulated annealing. Technical Report CS86-1, Department of Computer Science, University of New Mexico

Morton GC, Sangreacha R, Halina P, Loblaw A (2008) A comparison of anatomy-based inverse planning with simulated annealing and graphical optimization for high-dose-rate prostate brachytherapy. Brachytherapy 7(1):12–16

Moscato P, Fontanari JF (1990) Stochastic versus deterministic update in simulated annealing. Phys Lett A 146:204–208

Nissen V (1995) An overview of evolutionary algorithms in management applications. In: Biethahn J, Nissen V (eds) Evolutionary algorithms in management applications. Springer Verlag, New York, pp 44–97

Ogbu FA, Smith DK (1990) The application of the simulated annealing algorithm to the solution of the n/m/Cmax flowshop problem. Comput Oper Res 17:243–253

Osman IH (1993) Metastrategy simulated annealing and tabu search algorithms for the vehicle routing problem. Ann Oper Res 41:421–451

Outeiro MT, Chibante R, Carvalho AS, de Almeida AT (2008) A parameter optimized model of a proton exchange membrane fuel cell including temperature effects. J Power Sources 185(2):952–960

Pakhira MK (2003) A hybrid genetic algorithm using probabilistic selection. J Inst Eng (India) 84: 23–30

Paya I, Yepes V, Gonzalez-Vidosa F, Hospitaler A (2008) Multiobjective optimization of concrete frames by simulated annealing. Comput Aided Civil Infrastructure Eng 23(8):596–610

Pedamallu CS, Ozdamar L (2008) Comparison of simulated annealing, interval partitioning and hybrid algorithms in constrained global optimisation. In: Siarry P, Michalewicz Z (eds) Advances in metaheuristics for hard optimization 2008. Natural computing series. Springer, Berlin, pp 1–22

Penna TJP (2008) Travelling salesman problem and Tsallis statistics. Phys Rev E 51:R1–R3

Perea C, Alcaca J, Yepes V, Gonzalez-Vidosa F, Hospitaler A (2008) Design of reinforced concrete bridge frames by heuristic optimization. Adv Eng Software 39(8):676–688

Rodriguez-Tello E, Hao J-K, Torres-Jimenez J (2008) An effective two-stage simulated annealing algorithm for the minimum linear arrangement problem. Comput Oper Res 35:3331–3346

Ropke S, Pisinger D (2006) An adaptive large neighbourhood search heuristics for the pickup and delivery problem with time windows. Transportation Sci 40:455–472

Salamon P, Suibani P, Frost R (2002) Facts, conjectures and improvements for simulated annealing. SIAM Monographs on Mathematical Modeling and Computation 7, Society for Industrial and Applied Mathematics

Santé-Riveira I, Boullón-Magán M, Crecente-Maseda R, Miranda-Barrós D (2008) Algorithm based on simulated annealing for land-use allocation. Comput Geosci 34(3):259–268

Sechen D, Braun D, Sangiovanni-Vincetelli A (1988) Thunderbird: a complete standard cell layout package. IEEE J Solid State Circuits 23:410–420

Seçkiner SU, Kurt M (2007) A simulated annealing approach to the solution of job rotation scheduling problems. Appl Math Comput 188(1):31–45

Szu H, Hartley R (1987) Fast simulated annealing. Phys Lett A 122:157–162

Tavakkoli-Moghaddam R, Safaei N, Kah MMO, Rabbani M (2007) A new capacitated vehicle routing problem with split service for minimizing fleet cost by simulated annealing. J Franklin Inst 344(5): 406–425

Teitz MB, Bart P (1968) Heuristics methods for estimating the generalised vertex median of a weighted graph. Oper Res 16:955–961

Tewari S, Arnold J, Bhandarkar SM (2008) Likelihood of a particular order of genetic markers and the construction of genetic maps. J Bioinform Comput Biol 6(1):125–162

Thompson JM, Dowsland KA (1998) A robust simulated annealing based examination timetabling system. Comput Oper Res 25:637–648

Thompson JM, Dowsland KA (1996) General cooling schedules for a simulated annealing based timetabling system. In: Burke EK, Ross P (eds) Practice and theory of automated timetabling. Lecture notes in computer science, vol 1153. Springer-Verlag, Berlin

Tiourine S, Hurkens C, Lenstra JK (1995) An overview of algorithmic approaches to frequency assignment problems. Technical report, EUCLID CALMA project, Eindhoven University of Technology

Tovey CA (1988) Simulated simulated annealing. Am J Math Manag Sci 8:389–407

Triki E, Collette Y, Siarry P (2005) A theoretical study on the behavior of simulated annealing leading to a new cooling schedule. Eur J OR 166:77–92

Tsallis C, Stariolo DA (1996) Generalized simulated annealing. Phys A 233:395–406

Tuga M, Berretta R, Mendes A (2007) A hybrid simulated annealing with Kempe chain neighbourhood for the university timetabling problem. In: Lee R, Chowdhury M, Ray S, Lee T (eds) 6th IEEE/ACIS Conference Proceedings Computer and Information Science 2007, Melbourne, Australia, July 2007. IEEE Computer Society, Washington, pp 400–405

Vakharia AJ, Chang Y-L (1990) A simulated annealing approach to scheduling a manufacturing cell. Naval Res Logistics 37:559–577

Van Breedam A (1995) Improvement heuristics for the vehicle routing problem based on simulated annealing. Eur J Operational Res 86(3):480–490

Van Hentenryck P, Vergados Y (2007) Population-based simulated annealing for traveling tournaments. Proceedings of the Twenty-Second AAAI Conference on Artificial Intelligence, Vancouver, Canada, AAAI Press, pp 267–271

Van Laarhoven PJM, Aarts EHL (1987) Simulated annealing: theory and applications. Kluwer, Dordrecht, The Netherlands

Wales DJ, Scheraga HA (1999) Chemistry: global optimisation of clusters, crystals and biomolecules. Science 285:1368–1372

Wishart JD, Dong Z, Secanell MM (2006) Optimization of a PEM fuel cell system for low-speed hybrid electric vehicles. In: Proceedings of the ASME Design Engineering Technical Conference 2006, Philadelphia, PA, September 2006

Wong, DF, Leong HW, Liu HW (1998) Simulated annealing for VLSI design. The Springer International Series in Engineering and Computer Science, vol 42. Springer, Berlin

Wright M (1991) Scheduling English cricket umpires. J OR Soc 42:447–452

Wright M (1996) School timetabling using heuristic search. J OR Soc 47:347–357

Wright M (2001) Subcost-guided search – experiments with timetabling problems. J Heuristics 7:251–260

Yu P, Dai M-G, Wang J-L, Wu J-S (2008) Joint inversion of gravity and seismic data based on common gridded model with random density and velocity distributions. Chinese J Geophys 51(3):845–852

Zolfaghari S, Liang M (2002) Comparative study of simulated annealing, genetic algorithms and tabu search for solving binary and comprehensive machine-grouping problems. Int J Prod Res 40:2141–2158

50 Evolvable Hardware

Lukáš Sekanina
Faculty of Information Technology, Brno University of Technology, Brno,
Czech Republic
sekanina@fit.vutbr.cz

G. Rozenberg et al. (eds.), *Handbook of Natural Computing*, DOI 10.1007/978-3-540-92910-9_50,
© Springer-Verlag Berlin Heidelberg 2012

Abstract

This chapter surveys the field of evolvable hardware. After a brief overview of the reconfigurable devices used in evolvable hardware, elementary principles of evolvable hardware, corresponding terminology, and open problems in the field are introduced. Then, the chapter is divided into three main parts: extrinsic evolution, intrinsic evolution, and adaptive hardware. Extrinsic evolution (i.e., evolution using simulators) covers evolutionary design of digital circuits, analog circuits, antennas, optical systems, and microelectromechanical systems (MEMS). Intrinsic evolution conducted in field programmable gate arrays (FPGAs), field programmable transistor arrays (FPTAs), field programmable analog arrays (FPAAs), and some unconventional devices is discussed together with a description of the most successful applications. Examples of real-world adaptive hardware systems are also presented. Finally, an overview of major achievements and problems is given.

1 Introduction

At the beginning of the 1990s, researchers began using evolutionary algorithms (EAs) – population-based search algorithms (Eiben and Smith 2003) – to generate configurations for reconfigurable chips that could dynamically alter the functionality and physical connections of their circuits. This combination of EAs with reconfigurable devices spawned a new field called evolvable hardware (EHW) (Higuchi et al. 1993; de Garis 1993). Since that time, the EHW field has expanded and utilized many different combinations of EAs and biologically inspired algorithms with various reconfigurable devices including field programmable gate arrays (FPGAs), field programmable analog arrays (FPAAs), reconfigurable antennas, mirrors, MEMS, and special reconfigurable materials. Research in the field of EHW can be split into the two related areas of evolutionary hardware design and adaptive hardware. While evolutionary hardware design is the use of EAs for creating innovative (and sometimes patentable) physical designs, the goal of adaptive hardware is to endow physical systems with some adaptive characteristics in order to allow them to operate successfully in a changing environment or under the presence of faults.

❯ *Figure 1* shows the basic principle of the evolvable hardware method: Electronic circuits that are encoded as bit strings (chromosomes, in the parlance of EAs) are constructed and optimized by the evolutionary algorithm in order to obtain the circuit implementation satisfying the specification given by the designer. In order to evaluate the candidate circuit, the new configuration of a reconfigurable device is created on the basis of the chromosome content. This configuration is uploaded into the reconfigurable device and evaluated for a chosen set of input stimuli. The fitness function, which reflects the problem specification, can include behavioral as well as nonbehavioral requirements. For example, the correct functionality is a typical behavioral requirement. As a nonbehavioral requirement, we can mention the requirement for minimum power consumption or minimum area occupied on the chip. Once the evaluation of the population of candidate circuits is complete, a new population can be produced. That is typically performed by applying genetic operators (such as mutation and crossover) on existing circuit configurations. High-scored candidate circuits have a higher probability that their genetic material (configuration bitstreams) will be selected for future generations. The process of evolution is terminated when a perfect solution is obtained or when a certain number of generations is evaluated.

☐ **Fig. 1**
Evolvable hardware: Candidate configurations are generated by evolutionary algorithm, uploaded to a reconfigurable device and evaluated using the fitness function.

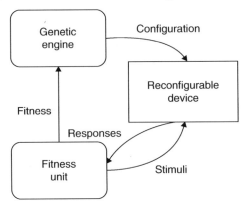

As the EA is a stochastic algorithm, the quality of resultant circuits is not guaranteed at the end of evolution. However, the method has two important advantages: (1) Artificial evolution can in principle produce intrinsic designs for electronic circuits, which lie outside the scope of circuits achievable by conventional design methods. (2) The challenge of conventional design is replaced by that of designing an evolutionary algorithm that automatically performs the design in a target place (e.g., in space). This may be harder than doing the design directly, but makes autonomy possible.

This chapter surveys the field of evolvable hardware. A more detailed explanation can be found in monographs (Zebulum et al. 2002; Thompson 1999; Greenwood and Tyrrell 2007; Higuchi et al. 2006; Sekanina 2004; Koza et al. 1999, 2003). The most important journals dealing with evolvable hardware are Genetic Programming and Evolvable Machines and IEEE Transactions on Evolutionary Computation. A lot of valuable material can be found in the conference proceedings: Evolvable Systems: From Biology to Hardware (Springer, 1996–), NASA/ESA Conference on Adaptive Hardware and Systems (IEEE, 2006–) and NASA/DoD Conference on Evolvable Hardware (IEEE, 1999–2005).

This chapter is organized as follows. ❷ Section 2 presents reconfigurable devices that can be used for evolvable hardware. In ❷ Sect. 3, elementary principles of evolvable hardware and corresponding terminology are introduced. ❷ Section 4 surveys the area of extrinsic evolution, i.e., evolutionary hardware design using simulators. ❷ Section 5 deals with the evolution conducted directly in physical hardware. Applications of evolutionary computing in the area of adaptive hardware are described in ❷ Sect. 6. Some other relevant research dealing with bioinspired algorithms and hardware is discussed in ❷ Sect. 7. Conclusions are given in ❷ Sect. 8.

2 Reconfigurable Technology

Common electronic chips such as processors or application-specific integrated circuits have an architecture that remains fixed during the lifespan of a system. It is impossible to change the architecture at runtime when faults occur in circuit components or even when a little modification introduced in a certain time could lead to a much better performance of the system. In recent years, we can observe a boom in the area of reconfigurable devices (Hauck and DeHon 2008). In comparison to fixed architectures, the structure and parameters of reconfigurable chips can be modified by writing configuration data to the configuration memory. While the first programmable devices (such as programmable logic arrays) used hundreds of bits to store the configuration, recent FPGAs require tens of megabytes to store their configuration data.

2.1   Why Reconfiguration?

Most reconfigurable devices consist of configurable blocks whose functions and interconnections are controlled by the configuration bitstream. Also, the connection of configurable blocks with I/O pins can be configured. ❷ *Figure 2* shows that in digital architectures, the configuration bits directly control the configurable switches and selection signals of multiplexers. Thus, a single chip can implement many different functions depending on its configuration. The main disadvantage of this approach is that the circuits allowing the "configurability" occupy a considerable area on the chip and make the whole system slower. The reasons for using reconfigurable chips are usually as follows:

- *Updating the firmware.* The reconfiguration extends the lifespan of a system. For example, when a new driver or peripheral device is introduced to a system, existing hardware could have a problem to communicate with it. However, if the system is implemented in a reconfigurable chip, the hardware can be updated by simply reprogramming the configuration memory. In this case, the reconfiguration is performed occasionally and only when the application is suspended.
- *Increasing the functional density.* The goal is to perform a complex task on a small chip and thus reduce the power consumption, size or weight of the application, even reduce the cost. The application has to be divided into modules whose configurations alternate on the chip. The reconfiguration is performed dynamically at runtime.
- *Increasing the reliability.* When a fault is detected and isolated, a system can be (in some cases) reconfigured to maintain its original function.

☐ **Fig. 2**
Multiplexer-based implementation of a reconfigurable functional unit (*left*) and a configurable interconnecting network implemented by configuration switches (*right*).

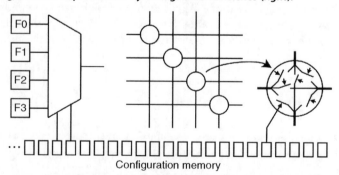

Configuration memory

- *Adapting hardware.* The goal is to dynamically create electronic circuits that are optimized for a given task, time, and location of the chip.
- *Shortening the design time.* Creating a configuration for a reconfigurable device usually takes much less time than building a new application-specific chip.

The possibility of reconfiguration is typical for digital architectures. However, reconfigurable devices are now available in the areas of analog circuits, antennas, mirrors, molecular electronics, and others. Major digital and analog reconfigurable devices used in the evolvable hardware field will be briefly described.

2.2 Field Programmable Gate Arrays

❯ *Figure 3* shows a typical architecture of a Xilinx FPGA (Xilinx Inc. 2009). It is a two-dimensional array of reconfigurable resources that include configurable logic blocks (CLB), programmable interconnect blocks (PIB), and reconfigurable I/O blocks (IOB). A CLB consists of so-called slices; each of them contains the function generators implemented using 3-, 4- or 6-input look-up tables, flip-flops, and some additional logic. The configuration bitstream is stored in the configuration SRAM memory.

The FPGA chips differ in the amount and type of resources available on the chip. The most advanced FPGAs contain more than 10 thousand CLBs and integrate, in addition to CLBs, various embedded hard cores such as SRAM memories, fast multipliers, processors, gigabit interfaces, and PCI interfaces (see PowerPC processors in ❯ *Fig. 3*). Because the existence of these cores has been identified as important to designers in the past, it is reasonable to integrate them as hard cores on the chip instead of implementing them using CLBs and other resources. Current FPGAs can compete with application-specific integrated circuits

◻ **Fig. 3**
FPGA Virtex II Pro architecture that contains two PowerPC processors.

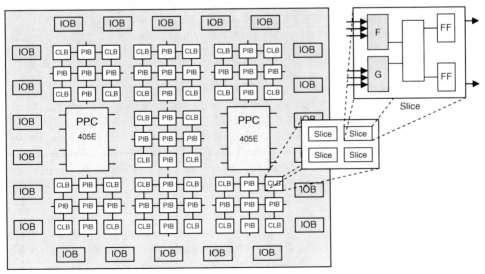

(ASICs) in many domains, for example, in applications of advanced signal processing or embedded systems.

Some FPGAs support a dynamic partial reconfiguration, which means that some parts of the FPGA can be reconfigured while remaining parts of the FPGA perform computation. In the following sections it will be seen that the possibility of partial reconfiguration is crucial for evolvable hardware. FPGAs can be configured either externally or internally. In the case of external reconfiguration, the configuration bit stream is copied to the configuration memory from an external (flash) memory. The internal reconfiguration is available in Xilinx Virtex FPGAs via the internal access configuration port (ICAP), which allows for reading and modifying the FPGA configurations by circuits created directly in the same FPGA.

The goal of digital circuit design is to provide such an implementation of the circuit which satisfies the user specification and is available in a reasonable time. As the conventional circuit design process with FPGAs is very similar to programming, the resultant system can be obtained relatively quickly. The designer has to describe the circuit structure or behavior using a hardware description language (such as VHDL or Verilog). Then, the source code is almost automatically transformed into the configuration bit stream for a particular FPGA. The transformation, which includes the synthesis, placement and routing is performed by CAD tools. This process can be constrained using various requirements, for example, the maximum delay of the circuit can be specified. Also, it is possible to simulate intermediate results of the transformation, modify the original source code when needed, and optimize the design. In most cases, the format of the configuration bit stream is not documented for the designer (with the exception of the XC6200 family). In this way, manufacturers protect their know-how and prevent the designers from physically destroying the FPGA chip.

2.3 Field Programmable Transistor and Analog Arrays

Reconfigurable analog circuits allow, in fact, a software control of analog circuits. In comparison with FPGAs, reconfigurable analog circuits contain fewer configurable blocks and operate at lower frequencies. The reconfiguration is usually based either on configurable transistor switches, analog multiplexers, switching capacitors, or operational transconductance amplifiers (OTAs). Reconfigurable analog chips were introduced much later than FPGAs.

2.3.1 FPTA-2

The field programmable transistor array (FPTA-2) developed at NASA JPL uses transistor switches to implement the reconfiguration (Stoica et al. 2002). The FPTA-2 can implement analog, digital, and mixed signal circuits. The architecture of the FPTA consists of an 8 × 8 array of reconfigurable cells. Each cell has a transistor set as well as a set of programmable resources, including programmable resistors and static capacitors. The reconfigurable circuitry consists of 14 transistors connected through 44 switches in each cell. ❷ *Figure 4* provides a detailed view of the reconfigurable transistor array cell. A total of 5000 bits are used to program the whole chip. The pattern of interconnection between cells is similar to the one used in commercial FPGAs: each cell interconnects with its north, south, east, and west neighbors. Another FPTA was developed at the University of Heidelberg (Langeheine 2005). This chip enables developing circuits directly at the transistor level; the designer can select the transistor type (P/N), parameters (channel length and size), and interconnection.

2.3.2 Anadigm FPAA

The reconfiguration of field programmable analog arrays (FPAA) is typically based on either switched capacitors or OTAs. Switched capacitors perform the function of configurable resistors. ❷ *Figure 5* shows the principle: capacitor C is connected between two switches controlled by signals S_1 and S_2. The switches are implemented using unipolar transistors and signals S_1 and S_2 are nonoverlapping clocks. The charge Q over one clock period transferred to the capacitor C is given by

$$Q = C(V_1 - V_2) \tag{1}$$

and the average current associated to this charge is

$$I_a = C(V_1 - V_2)/T \tag{2}$$

where T denotes the clock period. The value of the equivalent resistor can be calculated as

$$R = (V_1 - V_2)/I_a = T/C \tag{3}$$

◘ Fig. 4

A single cell of the FPTA-2 chip (Stoica et al. 2002). Configuration switches are shown as circles.

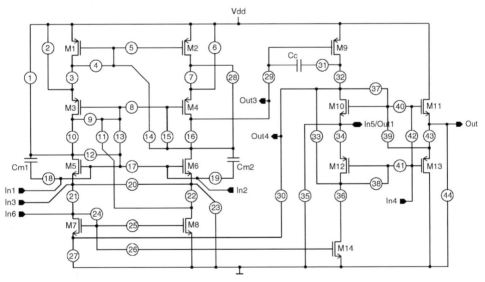

◘ Fig. 5

Switched capacitor and its equivalent resistor.

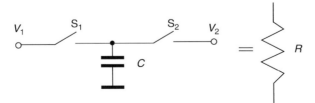

This resistor value can be controlled by the switching frequency $f = 1/T$. In comparison to conventional resistors, switching capacitors are advantageous in terms of linearity, dynamic range, precision, and size on the chip. As f can be controlled from software, analog circuits (such as filters and oscillators) can be easily tuned. A disadvantage might be that circuits containing switched capacitors operate in the discrete domain, that is, there is a limit on the possible operation frequency that is determined by f.

The commercially available Anadigm AN221E04 FPAA (Anadigm 2007), developed using switching capacitors, is an array of four configurable analog blocks (CAB), each of which contains two OpAmps, a comparator, and a successive approximation register (SAR) that performs 8-bit analog-to-digital conversion of signals. The device also contains one programmable look-up table that can be used to store information for the generation of arbitrary waveforms, and is shared among the CABs. The FPAA architecture is shown in ❯ *Fig. 6*. The configuration bit stream is stored in SRAM and the maximum switching frequency is 16 MHz.

❏ **Fig. 6**

Anadigm AN221E04 FPAA (Anadigm 2007).

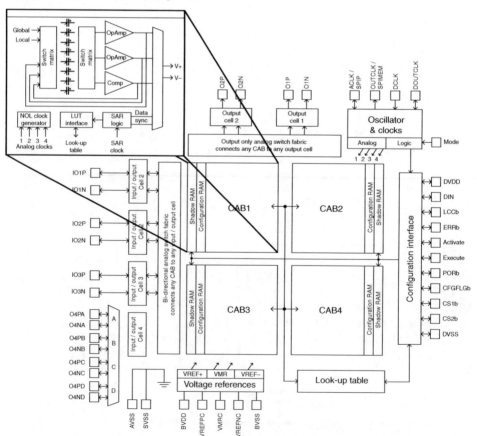

2.3.3 Operational Transconductance Amplifiers

Another approach to software control of analog circuits is based on operational transconductance amplifiers (OTAs). A typical OTA produces a current output I_0 that linearly depends on an input voltage present at both an inverting input (V_-) and non-inverting input (V_+):

$$I_0 = -g_m(V_+ - V_-) \tag{4}$$

where g_m is the transconductance of the circuit. The transconductance can be predefined or programmable using an external biasing current input I_{set} (see ❷ *Fig. 7*). Biasing currents for OTAs are generated using DACs. Ideally, the circuit has infinite values for both the input and output impedances. OTAs are the main building blocks of continuous time filters in which the transconductance (and thus the frequency characteristics) can be controlled externally. An example of an FPAA that utilizes configurable OTAs is that developed at the University of Freiburg (Henrici et al. 2007).

2.4 Other Reconfigurable Devices

Evolutionary design has been performed for many other reconfigurable devices. Special ASICs include reconfigurable chips: POEtic (Moreno et al. 2005), RISA (Greensted and Tyrrell 2007), PAMA (Zebulum et al. 2002), and REPOMO32 (Sekanina et al. 2009). Among more exotic examples, we find reconfigurable mirrors (Loktev et al. 2003), reconfigurable nanosystems (Tour 2003), reconfigurable antennas (Linden 2001) and reconfigurable liquid crystals (Harding et al. 2008). Whenever needed, detailed descriptions of these devices will be given in the sections below devoted to particular examples of evolvable systems.

3 First Approaches to Evolvable Hardware and Terminology

Practically, since the very beginning of the research in evolutionary computation, evolutionary algorithms have been applied in the area of hardware optimization. One of the earliest works was the optimization of wiring configurations for a subsystem on a ballistic missile, published in 1963 (Lohn and Hornby 2006). Various monographs (Drechsler 1998; Larsson 2005) summarize the applications from the areas of electronic circuits design, diagnostics, and testing. Later, evolutionary algorithms were applied to generate complete circuit structures, that is, not only to optimize parameters of existing circuits.

◘ **Fig. 7**
Operational transconductance amplifiers with fixed transconductance **(a)**, transconductance programmable by I_{set} **(b)**.

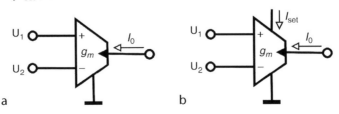

3.1   Higuchi's Experiment

In their pioneering work from 1992, Higuchi et al. used a genetic algorithm to create configurations for a physical electronic chip, the GAL16Z8 – a predecessor of modern FPGAs (Higuchi et al. 1993). This chip uses a programmable AND/OR array to implement digital circuits in the form of "sum of products." Only a part of the chip was utilized resulting in a chromosome size of 108 bits. A relatively simple combinational circuit (6-input/1-output multiplexer) was evolved only on the basis of its behavioral input/output specification. In order to obtain the fitness value, the candidate circuit response was measured for all possible input patterns (i.e., 2^6 combinations). For each correct output value, the fitness score was incremented, resulting in the maximum fitness value of 64 points. In this experiment, candidate configurations were evaluated using the GAL simulator. Only the resulting configuration was uploaded into the physical chip at the end of the evolution and tested physically. The scenario that employs a circuit simulator is known as *extrinsic evolution* (sometimes called offline evolution). This experiment is considered the birth of evolvable hardware.

3.2   Thompson's Experiment

In 1996 and the following years, Adrian Thompson performed a series of experiments, which clearly demonstrated that there can be a significant difference in the resulting behavior if candidate circuits are evaluated not in a circuit simulator but directly in a physical reconfigurable device (Thompson 1996). Thompson used an FPGA XC6216 chip to evolve a tone discriminator – a circuit discriminating between square waves of 1 and 10 kHz. With 10 × 10 configurable blocks of an XC6216 device, the circuit has to output 5 V for one of the frequencies and 0 V for the other. The problem is that the evolved circuit has to discriminate between input periods five orders of magnitude longer than the propagation time of each configurable block. No external components or synchronization signals were provided as is typical for conventional solutions to this problem. ❷ *Figure 8* shows the experimental setup.

The goal of the evolution is to maximize the difference between the average output voltage when the 1 and 10 kHz input signals are applied:

$$\text{fitness} = \frac{1}{5}|(k_1 \sum_{t \in S_1} i_t) - (k_2 \sum_{t \in S_{10}} i_t)| \tag{5}$$

where i_t represents the integrator output at the end of test tone t, S_1 is the set of 1 kHz tones, and S_{10} is the set of 10 kHz tones. k_1 and k_2 are experimentally determined constants.

Fortunately, the format of the configuration bit string of XC6216 was known to the users. Hence, Thompson could use a very efficient and fast dynamic reconfiguration to quickly generate new candidate circuits, though the experiment took 2–3 weeks because of time-consuming evaluation (around 5 s) of every individual. ❷ *Figure 8* shows one of the evolved solutions. Only 21 out of 10 × 10 blocks contribute to the actual circuit behavior. In particular, gray-labeled blocks are very important in the solution. Although they are not connected, they also influence the circuit behavior. According to Thompson, these blocks interact with the circuit in some nonstandard ways. The evolved circuit was carefully analyzed using simulators, tested under different conditions, and on other XC6216 chips (Thompson et al. 1999). Surprisingly, Thompson was not able to create a reliable simulation model of the evolved circuit. In addition, he observed that the circuit works only in the chip used during evolution.

◘ Fig. 8
Thompson's experiment setup (*left*) and utilization of logic blocks in evolved circuit (*right*, according to Thompson (1996)).

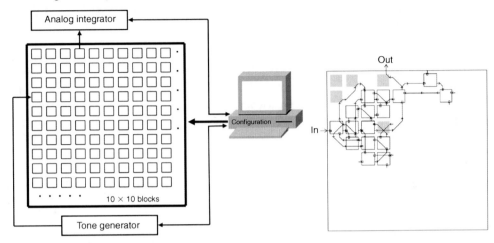

Although the circuit has a digital interface, its internal behavior is typical for analog circuits. It can be stated that the evolution was able to explore a full repertoire of behaviors available from the silicon resources provided to create the required behavior. This is an outstanding result, practically unreachable by means of conventional design methods. Hence, this experiment stimulated a lot of research in the evolvable hardware field.

The approach in which all candidate circuits are evaluated in a physical hardware is known as *intrinsic evolution* (sometimes called online evolution). The main advantage of the approach is that evolved circuits can be perfectly adapted to a given hardware substrate and external environment (temperature, radiation, ...) where the evolutionary design was carried out. On the other hand, intrinsically evolved solutions might not be robust and so portable to other reconfigurable devices. In order to avoid the portability problem, Stoica et al. introduced the so-called *mixtrinsic evolution*, in which some individuals are evaluated intrinsically and others extrinsically (Stoica et al. 2000).

3.3 Scalability Problems

The evolutionary circuit design is not competitive in all design problems because of the so-called *scalability problems*. From the viewpoint of the *scalability of representation*, the problem is that long chromosomes, which are usually required to represent complex solutions imply large search spaces that are typically difficult to search. In many cases, even a well-tuned parallel evolutionary algorithm running on a cluster of workstations fails to find an adequate solution in a reasonable time. In order to evolve large designs and simultaneously keep the size of the chromosome small, four main techniques have been developed:

- *Functional-level evolution* (Murakawa et al. 1996): Instead of gates and single-wire connections, the system is composed of complex application-specific functional blocks (such as adders, multipliers, and comparators) connected using multi-bit connections.

The selection of suitable functional blocks represents a domain knowledge that has to be included in the system.

- *Incremental evolution* (Torresen 1998, 2002): Target circuit is decomposed onto modules that are evolved separately. The decomposition strategy is a kind of domain knowledge that has to be supplied by the designer.
- *Development* (Kitano 1996; Koza et al. 1999): The above mentioned approaches employ a direct encoding of the target circuit (phenotype) in the chromosome (genotype). Hence, the size of the chromosome is proportional to the size of the circuit. Developmental approaches utilize indirect (generative) encodings that specify how to construct the target circuit. The phenotype is in fact constructed by a program that is encoded in the genotype. Designing these developmental encodings is not trivial and represents a domain knowledge that has to be supplied by the designer.
- *Modularization* (Walker and Miller 2008; Koza et al. 1999): Some EAs enable us to dynamically create and destroy reusable modules (subcircuits). The reuse of modules make the evolution easier even for large circuits.

Another problem is related to the fitness calculation time. In the case of the combinational circuit evolution, the evaluation time of a candidate circuit grows exponentially with the increasing number of inputs (assuming that all possible input combinations are tested in the fitness function). Hence, the evaluation time becomes the main bottleneck of the evolutionary approach when complex circuits with many inputs are evolved. This problem is known as the problem of *scalability of evaluation*. In order to reduce the time of evaluation, various techniques have been adopted:

- Only a subset of all possible input vectors is utilized. That is typical for the evolution of filters, classifiers, or robot controllers (Higuchi et al. 2006). Unfortunately, nobody can guarantee correct circuit behavior for those input combinations that were not used during evolution. Hence, evolved circuits have to be validated at the end of evolution using a test set – a representative set of input vectors that differ from the training set.
- In some cases, it is sufficient to evaluate only some structural properties (not the function) of candidate circuits that can be done with a reasonable time overhead. For example, because the testability of a candidate circuit can be calculated in a quadratic time, very large benchmark circuits (more than 1 million gates) with predefined testability properties were evolved (Pecenka et al. 2008).
- If the target system is linear, it is possible to perfectly evaluate a candidate circuit using a single input vector independently of the circuit complexity. Multiple-constant multipliers composed of adders, subtracters, and shifters were evolved for tens of outputs (Vasicek et al. 2008).

An obvious conclusion is that the perfect evaluation procedures are applicable only for small circuits or in very specific cases of large circuits. On the other hand, when more complex circuits have to be evolved, only an imperfect fitness calculation method may be employed due to time constraints.

4 Evolutionary Design Using Simulators

This section is devoted to extrinsic evolution of hardware. It covers the evolutionary design of digital circuits (at various levels), analog circuits, antennas, optical systems, and MEMS.

4.1 Gate-Level Evolution Using Cartesian Genetic Programming

Cartesian genetic programming (CGP), introduced by Miller and Thompson (2000), is a widely used method for extrinsic evolution of digital circuits and programs. CGP was originally defined for gate-level evolution; however, it can easily be extended for functional-level evolution. In its basic variant, candidate circuits are directly represented in the chromosome.

In CGP, a candidate entity (program or circuit) is modeled as an array of u (columns) × v (rows) of programmable elements (gates). The number of inputs, n_i, and outputs, n_o, is fixed. Each node input can be connected either to the output of a node placed in the previous L columns or to some of the program inputs. The L-back parameter, in fact, defines the level of connectivity and thus reduces/extends the search space. For example, if $L = 1$ only neighboring columns may be connected; if $L = u$, full connectivity is enabled. Feedback is not allowed. Each node is programmed to perform one of n_n-input functions defined in the set Γ (n_f denotes $|\Gamma|$). As ❯ *Fig. 9* shows, while the size of chromosome is fixed, the size of the phenotype is variable (i.e., some nodes are not used). Every individual is encoded using $u \times v \times (n_n + 1) + n_o$ integers. In order to make the search process easier, a common practice is to use some redundant elements in the CGP grid and a maximum value for the L-back parameter.

CGP operates with the population of $1 + \lambda$ individuals (typically, $\lambda = 5 - 20$). The initial population is randomly generated. Every new population consists of the best individual of the previous population and its λ offspring created by point mutation. If two or more individuals have received the same highest fitness score in the previous population, the individual that has not served as the parent in the previous population will be selected as the new parent. This strategy is used to ensure the diversity of population.

For the evolution of logic circuits, the fitness function is constructed in such a way that all possible input combinations are applied at the candidate circuit inputs, the outputs are collected and the goal is to minimize the difference between the obtained truth table and the required truth table. If evolution has found a solution that produces correct outputs for all possible input combinations, other parameters, such as the number of components or delay, can be optimized. The evolution is stopped when the best fitness value stagnates or the maximum number of generations is exhausted.

◻ **Fig. 9**
An example of a candidate program. CGP parameters are as follows: $L = 3$, $u = 4$, $v = 2$, $\Gamma = $ {AND (0), OR (1)}. Nodes 5 and 9 are not utilized. Chromosome: 1,2,1, 0,0,1, 2,3,0, 3,4,0, 1,6,0, 0,6,1, 1,7,0, 6,8,0, 6, 10. The last two integers indicate the outputs of the program.

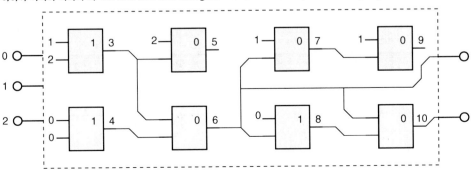

The design of small multipliers is the most popular benchmark problem for gate-level evolution. Because the direct CGP approach is not scalable it works only for 4-bit multipliers and smaller multipliers. ❯ *Table 1* summarizes the best-known results for various multipliers (Miller et al. 2000; Vassilev et al. 2000). For this class of circuits, CGP is able to create innovative designs, that is, circuits containing fewer gates than the best existing implementations, although only minimal domain knowledge has to be inserted into the representation.

CGP can be used to evolve compact implementations of such circuits whose efficient construction using conventional techniques is quite difficult. The construction of polymorphic circuits is a typical example (Stoica et al. 2001, 2002; Sekanina et al. 2008). Polymorphic circuits consist of conventional as well as polymorphic gates. Polymorphic gates are unconventional logic components that can switch their logic functions according to a changing environment. For example, a polymorphic gate controlled by the level of the power supply voltage exists, which operates as NAND when $V_{dd} = 3.3$ V and as NOR when $V_{dd} = 1.8$ V. Another gate operates as AND when the temperature is 27°C and as OR when the temperature is 125°C. Consider that the objective is to design a circuit that operates as a multiplier (the function will be denoted f_1) when polymorphic gates perform the NAND function and as a sorting network (the function will be denoted f_2) when polymorphic gates perform the NOR function. In order to use CGP for the design of polymorphic circuits with the minimum number of gates, the fitness function has to be modified so that a candidate circuit is evaluated in both modes. The new fitness value is defined as follows:

$$\text{fitness} = B_1 + B_2 + (u.v - z) \qquad (6)$$

where B_1 (resp. B_2) is the number of correct output bits for f_1 (resp. f_2) obtained as the response for all possible input combinations, z denotes the number of gates utilized in a particular candidate circuit, and $u.v$ is the total number of programmable gates available. The last term is considered only if the circuit behavior is perfect in both modes; otherwise $uv - z = 0$. ❯ *Table 2* summarizes the results obtained for various instances of the problem. In most cases, evolved circuits exhibit fewer gates than circuits created by multiplexing conventional implementations using polymorphic multiplexers (Sekanina et al. 2008).

4.2 Sequential Circuits

The evolution of sequential circuits is not as popular as the evolution of combinational circuits. However, evolutionary algorithms were successfully applied in the so-called state

❏ Table 1

The number of two-input gates in the best implementations of multipliers according to Miller et al. (2000) and Vassilev et al. (2000). CGP used with two-input gates AND, XOR, and AND with one input inverted, L-back is max., $\lambda = 4$

Multiplier	Best conventional	Best evolved	CGP array	Max. generations
2b × 2b	8	7	7 × 1	10k
3b × 2b	17	13	17 × 1	200k
3b × 3b	30	23	35 × 1	20M
4b × 3b	47	37	56 × 1	200M
4b × 4b	64	57	67 × 1	700M

assignment problem, the goal of which is to compute optimal binary codes for each symbolic state and construct the state transition table of a finite state machine (FSM). CGP is also used to implement combinational circuits computing the next state of the FSM. An improvement in the gate count (with respect to conventional techniques) was obtained for small circuits having up to 30 states (Nedjah and de Macedo 2005; Ali et al. 2004).

4.3 Incremental Evolution

The evolutionary design of larger gate-level circuits is usually performed using modular CGP (Walker and Miller 2008) or incremental evolution (Torresen 1998, 2002; Stomeo et al. 2006). A typical feature of circuits evolved modularly or incrementally is that they are fully functional, but no innovation is usually visible in their implementation. ❷ *Figure 10* shows basic approaches to the incremental evolution. The *n*-input/*m*-output circuit can be

❑ Table 2
Parameters of evolved polymorphic multipliers/sorters. Gates in "Gate set" are numbered as (1) NAND/NOR, (2) AND, (3) OR, (4) XOR, (5) NAND, (6) NOR, (7) NOT A, (8) NOT B, (9) MOV A, and (10) MOV B, where MOV denotes the identity operation. Population size is 15

Multiplier/sorter	$2b \times 2b/4b$	$3b \times 2b/5b$	$3b \times 3b/6b$	$4b \times 3b/7b$
$u \times v$	10 ×12	100 ×1	120 ×1	16 ×16
L-back	1	100	120	16
Mutation (genes)	1	2	4	4
Gate set	1, 2, 9, 10	1–4, 9, 10	1–10	1, 2, 9, 10
Runs	10	10	10	10
Successful runs	10	10	9	3
Generations (average)	52,580	854,900	26,972,648	62,617,151
Min. # of gates	23	30	52	113

❑ Fig. 10
(a) Original circuit, (b) input-oriented decomposition, (c) output-oriented decomposition.

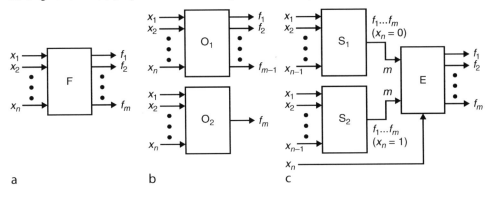

a
b
c

decomposed either according to the inputs or outputs. Both decompositions can be applied recursively and thus they can yield subcircuits that can be evolved directly using CGP, for example. Advanced variants of the incremental evolution further optimize evolved modules in order to maximize the number of gates shared among them and so to reduce the total number of gates. The most complex circuits evolved by the gate-level incremental evolution were reported in Stomeo et al. (2006): 6-bit multipliers, 17-input parity circuits, and some MCNC benchmarks (such as the alu4 with 14 inputs and 8 outputs). Incremental evolution was also used to design classifiers directly in the FPGA (see ❷ Sect. 5.1.2).

4.4 Developmental Approaches

As developmental approaches employ indirect encoding, the chromosome contains a program which is executed in order to construct the target circuit. The program can be implemented as an L-system, cellular automaton, if-then-else rules or, in general, as a program.

Arbitrarily large multipliers were constructed using evolved programs working in a grid of programmable nodes (Bidlo and Skarvada 2008). However, no innovation is observable in evolved multipliers in comparison to conventional multipliers. A similar approach was proposed to create arbitrarily large sorting networks, which exhibit slightly better properties (in the number of components and delay) than sorting networks created using conventional construction algorithms (Sekanina and Bidlo 2005). Papers of Gordon and Bentley (2002), Tufte and Haddow (2005), Tempesti et al. (2007), and Zhan et al. (2008) present more biologically plausible models of development for digital circuits. Although these approaches do not lead to innovative circuit designs, they are useful for investigation of principles of development, genetic regulatory networks, environmental interactions, and fault tolerance that could be useful for evolution and adaptation of large-scale digital circuits in future.

4.5 Functional-Level Evolution

CGP can easily be modified to evolve larger circuits. Instead of gates and single-wire connections, application-specific functions and multiple-bit connections are employed. The advantage is that while the size of the chromosome is similar to the gate-level evolution, the size of the phenotype can be arbitrarily large, depending on the building blocks used. Filters, predictors, and classifiers are typical circuits designed at this level. Because of the size of the candidate circuits, it is impossible to evaluate circuit responses for all possible input vectors. Hence, candidate circuits are evaluated using a training set during evolution and validated using a test set at the end of evolution. The goal of the evolution is to minimize the difference between the response of a candidate circuit and the target response. ❷ Section 5.1 will demonstrate real-world applications of functional-level evolution accelerated in FPGAs.

4.5.1 Digital Filter Evolution

Digital filter optimization and design is one of the most common applications of evolutionary computation in the hardware field. Digital filters are traditionally modeled using discrete-time systems theory (Ifeachor and Jervis 2002). A *discrete-time* system is essentially a mathematical algorithm that takes an input sequence, $x(n)$, and produces an output sequence, $y(n)$. A *digital filter* is an example of a discrete-time system. A discrete-time system may be linear or nonlinear, time invariant or time varying. *Linear time-invariant* (LTI) systems form an important class of systems used in digital signal processing. The input–output relationship of an LTI system is given by the convolution sum

$$y(n) = \sum_{k=-\infty}^{+\infty} h(k)x(n-k) \tag{7}$$

where $h(k)$ is the *impulse response* of the system. The values of $h(k)$ completely define the discrete-time system in the time domain. A general infinite impulse response (IIR) digital filter is described by the equation

$$y(n) = \sum_{k=0}^{N} b_k x(n-k) - \sum_{k=1}^{M} a_k y(n-k) \tag{8}$$

The output samples, $y(n)$, are derived from current and past input samples, $x(n)$, as well as from current and past output samples. The task of the designer is to propose the values of the coefficients a_k and b_k and the sizes of vectors N and M. In finite impulse response (FIR) filters (❷ *Fig. 11*), the current output sample, $y(n)$, is a function only of past and present values of the input, that is,

$$y(n) = \sum_{k=0}^{N} b_k x(n-k) \tag{9}$$

The stability and linear phase are the main advantages of FIR filters. On the other hand, in order to get a good filter, many coefficients have to be considered in contrast to IIR filters. In general, IIR filters are not stable (because of feedback). FIR filters are algebraically more difficult to synthesize.

Conventional design methods are well developed and represent the approach to digital filter design widely adopted by industry. Digital filters are usually implemented either on DSPs or as custom circuits. Their implementation is based on multiply-and-accumulate structures. The multiplier is the primary performance bottleneck when implementing filters in hardware as it is costly in terms of area, power, and signal delay. Hence, multiplierless filters were introduced in which multiplication is reduced to a series of bitshifts, additions, and subtractions.

Evolutionary algorithms have been utilized either to optimize filter coefficients (Harris and Ifeachor 1995) or to design complete filter structures. In particular, structures of multiplierless filters were sought by many authors (Wade et al. 1994; Hounsell et al. 2004; Erba et al. 2001; Gwaltney and Dutton 2005). Regardless of filter representation, the fitness function is usually constructed in the frequency domain. Typically, a candidate's filter frequency characteristics are measured at k frequencies f_i (distributed logarithmically) and compared with target frequency characteristics that are derived from specification (❷ *Fig. 11*). The fitness value is calculated as

$$\text{fitness} = \sum_{i=0}^{k-1} [W(d(f_i), f_i) * d(f_i)] \tag{10}$$

where $d(x)$ is the absolute value of the difference between the target and observed values at frequency x, and $W(y, x)$ is the weighting for difference y at frequency x. The goal is to approximate the target characteristics as accurately as possible and better-than-conventional

□ **Fig. 11**

FIR filter structure (*top*) and frequency characteristics of a low pass filter: ideal vs. real (*bottom*).

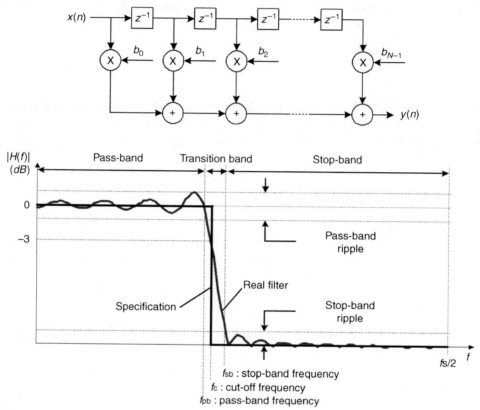

methods allow using the same resources. Additional objectives can be considered, for example, the circuit size or power consumption. The fitness calculation procedure is not trivial and can be time consuming.

4.5.2 Multiple Constant Multipliers

In digital filters, it is often required to multiply a single value x by N constants and so to generate N output products, that is, to implement a multiple constant multiplier (MCM). A multiplierless variant of MCM is such that it is composed of adders, subtractors, and shifters. Finding the optimal MCM, that is, the one with the fewest number of components is known to be NP-complete. A very efficient heuristic algorithm for the MCM problem was recently published (Voronenko and Püschel 2007).

Although MCMs are relatively complex circuits composed of high-level components, their interesting feature from the point of view of evolutionary design is that a candidate MCM can be perfectly evaluated by applying only one input vector (e.g., $x = 1$). The reason is that MCMs implement linear transforms. This feature is unique because it enables us to evolve very large circuits that are simultaneously perfectly evaluated. No training and test data sets are needed.

CGP was used to synthesize MCM, which generates N output values $c_1 x \ldots c_N x$, where $c_1 \ldots c_N$ are given constants and x is the only input variable (Vasicek et al. 2008). In addition to a perfect functionality, the number of components is optimized. CGP operates with parameters: $n_i = 1$, $n_o = N$, $\lambda = 7$ and Γ includes the addition, subtraction, shifts, and identity function. These functions as well as all connections are defined over b bits, where $b = 16$ in this case. ❷ *Figure 12* compares one of the best evolved solutions with the solution provided by the heuristics (Voronenko and Püschel 2007) for three selected constants. The evolved solution contains two shifters less and exhibits a shorter delay than the solution provided by the heuristics. CGP is able to handle MCMs with tens of outputs.

4.6 Evolutionary Approaches in Diagnostics and Testing

Similarly to the circuit design, digital circuit diagnostics and testing have profited from *evolutionary optimization* for a long time. Among typical problems that were attempted in this area, we mention test vectors selection, test scheduling, scan-chain optimization, and test

❏ **Fig. 12**
MCM with three coefficients (2,925, 23,111, 13,781): according to Voronenko and Püschel (2007) (*left*), the best evolved solution (Vasicek et al. 2008) (*right*).

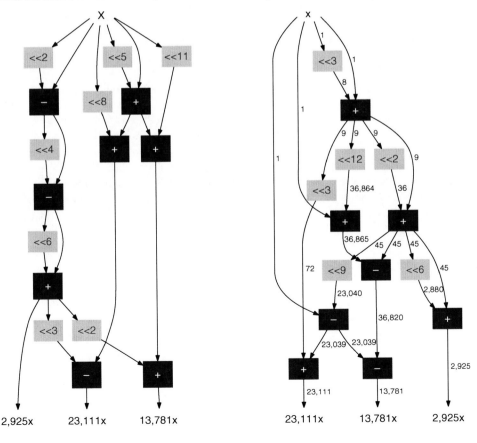

power consumption optimization under various constraints (Larsson 2005). Evolutionary design is applied in this area rarely. Two examples are briefly described below.

A method was described in Pecenka et al. (2008), which enables us to evolve large synthetic RTL benchmark circuits with a predefined structure and testability. These benchmark circuits are useful in a validation process of novel algorithms and tools in the area of digital-circuits testing. More specifically, as benchmark circuits are designed by means of the EA, they can contain constructions that do not usually appear in the circuits designed by classical design techniques. Thus, the use of evolved benchmark circuits in the process of evaluating new testability analysis tools can reveal problems of the tools that remain hidden when conventional benchmark circuits are used. With sizes up to 1.2 million gates, evolved benchmark circuits currently represent the most complex benchmark circuits with a known level of testability. Furthermore, these circuits are the largest that have ever been designed by means of evolutionary algorithms. This complexity was achieved because no requirement on the functionality was specified. Only some structural properties were measured in the fitness function that could be done in polynomial time.

Bernardi et al. presented a genetic programming-based approach to the construction of diagnosis-oriented software-based test sets for microprocessors (Bernardi et al. 2008). The methodology exploits existing manufacturing test sets designed for software-based self-test and improves them by using evolved test sets. Experimental results showed the feasibility, robustness, and effectiveness of the approach for diagnosing stuck-at faults on an Intel i8051 processor core. A similar approach was used to test peripheral cores.

4.7 Analog Circuits

Evolutionary algorithms can be used to optimize parameter values of circuit components in order to maximize circuit performance, reduce power consumption, or increase fabrication yield. However, in recent years, a lot of attention has been devoted in the evolvable hardware field to the evolutionary synthesis of complete structures of analog circuits. Analog circuits are evolved at various levels of description including passive circuits (composed of resistors, capacitors, and inductors), transistor-level circuits and circuits with operational amplifiers.

Candidate circuits are usually evaluated using the SPICE simulator, which is the most popular simulator for analog circuit designers. ❷ *Figure 13* shows a simple candidate circuit and its representation in a netlist – the input data format for SPICE. Before the circuit

❑ Fig. 13

A circuit and its description in netlist.

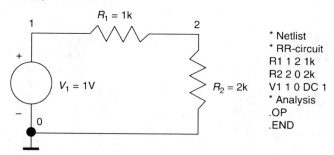

evaluation can begin, the chromosome has to be transformed to a netlist and the input stimuli have to be specified. Then, the circuit is simulated using methods supported by SPICE, for example, DC analysis, AC analysis, transient analysis, or noise analysis can be performed. The results of the simulation are available in a text file. A typical fitness function takes the form of ❯ Eq. 10, that is, the goal is to minimize the difference between the obtained signal and the target signal. The main disadvantage of the use of SPICE is that the fitness evaluation is relatively slow in comparison with measuring real circuits (several orders of magnitude reported in Stoica et al. (2002)).

Analog circuits can be encoded in the chromosome either directly or indirectly. Direct encoding is similar to the structure of the netlist. For each component, parameters (e.g., resistances of the resistors) are given together with the description of its terminal connections. The chromosome is represented as a linear variable-length array of integers or real values (for parameter values).

Innovative implementations of analog circuits were obtained using indirect encoding, so-called *developmental genetic programming* introduced by Koza et al. (1999). In his method, a target circuit is subsequently constructed from a simple initial circuit (called the embryo) according to a program that is encoded in the chromosome. ❯ *Figure 14* shows a typical embryonic circuit that contains fixed components (power supply, source resistor, and load resistor) and two modifiable wires Z0 and Z1. The developmental program (❯ *Fig. 14*), which is encoded as a tree, contains two subtrees, each of them specifying the construction process for one of the modifiable wires. The functions used in the circuit-constructing program tree can be divided into five categories (Koza et al. 2003):

- Topology-modifying functions (e.g., series division (S), parallel division (P), and flip (F)) that modify the topology of the developing circuit
- Component-creating functions that insert components into the developing circuit (insert a capacitor (C), insert an inductor (L))
- Development-controlling functions that control the developmental process by which the embryo and its successors are converted into a fully developed circuit (e.g., the no-operation function, the end of construction function (E))

◼ **Fig. 14**
Example of an embryonic circuit and a candidate program (*left*) used to create target circuit (*right*) (Koza et al. 1999).

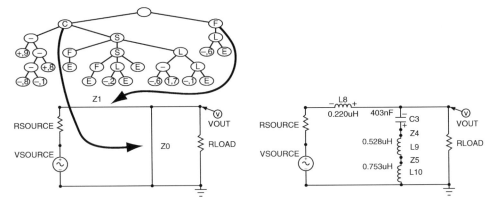

- Arithmetic-performing functions that may appear in a value-setting subtree
- Automatically defined functions that enable certain substructures to be reused (including parameterized reuse)

❯ *Figure 14* shows a fully developed circuit that was created from the embryo according to the given program. In order to obtain the fitness value of the circuit, this circuit has to be converted into a netlist and simulated.

Various books (Koza et al. 1999, 2003) present tens of circuits (passive analog circuits, circuits with transistors, circuits with operational amplifiers, controllers) evolved using the developmental genetic programming that can be classified as human-competitive designs, many times leading to rediscovering or even improving patented inventions. For example, an evolved analog circuit that duplicates the functionality of the low-voltage balun circuit (patented by Sang Gug Lee in 2001) is shown in ❯ *Fig. 15*. This design contains the key part of Lee's invention: a coupling capacitor that blocks DC (see capacitor C302 in ❯ *Fig. 15*).

Very high computational requirements are considered as the main disadvantage of the method. In contrast to CGP, it is typical for Koza's genetic programming that very large populations are used and only a few generations are produced. For example, Koza's team utilized two clusters of workstations, 1000 x Pentium II/350 MHz processor and 70 x DEC

◻ **Fig. 15**

Evolved analog circuit that duplicates the functionality of the low-voltage balun circuit that was patented by Sang Gug Lee in 2001 (according to Koza et al. 2003).

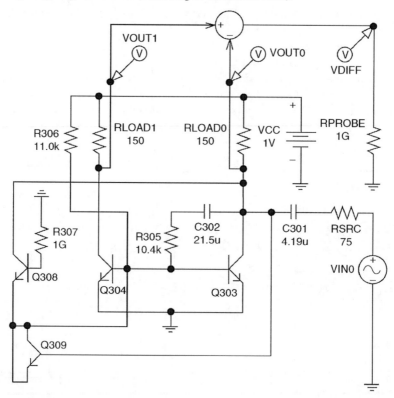

Alpha/533 MHz processor. For 36 tasks solved using GP on the clusters, the average population size is 3,350,000 individuals, 128.7 generations are produced on average and the average time to reaching a solution is 81.9 h (Koza et al. 2003).

Another criticism of Koza's genetic programming method for analog circuit synthesis comes from analog circuit designers who do not find evolved circuits trustworthy and suitable for fabrication. A solution to this problem was proposed by Gao, McConaghy, and Gielen who invented an analog circuit topology synthesis method called importance sampled circuit learning ensembles (ISCLEs). ISCLEs is a method that synthesizes circuit topologies that are novel in both functionality and topology, yet trustworthy, within reasonable CPU effort (Gao et al. 2008).

4.8 Antennas, Optical Systems, and MEMS

The evolutionary design of antennas, optical systems, MEMS, and other physical objects is strongly dependent on the existence of reliable simulators. With the increasing computer performance, sufficiently fast simulators have become available. These objects are usually evolved using generative encodings, similar to Koza's developmental genetic programming.

4.8.1 Antennas

Evolutionary optimization has been used in antenna design since the early 1990s (Linden 1997; Haupt and Werner 2007). Most of the work has been focused on optimizing parameters of a predetermined design as opposed to evolutionary design of the antenna's topology. One of the most famous examples of evolutionary design is the evolved antenna design for the flight hardware deployed on a NASA spacecraft for the Space Technology 5 (ST5) mission in 2006 (see ❏ *Fig. 16*). The antenna was evolved using generative encoding. The antenna's chromosome contains the instructions that control the construction of target shape. The process starts

❏ **Fig. 16**
One of the candidate trees (*left*) that defines the process of antenna construction in a 3D space (*middle*) and evolved antenna ST5-3-10 (Hornby et al. 2006) (*right*).

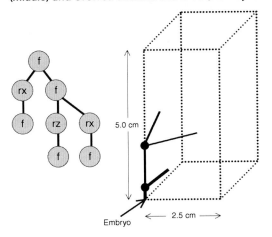

from a simple embryo – a small wire in this case. The instructions (such as F – move forward, R_i - rotate in direction i) can move and rotate the construction head in a 3D space of possible wire antennas. The paper by Hornby et al. (2006) explains that the fitness value of a candidate antenna is "the product of three scores involving the gain pattern, the VSWR, and pattern outliers. The gain pattern score compares the measured gain values across a range of elevation angles to the desired gain values and rewards antennas that exceed the requirements. VSWR is the ratio between the highest voltage and the lowest voltage in the signal envelope along a transmission line, with a ratio of 1 being ideal. The outlier score rewards antenna designs for having sample points with gains greater than zero." Candidate antennas were simulated using NEC software. The evolved antenna has a number of advantages, in particular, low power consumption, short fabrication time, and good performance.

4.8.2 Optical Systems

Evolutionary algorithms have been used to optimize the parameters of optical systems with a prespecified layout and prespecified number of components. Koza has applied developmental genetic programming to create novel topologies for optical systems. ❯ *Figure 17* shows one of the lens systems evolved using genetic programming that exhibits the same function as the patented solution of Koizumi and Watanabe, however its topology is different (Koza et al. 2005).

As the micron-meter resolution alignment of optical components usually takes a long time, to overcome this time problem the National Institute of Advanced Industrial Science and Technology in Japan (AIST) team proposed five systems that can automatically align the positions of optical components by evolutionary algorithms in very short times compared to conventional systems (Nosato et al. 2006): (1) an evolvable fiber alignment system, (2) an evolvable interferometer system, (3) an evolvable femtosecond laser system, (4) a wave-front correction system, and (5) a multiobjective adjustment system. In particular, the evolutionary algorithm was used to find positions of optical components that can be controlled by step motors. Proposed solutions have led to compacting the implementation of optical systems, reducing cost and lowering maintenance operations.

❑ Fig. 17
Evolved lens system according to Koza et al. (2005).

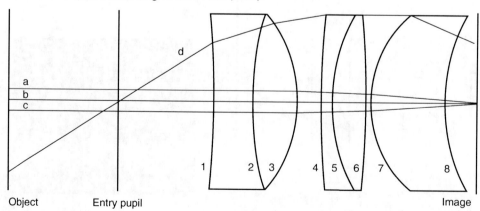

4.8.3 MEMS

Similarly to optical systems, evolutionary algorithms have been used to optimize parameters of MEMS, including the size, orientation, number of segments, etc. (Li and Antonsson 1998; Kamalian et al. 2002). Lohn, Kraus, and Hornby evolved a complete MEMS meandering resonator (i.e., topology and parameters) using a generative representation (Lohn et al. 2007). The JPL team has developed a tuning method for MEMS gyroscopes based on evolutionary computing to increase the accuracy of MEMS gyroscopes through electrostatic tuning. The method was successfully tested for the second generation of the JPL/Boeing post-resonator MEMS gyroscope (Keymeulen et al. 2006).

5 Intrinsic Evolution

This section surveys approaches to intrinsic evolution conducted directly in FPGAs, FPTAs, and FPAAs. Evolution on unconventional devices is also discussed.

5.1 Evolvable FPGAs

Nowadays, FPGAs are primarily used in the evolvable hardware field to accelerate circuit evolution. Most FPGA families can be configured only *externally*. The *internal reconfiguration* means that a circuit placed *inside* the FPGA can configure the programmable elements of the same FPGA. Although the internal configuration access port (ICAP) has been integrated into the recent Xilinx Virtex families (Virtex II, Virtex 4, Virtex 5) (Blodget et al. 2003), it is still too slow for many evolvable hardware applications. ❷ *Figure 18* shows an example of an evolvable platform utilizing ICAP (Upegui 2006). The system, which runs an EA on the MicroBlaze

■ Fig. 18
An evolvable system implemented in the FPGA using ICAP interface (Upegui 2006).

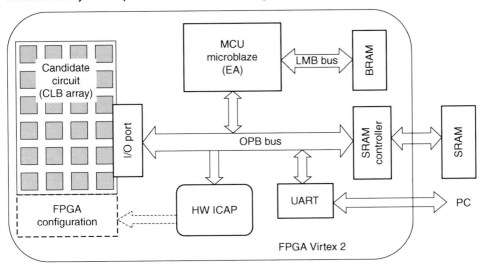

processor, reads a section of the configuration bitstream through ICAP, modifies the bitstream according to the genome currently evaluated in the MicroBlaze, sends the bitstream back through the ICAP for the partial reconfiguration of the FPGA, and evaluates the fitness of the current individual by interacting with the reconfigurable evolvable section through the standard OPB bus. The system was used to evolve the topology of Boolean networks (Upegui 2006).

In order to overcome the problem of slow reconfiguration, the concept of *virtual reconfigurable circuits* (VRC) was developed (Sekanina 2003, 2004). The VRC is, in fact, a second reconfiguration layer developed on top of the FPGA in order to obtain fast reconfiguration and application-specific programmable elements. ❯ *Figure 19* shows that the VRC consists of a 2D array of programmable elements E_i. The routing circuits are created using multiplexers. The configuration memory of the VRC is typically implemented as a register array. All bits of the configuration memory are connected to multiplexers that control the routing and selecting of functions in programmable elements. Because the array of programmable elements, routing circuits, configuration memory, style of reconfiguration, and granularity of VRC can be designed exactly according to the requirements of a given application, designers can create an optimized application-specific reconfigurable device. If the structure of the chromosome corresponds to the configuration interface of the VRC then a very fast reconfiguration can be achieved (e.g., consuming a few clock cycles only).

The FPGA-based implementations of evolvable hardware systems can be divided into two classes:

- *The FPGA serves in the fitness calculation only.* The evolutionary algorithm (which is usually executed on a personal computer) sends the configuration bitstreams representing candidate circuits to the FPGA in order to obtain their fitness values. The FPGA is configured externally. A typical example of this scenario is Thompson's experiment (see ❯ Sect. 3.2).
- *The entire evolvable system is implemented in the FPGA.* The idea of the complete hardware evolution for FPGAs was initially demonstrated in Tufte and Haddow (2000); however, the paper provides only a simple example of the optimization of a few FIR filter coefficients stored in a register. Current FPGA implementations are capable of evolving circuit structures. They utilize either ICAP or VRC.

❯ *Table 3* surveys examples of FPGA implementations of digital evolvable systems. One can identify the following components in all systems: the array of reconfigurable elements, evolutionary algorithm, fitness calculation unit, and controller. The problem domain determines the type of reconfigurable elements. In some cases, the evolution is performed directly with reconfigurable cells of the FPGA; in other cases, a kind of VRC is utilized. An evolutionary optimization of coefficients stored in registers represents the simplest example. The EA and fitness calculation unit can be implemented either as an application-specific circuit or as a program. The program is running either in a personal computer or in an embedded processor, which is integrated into the FPGA.

5.1.1 Evolution of Image Filters

An FPGA-based evolvable image filter architecture was proposed to create novel implementations of image filters (Sekanina 2004; Vasicek and Sekanina 2007). The system consists of the

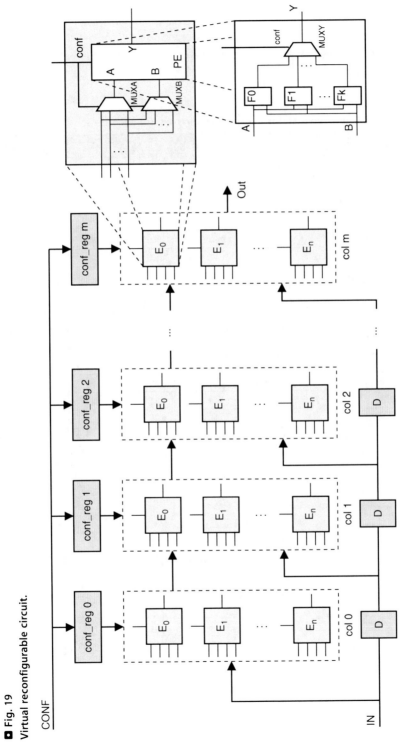

◨ Fig. 19
Virtual reconfigurable circuit.

genetic unit, an array of reconfigurable elements (implemented as VRC), and a fitness calculation unit. While candidate filters are created and evaluated using user logic available on the FPGA, the genetic operations are carried out in the on-chip PowerPC processor. Every image operator is considered as a digital circuit of nine 8-bit inputs and a single 8-bit output, which processes gray-scaled (8-bit/pixel) images (see ❷ *Fig. 20*).

◻ Table 3

Examples of FPGA implementations of evolvable digital systems

Reference	Application	Platform	EA	Fitness
External reconfiguration				
Thompson et al. (1999)	Tone discriminator	XC6216	PC	PC
Huelsbergen et al. (1999)	Oscillators	XC6216	PC	PC
Zhang et al. (2004)	Image filters	VRC	PC	PC
Gordon (2005)	Arithmetic circuits	Virtex CLB	PC	PC
Gwaltney and Gutton (2005)	IIR filters	VRC	DSP	DSP
Internal reconfiguration				
Tufte and Haddow (2000)	FIR filters	Register values	HW	HW
Martinek and Sekanina (2005)	Image filters	VRC	HW	HW
Vasicek and Sekanina (2007)	Image filters	VRC	PowerPC	HW
Sekanina and Friedl (2004)	Logic circuits	VRC	HW	HW
Vasicek and Sekanina (2008)	CGP accelerator	VRC	PowerPC	HW
Salomon et al. (2006)	Hash functions	VRC	HW	HW
Glette (2008)	Face recognition	VRC	MicroBlaze	HW
Glette et al. (2007)	Sonar spectrum class.	VRC	PowerPC	HW
Upegui and Sanchez (2006)	Cellular automaton	Virtex CLB	MicroBlaze	HW
Vasicek et al. (2008)	MCM	VRC	HW	HW

◻ Fig. 20

Image filter design problem.

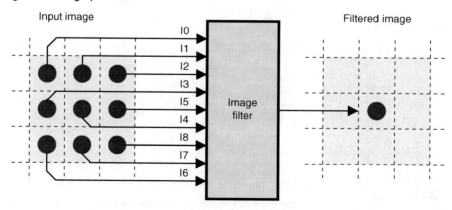

▶ *Figure 19* also shows a corresponding VRC, which consists of 2-input configurable functional blocks (CFBs), denoted as E_i, placed in a grid of eight columns and four rows. Any input of each CFB may be connected either to a primary circuit input or to the output of a CFB, which is placed anywhere in a preceding column. Every CFB can be programmed to implement one of the functions given in ▶ *Table 4*. All these functions operate with 8-bit operands and produce 8-bit results. The reconfiguration is performed column by column. The computation is pipelined; a column of CFBs represents a stage of the pipeline. The configuration bitstream of VRC, which is stored in a register array *conf_reg* consists of 384 bits. A single CFB is configured by 12 bits. The evolutionary algorithm directly operates with the configurations of the VRC.

The fitness calculation is carried out by the fitness unit (FU). The corrupted image, original image, and filtered image are stored in external SRAM memories, which are accessible via memory controllers implemented in the FPGA. The pixels of the corrupted image u are loaded from external SRAM1 memory and forwarded to the inputs of the VRC. The pixels of the filtered image v are sent back to the fitness unit, where they are compared with the pixels of the original image w that is stored in another external memory, SRAM2. The filtered image is simultaneously stored into the third external memory, SRAM3. The design objective is to minimize the difference between the filtered image and the original image, that is, the fitness value is calculated for an $M \times N$-pixel image as

$$\text{fitness} = \sum_{i=1}^{M-2} \sum_{j=1}^{N-2} |v(i,j) - w(i,j)| \tag{11}$$

As ▶ *Fig. 21* shows, the system is completely implemented in a single FPGA (except the SRAM memories). The program memory of the PowerPC as well as the population memory is implemented using on-chip Block RAM (BRAM) memories. The evaluation of the candidate configurations is pipelined in such manner as there are no idle clock cycles. Therefore, the time of evolution can be expressed as

$$t_{\text{evol}} = Q(M-2)(N-2)\frac{1}{f}$$

where Q is the number of evaluations, $N \times M$ is the number of pixels (note that border pixels are ignored), and f is the operation frequency. The system was implemented in a COMBO6X

◻ **Table 4**
List of functions implemented in each CFB

Code	Function	Description	Code	Function	Description
0	255	Constant	8	$x \gg 1$	Right shift by 1
1	x	Identity	9	$x \gg 2$	Right shift by 2
2	$255 - x$	Inversion	A	$swap(x, y)$	Swap nibbles
3	$x \vee y$	Bitwise OR	B	$x + y$	+ (addition)
4	$\bar{x} \vee y$	Bitwise $\bar{x}$ OR y	C	$x + {}^S y$	+ with saturation
5	$x \wedge y$	Bitwise AND	D	$(x + y) \gg 1$	Average
6	$\overline{x \wedge y}$	Bitwise NAND	E	$max(x, y)$	Maximum
7	$x \oplus y$	Bitwise XOR	F	$min(x, y)$	Minimum

■ **Fig. 21**
System for image filter evolution in FPGA.

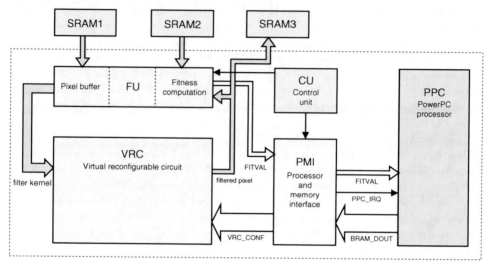

card, which is equipped with a Virtex-II Pro FPGA. It utilizes 20% slices. Experimental results show that approximately 6,000 candidate filters can be evaluated per second ($N = M = 128$), which is 44 times faster than the same algorithm running on a Celeron@2.4 GHz processor.

This platform was used to obtain the salt-and-pepper noise filters, Gaussian noise filters, edge detectors, and other image operators. By combining three evolved filters in a bank of filters, even better filtering properties can be obtained (Vasicek and Sekanina 2007). ❷ *Figure 22* compares evolved filters (a single filter and 3-bank filter) with conventional solutions (median filters (MF) and adaptive median filters (AMF)) for 25 images corrupted by the salt-and-pepper noise of various intensities. In comparison to conventional filters, evolved filters exhibit better filtering quality and lower implementation cost.

Recent work of Harding (2008) shows how the same application can be accelerated using a modern graphical processing unit that comes as a standard component of conventional PCs.

5.1.2 Incremental and Run-Time Evolution of Classifiers

Glette and Torresen proposed an application-specific architecture for online incremental evolution of classifiers (Glette 2008; Glette et al. 2007). The architecture consists of three main parts – the classification module, the evaluation module, and the CPU (PowerPC or MicroBlaze IP core) where the genetic algorithm is carried out. The reconfigurable part of the system is implemented as a VRC.

The classifier system consists of K category detection modules (CDMs), one for each category C_i to be classified (see ❷ *Fig. 23*). The CDM with the highest output indicates the resulting class. The input data to be classified are presented to each CDM concurrently. A CDM consists of M "rules," each of them is implemented using N functional units (FU) working concurrently over the input pattern (❷ *Fig. 24* left). The rule is activated when all outputs from the FUs are 1. The counter is used to count the number of activated FU rows. ❷ *Figure 24*

◻ Fig. 22

Comparison of standard image filters (MF, AMF) and evolved filters (single filter, 3-bank) for the salt-and-pepper noise of various intensity. The implementation cost is given in FPGA slices.

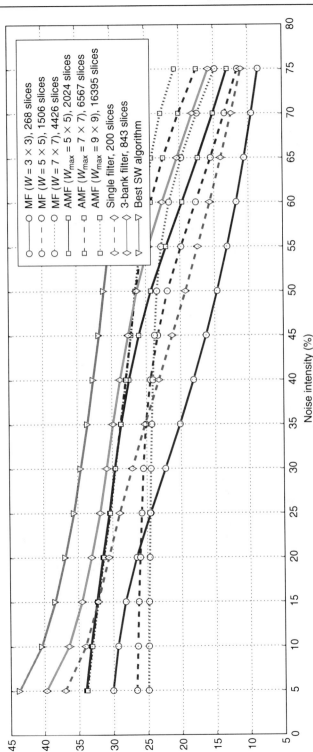

◘ Fig. 23

Evolvable classifier system. The pattern to be classified is input to all of the category detection modules (according to Glette (2008)).

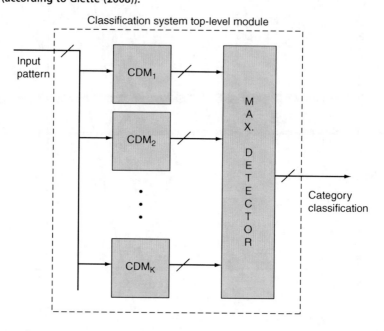

◘ Fig. 24

Category detection module (*left*) and function unit (*right*) according to Glette (2008).

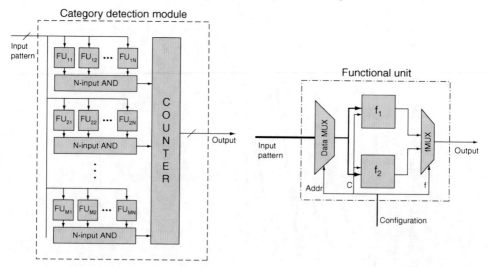

shows that each FU is controlled by configuration bits that determine some part of the input pattern to be processed by the FU, selection of function, and a constant value, C, which can be utilized for computations. This architecture is suitable for incremental evolution because each category detector, CDM_i, which is encoded using 100–200 bits, can be evolved separately.

According to Glette (2008), the fitness calculation is performed in the following way: Let V_t denote training vectors and V_v validation vectors utilized at the end of evolution. Each row of FUs is fed with the training vectors ($v \in V_t$), and fitness is based on the row's ability to give a positive (1) output for vectors v belonging to its own category ($C_v = C_i$), while giving a negative (0) output for the rest of the vectors ($C_v \neq C_i$). In the case of a positive output when $C_v = C_i$, the value A is added to the fitness sum. When $C_v \neq C_i$ and the row gives a negative output (value 0), 1 is added to the fitness sum. The other cases do not contribute to the fitness value. The fitness function F for a row can then be expressed in the following way, where o is the output of the FU row:

$$F = \sum_{v \in V_t} x_v \text{ where } x_v = \begin{cases} A \times o & \text{if } C_v = C_i \\ 1 - o & \text{if } C_v \neq C_i \end{cases} \tag{12}$$

The architecture was evaluated using nontrivial benchmark problems – face image recognition (Glette 2008), sonar spectrum classification (Glette et al. 2007) and electromyographic prosthetic hand control (Glette et al. 2008) – and compared with conventional as well as evolvable hardware approaches. The proposed solution exhibits fast classification and a short adaptation time. In comparison to existing evolvable hardware approaches, the quality of the classification is higher. In comparison to the state-of-the-art classification algorithms (such as support vector machines), the quality of classification is slightly worse.

5.2 Evolutionary Design in FPAA and FPTA

Although the previous section has dealt with the intrinsic approach, the goal of the presented applications was not to exploit the physical characteristics of FPGAs (as, e.g., Thompson et al. (1999)). The main objective was to make the evolutionary design faster. However, this is not always the case for analog circuits.

Full-custom design of analog circuits is time consuming and expensive. The aim of evolutionary design using FPTAs or FPAAs is to obtain circuits working directly in a real environment. The intrinsic evolution should help to overcome the "simulator vs reality" gap, enable designers to immediately verify the structure and parameters setting of the circuit under design, and provide a working prototype in a relatively short time.

❯ *Figure 25* shows a typical setup for analog circuit evolution in FPTA (or FPAA). The EA runs in a PC (or DSP) that generates candidate circuits (configurations) for FPTA. The PC also generates test stimuli that are converted to analog signals using a DAC. Circuit responses are converted using an ADC on digital data that are compared with the required target signals. The objective is to minimize the differences between the obtained signals and target signals. Alternatively, the fitness calculation can be based on a comparison of the measured frequency characteristics and target frequency characteristics. This approach was used to evolve various circuits including rectifiers, amplifiers, oscillators, filters, DA converters, controllers, regulators, and primitive gates (Aggarwal et al. 2006; Greenwood and Tyrrell 2007; Henrici et al. 2007; Stoica et al. 2002, 2006).

As far as the reconfigurable device is analog, it is natural for evolution to exploit physical properties of transistors, configurable switches, and other components and the environment to find the required behavior. The evolutionary process can escape the space of conventional implementations (which designers are able to represent in advance) and produce circuits that are difficult to understand or even reconstruct in a circuit simulator.

⬛ **Fig. 25**

A typical setup for circuit evolution in FPTA.

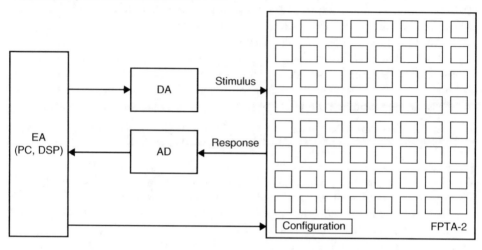

It was recognized that intrinsic evolution of analog circuits can lead to some serious problems. Circuits highly ranked during evolution may obtain low fitness when reevaluated individually at the end of evolution. It can be due to the transient behavior, which describes a configuration that is not stable as a function of time. Or, the circuit can be dependent on a previous configuration of the chip because of various parasitic capacitors in the chip. In practice, the transient behavior can be resolved by reevaluating the individuals for a longer time period. The operational range of the evolved circuit in the frequency domain is another potential pitfall, since in principle the circuit behavior should be evaluated for the overall frequency domain in which it is expected to work. Similar problems may appear when loads different from those used during evaluation are connected. Various techniques were proposed to eliminate these undesirable effects (Stoica et al. 2000, 2004; Thompson et al. 1999). In summary, evolved circuits may not be stable after some time, they may not work correctly when uploaded on another chip (or even to another area of the same chip) or used in a different environment.

A promising application for the evolution of analog circuits is adaptation in extreme environments. Future NASA missions will face extreme environments, including environments with large temperature differences (from −233°C to 460°C) and high radiation levels, upto a 5 Mrad total ionizing dose (TID) (Stoica et al. 2006). Conventional approaches to extreme environment electronics include *hardening-by-process* (i.e., fabricating devices using materials and device designs with higher tolerance to extreme environment) and *hardening-by-design* (i.e., the use of special design/compensation schemes). Both of these hardening approaches are limited, in particular for analog electronics, by the fact that current designs are fixed and, as components are affected by extreme environments, these drifts alter functionality. A recent approach pioneered by JPL is to mitigate drifts, degradation, or damage on electronic devices in extreme environments by using reconfigurable devices and an adaptive self-reconfiguration of circuit topology. This new approach is referred to as *hardening-by-reconfiguration*. As the only investigator in this area, JPL's evolvable hardware group has demonstrated for simple circuits created in the FPTA-2 that evolutionary approach can recover functionality lost (1) after a fault artificially injected into FPTA-2, (2) in extreme high and low temperatures (from −196°C to 320°C), and (3) in high radiation environments (up to 250 krad TID) (Keymeulen et al. 2000; Stoica et al. 2004, 2005, 2006; Zebulum et al. 2006).

5.3 Evolution on Unconventional Platforms

Inspired by Thompson's experiment, researchers began to use EAs to exploit physical properties of reconfigurable devices for computing. The aim was to obtain useful computations directly in materials, without formulating a mathematical model of the problem and without understanding "the rules of the game." In addition to common FPGAs and reconfigurable analog devices mentioned in previous sections, several platforms were proposed for these experiments, for example, an evolvable motherboard (Layzell 1998), RISA (Greensted and Tyrrell 2007) and REPOMO32 (Sekanina et al. 2009).

Recently, Miller has used the more general term, *evolvable matter*, to address the use of evolutionary algorithms for the design on any physical reconfigurable platform (e.g., chemical) in a real environment. The idea behind the concept is that applied voltages may induce physical changes that interact in unexpected ways with other distant voltage-inducted configurations in a rich physical substrate. In other words, it should theoretically be possible to perform the evolution directly *in materio* if the platform is configurable in some way. Promising technology for evolution in materio was reviewed in (Harding et al. 2008). In particular, liquid crystals were used as a platform to evolve various circuits including primitive logic functions, signal discriminators, and robot controllers (Harding 2006).

Tour's group has presented another example – the Nanocell (Tour 2003). The nanocell is a 2D network of self-assembled metallic particles connected by molecular switches (❯ *Fig. 26*). The nanocell is surrounded by a small number of lithographically defined access leads. The nanocell is not constructed as a specific logic gate – the logic is created in the nanocell by

❑ Fig. 26
Molecular switch used in NanoCell and its V/I characteristics (from Tour 2003).

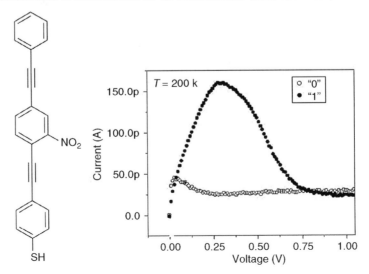

training it postfabrication by changing the states of the molecular switches. The training algorithm does not know the connections within the nanocell or the locations of the switches. However, the configuration can be changed by voltage pulses applied to the I/O pins. A genetic algorithm was utilized to generate these pulses in order to form the required logic gates.

Bartels et al. (2004) applied the evolutionary algorithm to configure a reconfigurable mirror used in a laser. The aim was to shape laser pulses in order to distort molecules in specific ways to catalyze chemical reactions, with the ultimate goal of manipulating large molecules, for example, proteins and enzymes. The method has also been used to create new quantum behaviors at the atomic level. Results completely unexpected from a physical point of view were reported. Evolved behaviors (such as anti-correlated attosecond harmonics in quantum systems) were not known to be physically possible. The ability to move beyond the nanoscale to the attoscale is considered as a major breakthrough. Potential applications of controlling the behavior of materials at the atomic level are enormous.

6 Adaptive Hardware

A promising approach to building adaptive hardware is to develop systems that combine a reconfigurable device with an intelligent mechanism for its reconfiguration. The reconfiguration mechanism is activated when the adaptation is needed. A straightforward solution for the adaptive system is to maintain a set of configurations designed a priori and use the most suitable one when the current configuration becomes unsatisfactory. An obvious weakness of this approach is that a suitable configuration may not be prepared. A more advanced approach, developed in the evolvable hardware field, tries to generate new configurations at run time using an evolutionary algorithm. A potential problem is that the EA is a stochastic algorithm. Nobody can guarantee that the resulting configuration really solves the problem. In addition, nobody can guarantee that the EA will provide an acceptable configuration on time. However, in many applications, a partial solution provided by the EA may be a very good result.

Examples of adaptive hardware systems will be presented according to the types of changes that usually lead to the requirement of system adaptation. These changes can be classified either as changes

- in the hardware platform itself (i.e., faults)
- in the input data characteristics, or
- in the specification

6.1 Self-repairing Hardware

The EA is used to recover the system function that was degraded/lost due to failure in the hardware platform (such as being stuck at zero, changes in the transistors' characteristics in extreme environments, etc.). Note that the requirement on the function remains unchanged and the input data are as expected. First of all, the failure has to be detected. Then, if no other mechanisms are capable of repairing the system, the EA is activated. In some cases, it is sufficient to tune only some circuit parameters (e.g., bias voltages) in order to repair the circuit. In more complicated cases, a new topology must be evolved.

6.1.1 SRAA

The self-reconfigurable analog array (SRAA) developed at NASA JPL provides the capability of *continual temperature compensation* (Stoica et al. 2008). The SRAA contains an array of 4×6 analog cells. Various specific circuits can be implemented in the SRAA for which mission-oriented ASICs were designed in the past. The SRAA cells are based on tunable OTAs used to compensate for the effects that come from temperature variations. A subset of cells, called reference cells, are continually compensated at the same time as other functional cells perform user operations. The result of compensation is then transferred to the main array. Elementary cells can be compensated for temperature-induced deviations within a $-180°C$ to $120°C$ range. The EA is implemented as a digital circuit either in an FPGA or as ASIC.

❷ *Figure 27* shows the SRAA chip that contains the array of cells divided into the functional and the reference analog arrays. The reference analog cells are individually probed to check for degradation. Both reference and functional cells include digital-to-analog converters (DACs) used to provide bias voltages or currents to tune the analog cell response. The switch box allows for the configuration of different functions in the functional analog array, as well as providing an interface between the reference analog cells and the test fixture. The SRAA contains 40 reconfiguration DACs for the 6×4 functional array.

The SRAA utilizes a hierarchical approach to calibration/compensation. First, a faulty configuration is replaced by a priori prepared configuration. If it fails, a second level takes place, which involves a less computationally expensive search algorithm. Finally, a third level is employed in case the previous levels do not succeed, and a global search is conducted using evolutionary algorithms.

Once an analog cell response goes out of specification (e.g., due to temperature or radiation effects), the FPGA will recalibrate the corresponding reference cell by reprogramming its

◨ **Fig. 27**
Self-reconfigurable analog array (SRAA) according to Stoica et al. (2008).

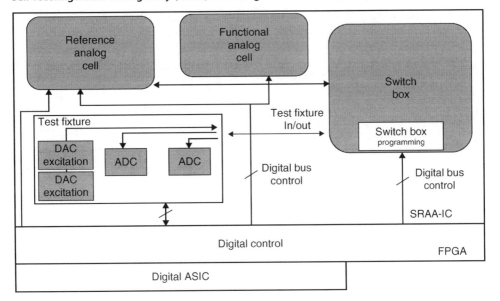

respective reconfiguration DACs. Once new values are found that recover the behavior of the reference cell, the same are used to reprogram the configuration DACs of the associated functional cells. This tuning process assumes that for each type of cell, the functional and reference versions will exhibit the same behavior. This assumption usually holds when the SRAA chip is exposed to extreme temperatures. Since the chip is fabricated in a rad-hard process, the SRAA will tolerate permanent radiation effects (total ionization dose [TID]) up to 300 krad. The SRAA is designed to maintain a stable operation over a wide temperature range of more than 300°C, from −180°C to 125°C. Over this range, parameters of interest may vary with < 1–5% deviation from their 27°C value, depending on the circuit.

6.1.2 Self-repairing FPGAs

An interesting evolutionary-repair approach for FPGAs was introduced by Garvie (2005). He employed triple module redundancy (TMR), which is a conventional scheme used to mitigate faults. A TMR system has three copies of a module and uses a voting circuit on their outputs so that the final output is an agreement between at least two modules. A TMR extension, TMR+Scrubbing, in addition, provides fault tolerance to single event upsets (SEUs) in the FPGA configuration data by regular reprogramming of the FPGA. Garvie extended TMR +Scrubbing by "Jiggling" to repair also permanent damage by using the two healthy modules to repair the faulty one (Garvie 2005). He recognized that the two healthy modules can provide a reference signal (i.e., training vectors) for evolutionary design of the configuration of the third module. Once a Jiggling repair is complete three healthy modules are again available (❷ *Fig. 28*). Permanent faults can be repaired until spare resources are exhausted. The Jiggling repair evolutionary module is implemented on the same FPGA and requires a small additional area on the chip. Hence, it can be itself repairable by another Jiggler. Reliability analysis for small benchmark circuits shows the Jiggling system using 2.8 times the overhead per module can survive 17 times longer than a TMR/Simplex approach.

6.2 Changes in the Specification or Input Data Characteristics

Consider an image-recognition system in which an image filter is used to enhance the input images coming from a camera. If the type of noise is variable, the filter may have problems

❑ **Fig. 28**
Triple module redundancy extended by a module repairing mechanism (Garvie 2005).

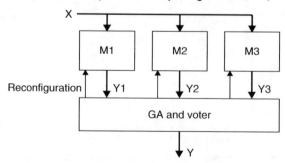

enhancing the images and the quality of recognition may decrease. Hence, the EA is employed to optimize the filter for a particular type of noise. In this case, the specification of the function remains unchanged; however, the system has to adapt because of changing environments. Also the specification can be changed either by an operator or autonomously, by the system itself which can, for example, continuously monitor its history and plan new actions (considered as circuit configurations). If the configuration that implements the required action is unknown in advance, the EA is activated to perform the adaptation. Several real-world applications developed in the AIST laboratory in Japan (Higuchi et al. 1999, 2006) in which this kind of adaptation has been employed will be shown.

6.2.1 Image Compression Chip

Adaptive image compression methods allow for modifying the compression algorithm according to the particular data that has to be compressed. In consequence, a higher compression ratio can be achieved. The AIST team has developed an evolvable chip for lossless image compression, which is suitable for electrophotographic printers (❖ *Fig. 29*) (Sakanashi et al. 2001). As the size of a typical image printed using this printer goes from tens to hundreds of MBs, a compression algorithm must be used to reduce the image size as well as the time needed to read the images from a hard disk. The conventional algorithm (JBIG) is based on predicting the next pixel value using an MQ coder according to its (typically 16) neighbors. The positions of the neighbors are determined by a so-called template, which is invariable in the conventional algorithm. Since the evolvable chip is able to find the optimal positions of pixels in the template for each block of the image, the predictions are better than in the conventional implementation and hence the compression rate is higher. The EA is executed for each image block separately with the aim of maximizing the compression ratio for the block. The incoming blocks can in fact be viewed as a changing environment for the prediction algorithm. The developed chip operates at 133 MHz and is 7 times faster than the reference implementation running on a Pentium 4.2 GHz processor. The compression rate is 1.7 times better than that achieved by the standard JBIG2 algorithm (Sakanashi et al. 2001, 2006). This is a very good result for lossless compression.

□ Fig. 29

Image compression chip developed at AIST that adapts pixel positions for prediction (Sakanashi et al. 2001).

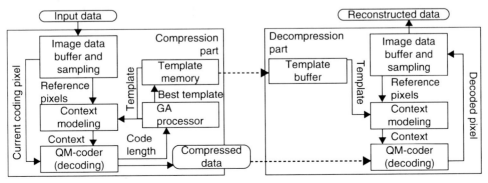

6.2.2 Adaptive Controllers

Another evolvable hardware chip from AIST was designed for a myoelectric prosthetic hand. The myoelectric hand is controlled by the signals generated with muscular movements (electromyography, EMG signals). It takes a long time, usually almost 1 month, before a disabled person is able to control a multifunction prosthetic hand freely using a fixed controller. This chip allows the myoelectric hand to adapt itself to the disabled person and thus significantly reduce the training period to several minutes (❯ *Fig. 30*).

The myoelectric hand described in Kajitani et al. (2001, 2006) is able to perform the six actions, which are paired (open–grasp, supination–pronation, and flexion–extension), with a separate motor control for each pair. The task for the evolvable hardware is to synthesize a circuit that can map input patterns (feature vectors extracted from EMG signals) to the desired actions of the hand.

The training procedure is as follows: First, training pattern data is obtained in an online mode, where the hand user makes one of the six actions from which ten EMG signal patterns are measured. For all six actions, a set of 60 signals is obtained as training data. Then, based on this obtained training data, a classifier is evolved in the PLA (16-bit input/6-bit output), which is available in the chip. GA is implemented on the same chip (partly in an on-chip processor and partly as a special circuit). The same chip was also used for an adaptive robot controller (Keymeulen et al. 1997).

First implementations of the chip achieved 81% classification accuracy. An experienced user is able to achieve 98% accuracy on the recent version of the chip, which uses more complex data preprocessing. The classifier evolution in hardware is 40 times faster than a 1.2 GHz Athlon processor (Kajitani et al. 2006). An incremental evolution-based scheme for

■ Fig. 30

AIST evolvable hardware chip used to implement adaptive controllers and classifiers (according to Kajitani et al. 1998).

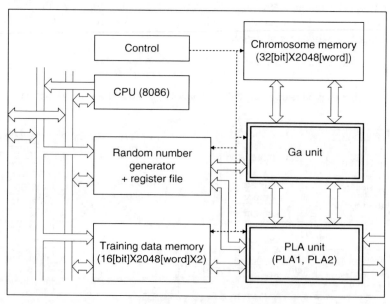

EMG signal classification achieving even a higher classification accuracy was also proposed (Glette et al. 2008).

6.2.3 Post-fabrication Adaptation

Post-fabrication calibration of ASICs can also be considered as a kind of hardware adaptation. Fabrication variance in the manufacture of chips can cause problems in some high-performance applications, especially when a cutting-edge manufacturing technology is employed. In other words, only some of the fabricated chips really meet the specification. Hence, an EA is used to tune selected circuit parameters with the aim of obtaining required circuit function even for the chips that do not work because of variance of the fabrication process. Although some area of the chip has to be used for implementing the reconfiguration logic and controller, the overall benefits are significant: It is possible to fabricate simpler, smaller, and low-power designs that do not contain large circuits needed to compensate fabrication variances, and thus increase the yield.

A typical chip in which the post-fabrication tuning was successfully applied is an intermediate filter (IF), commonly found in mobile telephones (Murakawa et al. 1998). In this case, a 1% discrepancy from the center frequency (455 kHz) is unacceptable. Chips that do not meet this requirement must be discarded lowering the yield. The AIST team used GA to control 18 OTAs that properly tune currents to compensate for the errors introduced by fabrication variances (see ❷ Fig. 31). The GA operates with a 20-member population and 234-bit chromosomes. The tuning is usually complete after 100 generations, which takes a few seconds. In the experiment, 29 out of 30 fabricated chips were successfully tuned. The method allowed us to reduce the area by 9% and the power consumption by 26%. The recent generation of the chip reduces the area by 49% (with respect to existing commercial designs) (Murakawa et al. 2006).

◼ **Fig. 31**
Tunable intermediate filter composed of programmable OTAs (from Murakawa et al. 1998).

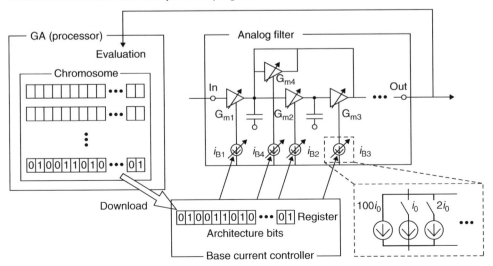

The same concept was used to compensate for out-of-specification timing delays between circuits, or "clock skew." AIST devised a clock timing adjustment architecture to eliminate the clock skew problem by using a genetic algorithm to make the clock timings perform within the intended design specifications. Simulation results showed yields rising from 2.9% to 51.1% using evolved clock-timing circuits (Takahashi et al. 2006).

A similar approach was applied to tune signal transmitters (of USB/IEEE 1394 standards) in order to compensate signal distortions at the end of the cable (Kasai et al. 2005). For a particular cable and its placement, the GA-based tuning of the transmitter has successfully achieved a transmission speed of 1.6 GHz, which is four times faster than the current IEEE1394 standard (400 MHz), and a cable length of 21 m, which is four times longer than the current USB standard (5 m).

The evolvable data-compression chip, evolvable controller, as well as the chips that enable post-fabrication parameters tuning were commercialized in Japan. The post-fabrication tuning seems to be the most attractive application of evolvable hardware from the commercial point of view.

7   Other Areas

Evolutionary design has been applied in many hardware-oriented areas, significantly extending the original intention of the evolvable hardware research – electronic circuit evolution and adaptation. For example, in evolutionary robotics, robot controllers are evolved together with robot physical bodies (Lipson and Pollack 2000). The specialized evolutionary robotics literature provides more details, for example, Nofli and Floreano (2000) and Gross et al. (2006). Evolving physical designs have become popular in the aerospace industry (Terrile et al. 2005). Other evolutionary design applications are well covered in Bentley (1999) and Bentley and Corne (2002).

Several research teams have developed adaptive and fault-tolerant research platforms inspired by embryonic processes (Durbeck and Macias 2001; Mange et al. 2000; Moreno et al. 2005; Tufte and Haddow 2005). These platforms implement arrays of artificial cells, which are able to divide (i.e., copy the configuration to neighboring cells) and differentiate (i.e., interpret the configuration according to, for example, location of the cell). These platforms contain various mechanisms for fault detection, localization, and repair. Circuit recovery is implemented using reconfiguration of spare cells and bridging over faulty cells. Fault-tolerance mechanisms are implemented hierarchically providing thus a really robust solution for digital circuits. These approaches will be useful for the massively parallel architectures that we expect in future with the progress in nanotechnology. However, the circuit overhead connected with the implementation of fault-tolerant mechanisms is impractically huge for current technology.

Finally, research is concentrated on the biological aspects and possibilities offered by modeling biological systems in hardware. This includes building biologically inspired structures such as virtual cells, artificial immune systems, and artificial brains using evolvable hardware. Research in this area focuses primarily on biological mechanisms, which can eventually lead to techniques for improved hardware design. Haddow speculates that modeling biological networks can be a killer-application for evolvable hardware (Haddow 2008).

8 Conclusions

The field of evolvable hardware has been surveyed. The status of its main subareas can be now summarized.

Digital circuits evolved so far contain from several gates to millions of gates. It is typical for the evolution of small circuits that all possible input vectors are used in the fitness function and the aim is to improve circuit parameters (area, delay) with respect to existing designs. Evolved mid-size circuits such as image filters or classifiers contain thousands of gates. As the training set has to be used in the fitness function, candidate circuits are evaluated imperfectly, that is, they may not work perfectly for some of the unseen input patterns. For million-component circuits evolved so far, requirements neither on function nor on size were specified; only some structural properties of circuits were evaluated. It can be stated that in all the circuit classes, the chromosomes have similar size, typically around one or two thousand bits. It seems that larger search spaces cannot be sought effectively using existing search algorithms and computational resources. In order to evolve a circuit larger than a small multiplier, a domain knowledge method has to be employed to define suitable components, possibilities of interconnection, and fitness function. Real-world circuits are usually evolved using function-level representation or incremental evolution. Although many developmental encodings were proposed and investigated to overcome the scalability problem, no really innovative designs have been obtained using developmental schemes so far.

In contrast to digital circuit design, analog circuit design requires more experience and intuition. Despite this fact, John Koza and his team have demonstrated high competitiveness of evolutionary algorithms (in particular, genetic programming) in the area of analog circuit synthesis. An important step toward evolving really trustworthy analog designs was done by McConaghy. Evolutionary design is particularly good at exploiting properties of physical systems in order to achieve target behaviors. This unique feature is important for fast tuning of analog circuits that are implemented in reconfigurable devices and operated in a real environment. Successful results were demonstrated for FPTAs running in various environments and FPAAs working as controllers. Similarly, it was demonstrated that design for other reconfigurable devices (e.g., antennas, optical systems, and nanosystems, including unconventional platforms such as liquid crystals and nanoparticles) can significantly benefit from the evolutionary approach.

The main contribution of the evolutionary hardware design can be seen in the areas where it is impossible to provide a perfect specification for target implementations, and conventional design methods are based on experience and intuition rather than on a fully automated methodology. Then, the EA can explore the "dark corners" of design spaces that humans have left unexplored. Also, the existence of a fast simulator for a particular domain is important for successful evolutionary designs.

Only few systems exist in which autonomous hardware adaptation is achieved using evolutionary computation. The reason is that there are some crucial problems associated with the use of EAs in these applications, namely, a solution is not always obtained in real-time, the quality of a solution may not be acceptable, and it is difficult/impossible for many applications to obtain training data during system operation. A typical adaptive/evolvable system uses an EA to optimize only a few parameters. In comparison to evolutionary circuit design, chromosomes are relatively short and the task is not difficult for EA (since the EA has to provide a sufficient solution almost anytime). However, with the development of reconfigurable technology, we can expect a boom in adaptive systems that will be tuned for a particular location, user, or time of operation. The evolutionary approach seems to be very promising in achieving the adaptive behavior.

Finally, evolvable hardware has an impact on the theory of computation. The evolved computational devices represent a distinct class of devices that exhibits a specific combination of properties, not visible and studied in the scope of all computational devices till now. Devices that belong to this class show the required behavior; however, in general, we do not understand how and why they perform the required computation. The reason is that the evolution can utilize, in addition to the "understandable composition of elementary components," material-dependent constructions and properties of environment (such as temperature, electromagnetic field, etc.) and, furthermore, unknown physical behaviors to establish the required functionality. Therefore, nothing is known about the mapping between an abstract computational model and its physical implementation. The standard notion of computation and implementation developed in computer science as well as in cognitive science has become very problematic with the existence of evolved computational devices (Sekanina 2007).

It can be concluded that evolvable hardware as a field is maturing. Real-world applications of evolutionary hardware synthesis as well as evolutionary-based adaptive hardware exist and bring advantages when compared to conventional solutions.

Acknowledgments

This work was partially supported by the Grant Agency of the Czech Republic under No. 102/07/0850 *Design and hardware implementation of a patent-invention machine* and the Research Plan No. MSM 0021630528 *Security-Oriented Research in Information Technology.*

References

Aggarwal V, Mao M, O'Reilly UM (2006) A self-tuning analog proportional-integral-derivative (PID) controller. In: AHS '06: Proceedings of the first NASA/ESA conference on adaptive hardware and systems. IEEE Computer Society, Washington, DC, USA, pp 12–19

Ali B, Almaini AEA, Kalganova T (2004) Evolutionary algorithms and their use in the design of sequential logic circuits. Genet Programming Evol Mach 5(1):11–29

Anadigm (2007) Anadigm, AN221E04 – field programmable analog arrays – user manual. URL http://www.anadigm.com_doc/UM021200-U007.pdf

Bartels RA, Murnane MM, Kapteyn HC, Christov I, Rabitz H (2004) Learning from learning algorithms: applications to attosecond dynamics of high-harmonic generation. Phys Rev A 70(1):1–5

Bentley PJ (ed) (1999) Evolutionary design by computers. Morgan Kaufmann, San Francisco, CA

Bentley PJ, Corne DW (2002) Creative evolutionary systems. Morgan Kaufmann, San Francisco, CA

Bernardi P, Sanchez E, Schillaci M, Squillero G, Reorda MS (2008) An effective technique for the automatic generation of diagnosis-oriented programs for processor cores. IEEE Trans Comput-Aided Des Integr Circuits Syst 27(3):570–574

Bidlo M, Skarvada J (2008) Instruction-based development: from evolution to generic structures of digital circuits. Int J Knowl-Based Intell Eng Syst 12(3):221–236

Blodget B, James-Roxby P, Keller E, McMillan S, Sundararajan P (2003) A self-reconfiguring platform. In: Proceedings of the 13th conference on field programmable logic and applications FPL'03, Lisbon, Portugal, LNCS, vol 2778. Springer Verlag, pp 565–574

Drechsler R (1998) Evolutionary algorithms for VLSI CAD. Kluwer Academic Publishers, Boston

Durbeck L, Macias N (2001) The cell matrix: an architecture for nanocomputing. Nanotechnology 12(3):217–230

Eiben AE, Smith JE (2003) Introduction to evolutionary computing. Springer, Berlin

Erba M, Rossi R, Liberali V, Tettamanzi A (2001) An evolutionary approach to automatic generation of VHDL code for low-power digital filters. In: Proceedings of the 4th European conference on genetic programming EuroGP2001, LNCS, vol 2038. Springer Verlag, Berlin, pp 36–50

Gao P, McConaghy T, Gielen G (2008) ISCLES: importance sampled circuit learning ensembles for trustworthy analog circuit topology synthesis. In: Proceedings of the 8th international conference on evolvable systems: from biology to hardware. LNCS, vol 5216. Springer Verlag, Berlin, pp 11–21

de Garis H (1993) Evolvable hardware – genetic programming of a Darwin Machine. In: International conference on artificial neural networks and genetic algorithms, Innsbruck, Austria. Springer Verlag

Garvie M (2005) Reliable electronics through artificial evolution. PhD thesis, University of Sussex

Glette K (2008) Design and implementation of scalable online evolvable hardware pattern recognition systems. PhD thesis, University of Oslo

Glette K, Torresen J, Yasunaga M (2007) An online EHW pattern recognition system applied to sonar spectrum classification. In: Evolvable systems: from biology to hardware, LNCS, vol 4684. Springer Verlag, pp 1–12

Glette K, Torresen J, Gruber T, Sick B, Kaufmann P, Platzner M (2008) Comparing evolvable hardware to conventional classifiers for electromyographic prosthetic hand control. In: Proceedings of the 2008 NASA/ESA conference on adaptive hardware and systems, Noordwijk. IEEE Computer Society, pp 32–39

Gordon T (2005) Exploiting development to enhance the scalability of hardware evolution. PhD thesis, Department of Computer Science, University College, London

Gordon TGW, Bentley PJ (2002) Towards development in evolvable hardware. In: Proceedings of the 2002 NASA/DoD conference on evolvable hardware. IEEE Computer Society Press, Washington, DC, pp 241–250

Greensted A, Tyrrell A (2007) RISA: a hardware platform for evolutionary design. In: Proceedings of 2007 IEEE workshop on evolvable and adaptive hardware, Long Beach, CA. IEEE, pp 1–7

Greenwood G, Tyrrell AM (2007) Introduction to evolvable hardware. IEEE Press, Los Alamitos, CA

Gross R, Bonani M, Mondada F, Dorigo M (2006) Autonomous self-assembly in Swarm-Bots. IEEE Trans Robot 22(6):1115–1130

Gwaltney D, Dutton K (2005) A VHDL core for intrinsic evolution of discrete time filters with signal feedback. In: Proceedings of the 2005 NASA/DoD conference on evolvable hardware. IEEE Computer Society, Washington, DC, USA, pp 43–50

Haddow PC (2008) Evolvable hardware: a tool for reverse engineering of biological systems. In: Proc. of the 8th int. conference on evolvable systems: from biology to hardware. LNCS, vol 5216. Springer Verlag, Berlin, pp 342–351

Harding S (2008) Evolution of image filters on graphics processor units using Cartesian genetic programming. In: 2008 IEEE world congress on computational intelligence. IEEE CIS, Hong Kong, pp 1921–1928

Harding SL (2006) Evolution in materio. Ph.D. thesis, University of York

Harding SL, Miller JF, Rietman EA (2008) Evolution in materio: exploiting the physics of materials for computation. J Unconventional Comput 4(2):155–194

Harris SP, Ifeachor EC (1995) Automating IIR filter design by genetic algorithm. In: Proceedings of the first IEE/IEEE international conference on genetic algorithms in engineering systems: innovations and applications (GALESIA'95), vol 414. IEE, London, pp 271–275

Hauck S, DeHon A (2008) Reconfigurable computing: the theory and practice of FPGA-based computation. Morgan Kaufmann, Seattle, WA

Haupt RL, Werner DH (2007) Genetic algorithms in electromagnetics. Wiley-IEEE Press, Hoboken, NJ

Henrici F, Becker J, Buhmann A, Ortmanns M, Manoli Y (2007) A continuous-time field programmable analog array using parasitic capacitance GM-C filters. In: Proceedings of the IEEE international symposium on circuits and systems. IEEE New Orleans, LA, pp 2236–2239

Higuchi T, Iwata M, Keymeulen D, Sakanashi H, Murakawa M, Kajitani I, Takahashi E, Toda K, Salami M, Kajihara N, Otsu N (1999) Real-world applications of analog and digital evolvable hardware. IEEE Trans Evolut Comput 3(3):220–235

Higuchi T, Liu Y, Yao X (2006) Evolvable hardware. Springer, Berlin

Higuchi T, Niwa T, Tanaka T, Iba H, de Garis H, Furuya T (1993) Evolving hardware with genetic learning: a first step towards building a Darwin machine. In: Proceedings of the 2nd international conference on simulated adaptive behavior. MIT Press, Cambridge, MA, pp 417–424

Hornby G, Globus A, Linden D, Lohn J (2006) Automated antenna design with evolutionary algorithms. In: Proceedings 2006 AIAA Space Conference. AIAA, San Jose, CA, pp 1–8

Hounsell BI, Arslan T, Thomson R (2004) Evolutionary design and adaptation of high performance digital filters within an embedded reconfigurable fault tolerant hardware platform. Soft Comput 8(5):307–317

Huelsbergen L, Rietman E, Slous R (1999) Evolving oscillators in silico. IEEE Trans Evolut Comput 3(3):197–204

Ifeachor E, Jervis B (2002) Digital signal processing: a practical approach (2nd edn). Prentice-Hall Upper Saddle River, NJ

Kajitani I, Hoshino T, Nishikawa D, Yokoi H, Nakaya S, Yamauchi T, Inuo T, Kajihara N, Iwata M, Keymeulen D, Higuchi T (1998) A gate-level EHW chip: implementing GA operations and reconfigurable hardware on a single LSI. In: Proceedings of the 2nd International conference on evolvable systems: from biology to hardware ICES' 98, Lausanne, Switzerland, LNCS, vol 1478. Springer, pp 1–12

Kajitani I, Iwata M, Higuchi T (2006) A GA hardware engine and its applications. In: Higuchi T, Liu Y, Yao X (eds) Evolvable hardware, Springer, Heidelberg, pp 41–63

Kajitani I, Sekita I, Otsu N, Higuchi T (2001) Improvements to the action decision rate for a multifunction prosthetic hand. In: The first international symposium on measurement, analysis and modeling of human functions. Sapporo, pp 84–89

Kamalian R, Zhou N, Agogino M (2002) A comparison of MEMS synthesis techniques. In: Proceedings of the 1st Pacific Rim Workshop on Transducers and Micro/Nano Technologies. Xiamen, China, pp 239–242

Kasai Y, Takahashi E, Iwata M, Iijima Y, Sakanashi H, Murakawa M, Higuchi T (2005) Adaptive waveform control in a data transceiver for multi-speed IEEE 1394 and USB communication. In: Evolvable systems: from biology to hardware, 6th International conference, ICES 2005, Sitges, Spain, LNCS, vol 3637. Springer, Berlin, 198–204

Keymeulen D, Durantez M, Konaka K, Kuniyoshi Y, Higuchi T (1997) An evolutionary robot navigation system using a gate-level evolvable hardware. In: Proceedings of the 1st International conference on evolvable systems: from biology to hardware ICES'96, LNCS, vol 1259. Tsukuba, Japan, Springer, Berlin, pp 195–209

Keymeulen D, Ferguson MI, Breuer L, Fink W, Oks B, Peay C, Terrile R, Kim Y-CD, MacDonald E, Foor D (2006) Hardware platforms for electrostatic tuning of MEMS gyroscope using nature-inspired computation. In: Higuchi T, Liu Y, Yao X (eds) Evolvable Hardware. Springer, Berlin, pp 209–222

Keymeulen D, Zebulum R, Jin Y, Stoica A (2000) Fault-tolerant evolvable hardware using field-programmable transistor arrays. IEEE Trans Reliability 49(3):305–316

Kitano H (1999) Morphogenesis for evolvable systems. In: Towards evolvable hardware: the evolutionary engineering approach. LNCS, vol 1062. Springer, Berlin, pp 99–117

Koza JR, Al-Sakran SH, Jones LW (2005) Automated re-invention of six patented optical lens systems using genetic programming. In: GECCO'05:

Proceedings of the 2005 conference on genetic and evolutionary computation. ACM, New York, NY, USA, pp 1953–1960

Koza JR, Bennett FH, Andre D, Keane MA (1999) Genetic programming III: Darwinian invention and problem solving. Morgan Kaufmann Publishers, San Francisco, CA

Koza JR, Keane MA, Streeter MJ, Mydlowec W, Yu J, Lanza G (2003) Genetic programming IV: routine human-competitive machine intelligence. Kluwer Academic Publishers, Norwell, MA

Langeheine J (2005) Intrinsic hardware evolution on the transistor level. Ph.D. thesis, Rupertus Carola University of Heidelberg

Larsson E (2005) Introduction to advanced system-on-chip test design and optimization. Springer, Dordrecht

Layzell PJ (1998) A new research tool for intrinsic hardware evolution. In: Proceedings of the evolvable systems: from biology to hardware conference. LNCS, vol 1478. Springer, Lausanne, Switzerland pp 47–56

Li H, Antonsson EK (1998) Genetic algorithms in MEMS synthesis. In: Proceedings of IMECE'98 1998 ASME International mechanical engineering congress and expositions, Anaheim, CA

Linden D (1997) Automated design and optimization of antennas using genetic algorithms. PhD thesis, MIT Cambridge

Linden DS (2001) A system for evolving antennas in-situ. In: EH'01: Proceedings of the 3rd NASA/DoD workshop on evolvable hardware, IEEE Computer Society, Washington, DC, USA, pp 249–255

Lipson H, Pollack JB (2000) Automatic design and manufacture of robotic lifeforms. Nature 406:974–978

Lohn JD, Hornby GS (2006) Evolvable hardware: using evolutionary computation to design and optimize hardware systems. IEEE Computat Intell Mag 1(1):19–27

Lohn JD, Kraus WF, Hornby GS (2007) Automated design of a MEMS resonator. In: Proceedings of the IEEE congress on evolutionary computation, Singapore, pp 3486–3491

Loktev M, Soloviev O, Vdovin G (2003) Adaptive optics – product guide. OKO Technologies, Delft

Mange D, Sipper M, Stauffer A, Tempesti G (2000) Towards robust integrated circuits: the embryonics approach. Proc IEEE 88(4):516–541

Martinek T, Sekanina L (2005) An evolvable image filter: experimental evaluation of a complete hardware implementation in FPGA. In: Evolvable systems: from biology to hardware, LNCS, vol 3637. Springer Verlag, Sitges, Spain, pp 76–85

Miller J, Job D, Vassilev V (2008) Principles in the evolutionary design of digital circuits – Part I. Genet programming evol Mach 1(1):8–35

Miller J, Thomson P (2000) Cartesian genetic programming. In: Proceedings of the 3rd European conference on genetic programming EuroGP2000, LNCS, vol 1802. Springer, Edinburgh, Scotland, pp 121–132

Moreno JM, Eriksson J, Iglesias J, Villa AEP (2005) Implementation of biologically plausible spiking neural networks models on the poetic tissue. In: Proceedings of evolvable systems: from biology to hardware, Sitges, Spain, LNCS, vol 3637. Springer, pp 188–197

Murakawa M, Kasai Y, Sakanashi H, Higuchi T (2006) Evolvable analog lSI. In: Higuchi T, Liu Y, Yao X (eds) Evolvable hardware. Springer, Berlin, pp 121–143

Murakawa M, Yoshizawa S, Kajitani I, Furuya T, Iwata M, Higuchi T (1996) Evolvable hardware at function level. In: Parallel problem solving from nature PPSN IV. LNCS, vol 1141. Springer, Berlin, pp 62–71

Murakawa M, Yoshizawa S, Adachi T, Suzuki S, Takasuka K, Iwata M, Higuchi T (1998) Analogue EHW chip for intermediate frequency filters. In: Evolvable systems: from biology to hardware, second International conference, ICES 98, Lausanne, Switzerland, LNCS, vol 1478. Springer, Heidelberg, pp 134–143

Nedjah N, de Macedo Mourelle L (2005) Evolutionary synthesis of synchronous finite state machines. In: Nedjah N, de Macedo Mourelle L (eds) Evolvable machines: theory and practice. Springer, Berlin, pp 103–127

Nofli S, Floreano D (2000) Evolutionary robotics: the biology, intelligence, and technology of self-organizing machines. MIT Press/Bradford Books, Cambridge, MA

Nosato H, Murakawa M, Kasai Y, Higuchi T (2006) Evolvable optical systems. In: Higuchi T, Liu Y, Yao X (eds) Evolvable Hardware. Springer, Heidelberg, pp 200–207

Pecenka T, Sekanina L, Kotasek Z (2008) Evolution of synthetic RTL benchmark circuits with predefined testability. ACM Trans Des Autom Electron Syst 13(3):1–21

Sakanashi H, Iwata M, Higuchi T (2001) A lossless compression method for halftone images using evolvable hardware. In: Evolvable systems: from biology to hardware, 4th International conference, ICES 2001 Tokyo, Japan. LNCS, vol 2210. Springer, Berlin, pp 314–326

Sakanashi H, Iwata M, Higuchi T (2006) EHW applied to image data compression. In: Higuchi T, Liu Y, Yao X (eds) Evolvable hardware. Springer, Berlin, pp 19–40

Salomon R, Widiger H, Tockhorn A (2006) Rapid evolution of time-efficient packet classifiers. In: IEEE congress on evolutionary computation, IEEE CIS, Vancouver, Canada, pp 2793–2799

Sekanina L (2003) Virtual reconfigurable circuits for real-world applications of evolvable hardware. In: Evolvable systems: from biology to hardware, fifth international conference, ICES 2003. LNCS, vol 2606. Springer, Trondheim, Norway, pp 186–197

Sekanina L (2004) Evolvable components: from theory to hardware implementations. Natural Computing, Springer Verlag, Berlin

Sekanina L (2007) Evolved computing devices and the implementation problem. Minds Mach 17(3):311–329

Sekanina L, Bidlo M (2005) Evolutionary design of arbitrarily large sorting networks using development. Genet Programming Evol Mach 6(3):319–347

Sekanina L, Friedl S (2004) An evolvable combinational unit for FPGAs. Comput Informatics 23(5):461–486

Sekanina L, Ruzicka R, Vasicek Z, Prokop R, Fujcik L (2009) Repomo32 – new reconfigurable polymorphic integrated circuit for adaptive hardware. In: Proceedings of 2009 IEEE workshop on evolvable and adaptive hardware. IEEE CIS, Nashville, TN, pp 39–46

Sekanina L, Starecek L, Kotasek Z, Gajda Z (2008) Polymorphic gates in design and test of digital circuits. Int J Unconventional Comput 4(2):125–142

Stoica A, Keymeulen D, Zebulum RS, Guo X (2006) Reconfigurable electronics for extreme environments. In: Higuchi T, Liu Y, Yao X (eds) Evolvable hardware. Springer, Heidelberg, pp 145–160

Stoica A, Keymeulen D, Zebulum RS, Katkoori S, Fernando P, Sankaran H, Mojarradi M, Daud T (2008) Self-reconfigurable mixed-signal integrated circuits architecture comprising a field programmable analog array and a general purpose genetic algorithm IP core. In: Evolvable systems: from biology to hardware, 8th International conference, ICES 2008. LNCS, vol 5216. Springer, Prague, pp 225–236

Stoica A, Wang X, Keymeulen D, Zebulum RS, Ferguson MI, Guo X (2005) Characterization and recovery of deep sub micron (DSM) technologies behavior under radiation. In: 2005 IEEE Aerospace Conference. IEEE, Montana, pp 1–9

Stoica A, Zebulum R, Keymeulen D (2000) Mixtrinsic evolution. In: Proceedings of the 3rd International conference on evolvable systems: from biology to hardware ICES'00, Edinburgh, Scotland, UK, LNCS, vol 1801. Springer, pp 208–217

Stoica A, Zebulum RS, Keymeulen D (2001) Polymorphic electronics. In: Proceedings of evolvable systems: from biology to hardware conference, LNCS, vol 2210. Springer, Tokyo, Japan, pp 291–302

Stoica A, Zebulum RS, Ferguson MI, Keymeulen D, Duong V (2002a) Evolving circuits in seconds: experiments with a stand-alone board-level evolvable system. In: Proceedings of the 2002 NASA/DoD conference on evolvable hardware (EH'02). IEEE Computer Society, Washington, DC, USA, pp 67–64

Stoica A, Zebulum RS, Keymeulen D, Lohn J (2002b) On polymorphic circuits and their design using evolutionary algorithms. In: Proceedings of IASTED international conference on applied informatics AI2002. Insbruck, Austria

Stoica A, Zebulum R, Guo X, Keymeulen D, Ferguson I, Duong V (2004a) Taking evolutionary circuit design from experimentation to implementation: some useful techniques and a silicon demonstration. IEE Proc Comp Digit Technol 151(4):295–300

Stoica A, Zebulum RS, Keymeulen D, Ferguson MI, Duong V, Guo X (2004b) Evolvable hardware techniques for on-chip automated reconfiguration of programmable devices. Soft Comput 8(5):354–365

Stomeo E, Kalganova T, Lambert C (2006) Generalized disjunction decomposition for evolvable hardware. IEEE Trans Syst Man Cybern Part B 36(5):1024–1043

Takahashi E, Kasai Y, Murakawa M, Higuchi T (2006) Post-fabrication clock-timing adjustment using genetic algorithms. In: Higuchi T, Liu Y, Yao X (eds) Evolvable hardware, Springer, Heidelberg, pp 65–84

Tempesti G, Mange D, Mudry PA, Rossier J, Stauffer A (2007) Self-replicating hardware for reliability: The embryonics project. JETC 3(2):1–21

Terrile R, Aghazarian H, Ferguson MI, Fink W, Huntsberger TL, Keymeulen D, Klimeck G, Kordon MA, Lee S, von Allmen P (2005) Evolutionary computation technologies for the automated design of space systems. In: 2005 NASA / DoD conference on evolvable hardware (EH 2005). IEEE Computer Society, Washington, DC, pp 131–138

Thompson A (1996) Silicon evolution. In: Proceedings of genetic programming GP'96. MIT Press, Cambridge, MA, pp 444–452

Thompson A (1999) Hardware evolution: automatic design of electronic circuits in reconfigurable hardware by artificial evolution. Springer, London

Thompson A, Layzell P, Zebulum S (1999) Explorations in design space: unconventional electronics design through artificial evolution. IEEE Trans Evolut Comput 3(3):167–196

Torresen J (1998) A divide-and-conquer approach to evolvable hardware. In: Proceedings of the 2nd International conference on evolvable systems: from biology to hardware ICES'98. LNCS, vol 1478. Springer, Lausanne, Switzerland, pp 57–65

Torresen J (2002) A scalable approach to evolvable hardware. Genetic programming and evolvable machines 3(3):259–282

Tour JM (2003) Molecular electronics. World Scientific, Singapore

Tufte G, Haddow P (2000) Evolving an adaptive digital filter. In: The second NASA/DoD workshop on evolvable hardware. IEEE Computer Society, Palo Alto, CA, pp 143–150

Tufte G, Haddow PC (2005) Towards development on a silicon-based cellular computing machine. Nat Comput 4(4):387–416

Upegui A (2006) Dynamically reconfigurable bio-inspired hardware. Ph.D. thesis, EPFL

Upegui A, Sanchez E (2006) Evolving hardware with self-reconfigurable connectivity in Xilinx FPGAs. In: The 1st NASA/ESA conference on adaptive hardware and systems (AHS-2006), IEEE Computer Society. Los Alamitos, CA, USA, pp 153–160

Vasicek Z, Sekanina L (2007a) An evolvable hardware system in Xilinx Virtex II Pro FPGA. Int J Innovative Comput Appl 1(1):63–73

Vasicek Z, Sekanina L (2007b) An area-efficient alternative to adaptive median filtering in FPGAs. In: Proceedings of 2007 conference on field programmable logic and applications. IEEE Computer Society, Los Alamitos, CA, pp 216–221

Vasicek Z, Sekanina L (2008) Hardware accelerators for Cartesian genetic programming. In: Proceedings of the 12th European conference on genetic programming, Naples, Italy. LNCS, vol 4971, pp 230–241

Vasicek Z, Zadnik M, Sekanina L, Tobola J (2008) On evolutionary synthesis of linear transforms in FPGA. In: Proceedings of the 8th International conference on evolvable systems: from biology to hardware. LNCS, vol 5216. Springer Verlag, Berlin, pp 141–152

Vassilev V, Job D, Miller J (2000) Towards the automatic design of more efficient digital circuits. In: Lohn J, Stoica A, Keymeulen D, Colombano S (eds) Proceedings of the 2nd NASA/DoD workshop on evolvable hardware. IEEE Computer Society, Los Alamitos, CA, USA, pp 151–160

Voronenko Y, Püschel M (2007) Multiplierless multiple constant multiplication. ACM Trans Algorithms 3(2):1–282

Wade G, Roberts A, Williams G (1994) Multiplier-less FIR filter design using a genetic algorithm. IEE Proc Vis Image Signal Process 141(3):175–180

Walker JA, Miller J (2008) The automatic acquisition, evolution and re-use of modules in Cartesian genetic programming. IEEE Trans Evolut Comput 12(4):397–417

Xilinx Inc. (2009) URL: http://www.xilinx.com

Zebulum R, Keymeulen D, Ramesham R, Sekanina L, Mao J, Kumar N, Stoica A (2006) Characterization and synthesis of circuits at extreme low temperatures. In: Higuchi T, Liu Y, Yao X (eds) Evolvable hardware. Springer, Berlin, pp 161–172

Zebulum R, Pacheco M, Vellasco M (2002) Evolutionary electronics – automatic design of electronic circuits and systems by genetic algorithms. The CRC Press International Series on Computational Intelligence. Boca Raton, FL

Zhan S, Miller JF, Tyrrell AM (2008) A developmental gene regulation network for constructing electronic circuits. In: Proceedings of the 8th international conference on evolvable systems: from biology to hardware. LNCS, vol 5216. Springer Verlag, Berlin, pp 177–188

Zhang Y, Smith S, Tyrrell A (2004) Intrinsic evolvable hardware in digital filter design. In: Applications of Evolutionary Computing, Coimbra, Portugal. LNCS, vol 3005. Springer Verlag, pp 389–398

51 Natural Computing in Finance – A Review

Anthony Brabazon[1] · *Jing Dang*[2] · *Ian Dempsey*[3] · *Michael O'Neill*[4] ·
David Edelman[5]

[1]Natural Computing Research and Applications Group, University
College Dublin, Ireland
anthony.brabazon@ucd.ie

[2]Natural Computing Research and Applications Group, University
College Dublin, Ireland
jing.dang@ucd.ie

[3]Pipeline Financial Group, Inc., New York, NY, USA
ian.dempsey@pipelinefinancial.com

[4]Natural Computing Research and Applications Group, University
College Dublin, Ireland
m.oneill@ucd.ie

[5]School of Business, UCD Michael Smurfit Graduate Business School,
Dublin, Ireland
david.edelman@ucd.ie

G. Rozenberg et al. (eds.), *Handbook of Natural Computing*, DOI 10.1007/978-3-540-92910-9_51,

Abstract

The field of natural computing (NC) has advanced rapidly over the past decade. One significant offshoot of this progress has been the application of NC methods in finance. This chapter provides an introduction to a wide range of financial problems to which NC methods have been usefully applied. The chapter also identifies open issues and suggests future directions for the application of NC methods in finance.

1 Introduction

Recent years have seen the application of multiple natural computing (NC) algorithms (defined in this chapter as computer algorithms whose design draws inspiration from phenomena in the natural world) for the purposes of financial modeling (Brabazon and O'Neill 2006). Particular features of financial markets including their dynamic and interconnected characteristics bear parallel with processes in the natural world and, prima facie, this makes NC methods "interesting" for financial modeling applications. Another feature of both natural and financial environments is the phenomenon of emergence, or the activities of multiple individual agents combining to coevolve their own environment.

The scale of NC applications in finance is illustrated by Chen and Kuo (2002) who list nearly 400 papers that had been published by 2001 on the use of evolutionary computation alone in computational economics and finance. Since then several hundred additional papers have been published underscoring the continued growth in this application area (see also Barbazon and O'Neill 2008, 2009; Chen 2002; Tsang and Martinez-Jaramillo 2004 and Wong et al. 2000 for additional examples of NC applications in finance).

Some of the major areas of financial applications using NC methods are: forecasting, algorithmic trading, portfolio management, risk management, derivatives modeling and market modeling. This chapter describes the utility of NC methods within each of these areas wherein their usage can be broadly categorized as optimization, model induction, and agent-based modeling.

1.1 Optimization

A wide variety of NC methodologies including genetic algorithms, evolutionary strategies, differential evolution, and particle swarm optimization have been applied for optimization purposes in finance. A particular advantage of these methodologies is that, if applied properly, they can cope with "difficult" search spaces. Examples of the use of optimization techniques in finance include optimal asset allocation, stock selection, risk management, pricing and hedging of options, and asset liability management (Zenios 2008).

1.2 Model Induction

While optimization applications of natural computing are important, the underlying model or data-generating process is not known in many real-world financial applications. Hence, the task is often to "recover" or discover an underlying model from a dataset. This is usually a difficult task as both the model structure and associated parameters must be uncovered.

Financial markets are affected by a myriad of interacting economic, political, and social events. The relationship between these factors and financial asset prices is not well understood and, moreover, is not stationary over time. Most theoretical financial asset pricing models are based on strong assumptions which are often not met in real-world asset markets. This offers opportunities for the application of model induction methodologies in order to recover the underlying data-generating processes. These methods can be applied, for example, to financial forecasting, credit risk assessment, and derivatives pricing.

1.3 Agent-Based Modeling

Agent-based modeling (ABM) has become a fruitful area of financial and economic research in recent years. ABM allows the simulation of markets which consist of heterogeneous agents, with differing risk attitudes and differing expectations to future outcomes, in contrast to traditional assumptions of investor homogeneity and rational expectations. ABM attempts to explain market behavior, replicate documented features of real-world markets, and allows one to gain insight into the likely outcomes of different regulatory policy choices.

A growing community of researchers are engaged in the application of natural computing methodologies in finance as illustrated by the number of conferences, workshops, and special sessions in this area. Examples of these include the annual track on Evolutionary Computation in Finance and Economics at the IEEE Congress on Evolutionary Computation, the IEEE Symposium on Computational Intelligence for Financial Engineering (CIFEr), the annual International Conference on Computational Intelligence in Economics & Finance (CIEF), and the European Workshop on Evolutionary Computation in Finance (EvoFIN) held annually as part of Evo*.

1.4 Structure of the Chapter

The rest of this chapter is organized as follows: ❷ Section 2 provides a concise overview of a number of key families of natural computing methods. ❷ Section 3 introduces various financial applications of natural computing methods and shows how these methodologies can add value in those applications. ❷ Section 4 concludes this chapter, suggesting multiple avenues of future work at the intersection of finance and natural computing.

2 Natural Computing

NC algorithms can be clustered into different groups depending on the aspects of the natural world upon which they are based. The main clusters that are relevant for finance applications illustrated in this chapter are neurocomputing, evolutionary computing, social computing, immunocomputing, physical computing, and developmental and grammatical computing (see ❷ *Fig. 1*).

Neurocomputing (or neural networks, NNs) typically draws inspiration from simplified models of the workings of the human brain or the nervous system. From a design perspective, neural networks can be characterized by a set of neurons (or nodes), the network structure which describes the pattern of connectivity between neurons, and the learning (or training)

□ Fig. 1
An overview of natural computing algorithms.

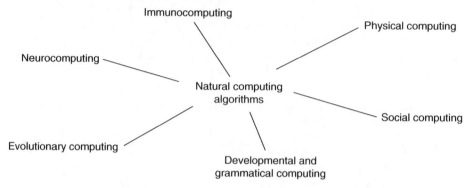

approach used. The predominant neurocomputing paradigms include *feedforward networks, recurrent networks, self-organizing networks, radial basis function networks, support vector machines* (Vapnik 1995; Cristianini and Shawe-Taylor 2000), etc. Financial firms worldwide are employing NNs to tackle difficult tasks involving intuitive judgment or requiring the detection of data patterns which elude conventional analytic techniques. For example, NNs are already being used to trade the securities markets, to forecast the economy, and to analyze credit risk. NNs were among the earliest NC methodologies to see widespread applications in finance (see, for example, Trippi and Turban 1993) but they do suffer from the practical drawback that their black-box nature makes their internal workings opaque to the user.

Evolutionary Computation (EC) is based upon neo-Darwinian principles of evolution. A population-based search process is used, whereby better (fitter) members of the population are preferentially selected for reproduction and modification, leading to a new population of individuals increasingly adapted to their environment. The main streams of EC are *genetic algorithms* (GA), *evolution strategies* (ES), *evolutionary programming* (EP), and *genetic programming* (GP). These methods have broad applications for optimization and model induction purposes. Recent extensions of the literature on EC to encompass dynamic, multi-objective, and constrained optimization problems has greatly increased the practical utility of these algorithms in finance.

Social Computing adopts a swarm metaphor and includes algorithms inspired by the flocking and schooling behavior of birds and fish. It also includes algorithms inspired by behaviors observed in social insects such as ants. These social systems exhibit a number of characteristics facilitating self-organization, flexibility, robustness, and direct or indirect communication among members of the population. Some examples of social computing include *ant colony, particle swarm,* and *bacterial foraging* algorithms. These algorithms are population-based like their evolutionary computation counterparts, and they operate by allowing the population of problem-solvers to communicate their relative success in solving the problem to each other, thereby biasing the future actions of the individuals in the population.

Immunocomputing encompasses a family of algorithms, which turn to the complex and adaptive biological immune system of vertebrates to inspire their design. The natural immune system represents an intricate network of specialized chemicals, cells, tissues, and organs with the ability to recognize, destroy, and remember an almost unlimited number of foreign bodies, and to protect the organism from misbehaving cells in the body. These properties are especially useful for tasks such as classification and optimization. Practical applications of immuno-computing include financial pattern-recognition such as the identification of potentially fraudulent credit card transactions, the identification of financially at-risk companies, and the identification of market "state."

Physical Computing draws inspiration from the physical processes of the natural world to design computational algorithms. These algorithms draw inspiration from phenomena such as simulated annealing and quantum mechanics. A claimed benefit of the quantum-inspired algorithms is that, because they use a quantum representation, they can maintain a good balance between exploration and exploitation. It is also suggested that they offer computa-tional efficiencies, as use of a quantum representation can allow the use of smaller population sizes than typical evolutionary algorithms. Computational efficiency is important in many financial applications such as real-time trading where systems have to deal with large data flows and a dynamic environment. Consequently, there is a continuing demand for optimiza-tion algorithms, which can potentially offer efficiency gains.

Developmental and Grammatical Computing borrows from both a developmental and a grammar metaphor. Grammatical computing refers to algorithms, which adopt concepts from linguistic grammars and are dominated by the generative form of grammars. Generative grammars are used to construct a "sentence" in the language specified by the grammar, and this generative process is metaphorically similar to the developmental process in biology in which "rules" govern the production of a complex, multicellular organism from a single embryonic cell. Generative grammars have been used in natural computing as a convenient representation by which developmental systems can be realized in-silico. The implementations of developmental and grammatical computing, such as *grammatical evolution* (GE) (O'Neill and Ryan 2001, 2003) (a grammatical variant of GP) may also embed an evolutionary algorithm typically used to drive the search process. GE has been already successfully applied to financial forecasting, credit rating assessment, and other financial applications.

These families of NC algorithms provide a rich set of tools for the development of quality optimization, model induction, and agent-based modeling applications, and all have seen application in finance. Readers requiring detailed information on these algorithms are referred to earlier chapters in this book. A review of these methods can also be found in Brabazon and O'Neill (2006), de Castro (2007), and Kari and Rozenberg (2008).

3 Financial Applications

This section introduces the application of NC methods across a range of financial areas including forecasting, algorithmic trading, portfolio management, risk management, deriva-tives modeling, and agent-based market modeling.

3.1 Forecasting

Financial forecasting applications may involve the prediction of future values of macroeco-nomic variables, individual stock market indices, commodity futures, the volatility of some financial products, etc. From an optimization perspective, NC methods can be applied either for variable identification or for parameter estimation (optimization). For example, NC methods can be used to select the explanatory variables from a large pool of candidates, which are then incorporated into the forecasting model. Even where the modeler knows the appropriate set of explanatory variables and model form, the selection of appropriate model parameters (coefficients) can be a difficult task, particularly for complex, nonlinear model structures. In the more difficult problem where the model form is not known, model induction methodologies such as GP or NNs can be used. This is potentially of considerable importance in financial applications, as many theoretical forecasting models are based on assumptions which are not met in real-world financial markets. This offers opportunities for the application of NC methods as model induction tools in order to "recover" the underlying data-generating processes directly from the data.

One family of NC methods which has seen extensive application for financial forecasting is NNs (e.g., Wong and Selvi 1998; Kaastra and Boyd 1996; Aiken 2000; Cao and Tay 2003a Edelman 2007). They offer particular advantages due to their ability to identify nonlinear models, handle noisy data, and embed a memory (recurrent NNs). A simple case of index prediction using a basic feedforward multilayer perceptrons (MLP) to construct a financial prediction model is illustrated in Brabazon and O'Neill (2006), where the MLP model is employed to predict the 5-day percentage change in the value of the FTSE 100 index. Ten inputs selected from a range of technical, fundamental, and intermarket data were included in the final model:

1. 5-Day lagged percentage change of the FTSE 100 index
2. 20-Day lagged percentage change of the FTSE 100 index
3. Ratio of the 10 vs 5-day moving average of the FTSE 100 index
4. Ratio of the 20 vs 10-day moving average of the FTSE 100 index
5. Bank of England Sterling index
6. S&P 500 composite index$_{(t)-(t-5)}$
7. LIBOR 1-month deposit rate
8. LIBOR 1-year deposit rate
9. Aluminum ($ per tonne)
10. Oil ($ per barrel)

In developing the final MLP models, a 11:6:1 structure was utilized, as illustrated in ❷ *Fig. 2*

$$y_t = L\left(\sum_{j=0}^{5} w_j L\left(\sum_{i=0}^{10} b_i w_{ij}\right)\right)$$

where b_i represents *input*$_i$ (b_0 is a bias node), w_{ij} represents the weight between input node$_i$ and hidden node$_j$, w_j represents the weight between hidden *node*$_j$ and the output node, and L represents the hyperbolic tangent function. Generally, NNs are developed through a trial-and-error approach guided by heuristics, the process is time consuming, and there is no guarantee that the final network structure is optimal. One approach is to automate the construction of the NN using evolutionary approaches (e.g., Aiken and Bsat 1999; Armano et al. 2005; and Cao and Tay 2003b).

GP can also be applied for forecasting. One of the best-known examples, EDDIE (which stands for *Evolutionary Dynamic Data Investment Evaluator*), was developed as an interactive decision tool (Tsang and Lee (2002); Li (2001)). It is a *genetic-programming-* based system for channeling expert knowledge into forecasting. Given a set of variables, EDDIE attempts to find interactions among variables and discover nonlinear functions.

☐ Fig. 2

Index prediction using MLP model, with 11 input nodes, six hidden nodes, and a single output node.

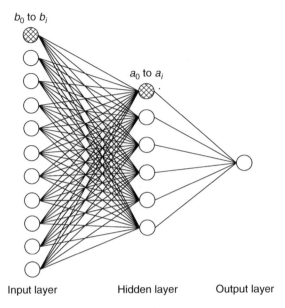

b_0 to b_i

a_0 to a_i

Input layer   Hidden layer   Output layer

Other examples of forecasting in the financial literature include the prediction of takeover targets (Harris et al. 1982; Belkaoui 1978; Palepu 1986; Rege 1984; Hickey et al. 2006), the prediction of auditor qualification of the financial statements of a company (Mutchler 1985; Dopuch et al. 1987; Thompson et al. 2007), the prediction of earnings and the prediction of IPO underpricing. In earnings prediction, Trigueros (1999) uses a GA to select explanatory variables from financial statements in order to predict corporate earnings. In the prediction of IPO underpricing, Quintana et al. (2005) employs GA for rule-based prediction. A less common, but nonetheless important, application in the literature is the prediction of volatility. Neely and Weller (2002) use GP to predict exchange rate volatility, where GP is applied for producing forecasting rules of the out-of-sample daily volatility in the foreign exchange market.

Typically, studies applying NNs, GA, or GP methods, for financial time series forecasting, use measures of goodness of fit drawn from statistics such as *mean squared error, sum of squared error, mean absolute percentage error,* etc., as their error (or fitness) function. The aim is to uncover or train a model using historical data, which "fits" that data well. Unsurprisingly, the choice of fitness function usually has a critical impact on the behavior of resulting model, hence a model constructed using one fitness metric would not necessarily perform well on another. While many forecasting studies applying NC methods to financial data indicate that models could be constructed to fit historic data fairly well, a common finding is that the quality of the forecasts diminishes, over time, out of sample.

3.2 Algorithmic Trading

Algorithmic trading is defined here as the use of computer programs to assist with any aspect of the trading of financial assets. It can therefore encompass systems which decide on certain aspects of the order such as the timing, price, or even the final quantity of the order. Hence, algorithmic trading can be combined with any investment strategy. A few related processes of algorithmic trading where NC methods can be applied, namely, investment analysis, arbitrage, and trade execution, are illustrated below.

3.2.1 Investment Analysis

Financial trading has seen a large number of applications of NC methods. Typically, these studies take one of two approaches of investment analysis, using either fundamental data (*fundamental analysis*) or market data, primarily price and volume (*technical analysis*).

Fundamental Analysis

Taking the example of investing in stocks, fundamental investment concentrates on the use of accounting information about the company, as well as industry and macroeconomic data, in order to identify companies which are mispriced by the market. In other words, the objective is to identify stocks which are good value (underpriced by the market), or stocks which are overpriced by the market (and therefore are candidates for "shorting"). In this approach, the investor needs to develop stock screening rules in order to decide which stocks to invest in. These rules were formulated manually in decades before computers. With a natural computing algorithm such as the GA, a large range of stock filter rules can be searched efficiently in order to find the highest-quality rules. In this approach, each individual in the population corresponds to a potential stock filter rule. The utility of these rules are tested using historical data, with the best rule (or set of rules) then being used for investment purposes (❷ *Fig. 3*). More generally, GP methods can be applied to evolve the structure of the filter rules.

Technical Analysis

In contrast to investors using a fundamental investment approach, technical analysts attempt to identify imbalances in the supply and demand for a financial asset using information from the time-series of the asset's trading (such as historical price, volume, and volatility data). Usually, investors who adopt a technical analysis approach look to combine technical indicators (preprocessed price and volume time series data about a financial asset), in order to produce a "trading signal". For example, a "technical indicator" could be the *moving average convergence-divergence* (MACD) oscillator, calculated by taking the difference of a short-run and a long-run moving average. If the difference is positive, it may indicate that the market is

■ Fig. 3
String encoding of a number of fundamental indicators. Each indicator can be coded as a 0 (no) or 1 (yes).

High sales growth relative to industry average?	High debt level relative to industry average?	High level of cash flow from operations relative to industry average?	High level of liquidity relative to industry average?	High profit level relative to industry average?

trending upward. For example, a buy signal could be generated when the shorter moving average crosses the longer moving average in an upward direction. A sell signal could be generated in a reverse case. Hence, a sample MACD trading rule could be:

$$IF \ x\text{-day MA of price} \geq y\text{-day MA of price}$$
$$THEN \ Go \ Long \ ELSE \ Go \ Short$$

where $x < y$. The optimal value of x and y can be evolved through a genetic algorithm, in order to maximize the trading profit (the fitness measure). A candidate solution encoded as a binary string of length 8 is illustrated below.

x:	0	0	0	0	1	0	1	0
y:	0	0	1	1	0	0	1	0

This solution indicates that $x = 10$ and $y = 50$. The MACD oscillator is a crude band-pass filter, removing both high-frequency price movements and certain low-frequency price movements, depending on the precise moving average lags selected. In essence, the choice of the two lags produces a filter, which is sensitive to particular price-change frequencies. In a recursive fashion, more complex combinations of moving averages of values calculated from an MACD oscillator can themselves be used to generate trading rules. In the past decades, the search for apparently useful technical indicators (or combinations of these) was undertaken manually by investors who back tested various indicators on historical financial data. GP allows the automation of this process with the concurrent vast expansion of the search space, which can be feasibly searched (Dempster and Jones 2001).

A trading rule should specify the entry, profit-taking, and stop-loss or exit strategy. NC methods can be used for rule optimization (to find optimal parameters for a fixed rule) or rule induction (to find optimal combination of diversified rules). Early applications of EC to uncover trading rules include Bauer (1994), Mutchler (1985), Allen and Karjalainen (1999), and Fyfe et al. (1999). The flexibility to implement different fitness functions is one particular advantage of EC approaches. However, there are also some issues related to the fitness functions as to include transaction cost and to consider different types of risks (Pavlidis et al. 2007). While markets exhibit periods in which a static trading rule can work, it is hard to find evidence of rules which are successful over long time periods. Of course, as financial markets comprise a dynamic system, the utility of any static trading system can be expected to degrade over time (Dempsey et al. 2009). One basic way of examining the characteristics of a trading system is to use an equity curve (❯ *Fig. 4*). The use of an adaptive trading strategy seems more plausible (Ghandar et al. 2008; da Costa Pereira and Tettamanzi 2008).

Recent work has seen a broadening of the information sources used as inputs in trading models. Instead of typical data drawn from the market, financial statements or macroeconomic data, for example, Thomas and Sycara (2002) and Larkin and Ryan (2008), used text data drawn from either internet message boards/the financial press in the creation of trading rules.

A wide range of forecasting approaches have also been employed to support making trade decisions, such as support vector machines (Edelman and Davy 2004), and hybrid methods like neuro-fuzzy hybrids (Zaiyi et al. 2006), neuro-genetic hybrids, geno-fuzzy (Ghandar et al. 2008) and ensemble methods (combining multiple models) (Kwon and Moon 2004).

◻ **Fig. 4**

Sample equity curves showing cumulative returns on the y axis and time on the x axis. The left-hand graph exhibits gradual return accumulation, whereas the right-hand graph suggests that the model is working less well on the second half of the time period.

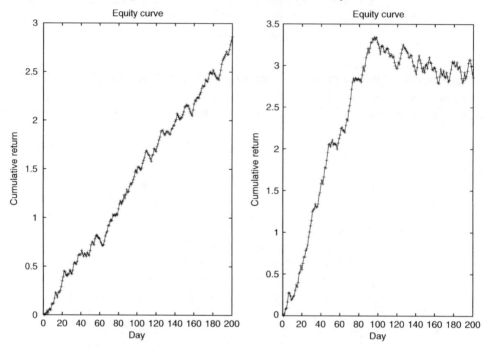

3.2.2   Arbitrage

Arbitrage trading can be defined in a variety of ways, but, broadly speaking, these trades seek to make profits by exploiting price differences of identical or similar financial instruments, on different markets or in different forms (e.g., buying a share at $23.78 on one exchange, and selling it immediately on another for $23.82, thereby exploiting a pricing difference for the same asset between the two exchanges). As would be expected, arbitrage opportunities tend to be closed very quickly and transaction cost can negate apparent arbitrage possibilities.

A simple example of an arbitrage play based on *Put-Call Parity* is illustrated in Tung and Quek (2005). In essence, the concept underlying this trade is that the price of a "long" position on an asset and an associated put option (the right to sell that asset in the future at a specified price, see ❷ Sect. 3.5 for a detailed description of options) must be equal to the price of a long call option on the same asset and a long position in a risk-free bond. Specifically, for example, for a European option:

$$S_0 + P_E(S_0, T, K) = C_E(S_0, T, K) + K(1 + r)^{-T}$$

where S_0 is the current price of the underlying asset, P_E is the current price of a European put, C_E is the current price of a European call, r is the risk free rate of return, K is the strike price for both options, and T is the time to maturity of the options. If either the put or call option are mispriced, the investor can, in theory, make a risk-free gain by constructing a portfolio of the four financial instruments.

The above example describes an arbitrage opportunity between the cash market (for the asset) and the option market. More generally, arbitrage opportunities can also exist between cash and future markets and between future and options markets. In its purest form, arbitrage is a risk-free transaction but in reality, most arbitrage trades are exposed to some risk such as liquidity risk or credit risk, and more significantly, execution risk where prices move before all elements of the trade can be completed.

An alternative approach that wait for arbitrage opportunities to emerge and then try to trade on them, is to anticipate or forecast opportunities in advance of the actual mispricing occurring. Markose et al. (2002) adopted this approach and developed a GP model to predict arbitrage opportunities between the FTSE 100 index futures and options market up to 10 min in advance of the arbitrage opportunity arising. Another example of the application of computational intelligence for uncovering arbitrage opportunities is provided in Tung and Quek (2005) who used self-organising, fuzzy NNs to identify mispriced American style currency options between the USD and the GBP, and then use a Delta hedge trading strategy to execute the arbitrage play.

3.2.3 Trade Execution

An important issue in trading financial assets is the efficient execution of large institutional size orders, and applications of EC for this task have begun to emerge in recent years. Typically, orders to buy or sell a stock can be either *market orders* (the transaction is undertaken immediately in the market at current prices) or limit orders (the purchase/sale must occur at a price which is no greater than/or less than a specific price). When a market order is placed, the customer does not have control over the final price, and in a limit order, while the customer has some price control, there is no guarantee that the order will actually be executed. So for example, if a customer places a limit order to buy a stock at $25 per share, the transaction will only take place if the market price falls to $25 or less.

Over the last number of years, trading algorithms have been executing an ever-increasing number of trades on markets. In the USA their rise has been brought about through a series of technological and regulatory changes. Since 2001, with the move to decimalization of the US equity markets and the widespread acceptance of electronic market places, the average trade size has declined from 1,200 shares per transaction in 2000 to 300 shares in 2008 (NYSE Euronext). This in turn has led to an explosion in the number of trades executed and a narrowing of spreads, with large institutional orders taking longer to execute. As a result, investors wishing to trade large blocks face trade-offs in balancing the risk of tipping their hand and providing information to the marketplace, thereby suffering market risk as the trade is executed. Trading algorithms seek to optimally execute these orders, using the vast amounts of data produced by the market place and submitting appropriately sized smaller orders to various destinations with the aim of achieving best execution. In reaching this goal, an entire ecology of different trading algorithms have been designed to perform under different market conditions, with recent innovations intelligently switching between these algorithms depending on current market conditions.

When trading shares, particularly when an investor is looking to buy or sell a large quantity of stock, the problem of *market impact* arises. Market impact occurs when the actions of an investor start to move the price adversely against themselves. Hence, market impact is the difference between a transaction price and what the market price would have been in the absence of the transaction. For example, the order may be executed as quickly as possible through sweeping any orders posted to the limit-order book, however, this would incur significant cost and drive the price of the stock against the investor. In this case, the investor avoids market risk but, by demanding instantaneous liquidity, incurs significant market impact costs. The obvious strategy to minimize market impact is to break the order into smaller lots and spread it over several purchases. While this will reduce the market impact, it incurs the risk of suffering *opportunity cost*, that market prices may start moving against investor during the multiple purchases. Added to this, the steady flow of small orders over time will inform other market participants of the presence of a large institutional order and so encourage competitors to run ahead of the investor. Hence, the design of trade execution strategies is intended to balance out the total cost of market impact and opportunity cost while maintaining a tight control over information leakage.

In selecting a trade execution strategy, the investor must not only balance her preferences but must also be prepared to adapt and change strategy as market conditions evolve. NC techniques provide ample scope to assist in uncovering information that can help optimize trading algorithm selection and/or adaptation. Various rules and heuristics can be evolved and adapted using (for example) GAs that can provide a trading strategy with predictive capability (Stephens and Sukumar 2006) with the aim of selecting best trading tactic under current market conditions. For institutional sized orders, which can be on the order of millions of shares, the reduction in average price by a couple of pennies can lead to significant savings.

Despite the importance of optimizing trade execution, there has been relatively little attention paid in the literature to the application of evolutionary methodologies for this task. One interesting exception is Lim and Coggins (2005) who used a GA to evolve a trading strategy in order to optimize trade execution performance using order book data from the Australian Stock Exchange. In this study, the approach taken was to initially split each trade into a series of *N* equal sized orders, and the objective was to evolve the timing strategy for the execution of each of these *N* orders during a single trading day. Each order was submitted as a limit order at the best ask or bid prevailing at the time the order was submitted, depending on whether the investor was seeking to buy or sell shares. A simple traditional trading strategy could be to submit one of these orders every 10 min. However, there is no guarantee that a 10-min order spacing would produce good results in terms of minimizing market impact. Lim and Coggins (2005) used a GA to uncover good quality timings for each order by evolving a chromosome of *N* genes, where each gene encoded the maximum lifetime that the order would remain on the order book (if it had not already been executed) before it was automatically ticked over the spread (e.g., a limit buy order being repriced to the current ask) to close out the trade. Any uncompleted trade at the end of the day were closed out the same way. Hence, the GA evolved the maximum time that each order would be exposed to the market before being crossed over the spread.

A variety of fitness functions could be designed to drive the evolution of the trading strategy but a common metric of trade execution performance is its *volume weighted average price* (VWAP):

$$VWAP = \frac{\sum(Price \cdot Volume)}{\sum(Volume)}$$

The VWAP of a strategy can be calculated and benchmarked against (for example) the overall VWAP for that share during the period of the trading strategy's execution. The aim is to evolve a strategy which produces as competitive a VWAP as possible.

In the above approach, the basic structure of the execution rule is determined in advance (number of trades, etc.) and the task of GA is to parameterize the rule. Another approach which could be applied is to use GP to evolve the structure of the execution rule, as well as its parameters.

In real-world trading, a number of interesting additional issues arise. While the order book provides an indication of the current state of supply and demand for a share, it does not always present a true reflection of the investor's trading intentions. For example, market participants can attempt to "game" the order book by placing limit orders which are subsequently canceled or amended, and (as described above) the order book may contain *iceberg orders* (a large order which has been split into several smaller orders in order to disguise the investor's trading intent). The dynamic nature of the order book suggests that an agent-based modeling approach could be used in order to uncover robust trade execution strategies.

3.3 Portfolio Management

In finance, a *portfolio* refers to a basket of financial assets such as stocks, bonds, and cash equivalents. Portfolio management involves the art and science of making decisions about investment mix and policy, matching investments to objectives, asset allocation, and balancing risk and return. An overview of the portfolio management process is illustrated in ❷ *Fig. 5*. For a portfolio manager, the first and foremost part of the investment process is understanding the client's needs, the client's tax status and his or her risk preferences. The next part of the process is the actual construction of the portfolio, which involves asset class allocation and security selection decisions. Asset class allocation refers to the allocation of the portfolio across different asset classes defined broadly as equities, fixed income securities, and real assets (such as real estate, commodities, and other assets). The security selection decision refers to the selection of specific securities under each asset class. The final component of portfolio management is trade execution, the efficient purchase or sale of the relevant assets in the marketplace. An important and open question is how best to measure the performance of the resulting portfolio.

"*Optimization is the engineering part of portfolio construction*," as mentioned by Fabozzi et al. (2007) in their recent survey for quantitative equity portfolio management: "*Most portfolio construction problems can be cast in an optimization framework, where optimization is applied to obtain the desired optimal risk-return profile.*" Multiple elements of the portfolio management process, such as asset allocation, security selection, index tracking, etc., where optimization is crucial, are amenable to NC methodologies. Applications of NC methods for the purpose of asset allocation and index tracking are illustrated below.

3.3.1 Asset Allocation

Asset allocation is the selection of a portfolio of investments where each component is an asset class rather than an individual security. The aim of asset allocation is to invest a fixed amount of money in a diverse set of assets so as to maximize return while minimizing a risk measure. The solution to this is a Pareto frontier (usually referred to as the *efficient frontier*) shown in ❷ *Fig. 6*, as for a given level of risk, there should not be a portfolio with a higher rate of return, or for a given level of return, there should not be a portfolio with a lower level of risk.

☐ **Fig. 5**

An overview of the portfolio management process (from Damodaran (2003)).

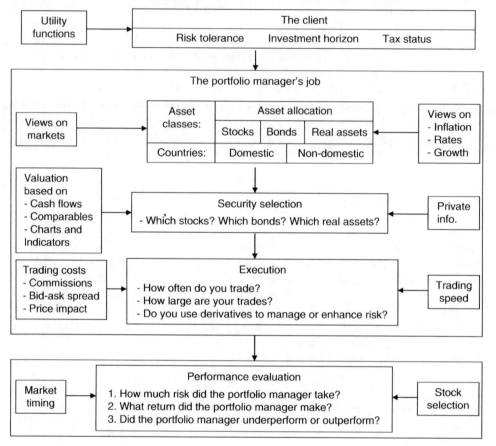

☐ **Fig. 6**

Pareto frontier. The points that correspond to the risk-return of the set of portfolios that are Pareto optimal.

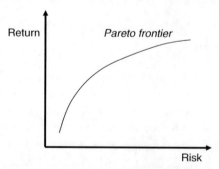

Once the frontier is uncovered, the final choice of portfolio is determined by the individual investor's risk preference.

A classical approach used for asset allocation is the Markowitz mean-variance model (Markowitz 1952, 1959). It assumes that investors wish to maximize their return (measured as mean or expected return) and minimize their risk (measured as variance or the standard deviation of their return). This produces the risk-return trade-off. The goal of the Markowitz model is therefore to find an optimal portfolio p of N assets, each with a weighting w_i (in percentage), such that the return E_p is maximized:

$$E_p = \sum_{i=1}^{N} w_i \cdot \mu_i$$

while minimizing the variance of the return:

$$V_p = \sum_{i=1}^{N} \sum_{\substack{j=1 \\ i \neq j}}^{N} w_i \cdot w_j \cdot \sigma_{ij}$$

subject to

$$\sum_{j=1}^{N} w_i = 1$$

$$w_i \geq 0; i = 1, \ldots, N$$

where σ_{ij} is the covariance of return between asset i and j, the constraints are used to ensure that all the money is invested and all investments are positive (assuming that short selling is not allowed). It produces a multi-objective optimization problem (maximize return, minimize risk) and there are two equivalent formulations which are the dual of each other (first, fix a value of expected return and find the portfolio that minimizes the risk, second, select a level of risk and find the portfolio that maximizes the expected return). Either of the formulations produces a quadratic programming problem and several algorithms exist, which can be applied to uncover good-quality portfolios.

Adding Real-World Constraints

Quadratic programming relies on a number of assumptions, including a single quadratic objective function, linear constraints, and the existence of a positive definite covariance matrix between the asset returns. However, these assumptions are typically breached in the real world. For example, the constraints can include *cardinality constraints* (a limit on the number of assets which can be held in the portfolio), i.e., $\sum_{i=1}^{N} sign(w_i) = K$, or there may be *threshold limits* on the amount of investment in any single asset, i.e., $w_i \leq m_i$ $i = 1, \ldots, N$, where m_i is the threshold amount for asset i. Other constraints may include industry or sector (or *concentration*) holding constraints, round lot constraints, and transaction cost may have both fixed and nonlinear variable cost elements. These constraints can lead to non-convex, non-differential models. In addition, some constraints may be hard and others may be soft. Hence, real-world portfolio selection can present a difficult, high-dimensional, constrained optimization problem, which is beyond the capabilities of traditional optimization methods. In this setting, heuristic approaches such as evolutionary computing methods are of particular interest, because of their ability to find good solutions, even if optimality is not assured.

MOEA and Portfolio Selection

An extensive literature on multi-objective evolutionary algorithms (MOEA) has developed over the past 20 years (see Schaffer 1984; Deb 2001; Coello et al. 2002 for a detailed review). MOEA have an advantage of maintaining a population of solutions and therefore offer the potential to uncover multiple points on the Pareto frontier. A wide range of approaches have been offered to deal with different types of constraints including penalty function approaches, repair mechanisms, the design of appropriate representations, and diversity-generation operators. A stream of literature also exists which has used hybrid evolutionary algorithms or local search techniques for MOEA. Many of these approaches have been applied for portfolio selection.

The earliest papers to apply EAs for portfolio selection include Arone et al. (1993), Loraschi et al. (1995), Shoaf and Foster (1998), and Vedarajan et al. (1997). In the case of Arone et al. (1993) multiple GA populations were used to identify the Pareto frontier. Rather than use the standard Markowitz model, the authors used a downside risk measure. The multi-objective problem was converted into a single objective using a trade-off function, with each population using a different trade-off coefficient and therefore producing a different portion of the Pareto frontier. The formulation of the portfolio problem includes cardinality and buy-in constraints, and a repair mechanism was applied in order to ensure that generated solutions were feasible. The utility of differing crossover operators and differing genotypic representations for the portfolio selection problem was examined by Streichert et al. (2004a, b), and the application of EC hybrids was examined by Subbu et al. (2005) and Streichert et al. (2006). The impact of cardinality constraints was examined in Fieldsend et al. (2004) and Moral-Escudero et al. (2006) (the latter also adopted an EC hybrid approach). A number of other applications including stock ranking, and credit portfolio optimization have been reported in the literature (see Castillo Tapia and Coella (2007) and Schlottmann and Seese (2004) for a review of MOEA applications in finance).

In practical settings, investment managers are concerned with a variety of risk and return measures, not just expected return and its variance. For example, *value at risk* (VaR), the risk that a portfolio could lose a significant amount of its value over a defined time window (more precisely, VaR at level $(1-\alpha)$ is the α-quantile of the loss distribution), has gained importance especially for regulatory purposes. VaR is typically nonlinear and non-convex, making optimization in models, which use this metric difficult. More generally, a portfolio manager may be concerned with more than one risk constraint. A variety of papers have applied MOEA to non-Markowitz risk metrics, including Lipinski (2008) which uses a compound risk metric. Hochreiter (2008) introduces an evolutionary stochastic portfolio optimization methodology and illustrates its application using a set of structurally different risk measures, which include, standard deviation, mean-absolute downside semi deviation, value-at-risk, and expected shortfall. Recent work has also seen the application of coevolutionary MOEAs for portfolio optimization (Dreżewski and Siwik 2008).

3.3.2 Index Tracking

There are two common types of portfolio management strategies: passive and active. Active portfolio management consists of picking assets which are expected to outperform the market. In contrast, passive management simply tracks a market index, where the objective is to form a portfolio that replicates the performance of an index as closely as possible.

Passive portfolio management strategies have become very common in recent decades. Just like general portfolio optimization, the construction of an index tracking portfolio is a constrained optimization problem, where the objective is to minimize a measure of *tracking error* (or difference between the return to the portfolio and the return to the index), subject to a variety of constraints, similar to those in the general portfolio optimization problem. The solution space is non-convex suggesting a useful role for population-based, global optimization heuristics.

At first glance, the construction of an index tracking portfolio appears trivial, merely requiring the purchase of the same basket of assets that make up the index, using the same weight that each asset has in the index. However, the creation of a perfect replica portfolio is difficult for several reasons including a requirement for frequent portfolio rebalancing (with associated transactions costs), integer constraints on asset purchases, mandate limits on the maximum holding in any individual asset, etc. Another practical issue is that not all assets making up market indices have equal liquidity. Hence, there may be good reason to seek to track the performance of a broad market index using a portfolio which comprises of a subset (rather than all) of the assets making up the index. Another feature of this problem is that the investor's risk attitudes to tracking errors are not symmetric. While investors will not wish to underperform the index, they will not object if the portfolio outperforms the index. EC applications to the index tracking problem include Shapcott (1992), Streichert and Tanaka-Yamawaki (2006), Orito et al. (2007), and Maringer (2008) who also examine the impact of investor loss-aversion preferences on tracking portfolio construction.

3.4 Risk Management

Risk management is a critical aspect of investing. Some of the main types of risks faced by investors are illustrated in ❯ *Fig. 7* below.

◘ Fig. 7
Risk illustration (from Sound Practices for Hedge Fund Managers (2000)).

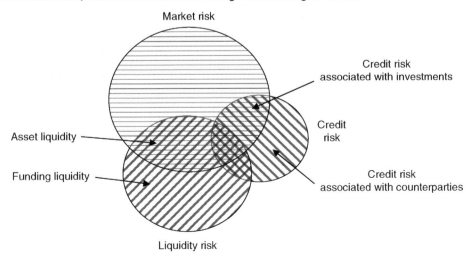

3.4.1 Market Risk Computation

Market risk refers to the risk faced by an investor arising from changes in financial market prices. The degree of risk of loss faced by an investor will vary depending on the price volatility of the assets they hold. Not all assets will have the same degree of volatility. There are various techniques to measure market risk (Marrison 2002), applications of NC methodology include the calculation of value-at-risk (VaR) and the sensitivities. The VaR approach measures the worst expected loss under normal market conditions over a specific time interval. The loss distribution is usually assumed to be normal when calculating VaR. Evolutionary algorithms do not need to embed this assumption and can incorporate any preferred loss distribution type (Uludag et al. 2007). NNs have also been used for other market risk measures, such as conditional VaR estimates (Lee et al. 2005), and expected shortfall (Diagne 2002). The sensitivity analysis approach measures how much the portfolio's (or a specific financial instrument's) value is expected to change, if there is a small change in one of the market-risk factors, such as interest rates, equity prices, commodity prices, etc. The sensitivities are the so-called *greeks* for derivatives; *duration* and *convexity* for fixed income products. Keber and Schuster (2001) has used GP to calculate greeks of the American-type put options (see ❷ Sect. 3.5 for an explanation of derivative products).

3.4.2 Credit Risk Assessment

Credit risk assessment is an important component of the lending decision of financial institutions and other commercial companies. Examples of decisions of where credit scoring could be useful include, decisions such as should a loan be extended to a firm or to an individual, should a customer be allowed to purchase goods on credit, or what credit limit should be offered to a customer on their credit card?

Over the past several decades, an extensive literature has amassed on modeling creditworthiness and default risk. Traditional statistical methods including linear discriminant analysis (Altman 1968) and logit (Ohlson 1980) have been applied. In all of these applications, the objective is to develop a model, which will provide a metric of creditworthiness from a series of explanatory variables. Typically, in assessing corporate creditworthiness, explanatory variables can include numbers drawn from the financial statements of the firm, from financial markets, general macroeconomic variables, and nonfinancial, firm-specific information). In assessing personal consumer creditworthiness, explanatory variables can include income, age, occupation, current employment status, past borrowing record, etc. (West 2000; Yobas et al. 2000; Leung et al. 2007). Closely associated streams of academic literature include corporate failure or bankruptcy prediction (Altman 1968), and the reverse engineering of the bond-rating models used by rating firms such as Standard & Poor's (S&P), Moody's, Fitch's, or Dominion Bond Rating Service (Ederington 1985; Dutta and Shekhar 1988; Gentry et al. 1988). Assessments of credit default probability could also form a useful input into a stock- or bond-trading model. A practical problem in constructing risk-assessment models is that there is no clear theoretical framework for guiding the choice of explanatory variables or model form. In the absence of an underlying theory, most published work on credit rating employs a data-inductive modeling approach. This produces a high-dimensional combinatorial problem, as the modeler is attempting to uncover a good set of explanatory variables and model form.

An illustration of an early credit risk assessment model is provided by Altman's (1968) classic study in which five ratios were combined to produce a linear discriminant classification model for corporate bankruptcy. A Z score was calculated for each company, and this value determined whether the company was classified as bankrupt or solvent:

$$Z = 0.012X_1 + 0.014X_2 + 0.033X_3 + 0.006X_4 + 0.999X_5$$

where

X_1 = working capital to total assets

X_2 = retained earnings to total assets

X_3 = earnings before interest and taxes to total assets

X_4 = market value of equity to book value of total debt

X_5 = sales to total assets

As the range of NC techniques have expanded over the past 20 years, each new technique has been applied to credit scoring and corporate failure prediction. Examples include feedforward NNs (Wilson et al. 1995; Atiya 2001), self-organizing maps (Serrano-Cina 1996; Kiviluoto and Bergius 1998), GAs (Kumar et al. 1997; Varetto 1998), Ant models (Wang et al. 2004; Brabazon and O'Neill 2006), GP and GE (McKee and Lensberg 2002; Brabazon and O'Neill 2006; O'Neill and Ryan 2001; Alfaro-Cid et al. 2008). The domain offers particular potential for evolutionary automatic programing methodologies such as GP or GE as these methods can produce human-readable credit decision rules. This can be important in some countries where lenders can be required to justify decisions not to grant loans. Another advantage of GP and GE is that the rule-evolution process can be seeded using domain knowledge.

Another closely related application is the prediction of bank failure (Lee et al. 2006), with many regulatory authorities using risk models in order to assess which financial institutions require the closest scrutiny. Obviously, for these applications it is important that the regulatory authority can verify the correctness of the underlying prediction model, hence, methodologies which can incorporate expert knowledge and produce interpretable decision rules, such as fuzzy systems and GP, are of particular interest.

Of course, there are other types of risks which need to be quantified in practice, such as liquidity risk and operational risk (arising due to poor or inadequate management control systems or due to human error). However, as yet, there is little literature concerning the application of NC methods in these areas.

3.5 Derivatives Modeling

Derivatives are contracts whose value is derived from the value of the underlying assets, such as equities, interest rates, currencies, market indices, commodities, etc. Two of the best known forms of derivatives are futures and options. A futures contract is an agreement to buy or sell goods, currency or securities on an agreed future date and for a price fixed in advance. An option is a financial instrument that simply gives the holder (buyer) the right, but not the obligation, to buy (a call option), or sell (a put option), a specified underlying asset at a pre-agreed price on or before a given date. A European style option refers to an option that may only be exercised on expiration; while an American style option can be exercised on any trading day on or before expiration.

The key issue for investors wishing to trade in derivatives is the determination of the fair price for the derivative. For some standard derivatives (based on specific assumptions such as continuous time finance theory), closed-form pricing equations have been determined (e.g., the Black–Scholes model (Black and Scholes 1973; Merton 1973) for pricing European options, the Cox et al. (1979) binomial model, etc.). The traditional approach to pricing a derivative is (Noe and Wang 2002):

- Specify a stochastic process for the underlying asset(s)
- Derive the pricing equation for the derivative (using a no-arbitrage argument) and
- Price the derivative by solving the pricing equation

Of course, this approach can be difficult to implement, as the relevant stochastic process may be imperfectly understood, and the pricing equation may be too difficult to solve analytically. In the latter case, there is scope to use tools such as Monte Carlo (MC) simulation to estimate the expected payoff and the associated payoff risk for the derivative. In valuing a complex derivative using MC, the typical approach is to randomly generate a set of independent price paths for each security underpinning the derivative, then compute the present value of the payoff to the derivative under each set of these price paths. The simulation process is repeated multiple times and the distribution of the payoffs is considered to characterize the derivative. An example of this approach is illustrated in Kim and Byun (2005). A critical issue in applying an MC approach is the correct design of the theoretical pricing model.

There have been two main avenues of application of NC methods in pricing financial derivatives, namely,

- Model calibration and
- Model induction

In model calibration, the objective is to estimate the parameters of (or "calibrate") a theoretical pricing model. The parameters are estimated by fitting the model to the relevant return time series. Typically, the pricing model will have a complex, nonlinear structure with multiple parameters. Hence, global search heuristics such as the genetic algorithm can have utility in uncovering a high-quality set of parameters. Examples of the use of NC algorithms for model calibration include Dang et al. (2008) and Fan et al. (2007).

In model induction, NC approaches such as GP would have particular utility when little is known about the underlying asset-pricing dynamics as both the structure and the parameters of the pricing model are estimated directly from the data, thereby extracting the pricing model implicitly. Even where theory does exist, model induction methodologies allow one to investigate whether other plausible theories may exist to explain observed prices. Applications of such methods include NNs (Malliaris and Salchenberger 1993; Hutchinson et al. 1994; White 1998), self-organizing, fuzzy NNs (Tung and Quek 2005), or GP (Chen et al. 1999; Chidambaran et al. 1998; Keber 2000; Chidambaran 2003; Yin et al. 2007) to recover a proxy for the price-generating model directly from the data.

An example of using GP to generate an option pricing model is illustrated below. One advantage noted by Chidambaran (2003) is that the GP process can be seeded with the Black–Scholes equation, with the final resulting model being an adaptation of the Black–Scholes equation for conditions which violate its underlying assumptions. For example, in the

Black–Scholes setting for a non-dividend-paying European call option, there are five factors that affect the price of the option (assuming no dividends):

1. S_0 – the underlying asset price
2. K – the exercise price of the option
3. T – the time to maturity (of the option)
4. r – the risk free rate of return (of the underlying asset)
5. σ – the expected volatility of the asset price

Items 1 and 2 can be combined to give $(S_0 - K)$ or a measure of the *moneyness* of the option. An option is said to be "in the money" when this value is greater than zero, and "out of the money" when it is less than zero. In developing a pricing model for options from these factors, the Black–Scholes model embeds several critical assumptions. It is assumed that the stock price undergoes a diffusion process that is lognormally distributed, with an instantaneous drift and volatility given by μ and σ, respectively. The volatility σ and the risk-free rate r are also assumed to be constant during option's life. This implies that:

$$\ln \frac{S_T}{S_0} \sim N\left(\left(\mu - \frac{\sigma^2}{2}\right)T, \sigma\sqrt{T}\right)$$

and in turn this leads to:

$$S_T = S_0 e^{\eta\, T}, \eta \sim N\left(\left(\mu - \frac{\sigma^2}{2}\right), \frac{\sigma}{\sqrt{T}}\right)$$

where μ is the instantaneous expected return on the stock, σ is the instantaneous volatility of stock price return, S_T is the stock price at a future time T, S_0 is the stock price at time zero, $N(m,s)$ denotes the normal density function with mean m and standard deviation s and η is defined as the continuously compounded rate of return per annum realized between time zero and T. The Black–Scholes formula for the price at time zero of an European call option on a non-dividend-paying stock is therefore:

$$C_0 = S_0 N(d_1) - K e^{-rT} N(d_2)$$

where

$$d_1 = \frac{\ln(S_0/K) + (r + \sigma^2/2)T}{\sigma\sqrt{T}}$$

$$d_2 = d_1 - \sigma\sqrt{T}$$

$N(\chi)$ is the cumulative probability distribution function for a standardized normal distribution, C_0 is the price of the European call option at time 0.

Of course, assuming that the Black–Scholes model did precisely value options, model induction techniques such as neural networks or GP could be used to recover the structure of the option pricing model directly from a historical time series of option prices (C_0) and the five factors that influence option prices. ❍ *Figure 8* illustrates a tree representation of the Black–Scholes Model which could (potentially) be recovered by GP.

In reality, some of the key assumptions in the Black–Scholes model do not hold in real-world option markets, and hence the model does not explain observed option prices correctly. For example, prices can experience discontinuous jumps, the distribution of price changes has fatter tails than those implied by a lognormal distribution, and asset price volatility changes over time. (There is a long line of literature which examines alternative (non-normal) stock

⬛ **Fig. 8**

Stylized illustration of a tree representation of the Black–Scholes model (not all sub-trees are shown).

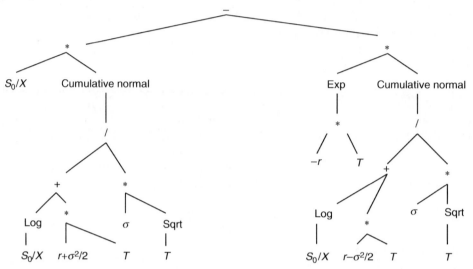

return models including Poisson jump-diffusion return processes and GARCH processes. However, closed-form solutions for the option price cannot be obtained for all these models.) The latter issue can be easily seen if market prices are substituted into the Black–Scholes model, in order to calculate the "implied volatility" for that option. If the assumptions underlying the Black–Scholes option pricing model were correct, the implied volatilities for options on the same underlying asset would be constant for different strike prices and maturities. However, in practice, the Black–Scholes implied volatilities are varying over strike price and maturity. Keber (2002) has used GP to generate formulae for determining the implied volatility based on American put options.

3.6 Agent-Based Market Modeling

The essence of agent-based modeling (ABM) lies in the notion of autonomous agents whose behavior evolves endogenously leading to complex, emergent, system dynamics, which are not predictable from the properties of the individual agents. ABM is an exciting new tool for exploring behavior in financial markets that are far from traditional notions of equilibrium, and where agents exhibit behavior that is less than fully rational at times. Financial markets are particularly appealing applications for agent-based methods especially considering the following: issues of price and information aggregation tend to be sharper in financial settings where agent objectives tend to be clearer; financial markets are rich in data sets (such as price and volume data) at many different frequencies, that can be used for testing and calibrating agent-based models; financial markets are well organized, centralized, and trade homogeneous products in a generally efficient fashion relative to markets for other goods and services; there are continuing developments in the area of experimental financial markets, which give carefully controlled environments that can be compared with agent-based experiments; the key debates in finance about market efficiency and rationality are still unresolved; many puzzles of the financial time series (such as the volatility persistence) are still not well understood. In designing ABMs of financial markets, modelers face a daunting list of design choices which can critically impact on the system's behavior. Important design questions include:

- Representation and structure of the actual trading agents. Agents can vary from simple budget constrained *zero intelligence agents* (refers to a type of trader that randomly makes price bids (offers to buy) and/or price asks (offers to sell) subject only to a budget constraint), as in Gode and Sunder (1993), to sophisticated learning agents as in Chen and Yeh (2001).
- The actual mechanism that governs the trading of assets. Ways of designing this include assuming a simple price response to excess demand, building the market, such that a kind of local equilibrium price can be found easily, or explicitly modeling the dynamics of trading to mimic the continuous trading of real-world markets.
- Types of securities to be incorporated into the agent-based market model, where typically simple securities (such as stocks) are considered.

In designing agent-based models (ABMs) of financial markets, NC methods can be used to model the information processing and storage by agents, the process of adaptive learning by agents, or to model the trading mechanism. One example of the use of ABM to simulate a financial market is provided by LeBaron (2002) in building the well-known Santa Fe Artificial Stock Market. This model simulates price generation and trading in a market made up of artificial adaptive agents. This is a prototype example of a complex system and is thought to illustrate the benefits of simulation modeling. The Santa Fe Artificial Stock Market consists of a central computational market and a number of artificially intelligent agents. The agents choose between investing in a stock and leaving their money in the bank, which pays a fixed interest rate. The stock pays a stochastic dividend and has a price which fluctuates according to agent demand. The agents make their investment decisions by attempting to forecast the future return on the stock, using GA to generate, test, and evolve predictive rules. Other applications of ABM include the simulation of a foreign exchange market (Izumi 1999), the modeling of an artificial stock option market (Ecca et al. 2008) and the modeling of an artificial payment card market (Alexandrova-Kabadjova 2008).

A key output from the ABM literature on financial markets is that it illustrates that complex market behavior can arise from the interaction of quite simple agents. Carefully constructed, ABM can help increase one's understanding of market processes and can potentially provide insights for policy makers and regulators, where, unlike laboratory sciences, one cannot rerun a real market under different regulations "to see what would happen." Of course, issues of model validation are important in all ABM applications, including those in financial markets.

4 The Future

Though a plethora of academic literature on NC applications in finance exists, it is notable that many papers have been concerned with proof of concept rather than robust, industry-strength, applications. While NC methods offer potential in multiple areas in finance, the maturing of their application requires future work focusing on complex real-world problems. This will require the construction of multidisciplinary research teams, drawing academic expertise as necessary from finance, computer science, mathematics, biology, etc., and combining this with industrial collaborators. For example, quality work on trading systems requires deep knowledge of market microstructure, the regulatory environment, available financial instruments, and the technology available to traders. Indicated below are some promising future directions for research at the nexus of natural computing and finance.

- *Forecasting*: While forecasting models applying NC methods can typically be constructed to fit historical data fairly well, a common finding is that the quality of the out-of-sample forecasts degrades over time. Hence, one can expect to see increased use of sophisticated methods for preprocessing the raw time-series inputs, and for the adaptation of the resulting models in response to changing environmental conditions. Another area of growing interest is the incorporation of information from text mining (e.g., from the financial press) into forecasting models.
- *Algorithmic Trading*: While many published works have focused on the development of simplified trading systems, successful real-world applications have often focused on the support of specific elements of the investment process. In the medium term, one can expect to see a notable increase in the rigor of the published works in this area as computer scientists form teams with finance academics and practitioners, incorporating realistic models of market microstructure. The area of trade execution has seen relatively little published application of NC methodologies, despite its real-world significance. Model induction tools such as GP offer interesting potential here, as do agent-based modeling approaches. The latter could be used to uncover robust trade execution strategies.
- *Portfolio Optimization*: There has already been an extensive application of NC methods for portfolio optimization, but there is still a need for further systematic investigation of portfolio constraints for special sub-application areas including hedge funds, pension funds, and insurance. The extension of dynamic portfolio optimization needs further development, which may involve the simulation or forecasting of extreme market situations, and the solution of multistage constraint optimization problems.
- *Risk Management*: Recent events on the financial markets have underscored the importance of risk management, and the weakness of existing theoretical models in this area. It is interesting to note that applications of NC methods in risk management have not attracted as much attention as might be expected in the literature and this remains an open research area. For example, NC methods could be applied to assist in the development of enterprise-wide risk management systems, and to improve the flexibility and efficiency of large-scale multistage *asset liability management* (ALM) models.
- *Derivatives Modeling*: In spite of the vast array of derivative products available, and the weakness of financial theory, once one moves beyond vanilla products there have only been a relatively limited number of applications of NC for model calibration or induction in this area. Possibilities also exist to hybridize NC methods with traditional numerical methods, and to develop dynamic derivative pricing models.
- *Agent-Based Market Modeling*: The field of ABM is attracting significant attention with the increasing questioning of agent homogeneity which underlies classical financial economics. ABM allows one to examine the effect of the differing forms of market structure on market behavior. Doubtless, the next few years will see increased focus on this, given the failures of market regulation during the recent financial crisis. The coevolutionary element of markets lends itself well to NC approaches in terms of modeling of agent behavior and strategy adaptation.

A practical issue that arises in the application of NC to finance is that the underlying algorithms are themselves undergoing a process of maturation. Recent years have seen extensive research in order to extend canonical NC algorithms into high-dimensional environments (enhancing algorithmic scalability), to develop efficient algorithms for constrained optimization, and to develop practical application of the algorithms in dynamic problem environments. Meanwhile, we have also seen developments in computer hardware, and in our ability to implement parallel versions of NC algorithms (e.g., using *graphics processing unit* (GPU) implementations). These two strands of development are creating an ever more powerful toolbox of NC algorithms for financial modelers.

Acknowledgments

This chapter has emanated from research conducted with the financial support of Science Foundation Ireland under Grant Number 08/SRC/FM1389.

References

Aiken M (2000) Forecasting the United States gross domestic product with a neural network. J Int Inform Manag 9(1):11–21

Aiken M, Bsat M (1999) Forecasting market trends with neural networks. Inform Syst Manag 16(4):42–48

Alexandrova-Kabadjova B, Tsang E, Krause A (2008) Evolutionary learning of the optimal pricing strategy in an artificial payment card market. In: Brabazon A, O'Neill M(eds) Natural computing in computational finance. Springer, Berlin, pp 233–251

Alfaro-Cid E, Cuesta-Canada A, Sharman K, Esparcia-Alcazar A (2008) Strong typing, variable reduction and bloat control for solving the bankruptcy prediction problem using genetic programming. In: Brabazon A, O'Neill M (eds) Natural computing in computational finance. Springer, Berlin, pp 161–186

Allen F, Karjalainen R (1999) Using genetic algorithms to find technical trading rules. J Financ Econ 51:245–271

Altman E (1968) Financial ratios, discriminant analysis and the prediction of corporate bankruptcy. J Financ 23:589–609

Armano G, Marchesi M, Murru A (2005) A hybrid genetic-neural architecture for stock indexes forecasting. Inform Sci 170:3–33

Arone S, Loraschi A, Tettamanzi A (1993) A genetic approach to portfolio selection. Neural Netw World Int J Neural Mass-Parallel Comput Inform Syst 3:597–604

Atiya A (2001) Bankruptcy prediction for credit risk using neural networks: a survey and new results. IEEE Trans Neural Netw 12(4):929–935

Bauer R (1994) Genetic algorithms and investment strategies. Wiley, New York

Belkaoui A (1978) Financial ratios as predictors of Canadian takeovers. J Bus Financ Account 5 (5):93–108

Black F, Scholes M (1973) The pricing of options and corporate liabilities. J Polit Econ 81:637–659

Brabazon A, O'Neill M (2006) Biologically inspired algorithms for financial modelling. Springer, Berlin

Brabazon A, O'Neill M (eds) (2008) Natural computing in computational finance. Springer, Berlin

Brabazon A, O'Neill M (eds) (2009) Natural computing in computational finance, vol II. Springer, Berlin

Castillo Tapia MG, Coello Coello C (2007) Applications of multi-objective evolutionary algorithms in economics and finance: a survey. In: CEC 2007: Proceedings of the IEEE international conference on evolutionary computation, Singapore, September 2007. IEEE Press, Piscataway, NJ, pp 532–539

de Castro LN (2007) Fundamentals of natural computing: an overview. Phys Life Rev 4(1):1–36

Cao LJ, Tay FEH (2003a) Support vector machine with adaptive parameters in financial time series forecasting. IEEE Trans Neural Netw 14(6):1506–1518

Cao LJ, Tay FEH (2003b) A hybrid neurogenetic approach for stock forecasting. IEEE Trans Neural Netw 18(3):851–864

Chen S-H (eds) (2002) Genetic algorithms and genetic programming in computational finance. Kluwer Academic, Norwell, MA

Chen S-H, Kuo T-W (2002) Evolutionary computation in economics and finance: a bibliography. In: Chen S-H (ed) Evolutionary computation in economics and finance. Physica-Verlag, Heidelberg

Chen SH, Yeh CH (2001) Evolving traders and the business school with genetic programming: a new architecture of the agent-based stock market. J Econ Dyn Control 25(3–4):363–393

Chen S-H, Lee W-C, Yeh C-H (1999) Hedging derivative securities with genetic programming. Int J Intell Syst Account Financ Manag 8(4):237–251

Chidambaran N (2003) Genetic programming with Monte Carlo simulation for option pricing. In: Proceedings of the 2003 IEEE winter simulation conference, New Orleans, LA, December 2003. IEEE Press, Piscataway, NJ, pp 285–292

Chidambaran N, Lee C, Trigueros J (1998) Adapting Black-Scholes to a non-Black-Scholes environment via genetic programming. In: CIFEr: Proceedings of the IEEE/IAFE/INFORMS 1998 conference on computational intelligence for financial engineering, New York, March 1998. IEEE Press, Piscataway, NJ, pp 197–211

Coello C, Van Veldhuizen D, Lamont G (2002) Evolutionary algorithms for solving multi-objective problems. Kluwer Academic, New York

da Costa Pereira C, Tettamanzi A (2008) Fuzzy-evolutionary modeling for single-position day trading. In: Brabazon A, O'Neill M (eds) Natural computing in computational finance. Springer, Berlin, pp 131–159

Cox J, Ross S, Rubinstein M (1979) Option pricing: a simplified approach. J Financ Econ 7:229–264

Cristianini N, Shawe-Taylor J (2000) An introduction to support vector machines and other kernel-based learning methods. Cambridge University Press, New York

Damodaran A (2003) Investment philosophies: successful investment philosophies and the greatest investors who made them work. Wiley, Hoboken, NJ, p 8

Dang J, Brabazon A, O'Neill M, Edelman D (2008) Option model calibration using a bacterial foraging optimisation algorithm. In: EvoFin 2008: Proceedings of the 2nd European workshop on evolutionary computation in finance and economics, Napoli, Italy, March 2008. Lecture notes in computer science, vol 4974. Springer, New York, pp 133–143

Deb K (2001) Multi-objective optimization using evolutionary algorithms. Wiley, Chichester, UK

Dempsey I, O'Neill M, Brabazon A (2009) Foundations in grammatical evolution for dynamic environments. Springer, Berlin

Dempster M, Jones C (2001) A real-time adaptive trading system using genetic programming. Quant Financ 1:397–413

Diagne M (2002) Financial risk management and portfolio optimization using neural networks and extreme value theory. Ph.D. thesis, University of Kaiserslautern

Dopuch N, Holthausen RW, Leftwich RW (1987) Predicting audit qualifications with financial and market variables. Account Rev LXII(3):431–454

Dreżewski R, Siwik L (2008) Co-evolutionary multi-agent system for portfolio optimization. In: Brabazon A, O'Neill M (eds) Natural computing in computational finance. Springer, Berlin, pp 271–299

Dutta S, Shekhar S (1988) Bond rating: a non-conservative application of neural networks. In: Proceedings of IEEE international conference on neural networks, II, San Diego, CA, July 1988, IEEE Press, Piscataway, NJ, pp 443–450

Ecca S, Marchesi M, Setzu A (2008) Modeling and simulation of an artificial stock option market. Comput Econ 32(1):37–53

Edelman D (2007) Adapting support vector machine methods for horserace odds prediction. Ann OR 151(1):325–336

Edelman D, Davy P (2004) Adaptive technical analysis in the financial markets using machine learning: a statistical view. In: Fulcher J, Jain LC (eds) Applied intelligent systems new directions series studies in fuzziness and soft computing. Springer, Berlin, pp 1–15

Ederington H (1985) Classification models and bond ratings. Financ Rev 20(4):237–262

Fabozzi FJ et al. (2007) Trends in quantitative equity management: survey results. Quant Financ 7(2):115–122

Fan K, Brabazon A, O'Sullivan C, O'Neill M (2007) Quantum-inspired evolutionary algorithms for calibration of the VG option pricing model. In: EvoFin 2007: Proceedings of the 1st European workshop on evolutionary computation in finance and economics, Valencia, Spain, April 2007. Lecture notes in computer science, vol 4447. Springer, Berlin, pp 186–195

Fieldsend J, Matatko J, Peng M (2004) Cardinality constrained portfolio optimisation. In: IDEAL 2004: Intelligent data engineering and automated learning, Exeter, UK, August 2004. Lecture notes in computer science, vol 3177. Springer, New York, pp 788–793

Fyfe C, Marney J, Tarbert H (1999) Technical analysis versus market efficiency – a genetic programming approach. Appl Financ Econ 9(2):183–191

Gentry J, Whitford D, Newbold P (1988) Predicting industrial bond ratings with a probit model and funds flow components. Financ Rev 23(3):269–286

Ghandar A, Michalewicz Z, Schmidt M, Tô T-D, Zurbrugg R (2008) Computational intelligence for evolving trading rules. IEEE Trans Evol Comput 13(1):71–86

Gode DK, Sunder S (1993) Allocative efficiency of markets with zero intelligence traders. J Polit Economy 101:119–137

Harris R, Stewart J, Guilkey D, Carleton W (1982) Characteristics of acquired firms: fixed and random coefficients probit analyses. South Econ J 49(1):164–184

Hickey R, Little E, Brabazon A (2006) Identifying merger and takeover targets using a self-organising map. In: ICAI '06: Proceedings of the 2006 international conference on artificial intelligence, Las Vegas, NV, June 2006. CSEA Press

Hochreiter R (2008) Evolutionary stochastic portfolio optimization. In: Brabazon A, O'Neill M (eds) Natural computing in computational finance. Springer, Berlin, pp 67–87

Hutchinson J, Lo A, Poggio T (1994) A non-parametric approach to pricing and hedging derivative securities via learning networks. J Financ 49:851–889

Izumi K (1999) An artificial market model of a foreign exchange market. Ph.D. Dissertation, Tokyo University

Kaastra I, Boyd M (1996) Designing a neural network for forecasting financial and economic time series. Neurocomputing 10:215–236

Kari L, Rozenberg G (2008) The many facets of natural computing. Commun ACM 51(10):72–83

Keber C (2000) Option valuation with the genetic programming approach. In: Proceedings of the sixth international conference. MIT Press, Cambridge, MA, pp 689–703

Keber C (2002) Evolutionary computation in option pricing: determining implied volatilities based on American put options. In: Chen S-H (eds) Evolutionary computation in economics and finance. Physica-Verlag, New York, pp 399–415

Keber C, Schuster M (2001) Evolutionary computation and the Vega risk of American put options. IEEE Trans Neural Netw 12(4):704–715

Kim J, Byun S (2005) A parallel Monte Carlo simulation on cluster systems for financial derivatives pricing. In: CEC 2005: Proceedings of the IEEE international conference on evolutionary computation, Edinburgh, UK, September 2005. IEEE Press Piscataway, NJ

Kiviluoto K, Bergius P (1998) Maps for analysing failures of small and medium-sized enterprises. In: Deboeck G, Kohonen T (eds) Visual explorations in finance with self-organizing maps. Springer-Verlag, Berlin, pp 59–71

Kumar N, Krovi R, Rajagopalan B (1997) Financial decision support with hybrid genetic and neural based modeling tools. Eur J Oper Res 103(2):339–349

Kwon Y-K, Moon B-R (2004) Evolutionary ensemble for stock prediction. In: GECCO 2004: Proceedings of the genetic and evolutionary computation conference, Seattle, WA, June 2004. Lecture notes in computer science, vol 3103. Springer, New York, pp 1120–1113

Larkin F, Ryan C (2008) Good news: using news feeds with genetic programming to predict stock prices. In: O'Neill M et al. (eds) EuroGP 2008: Proceedings of the 11th European conference on genetic programming, Napoli, Italy, March 2008. Lecture notes in computer science, vol 4971. Springer, Berlin, pp 49–60

LeBaron B (2002) Building the Santa Fe artificial stock market. Working Paper, Brandeis University, June 2002

Lee H, Lee J, Yoon Y, Kim S (2005) Coherent risk measure using feedforward neural networks. In: ISNN 2005, Chongqing, China, May–June 2005. Lecture notes in computer science, vol 3497. Springer, Berlin, pp 904–909

Lee C, Quek C, Maskell D (2006) A brain-inspired fuzzy neuro-predictor for bank failure analysis. In: CEC 2006: Proceedings of the IEEE international conference on evolutionary computation, Vancouver, BC, Canada, July 2006. IEEE Press, Piscataway, NJ, pp 7927–7934

Leung K, Cheong F, Cheong C (2007) Consumer credit scoring using an artificial immune system algorithm. In: CEC 2007: Proceedings of the IEEE international conference on evolutionary computation, Singapore, September 2007. IEEE Press, Piscataway, NJ, pp 3377–3384

Li J (2001) FGP: a genetic programming based tool for financial forecasting. Ph.D. Thesis, University of Essex

Lim M, Coggins R (2005) Optimal trade execution: an evolutionary approach. In: CEC 2005: Proceedings of the IEEE international conference on evolutionary computation, Edinburgh, UK, September 2005. IEEE Press, Piscataway, NJ, pp 1045–1052

Lipinski P (2008) Evolutionary strategies for building risk-optimal portfolios. In: Brabazon A, O'Neill M (eds) Natural computing in computational finance. Springer, Berlin, pp 53–65

Loraschi A, Tettamanzi A, Tomassini M, Verda P (1995) Distributed genetic algorithms with an application to portfolio selection problems. In: Pearson D, Steele N, Albrecht R (eds) Artificial neural networks and genetic algorithms. Springer, Berlin, pp 384–387

Malliaris M, Salchenberger L (1993) A neural network model for estimating option prices. Appl Intell 3(3):193–206

Maringer D (2008) Constrained index tracking under loss aversion using differential evolution. In: Brabazon A, O'Neill M (eds) Natural computing in computational finance. Springer, Berlin, pp 7–24

Markose S, Tsang E, Er H (2002) Evolutionary decision trees for stock index options and futures arbitrage. In: Chen S-H (eds) Genetic algorithms and genetic programming in computational finance. Kluwer Academic, Norwell, MA, pp 281–308

Markowitz H (1952) Portfolio selection. J Financ 1(7):77–91

Markowitz H (1959) Portfolio selection: efficient diversification of investments. Wiley, New York

Marrison C (2002) The fundamentals of risk measurement. McGraw-Hill, New York

McKee T, Lensberg T (2002) Genetic programming and rough sets: a hybrid approach to bankruptcy classification. Eur J Oper Res 138:436–451

Merton R (1973) Rational theory of option pricing. Bell J Econ Manag Sci 4:141–183

Moral-Escudero R, Ruiz-Torrubiano R, Suarez A (2006) Selection of optimal investment portfolios with cardinality constraints. In: CEC 2006: Proceedings of the IEEE international conference on evolutionary computation, Vancouver, BC, Canada, July 2006. IEEE Press, Piscataway, NJ, pp 8551–8557

Mutchler JF (1985) A multivariate analysis of the auditor's going-concern opinion decision. J Account Res 23(2):668–682

Neely C, Weller P (2002) Using a genetic program to predict exchange rate volatility. In: Chen S-H (eds) Genetic algorithms and genetic programming in computational finance. Kluwer Academic, Norwell, MA, pp 263–278

Neely C, Weller P, Dittmar R (1997) Is technical analysis in the foreign exchange market profitable? A genetic programming approach. J Financ Quant Anal 32(4):405–428

Noe T, Wang J (2002) The self-evolving logic of financial claim prices. In: Chen S-H (eds) Genetic algorithms and genetic programming in computational finance. Kluwer, Norwell, MA, pp 249–262

Ohlson J (1980) Financial ratios and the probabilistic prediction of bankruptcy. J Account Res 18 (1):109–131

O'Neill M, Ryan C (2001) Grammatical evolution. IEEE Trans Evol Comput 5(4):349–358

O'Neill M, Ryan C (2003) Grammatical evolution: evolutionary automatic programming in an arbitrary language. Kluwer, Boston, MA

Orito Y, Takeda M, Iimura K, Yamazaki G (2007) Evaluating the efficiency if index fund selections over the fund's future period. In: Chen S-H, Wang P, Kuo T-W (eds) Computational intelligence in economics and finance. Springer, New York, pp 157–168

Palepu K (1986) Predicting takeover targets: a methodological and empirical analysis. J Account Econ 8:3–25

Pavlidis N, Pavlidis E, Epitropakis M, Plagianakos V, Vrahatis M (2007) Computational intelligence algorithms for risk-adjusted trading strategies. In: CEC 2007: Proceedings of the 2007 congress on evolutionary computation, Singapore, September 2007. IEEE Press, Piscataway, NJ, pp 540–547

Quintana D, Luque C, Isasi P (2005) Evolutionary rule-based system for IPO underpricing prediction. In: GECCO 2005: Proceedings of the genetic and evolutionary computation conference, Washington DC, June 2005. ACM, New York, pp 983–989

Rege U (1984) Accounting ratios to locate take-over targets. J Bus Financ Account 11(3):301–311

Schaffer J (1984) Multiple objective optimization with vector evaluated genetic algorithms. Ph.D. Thesis, Vanderbilt University

Schlottmann F, Seese D (2004) Financial applications of multi-objective evolutionary algorithms: recent developments and future research directions. In: Coello Coello C, Lamont G (eds) Applications of multi-objective evolutionary algorithms. World Scientific, Singapore, pp 627–652

Serrano-Cina C (1996) Self organizing neural networks for financial diagnosis. Decis Support Syst 17(3): 227–238

Shapcott J (1992) Index tracking: genetic algorithms for investment portfolio selection. Technical report, EPCC-SS92-24, Edinburgh Parallel Computing Centre

Shoaf J, Foster J (1998) The efficient set GA for stock portfolios. In: CEC 98: Proceedings of the IEEE international conference on evolutionary computation, Anchorage, AK, May 1998. IEEE Press, Piscataway, NJ, pp 354–359

Sound Practices for Hedge Fund Managers (2000) the Managed Funds Association (MFA), New York, p 16

Stephens CR, Sukumar R (2006) An introduction to datamining. In: Grover R, Vriens M (eds) The handbook of market research do's and don'ts. Sage, Thousand Oaks, CA

Streichert F, Tanaka-Yamawaki M (2006) The effect of local search on the constrained portfolio selection problem. In: CEC 2006: Proceedings of the IEEE international conference on evolutionary computation, Vancouver, BC, Canada, July 2006. IEEE Press Piscataway, NJ, pp 8537–8543

Streichert F, Ulmer H, Zell A (2004a) Evaluating a hybrid encoding and three crossover operators on the constrained portfolio selection problem. In: CEC 2004: Proceedings of the IEEE international conference on evolutionary computation, Portland, OR, June 2004. IEEE Press, Piscataway, NJ, pp 932–939

Streichert F, Ulmer H, Zell A (2004b) Comparing discrete and continuous genotypes on the constrained portfolio selection problem. In: GECCO 2004: Proceedings of the genetic and evolutionary computation conference, Seattle, WA, June 2004. Springer, Berlin, pp 1239–1250

Subbu R, Bonissone P, Eklund N, Bollapragada S, Chalermkraivuth K (2005) Multiobjective financial portfolio design: a hybrid evolutionary approach. In: CEC 2005: Proceedings of the IEEE international conference on evolutionary computation, Edinburgh, UK, September 2005. IEEE Press, Piscataway, NJ, pp 2429–2436

Thomas J, Sycara K (2002) GP and the predictive power of internet message traffic. In: Chen S-H (eds) Genetic algorithms and genetic programming in computational finance. Kluwer Academic, Norwell, MA, pp 81–102

Thompson D, Thompson S, Brabazon A (2007) Predicting going concern audit qualification using neural networks. In: ICAI'07: Proceedings of the 2007 international conference on artificial intelligence. CSEA Press, Las Vegas, NV, June 2007

Trigueros J (1999) Extracting earnings information from financial statements via genetic algorithms. In: Proceedings of the 1999 IEEE international conference on computational intelligence for financial engineering, New York, March 1999. IEEE Press, Piscataway, NJ, pp 281–296

Trippi RR, Turban E (1993) Neural networks in finance and investing. McGraw-Hill, New York

Tsang E, Li J (2002) EDDIE for financial forecasting. In: Chen S-H (ed) Genetic algorithms and genetic programming in computational finance. Kluwer Academic, Norwell, MA, pp 161–174

Tsang E, Martinez-Jaramillo S (2004) Computational finance. IEEE Computational Intelligence Society Newsletter, August 8–13, 2004

Tung W, Quek C (2005) GenSoOPATS: a brain-inspired dynamically evolving option pricing model and arbitrage system. In: CEC 2005: Proceedings of the IEEE international conference on evolutionary computation, Edinburgh, UK, September 2005. IEEE Press, Piscataway, NJ, pp 1722–1729

Uludag G et al. (2007) Comparison of evolutionary techniques for value-at-risk calculation. In: Proceedings of EvoWorkshops 2007, Valencia, Spain, April 2007. Lecture notes in computer science, vol 4448, pp 218–227

Vapnik V (1995) The nature of statistical learning theory. Springer-Verlag, New York

Varetto F (1998) Genetic algorithms in the analysis of insolvency risk. J Bank Financ 22(10):1421–1439

Vedarajan G, Chan L, Goldberg D (1997) Investment portfolio optimization using genetic algorithms. In: Koza J (ed) Late breaking papers at the genetic programming 1997, Stanford, CA, July 1997, pp 256–263

Wang C, Zhao X, Kang Li (2004) Business failure prediction using modified ants algorithm. In: Chen S-H, Wang P, Kuo T-W (eds) Computational intelligence in economics and finance. Springer, Berlin

West D (2000) Neural network credit scoring models. Comput Oper Res 27:1131–1152

White A (1998) A genetic adaptive neural network approach to pricing options: a simulation analysis. J Comput Intell Financ 6(2):13–23

Wilson N, Chong K, Peel M (1995) Neural network simulation and the prediction of corporate outcomes: some empirical findings. Int J Econ Bus 2(1):31–50

Wong BK, Selvi Y (1998) Neural network applications in finance: a review and analysis of literature. Inform Manag 34(3):129–140

Wong B, Lai V, Lam J (2000) A bibliography of neural network business applications research: 1994–1998. Comput Oper Res 27:1045–1076

Yin Z, Brabazon A, O'Sullivan C (2007) Adaptive genetic programming for option pricing. In: GECCO 2007: Proceedings of the genetic and evolutionary computation conference, London, England, July 2007. ACM, New York, pp 2588–2594

Yobas M, Crook J, Ross P (2000) Credit scoring using neural and evolutionary techniques. IMA J Math Appl Bus Ind 11:111–125

Zaiyi G, Quek C, Maskell D (2006) FCMAC-AARS: a novel FNN architecture for stock market prediction and trading. In: CEC 2006: Proceedings of the IEEE international conference on evolutionary computation, Vancouver, BC, Canada, July 2006. IEEE Press, Piscataway, NJ, pp 8544–8550

Zenios SA (2008) Practical financial optimization. Wiley-Blackwell, Chichester, UK

52 Selected Aspects of Natural Computing

David W. Corne[1] · *Kalyanmoy Deb*[2] · *Joshua Knowles*[3] · *Xin Yao*[4]
[1]School of Mathematical and Computer Sciences, Heriot-Watt
University, Edinburgh, UK
dwcorne@macs.hw.ac.uk
[2]Department of Mechanical Engineering, Indian Institute of Technology,
Kanpur, India
deb@iitk.ac.in
[3]School of Computer Science and Manchester Interdisciplinary Biocentre
(MIB), University of Manchester, UK
j.knowles@manchester.ac.uk
[4]Natural Computation Group, School of Computer Science, University
of Birmingham, UK
x.yao@cs.bham.ac.uk

G. Rozenberg et al. (eds.), *Handbook of Natural Computing*, DOI 10.1007/978-3-540-92910-9_52,
© Springer-Verlag Berlin Heidelberg 2012

Abstract

In this chapter we will discuss a selection of application areas in which natural computation shows its value in real-world enterprises. For the purposes of demonstrating the significant impact and potential of natural computation in practice, there is certainly no shortage of documented examples that could be selected. We present just ten applications, ranging from specific problems to specific domains, and ranging from cases familiar to the authors to highlights known well in the general natural computation community. Each displays the proven promise or great potential of nature-inspired computation in high-profile and important real-world applications, and we hope that these applications inspire both students and practitioners.

1　Introduction

The study of natural computation has borne several fruits for science, industry, and commerce. By providing exemplary strategies for designing complex biological organisms, nature has suggested ways in which design spaces can be explored and developed into innovative new products. By exhibiting examples of effective cooperation among organisms, nature has hinted at new ideas for search and control engineering. By showing how highly interconnected networks of simple biological processing units can learn and adapt, nature has paved the way for the development of computational systems that can discriminate between complex patterns and improve their abilities over time. And the list goes on.

It is instructive to note that the methods used that have been inspired by nature are far more than simply "alternative approaches" to the problems and applications that they address. In many domains, nature-inspired methods have broken through barriers in the erstwhile achievements and capabilities of "classical" computing. In many cases, the role of natural inspiration in such breakthroughs can be viewed as that of a strategic pointer, or a kind of "tiebreaker." For example, there are many, many ways that one might build complex multiparameter statistical models for general use in classification or prediction; however, nature has extensive experience in a particular area of this design space, namely neural networks – this inspiration has guided much of the machine learning and pattern recognition community toward exploiting a particular style of statistical approach that has proved extremely successful. Similar can be said of the use of immune system metaphors to underpin the design of techniques that detect anomalous patterns in systems, or of evolutionary methods for design.

Moreover, it seems clear that natural inspiration has in some cases led to the exploration of algorithms that would not necessarily have been adopted, but have nevertheless proven significantly more successful than alternative techniques. Particle swarm optimization, for example, has been found enormously successful on a range of optimization problems, despite its natural inspiration having little to do with solving an optimization problem. Meanwhile, evolutionary computation, in its earliest days, was subjected to much skepticism and general lack of attention – why should a method be viable for real-world problems when that method, in nature, seems to take millions of years to achieve its ends? What need is there for slow methods that rely on random mutation, when classical optimization has a mature battery of sophisticated techniques with sound mathematical bases? Nevertheless, evolutionary methods are now firmly established, thanks to a long series of successful applications in which their performance is unmatched by classical techniques.

The idea of this chapter is to present and discuss a collection of exemplars of the claims made in this introduction. A handful of selected applications of natural computation will be looked at, each chosen for a subset of reasons, such as level of general interest, or impact. Some classic applications will be considered, which still serve as inspirational to current practitioners, and some newer areas will be looked at, with exciting or profound prospects for the future.

The applications are loosely clustered into four themes as follows. The first theme deals with applications under the banner of "Strategies," where three examples in which natural computing methods have been used to produce novel and useful strategies for different enterprises are studied in detail. These include an evolutionary/neural hybrid method that led to the generation of an expert checkers player, the use of genetic programming (GP) to discover rules for financial trading, and the exploitation of a learning classifier system to generate novel strategies for fighter pilots. The next theme is "science and engineering" where applications that have wider significance for progress in one or more areas of science and engineering, in areas (or in ways) that may not be traditionally associated with natural computing, are considered. The two exemplars in this area are the use of multi-objective evolutionary computation for a range of areas (often in the bio and analytical sciences) for *closed-loop* optimization, and the concept of *innovization*, which exploits multi-objective evolutionary computation in a way that leads to generic design insights for mechanical engineering (and other) problems. The next theme, "logistics," exemplifies how natural computing (largely, learning classifier systems and evolutionary computation) has provided us with successful ways to address difficult logistics problems (the case of a real-world truck scheduling problem is looked at), as well as a way to design new fast *algorithms* for a range of logistics and combinatorial problems, via approaches referred to as "super-heuristics" and/or "hyper-heuristics." Finally, the theme of "Design" is considered and three quite contrasting examples are discussed. These are, in turn, antenna design, Batik pattern design, and the emerging area of software design using natural computing methods.

2 Strategies: Generating Expert Pilots, Players, and Traders

Many problems in science and industry can be formulated as an attempt to find a good strategy. A strategy is, for practical purposes, a set of rules (or an algorithm, or a decision tree, and so forth) that sets out what to do in a variety of situations. Expert game players are experts, presumably, because they use good strategies. Similar is true for good pilots, and successful stock market traders, as well as myriad other professionals who are experts in their particular domains. It may well come as a surprise to some that humans do not have the last word on good strategy – strategies can be discovered by software, which, in some cases, can outperform most or even all human experts in particular fields. In this section, three examples of applications in which strategies have been developed via natural computing techniques, respectively, for piloting fighter aircraft, for playing expert-level checkers, and for trading on the stock market will be looked at.

2.1 Discovery of Novel Combat Maneuvers

In the early 1990s and beyond, building on funding support from NASA and the USAF, a diverse group of academics and engineers collaborated to explore the automated development of strategies for piloting fighter aircraft. A broad account of this work (as well as herein) is available in Smith et al. (2002). The natural computing technology employed is termed "genetics-based machine learning" (GBML), the most common manifestation of which is the *learning classifier system* – essentially a rule-based system that adapts over time, with an evolutionary process central to the rule adaptation strategy. In this work, such an adaptive rule-based system takes the role of a test pilot. In the remainder of this section, some of the background and motivation for this application are covered, the computational techniques used are explained, and some of the interesting and novel results that emerged from this work are presented.

2.1.1 Background: New Aircraft and Novel Maneuvers

As explained in Smith et al. (2002), a standard approach, when developing a new fighter aircraft, is to make a prototype for experimentation by test pilots, who then explore the performance of the new aircraft and, importantly, are then able to develop combat maneuver strategies in simulated combat scenarios. Without such testing, it is almost impossible to understand how a new aircraft will actually perform in action, which in turn depends, of course, on how it will be flown by experienced pilots. In particular, it is very important that pilots are able to develop effective and innovative combat strategies that exploit the technology in the new craft. Following such testing, issues in performance are then fed back into the design process, and perhaps the prototype will need to be reengineered, and so forth. This testing process is obviously very expensive – costing the price of at least one prototype craft, and the time of highly skilled pilots. One way to cut this cost includes using a real pilot, but to "fly" a simulation of the new craft; another is to resort to entirely analytical methods; however, both of these approaches are problematic for different reasons (Smith et al. 2002). The idea of Smith et al.'s research is to explore a third approach, in which a machine learning system takes the place of a test pilot and operates in the context of sophisticated flight simulations.

To help better understand the motivation for this work and grasp the importance of developing novel maneuvers, it will be useful to recount some background in fighter aircraft piloting. This is adapted next from Smith et al. (2002), while a comprehensive account is in Shaw (1998). A relatively new aspect of modern fighter aircraft is the use of post-stall technology (PST). This refers to systems that enable the pilot to fly at extremely high angles of attack (the angle between the aircraft's velocity and its nose–tail axis). Pilots have developed a range of combat maneuvers associated with PST flight, including, for example, the Herbst Maneuver, in which the aircraft quickly reverses direction via a combination of rolling and a high angle of attack. In another example, the Cobra Maneuver, the aircraft makes a very quick pitch-up from horizontal to 30° past vertical; the pilot then pitches the aircraft's nose down and resumes normal flight angles. This causes dramatic deceleration, meaning that a pursuing fighter will overshoot. The technologies that allow PST flight have led to the invention of these and several other maneuvers, as well as the prevalence of tactics that involve "out-of-plane" maneuvering, where the attacking aircraft flies in a continually changing maneuvering plane, invariably different from the plane of the target craft. This link between new technologies and new maneuvers is critical in the design and deployment of new aircraft and is the focus of Smith et al.'s work. The results that are described later involve experiments in which the attacking craft was an X-31 experimental fighter plane, with sophisticated PST capability, and where the target craft in the simulations was an F-18.

2.1.2 Learning Classifier Systems

Learning classifier systems (LCSs: Holland et al. 1986; Grefenstette 1988; Goldberg 1989; Holland 1992) use a collection of rules called *classifiers* in the form of state/action pairs. Each such pair indicates an action to take if the environment currently matches the "state" part of the rule. An LCS operates in an environment according to its current set of classifiers, and uses reinforcement learning and other adaptation methods, in particular, including genetic algorithms (GAs) to gradually adapt the rules over time. Classically, a classifier in an LCS represents states and actions as binary strings, but states may also contain "don't care" characters (#s). For example, the string: "0 1 0 # 1/0 1 0" is a classifier with the meaning: "If the environment is in state 0 1 0 0 1 or state 0 1 0 1 1, then perform action 0 1 0."

In a typical LCS, each cycle begins with a message representing the state of the environment (as will be seen, the environment in the fighter plane combat context is simply a characterization of the relative positions and velocities of the aircraft in the simulation). The LCS then sees which of its classifiers match this environmental message. There may of course be several, and some form of conflict resolution method must then be invoked to decide which classifier's action will be executed. The action eventually chosen is then performed. This action may lead to a reward – that is, some aspect of the environment becomes (more) favorable, and the classifier that led to this action receives an increment to its "fitness" score. In some classifier systems, sophisticated credit allocation systems are in place to ensure that the most recent action does not necessarily receive all of the credit. After some specified number of cycles, the genetic algorithm is invoked. The GA population is formed from a subset of the classifiers focusing on those with higher fitness. New classifiers are produced by standard genetic operations on selected classifiers (where, naturally, selection is biased by fitness), and these are then incorporated into the LCS, overwriting some of the less fit existing classifiers. Clearly, an LCS operates in a way that attempts to find – via the GA, which is in turn informed by the fitnesses of classifiers, which in turn are informed by experience in the environment – a good set of classifiers that achieves continual rewards in its environment.

2.1.3 Implementation, Experimentation Details, and Results

The way that LCS technology has been used, with considerable and long-established success in the domain of combat maneuver discovery (Smith and Dike 1995; Smith et al. 2002), is basically as follows. The task faced by the LCS is (typically) a one-on-one engagement for a specific amount of time, such as 30 s. There is a specified initial configuration of positions and velocities, and the period is divided into periods of 0.1 s (i.e., each action of the classifier pilot must last for at least 0.1 s before another action can be performed). At each of the (typically) 300 timesteps during a simulation, each aircraft observes the current configuration, and decides on an action. At the end of an engagement, a score can be calculated based on the relative probabilities of the two aircrafts having damaged their opponents.

All Smith et al.'s experiments employed AASPEM, the Air-to-Air System Performance Evaluation Model developed by the U.S. Government for computer simulation of air-to-air combat. The encoding of the state/action parts of a classifier were as follows. The state part of a classifier comprised 20 bits: six bits were used to encode the two "aspect angles" that gave the current relative positions of the aircraft in terms of their lines of sight. The remaining 14 bits were used to encode seven parameters (hence, each discretized into four bins), namely: range, speed, delta speed, altitude, delta altitude, climb angle, and opponent's climb angle. The action part of a classifier comprised eight bits, encoding three parameters: a relative bank angle (three bits), an angle of attack (three bits), and a speed (two bits). Speed, for example, was either 100 knots (00), 200 knots (01), 350 knots (10), or 480 knots (11). The meaning of an action that specified, for example, relative bank angle of 30° and speed of 200 knots, was to aim for these as desired targets. In all cases, the simulation environment (i.e., the AASPEM system) would automatically interpret these aims into realistic actions.

A set of such classifiers, therefore, represents a general strategy, and subsets of related classifiers can potentially encode entire novel maneuvers. During a simulated engagement, the classifiers are run for 300 cycles, as described; when no classifier matches the current environmental configuration, a default action for straight, level flight is used. When more than one classifier is matched, the fittest one is chosen as the provider of the action. At the end of the 300 cycles, a fitness measure is calculated. Following experiment with various approaches, the most promising fitness measurement was found to be based on the difference between the self and the opponents' "aspect angles." This basically gives a score that is a linear function along the continuum from: *self is aiming directly at opponent's tail* to *opponent is aiming directly at self's tail*, with the former obviously preferred. The fitness assigned after an engagement was based on the average of this value over the entire engagement, and is assigned to every individual classifier that was active at any point during the engagement.

Following a full engagement, the GA then operates over the whole population of classifiers. Using a moderate selection pressure for parents, and standard crossover and mutation operations, a collection of new classifiers is generated. The fitness assigned to a new classifier is simply the average of its parents' fitnesses. Typically, about half of the classifiers in a population were replaced with new classifiers. A learning run would continue with repeated such engagements (perhaps ~500), each resulting in fitness assignment and operation of the GA, leading to a revised set of classifiers for use in the next engagement. The starting configuration for all engagements in a single run was always one from the small set of tactically interesting situations shown in ❷ *Fig. 1*.

The conditions in ❷ *Fig. 1* were designed to generate X-31 tactics results for a balanced set of relevant situations.

The early experiments described in Smith and Dike (1995) were quite similar, involving one-on-one combat, in which the LCS attempted to find novel maneuvers for the X-31, but the opponent F/A-18 aircraft used a fixed (although suitably reactive and challenging) set of standard maneuvers embedded in AASPEM. In short, the opponent would always attempt to execute the fastest possible turn that would leave it pointing directly at its opponent, at the same time attempting to match the opponent's altitude.

As reported, in considerably more detail, in Smith and Dike (1995), this setup led to the discovery of a wide variety of new and novel fighter maneuvers, which were evaluated in positive terms by real fighter test pilots. An example of such a maneuver is shown in ❷ *Fig. 2*.

The strategy discovered by the LCS in ❷ *Fig. 2* involves pitching upward sharply, stalling, tipping over, and then engaging the opponent with a favorable relative position. This turns out to be a variation on the "Herbst maneuver" mentioned earlier – in fact it was common for the LCS to rediscover existing maneuvers, as well as discover novel variations.

In a later work (Smith et al. 2000), both opponents were controlled by a separate LCS. As Smith et al. (2002) describe, in this scenario, reminiscent of the continuous iterated prisoners' dilemma (IPD), the resulting dynamic system has four potential attractors, the most attractive of which is an "arms race" dynamic, in which each pilot continuously improves his or her

▢ Fig. 1

The matrix of initial conditions for the combat simulations. Reprinted from Smith et al. (2002), with permission from Elsevier.

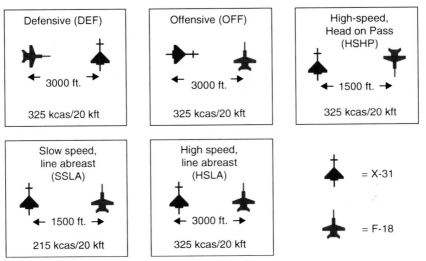

▢ Fig. 2

An example of a novel maneuver evolved by the learning classifier system under the HSHP starting condition (see ❷ *Fig. 1*). The aircraft on the *left* is following a fixed, but reactive strategy; the aircraft on the *right* is following a strategy evolved by the LCS, which in turn is a new variation on the Herbst maneuver. Reprinted from Smith et al. (2002), with permission from Elsevier.

strategies. Smith et al. (2000) explored various setups and indeed found that an arms race effect reliably occurs under some conditions.

2.1.4 Findings and Impact

Smith and Dike (1995) and Smith et al. (2000) contain and discuss several more examples of discovered maneuvers, including some revealing expositions of arms races that develop under the conditions of two LCSs in combat with each other. One clear result of this (still ongoing) work is the real impact it has had on its industrial collaborators. In general, the aerodynamics of a new aircraft can be understood before the first prototype is flown; but, the complexities of piloting and combat, and consequently any real knowledge about the potential combat performance with skilled pilots, are much more difficult to predict. Discovering successful combat maneuvers in the way described has many advantages – in particular, without the cost of test pilot time or prototype construction, LCS experiments generate rich sources of information on combat advantages (or disadvantages) that can be fed back to designers, pilots, and customers. The system described briefly here, and more fully in Smith et al. (2000, 2002), and Smith and Dike (1995), has resulted in several novel strategies that have been approved by test fighter pilots and continue to provide useful results in a highly complex, real-world domain.

2.2 Developing an Expert Checkers Player

The next example comes from the area of "computational intelligence." The term "computational intelligence" has come to be associated largely with the major fruits of nature-inspired computing, particularly, evolutionary, neural, and fuzzy techniques. This is not to be confused with the older, more well-established term "artificial intelligence," which stands for the much wider enterprise of, by whatever means, designing algorithms and systems that perform functions that can be called "intelligent." Artificial intelligence (AI) includes classic areas and techniques such as expert systems, heuristic tree search, machine vision, natural language processing, planning, and so forth, as well as the growing range of nature-inspired techniques. AI is concerned with everything from full-scale intelligent systems, through to the details of appropriate heuristics for edge detection in images from a narrow domain.

Basically, almost any activity, other than those that are "easy" for computers to handle with standard techniques, can be labeled with the adjective "intelligent." However, via natural computing, achievements have been made that will seem genuinely surprising to many people. It is no great surprise, for example, that computers can design, more successfully than humans, effective production schedules for factories with thousands of jobs per day. However, it perhaps is surprising that one can produce a software that plays checkers at the level of an expert, without encoding any expert knowledge of the game.

2.2.1 Blondie24

In 1999, on an Internet gaming site called "The Zone," an online checkers player with the screen name *Blondie24* regularly played against a pool of 165 human opponents, and achieved a rating of 2,048, placing it well into the top half a percent of checkers players using that site. Blondie24 learned to play well at checkers, as did all of the good human players using that site (or otherwise). However, "she" was (and still is) a computer program.

In common with many successful artificial intelligence game playing programs, Blondie24 (Chellapilla and Fogel 1999a, b, 2001; Fogel 2002) incorporates a minimax algorithm (Russell and Norvig 2003) to traverse the game tree induced by the available moves from the current position. However, individual nodes in the tree are evaluated by an artificial neural network (ANN). The input to this ANN is a specialized representation of the current state of the game, and the output is a single value that is then used by the minimax algorithm. So far so clear – one can perhaps imagine that a well-trained or well-designed ANN could be capable of returning values in this context that would translate to competent checkers playing. But how can one design, or train, such an ANN? In Blondie24's case, training was accomplished by using an evolutionary algorithm. A population of such ANNs played against each other, accumulating points over many games. The result of a game between two such ANNs comes down to a single value (per ANN) – either 1 (win), 0 (draw), or −2 (lose) – and the overarching evolutionary algorithm operates by regarding the fitness of an ANN as its total score after a number of games. In each "generation" of this evolutionary algorithm, the ANNs with the lowest scores are eliminated, and new ones are generated by making mutant copies of the better performers, and so it continues.

For several reasons, Blondie24's design and its success are both surprising and significant. Its prowess at checkers does not rely on tuition by human experts. Instead, it emerges from the evolutionary algorithm process, guided only by the bare, raw total of points earned after playing several games. If an individual had a fitness of six, for example, it was considered better (and hence had more chance for selection as a parent) than an individual with fitness four. However, this takes no account of the distribution of wins, losses, and draws. The individual with fitness six may have won six games and drawn four, while the individual with fitness four may have won eight games and lost two.

Guided only by this summary measure of performance, an evolutionary algorithm was able to traverse the space of checkers-playing ANNs (or, more correctly, ANNs for evaluating game positions in the context of minimax search) and emerge with expert-level players. It is worth covering in more detail the approach taken to generate Blondie24, which is done next, following the treatment provided in Chellapilla and Fogel (2001).

2.2.2 Checkers: The Game

Checkers, known in some countries as "draughts," involves an eight-by-eight board with squares of alternating colors, equivalent to a chessboard. Each player has 12 identical pieces, and the initial game position is as detailed in ❷ *Fig. 3*.

When it is a player's turn to move, the allowed moves are as follows: an individual piece can move diagonally forward by one square; or an individual may jump over an opponent's checker into an empty square. Such a "jump" is only allowed if it takes two diagonal steps in the same direction, the first such step is occupied by an opponent's piece, and the second step is currently empty. After a jump, the opponent's piece is removed from play.

If one or more jump moves are available, then it is mandatory for the player to make such a move. If an opponent manages to find itself in the final row (from their side's viewpoint), it becomes a "king" piece. It is then able to move either forward or backward, but otherwise

◻ **Fig. 3**
The initial position in a game of checkers. The White player moves upward, and the Black player moves downward.

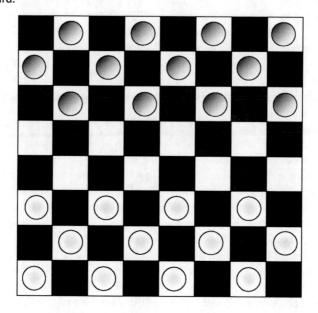

follow the same rules. The object of the game is to reach a position in which the opponent has no possible moves – a common way in which this happens is for the opponent's pieces to all have been removed.

2.2.3 Representing the Board and Evaluating Moves

Chellapilla and Fogel (2001) used a straightforward and sensible approach to encoding a board position. The current state of the game is simply represented by a vector of 32 numbers, one for each board position. The numbers in a position are either $-K$, -1, 0, 1, or K, where K represents a value assigned to a king. From the viewpoint of a given player, a 1 or a K at a given board position represent, respectively, either a standard piece or a king at that position, while the negative values are used to represent the opponent's pieces, and 0 indicates an empty position. In Chellapilla and Fogel's work, K was not preset. Rather than bias the process toward giving a king any particular relative value over an ordinary piece, the value of K was itself subject to evolution.

When a move is to be made, Blondie24 operates by evaluating, in turn, each of its possible moves. Any such move leads to a future board position, and this future board position is evaluated by the ANN. The input to the ANN is therefore this 32-dimensional vector. As is well known from ANN theory, any reasonable ANN architecture (in terms of the number of hidden layers and the numbers of nodes in each layer) might suffice in being capable of then performing the appropriate mapping from input vector to appropriate, useful output. The difficulty, as ever, is in choosing an appropriate training regime, which promotes learning of suitable features and components of the problem state that are useful guides toward a proper evaluation. After initial experiments with a more straightforward neural network architecture (which did not encapsulate the spatial information that human players take for granted), Chellapilla and Fogel designed the architecture of the ANN in a way that highlighted potentially appropriate features. This was done as follows.

Each 3×3 block on the board was represented by its own unit in the first hidden layer. That is, given any specific 3×3 block, one of the units in the first hidden layer received incoming connections from that specific set of nine inputs from the input layer (from the nine parts of the vector corresponding to the component positions of that 3×3 block), and had no incoming connections from any of the other units. In this way, a specific signal emerging from this unit, for later processing in subsequent layers, summarizes the state of play in that specific 3×3 block. The first hidden layer contained such a unit for each of the 36 different 3×3 blocks on the board. In just the same way, each 4×4 block, 5×5 block, 6×6 block, 7×7 block, and 8×8 block (of which there was of course just one) was represented in the first hidden layer by its own unit. This resulted in a set of 91 units which comprised the first hidden layer.

The complete picture of the ANN's architecture is given in ❯ *Fig. 4*. Between the input layer, which simply carries 32 units, one per board position, and the first hidden layer, the connections are arranged according to the specific feature encoded by each of the units in hidden layer 1. Between the pairs of layers, the connections are all complete – for example, each unit in hidden layer 1 has a feedforward connection to each unit in hidden layer 2, and similarly for hidden layers 2 and 3, while every unit in hidden layer 3 is connected to the single output unit. The output unit receives an additional input, which is the sum of the 32 board positions.

In total, including bias weights, there are 5,046 connections in this network, each of which is a real-valued weight subject to the evolution process. In addition, every hidden layer unit has a bias input, which means an additional weight to be evolved. Each unit in the hidden layers operates in the standard way, common in most ANN applications, by calculating the weighted sum of its inputs and applying the hyperbolic tangent function, resulting in an output signal

◻ **Fig. 4**
The architecture of the Blondie24 artificial neural network.

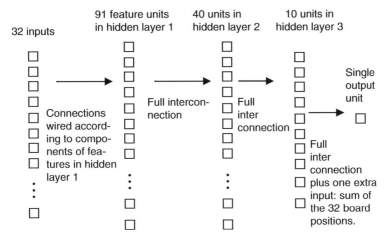

strictly between −1 and 1. From the perspective of the ANN "player," this ultimate scalar value is directly used as an estimate of the value of this board position: the closer it is to 1, the better for the ANN. However, where the board position was actually a win for the ANN, the value was taken to be precisely 1, and if it was a win for the opponent, the value was taken to be −1.

2.2.4 Evolving Checkers Players

The process begins with a population of 15 such ANNs, which are initialized randomly. Every connection weight and bias value is given a value chosen uniformly at random from the interval $[-0.2, 0.2]$, and with K set initially at 2.0. In common with the practice of evolutionary programming and evolution strategies, each individual in the population also contained a vector of step-size parameters. For every connection weight, and every corresponding bias unit, there was also a step-size parameter governing the range of mutations that would be applied to that parameter. That is, when a weight or bias parameter was mutated, this was done by adding a Gaussian perturbation whose mean was 0 and whose variance was provided by the associated step-size parameter in the chromosome. The step-sizes were initially all set at 0.05, and then subject to evolution along with the other parameters.

Whenever an ANN was selected as a parent, its offspring was generated as follows: first, each of the step-size parameters was mutated, by multiplying it with a random number from a specific exponential distribution, and every weight and bias parameter was mutated by adding a Gaussian perturbation whose step size was the associated step-size parameter, as indicated. Finally, recall that each individual also carries its own value for K, which is also subject to evolution. This was mutated by adding a perturbation chosen uniformly at random from the set $\{-0.1, 0, 0.1\}$, but was protected from moving below 1 or above 3.

During the evolution process, each ANN played one game each against five opponents, selected uniformly from the population. With the scoring for individual games as indicated, the ANN would therefore accumulate a score over these five games ranging from ten (all losses) to five (all wins). A game was declared as a draw (zero points) if it lasted for 100 moves. Essentially, in each generation, each ANN took part in around ten games, and the top 15 (in terms of points received) became parents for the next generation. Each individual game was played using a minimax alpha–beta search set to four-ply (with extended ply in a number of special cases). After 840 generations in which evolving ANNs played against each other, the best-resulting ANN was then harvested and recruited to play against human opponents on the Internet gaming site "The Zone." The next subsection summarizes the surprising and remarkable resulting performance of this ANN.

2.2.5 Humans Versus Evolved ANNs

Over a 2-month period, the evolved ANN, eventually named "Blondie24" (which was successful in attracting opponents) played 165 games against human opponents at "The Zone," although opponents were not aware they were playing against a computer program. In these games, the ANN used an eight-ply search, and faced a variety of opponents. The ANN's performance placed it at better than 99.6% of all the (rated) players using the site. On one occasion, the ANN beat an expert-level player (with a rating of 2,173, just below the master level of 2,200) who was ranked 98th of over 80,000 registered players.

Chellapilla and Fogel (2001) performed some comprehensive control experiments, which showed that the evolved ANN operated with a clear advantage over a system that simply used the piece differential as the basis for choosing moves in an eight-play approach. In particular, they compared the ANN with a piece-differential-based player, on the basis of using equal CPU time in their look-ahead search at each move; this disadvantages the ANN, since it has over 5,000 weight parameters involved in its heuristic calculation, so the piece-differential player can look further ahead in the time available. These experiments showed conclusively that the evolved ANN was a significantly better player in both equal-ply and equal CPU-time conditions.

The achievement of Blondie24 is remarkable from many viewpoints: particularly the essential simplicity of its approach, the fact that the search landscape for the evolutionary algorithm was so huge, and the fact that fitness assessment was a relatively coarse measure of a network's performance. A straightforward assessment of Blondie24's "message" is that it exemplifies the flexibility and potential of evolutionary search, even when this is recruited to search a coarse-grained 10,000-dimensional landscape (the evolution strategy that was employed optimized both a weight and a step-size parameter for each connection). Achieving expert-level performance (over 2,000 points) is considerably superior to most humans. Perhaps not surprisingly, this is also certainly superior to a simpler (but seminal) approach in this area by Samuel (1959), which attempted to derive, by an iterative learning process, a polynomial board rating function. Chellapilla and Fogel (2001) note that this was considered to rate below 1,600 in the opinion of the American Checkers Federation games editor.

The world champion checkers program, Chinook (Schaeffer et al. 1996), is rated at over 2,800, over 100 points above its closest human competitors (Schaeffer 1996). In fact, it is now known that Chinook can never be defeated in "go-as-you-please" checkers, in which there are no restrictions on the initial moves. The chief difference between Blondie24 and Chinook is the amount of built-in specialized knowledge. In Chinook, the level of such knowledge is very substantial indeed; in Blondie24, it is virtually none. Along with many other elements informed by careful expert knowledge and tuning, Chinook incorporates a database of games from previous grand masters and a complete endgame database for *all* cases that start from ten pieces or fewer. Blondie24 and Chinook represent entirely different artificial intelligence approaches to designing a game-playing program. It is not difficult to argue that the approach taken by Blondie24 is the more interesting and impactful – from no prior knowledge, other than a built-in awareness of the rules of the game, an expert-level player emerged from the evolutionary process, providing a very tough, usually insurmountable challenge to all but a very small percentage of human players.

Finally, since the checkers research, Chellapilla and Fogel's approach was extended to address chess, by combining the coevolutionary spatial neural network approach with domain-specific knowledge (Fogel et al. 2004, 2006). The result was an evolved chess player that earned wins over Fritz 8, which was the fifth best computer program in the world at that time.

2.3   Discovering Financial Trading Rules

Financial markets are complex and ever-changing environments in which groups of individuals, companies, and other investors are always competing for profit. There are many opportunities in this area for machine learning and optimization methods, and consequently a variety of natural computation approaches to be exploited, and a chapter in this volume is indeed devoted to this topic. This section focuses on one specific thread of research in this area – which has a simply grasped approach and a straightforward task to solve. This is the use of GP to discover new and valuable rules for financial trading.

It is now common to see applications of evolutionary computation applied to the financial markets (Brabazon and O'Neill 2005, this volume). GP (Koza 1992; Angeline 1996; Banzhaf et al. 1998) is particularly prominent in terms of the degree to which it has recently been applied in finance (Chen and Yeh 1996; Fyfe et al. 1999; Allen and Karjalainen 1999; Marney et al. 2001; Chen 2002; Cheng and Khai 2002; Farnsworth et al. 2004; Potvin et al. 2004). This section focuses on the specific area in finance known as *technical analysis* (Pring 1980; Ruggiero 1997; Murphy 1999; Lo et al. 2000). Technical analysis is a set of techniques that forecast the future direction of stock prices via the study of historical data. Many different methods and tools are used, all of which rely on the principle that price patterns and trends exist in markets, and that these can be identified and exploited.

Common tools in technical analysis include indicators such as moving averages (the mean value of the price for a given stock or index over a given recent time period), relative strength indicators (a function of the ratio of recent upward movements to recent downward movements). There have been a number of attempts to use GP in technical analysis for learning technical trading rules, and a typical strategy is for such a GP-produced rule to be a combination of technical indicator "primitives" with other mathematical operations. Such a rule is often called a "signal." For example, GP may be employed to find both a good buy signal and a good sell signal – that is, one rule that, if its output is above 0, indicates that it is a good time to buy, and a different rule indicates when it is a good time to sell.

Early attempts to use GP in technical trading analysis were by Chen and Yeh (1996) and Allen and Karjalainen (1999). However, although GP could produce profitable rules for the stock exchange markets, their performance did not show any benefit when compared to the standard buy-and-hold approach. "Buy-and-hold" simply means, for a given period, buying the stock at the beginning of the period, and selling at the end – hence, always a good idea in a market that generally moves up during the period.

More recent applications of GP in this context have been more encouraging (Marney et al. 2000, 2001; Neely 2001). Becker and Seshadri's (2003a, b, c) work is looked at in particular, which found GP-evolved technical trading rules that outperformed buy-and-hold (at least if dividends are excluded from stock returns). In turn, their approach was founded in Allen and Karjalainen (1999), with various modifications. After giving some detail of the overall approach, further experiments from Lohpetch and Corne (2009, 2010), which probed certain boundaries of the technique and examined its robustness, are summarized.

2.3.1 Becker and Seshadri's Approach to Evolving Trading Rules

Becker and Seshadri (2003a, b, c), based on Allen and Karjeleinen (1999), used a fairly standard GP approach and found rules that significantly outperformed buy-and-hold on average over a 12-year test period of trading with the Standard & Poors (S&P 500) index. Their GP's function set contained the standard arithmetic, Boolean, and relational operators, and the terminal set included some basic technical indicators. An example of a specific rule found by their method is in ❷ *Fig. 5.*

☐ Fig. 5
Example of a trading rule found by Becker and Seshadri's GP approach.

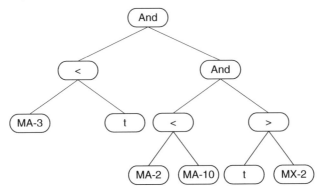

The rule in ❷ *Fig. 5* has the following basic interpretation "the 3-month moving average (MA-3) is less than the lower trend line (t) and the 2-month moving average (MA-2) is less than the 10-month moving average (MA-10) and the lower trend line (t) is greater than the second previous 3-month moving average maxima (MX-2)." This signals trading behavior in the following way: If the trader is currently out of the market (no stocks invested in the S&P 500), and the rule evaluates to *true*, then *buy*; if the trader is currently in the market, and the rule becomes *false*, then sell. This procedure assumes a fixed amount to invest (e.g., $1,000) whenever there is a buy signal.

The remainder of this subsection explains the approach in more detail, and tries to emphasize the key points that are necessary to replicate similar performance. In passing, the main ways in which Becker and Seshadri modified the original approach of Allen and Karjeleinen are noted. These were: monthly trading decisions rather than daily trading; a reduced function set in the GP approach; a larger terminal set in the GP approach (with more technical indicators); the use of a complexity-penalizing element to avoid over-fitting; and finally, modifying fitness function to consider the number of periods with well-performing returns, rather than just the total return over the test period. In combination, these methods enabled Becker and Seshadri to find rules that outperformed buy and hold for the period they tested, when trading on the S&P 500 index. It is an open question as to which modifications were most important to this achievement; however, Lohpetch and Corne (2010) begin to answer that question, as will be seen, by showing (as is intuitively the case) that it is increasingly easier to find good rules as one changes the trading interval from daily to weekly, and then to monthly.

In the following, we exclusively use S&P 500 data (as did Allen and Karjeleinen 1999 and Becker and Seshadri 2003a, b, c); so, the "portfolio" is the fixed set of 500 stocks in the S&P 500 index, which aggregate to provide daily price indicators.

2.3.2 Function and Terminal Sets Used by Becker and Seshadri

In Becker and Seshadri's GP approach, the function set comprises simply the Boolean operators and, or, and not, and the relational operators > and <. The terminal set comprises the following, in which "period" was always *month* in Becker and Seshadri's work, but later Lohpetch and Corne (2010) will be discussed in which it could be day, week, or month in different experiments.

- Opening, closing, high and low prices for the current period;
- 2-, 3-, 5-, and 10-period moving averages;
- Rate of change indicator: 3-period and 12-period;
- Price Resistance indicators: the two previous 3-period moving average minima, and the two previous 3-period moving average maxima;
- Trend Line indicators: a lower resistance line based on the slope of the two previous minima; an upper resistance line based on the slope of the two previous maxima.

The n-period moving average at period m is the mean of the closing prices of the n previous months (included m). The n-period rate of change indicator measured at period m is: $(c(m) - c(m - (n - 1))) \times 100)/c(m - (n - 1))$, where $c(x)$ indicates the closing price for period x. Previous maxima MX1 and MX2 are obtained by considering the 3-period moving averages at each point in the previous 12 periods. Of the two highest values, that closest in time to the current period is MX1, and the other is MX2. The two previous minima are similarly defined. Finally, to identify trend line indicators, the two previous maxima are used to define a line in the obvious way, and the extrapolation of that line from the current period becomes the upper trend line indicator; the lower trend line indicator is defined similarly, using the two previous minima.

2.3.3 Becker and Seshadri's Fitness Function

The fitness function has three main elements. First is the so-called "excess return," indicating how much would have been earned by using the trading rule, in excess of the return that would have been obtained from a buy-and-hold strategy. The other elements of the fitness function were introduced by Becker and Seshadri to avoid over-fitting. These were as follows: a factor that promoted fitness for trading rules that were less complex (e.g., with reference to ❯ *Fig. 5*, a less complex rule is one in which the tree has smaller depth); and, a factor that considered "performance consistency" (PC), favoring rules that generally were used often, each time providing a good return, rather than rules that were very fortunate in only brief periods.

   In more detail, the excess return is simply $E = r - r_{bh}$, where r is the return on an investment of \$1,000, and r_{bh} is the corresponding return that would have been achieved using a buy-and-hold strategy. To calculate r, Allen and Karjeleinen (1999) and Becker and Seshadri (2003a, b, c) used:

$$r = \sum_{t=1}^{T} r_t I_b(t) + \sum_{t=1}^{T} r_f(t) I_s(t) + n \ln\left(\frac{1-c}{1+c}\right)$$

in which: $r_t = \log P_t - \log P_{t-1}$, indicating the continuously compounded return, where P_t is the price at time t. The term $I_b(t)$ is the buy signal, either 1 (the rule indicates *buy* at time t) or 0. The sell signal, $I_s(t)$, is analogously defined. So, the first component gives the return on investment from the times when the investor is in the market, and the second component, $r_f(t)$, indicates the risk-free return which would otherwise be available, which is taken for any particular day t from the published US Treasury bill data (these data are available from http://research.stlouisfed.org/fred/data/irates/tb3ms). The second component therefore represents time out of the market, in which it is assumed that the investor's funds are earning a standard risk-free interest. Finally, the third component is a correction for transaction costs, estimating the compounded loss from the expenditure on transactions; a single transaction is assumed to cost 0.25% of the traded volume – for example, $2.50 for a transaction of volume $1,000. The number of transactions sanctioned during the period by the rule is n.

The second main part of the fitness function, r_{bh}, is calculated as:

$$r_{bh} = \sum r_t + \ln\left(\frac{1-c}{1+c}\right)$$

in which r_t is as indicated above. This calculates the return over the period from risk-free investment in US Treasury bills, involving a single buy transaction.

Becker and Seshadri's complexity-penalizing adjustment works as follows: Given a rule that has *depth* and fitness value (excess return) f, the adjusted fitness becomes $5f/\max(5, depth)$. This involves the constant 5 as a relatively arbitrary desired maximal depth, and in the trading rule evolution context, there has been little parametric investigation around this value so far. The other of Becker and Seshadri's modification to the excess return fitness function, PC works as follows. The excess return E is calculated for each successive group of K windows of a certain length covering the entire test period.

The value returned is simply the number of these periods for which E was greater than both the corresponding buy-and-hold return (from investing in the index over that period) and the risk-free return during that period. For example, if the rule is evaluated over a 5-year test period, the PC version of the fitness function might use 12-month windows. Clearly, there are five such windows in the test period, and the fitness value returned will simply be an integer between 0 (the rule did not outperform buy-and-hold and risk-free investment in any of the five windows) and 5 (the rule was more successful than both buy-and-hold and risk-free returns in all of the windows).

At last, the above background enables one to state the fitness function used (with minor variations in each case) in Becker and Seshadri (2003a, b, c) and Lohpetch and Corne (2009, 2010). The fitness of a GP tree of depth d in these studies was the performance-consistency-based fitness (i.e., a number from 0 to X, where there were X windows covering the test data period) adjusted to penalize undue complexity by $5f/\max(5, d)$, in which f is the number of the X windows in which both the corresponding buy-and-hold return and the risk-free return were outperformed by the rule.

2.3.4 Some Illustrative Results

Reported here are some results that show how this approach performs on various windows of time when trading with the S&P 500 index. The results shown are some of those from Lohpetch and Corne (2010), and the subset of those that were obtained under the same test conditions as used by Becker and Seshadri (i.e., monthly trading for a specific training and test window) are quite indicative of Becker and Seshadri's own results. However, it is worth first discussing some further details of the way that the GP method was set up for the experiments.

Although perhaps not always the case, it seems that the precise choice of mutation and crossover methods makes little real difference in this application; the chance of evolving effective trading rules seems clearly related to a good choice of function and terminal sets for the expression trees, as well as a wise choice of fitness function. Although, as will be seen later, the frequency of trading is a significant factor. Meanwhile, Lohpetch and Corne (2009, 2010) used standard mutation operators, as described by Angeline (1996), namely, *grow, switch, shrink*, and *cycle* mutation, and used standard subtree-swap crossover (Koza 1992). Finally, it is noted that, in the experiments whose results are summarized next, the population was initialized by growing trees to a maximum depth of 5; however, no constraint was placed on tree size beyond the initial generation, other than the pressure toward less complex trees which is a part of the fitness function.

Presented now are some results that indicate the performance achievable by such a GP system as described in the last section. In the experiments summarized here, from Lohpetch and Corne (2009, 2010), a population size of 500 was used, and other relatively standard GP settings, with a run continued for 50 generations. Here results for each of the daily, weekly, and monthly trading are shown, and it is found that outperformance of buy-and-hold can indeed be achieved even for daily trading, but moving from monthly to daily trading, the performance of evolved rules becomes increasingly dependent on prevailing market conditions. The data used is the S&P 500 index from 1960 onward. In Becker and Seshadri's demonstration of outperforming buy-and-hold, only monthly trading was used, and their results arise from training the rules over the period 1960–1991, and evaluating them on a test period spanning 1992–2003. This corresponds to "MonthlySplit1" in the following; however, it is clear from Lohpetch and Corne (2009) that more robust performance is obtained when a validation period is used. The following illustrative results therefore reflect a training/validation/test regime in which the GP training run evaluated fitness on the training period only, but the rule that achieved the best performance on a validation period was harvested, and this was the rule evaluated on the test period.

Results for four different monthly trading data splits are summarized below. The splits themselves are as follows, in which N gives the length of the validation period in years, immediately following the training period, and K gives the length of the test period in years, again immediately following the validation period.

- MonthlySplit1: 31 years training; $N = 12$, $K = 5$
- MonthlySplit2: 31 years training, $N = 8$, $K = 8$
- MonthlySplit3: 31 years training, $N = 9$, $K = 9$
- MonthlySplit4: 25 years training, $N = 12$, $K = 12$

Corresponding splits for the weekly and daily trading experiments are also summarized here (for details see Lohpetch and Corne 2010). Four different weekly trading and daily trading data splits were also investigated, roughly corresponding to the monthly data splits in terms of the number of data points in each split. For example, WeeklySplit1 involved 366 weeks trading, 158 weeks validation, and 157 weeks testing. Similarly, the training periods for the daily splits were approximately 1 year in length. The four different weekly and daily splits started at different times, spread evenly between 1960 and 1996.

❯ *Figure 6* shows the four monthly data splits aligned against the S&P 500 index for the period 1960–2008. Note that the market movements were net positive in each part of each split, indicating that outperforming buy-and-hold was in all cases a challenge.

In Lohpetch and Corne's (2010) experiments, they also explored different lengths of window for the PC element of the fitness function. In Becker and Seshadri's work, the PC approach clearly results in improved performance, however, they only reported on the use of

■ Fig. 6
The S&P 500 index over the period 1960–2008, illustrating the four data splits for the case of monthly trading.

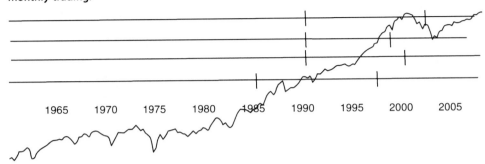

■ Table 1
Summary of results for monthly trading

Data split	PC period	Trials outperforming buy-and-hold	PC period	Trials outperforming buy-and-hold
Monthly Split1	6	10 out of 10	12	10 out of 10
Monthly Split2	6	9 out of 10	12	10 out of 10
Monthly Split3	6	10 out of 10	12	9 out of 10
Monthly Split4	6	10 out of 10	12	10 out of 10
Monthly Split1	18	10 out of 10	24	10 out of 10
Monthly Split2	18	8 out of 10	24	10 out of 10
Monthly Split3	18	8 out of 10	24	7 out of 10
Monthly Split4	18	10 out of 10	24	10 out of 10

12-month windows. Lohpetch and Corne experimented with different lengths for these windows for each trading situation, namely: 6-, 12-, 18-, and 24-month periods for monthly trading; 12 and 24 weeks for weekly trading, and 12 and 24 days for daily trading.

For each trading period (monthly, weekly, daily), Lohpetch and Corne did ten runs for each combination of data split and consistency of performance period. The outcome of the ten runs is summarized in ❷ *Tables 1–3*, simply as the number of times that the result outperformed buy-and-hold.

As ❷ *Table 1* shows, monthly splits 1 and 4 were clearly well disposed to good performance, but performance was also rather robust on the other monthly splits. Note that outperforming buy-and-hold would seem to be more likely, according to a priori intuition, when the performance of buy-and-hold in the test period is relatively weak, but this is not the case for monthly splits 1 and 4 (see ❷ *Fig. 6*). The results are quite impressive from many points of view. In many cases, ten tests out of ten showed that a simple trading rule evolved by GP was able to outperform buy-and-hold in an upwardly moving market.

❷ *Table 2* shows the results, summarized in the same way, for the case of weekly trading, and ❷ *Table 3* presents the corresponding results for the case of daily trading. These clearly

◻ Table 2

Summary results for weekly trading

Data split	PC period	Trials outperforming buy-and-hold	PC period	Trials outperforming buy-and-hold
Weekly Split1	12	2 out of 10	24	7 out of 10
Weekly Split2	12	10 out of 10	24	5 out of 10
Weekly Split3	12	4 out of 10	24	4 out of 10
Weekly Split4	12	10 out of 10	24	10 out of 10

◻ Table 3

Summary of results for daily trading

Data split	PC period	Trials outperforming buy-and-hold	PC period	Trials outperforming buy-and-hold
Daily Split1	12	0 out of 10	24	0 out of 10
Daily Split2	12	0 out of 10	24	0 out of 10
Daily Split3	12	10 out of 10	24	9 out of 10
Daily Split4	12	2 out of 10	24	4 out of 10

show increasingly less robust results. It certainly seems that this relatively straightforward GP method can find robust rules for weekly trading that outperform buy-and-hold in some circumstances (splits 2 and 4), with less reliable performance in other cases. However, Lohpetch and Corne (2009, 2010) were not able to discern any pattern that explains this from analyses of the data splits. Finally, for daily trading, ❷ *Table 3* shows that outperforming buy-and-hold is less likely, with strong performance in only one of the four data splits, and very poor performance in two of the data splits.

2.3.5 A Brief Discussion

The investigation of GP in financial applications and, in particular, the use of it to discover technical trading rules, remains an active thread of research in both industry and academia. In the published academic research, it was commonly found in earlier studies that rules found by GP were profitable, but usually not competitive with straightforward "buy-and-hold" strategies. However, as seen earlier, the situation is changing and it now seems that progress is being made in finding ways to use GP to produce effective and interesting rules that might be used by individual traders. There are several caveats, and of course this enterprise is only one thread of work in a wide area that also involves natural language understanding and many other areas of machine learning (e.g., to spot ideal trading opportunities based on the latest online news). However, this work represents another example of the way in which natural computation can help generate strategies for complex situations which are competitive with those that are self-designed.

It should also be noted that the approach described in this section is far from the last word in the application of GP to the specific area of technical trading. Pains have been taken to describe a classic approach, and it has been shown that it can indeed find robustly profitable trading rules under a range of conditions – however, several more sophisticated ways to use GP in this area also exist. For example, rather than simply evolve a single rule that encapsulates both a buy and a sell signal, different rules can be evolved separately for buying and selling. Also, it is noted that interested researchers may pick up code for evolving technical trading rules (written by Dome Lohpetch) from the following site: http://www.macs.hw.ac.uk/~dwcorne/gptrcode/.

It is also worth mentioning alternative directions which attempt to gain on buy-and-hold by including risk metrics in the rules (or in their evaluation). Typically, a risk measure such as the Sharpe ratio (Sharpe 1966) is used to normalize the estimate of financial return, effectively downgrading the performance of rules that promote trading in volatile conditions, promoting rules more likely to be applied by investors. For example, in attempting to build on work by Fyfe et al. (1999), Marney et al. (2000, 2001) included the use of metrics for calculating risk, although still did not outperform buy-and-hold. More recently, Marney et al. (2005) used the Sharpe ratio and found that a technical trading rule easily outperformed simple buy-and-hold in terms of unadjusted returns, but not in terms of risk-adjusted returns. There is clearly much work still to do until techniques exist in the research literature that can robustly outperform buy-and-hold in a way that satisfies risk-conscious traders, although the progress and effort in this direction makes it clear that this will be achieved, as well as suggesting that private and unpublished research in commercial organizations has almost certainly achieved this already with appropriate use of GP and similar technologies.

3 Examples of Natural Computing's "Outreach" Elsewhere in Science and Engineering

In this section, two areas of natural computation are selected that have wider implication for significant areas of science and technology. Mostly, an application of a natural computing technique may produce excellent results in its domain, and the impact of those results, though potentially significant, tend to remain solidly within that domain. Progress in general financial mathematics, for example, will not be revolutionized by the trading application discussed in ❷ Sect. 2. However, sometimes an exemplar application will open up previously unconsidered possibilities in a whole subfield of science. In this section, two examples are discussed in which such broader consequences can be seen. The first is the use of (mostly) multi-objective evolutionary computation in the area of *closed-loop* optimization, in order to optimize a range of processes and products in the biosciences, process industries, and other areas. In this arena, evolutionary computing was never an "obvious" technique to try, given the potential cost in time; however, its use has time and again proven worthy, and this in turn leads directly to better and faster processes and products emerging from, for example, the use of the instruments that have been configured via evolutionary techniques. The second example area looked at in this section is the concept of *innovization*, which exploits multi-objective evolutionary computation in a way that leads to generic design insights for mechanical engineering (and other) problems. In multi-objective problems (see Deb 2001; Corne et al. 2003a), the result of solving the problem is (usually) a large collection of diverse solutions, each optimal in a sense, but traversing a Pareto surface of optimal from (for example) highly reliable and high-cost solutions to exceptionally cheap but less reliable ones. The notion of innovization is to exploit the prowess of evolutionary computing in obtaining such a diverse set, by further analyzing this collection of designs to find, as it turns out, previously unknown generic design rules, which seem to be true of all "optimal" designs, wherever they sit on this Pareto surface. A well-designed natural computing approach to a specific problem in mechanical engineering, for example, thereby leads to new design principles that can have much wider impact than simply solving the given problem.

3.1 Applications in Analytical Science: Closed-Loop Evolutionary Multi-objective Optimization

Knowles (2009) provides a detailed and comprehensive summary of historical origins and current work in the broad area of closed-loop optimization using evolutionary multi-objective algorithms. A similar but briefer treatment is provided here, including a summary of two of the several interesting modern case studies covered in Knowles (2009).

As Knowles (2009) points out, the idea of using an evolution-inspired technique for producing solutions to optimization problems has been explored for around 60 years so far, starting in the 1950s. The celebrated British statistician George E.P. Box used the term "closed-loop" in describing the kind of evolution experiments that were first investigated, while Ingo Rechenberg (a pioneer in evolutionary computation) used the phrase "evolutionary experimentation." In closed-loop evolution-inspired optimization, the evolution process is a combination of computation and physical experiment. The evaluation of candidate design solutions is done in the real world by conducting physical experiments. Much of the pioneering work in evolutionary computation (by Rechenberg and his team) was of this kind. In much more recent times, the closed-loop approach has been used, commonly with much success, in evolvable hardware research (see the chapter ❷ Evolvable Hardware in this volume), in evolutionary robotics research, as well as in microbiology and biochemistry. In this section, some brief example case studies are described to illustrate the increasingly wide emerging impact of this technique at the evolution/engineering interface.

With a focus on closed-loop evolutionary multi-objective optimization (CL-EMO) in particular, two cases are looked at: (1) instrument optimization in analytical biochemistry and (2) on-chip synthetic biomolecule design; these are described in greater detail in Knowles (2009) as well as further references detailed later, along with other quite different examples. However, before these case notes, historical development and fundamental concepts in closed-loop optimization and CL-EMO will briefly be looked at.

3.1.1 Historic Highlights in Closed-Loop Optimization

In Berlin in the 1960s, Rechenberg, Schwefel, and Bienert conducted a series of studies in engineering and fluid dynamics, in which they tested the idea of using a process inspired by evolution to search for new and successful designs. Their work clearly demonstrated that complex design engineering problems (including: the optimal shape of a fluid-bearing pipe, and the design of a supersonic jet nozzle) could be addressed in this way with rampant success (see Chap. 8 of Fogel 1998, as well as Rechenberg 1965, 2000). The design process itself was found to be efficient and scalable, and the results were highly effective. Rechenberg and his team were using an early example of an evolutionary algorithm, but in which only the selection and variation steps were done by a microprocessor; the rest, the evaluation of candidate designs, was done by constructing prototype designs and performing experiments to test their properties. Innovative solutions were found to all of the engineering design problems that they studied.

Predating Rechenberg's work, a similar principle was used by George Box, who introduced "evolutionary operation" (EVOP) in 1957. This was also an experimental method of optimization, which Box (1957) envisaged being used regularly in factories and similar processing facilities. Box's "closed-loop" scheme involved some human input, and was somewhat more deterministic than the approach taken in Berlin, but, just as Rechenberg's work, was inspired by principles from natural evolution. Box's methods were both successful and very influential (Hunter and Kittrell 1966), remaining in use today. Meanwhile, the work of Rechenberg's team was the beginning of the field of evolution strategies, one of the foundation stones of the current field of evolutionary computation.

Since these early studies, however, evolutionary computation as a whole has largely been concerned with entirely in silico optimization. The great majority of growth in this research area, as well as in industrial practice, concerns applications that involve convenient and entirely computational estimations of the fitness of computational abstractions of solutions. This is fine for a vast collection of scenarios, but there remains a need – in fact a quickly expanding one – for applications in which it makes sense for designs to be realized and evaluated physically throughout the simulated evolution process.

Research in evolvable hardware shows that, if the evolution process is given direct access to a complex physical structure, designs can be evolved that use entirely different principles than would be used by human designers, often exploiting aspects of the physics of the structures involved that are unfamiliar to human experts, or simply too difficult to use as part of the design process. Thompson and Layzell's work with field programmable gate arrays is exemplary of this. Meanwhile, evolutionary robotics projects have often relied upon the controllers being evolved in real time within physical robots, while they are performing real tasks in a real environment (Nolfi and Floreano 2004; Trianni et al. 2006). The benefits of such evolution experiments, exposed to and exploiting the true physics of the designs being evolved, are not just confined to evolutionary robotics (e.g., Davies et al. 2000; Evans et al. 2001).

Later, three further and recent uses of closed-loop evolutionary optimization are described, from recent work in which the third author (JDK) has been involved. Each of them is a scenario in which direct experimental evaluation of solutions is either the only option or is clearly preferable to simulation. Also each of them involves multi-objective evolution, a notable advance of the last 20 years (Fonseca and Fleming 1995; Coello 2000, 2006; Deb 2001; Corne et al. 2003a) which was not available to Box or Rechenberg. One of the several benefits of a multi-objective approach in these scenarios is that the different design objectives may simply be stated, without any need to define normalizations, weights, or priorities that mangle them into a single scalar (and usually misleading) measure of quality.

At this point, it is worth noting that there is widespread use of certain statistical methods in industry, for the types of problems that are considered in the "closed-loop" setting. The techniques employed are referred to as design of experiments (DoE) approaches, or sometimes experimental design (ED)-based approaches (Fisher 1971; Chernoff 1972; Myers and Montgomery 1995; Box et al. 2005). Such methods emphasize rational reasoning from all the information obtained so far, as opposed to more randomized exploration. Standard DoE is typically used for probing low-dimensional parameter spaces using few experiments, while evolutionary algorithms are typically used for optimization in high-dimensional spaces, using many evaluations, and optimizing many different types of structure, including permutations, graphs, networks, and so on. However, there is an increasingly disappearing divide between the two types of approach, especially since the advent of sequential DoE, which incorporates aspects of evolutionary computing. The closed-loop optimization scenarios considered in this section lie between these niches, and benefit from aspects of both approaches.

3.1.2 Fundamentals of Closed-Loop Evolutionary Multi-objective Optimization

In closed-loop EMO, candidate solutions to a problem are generated by an algorithm in computer simulation, but their evaluation is achieved by physical experiment. Evaluations are fed back to the algorithm and its generation of subsequent solutions is a function of these. Thus, the process has the form of a closed loop, being at least partially sequential. Closed-loop problems can be defined generally as multi-objective optimization problems in which, essentially, one needs to find some ideal solution vector x, which simultaneously minimizes each of a collection of k objective functions $f_1(x), f_2(x), \ldots f_k(x)$. Typically, a single physical experiment $g(x)$ yields the k measurements $f_1(x), f_2(x), \ldots f_k(x)$. That is, the k objectives are k different measurements that are made as the result of a single experiment, all of which need to be optimized in some way. Typically, at least some of the objectives will be in conflict (Brockhoff and Zitzler 2006), and no single solution is a minimizer of all functions. Rather, the improvement of one objective is only possible by sacrificing, or trading off, quality in some other objective. The solutions corresponding to optimal values of the k objectives are known as the Pareto set, and when plotted in objective space, form the Pareto front (see ❷ *Fig. 7*).

❑ **Fig. 7**

An illustration of a Pareto front for a typical optimization problem with two objectives, both of which have to be minimized. Each of the solutions on the Pareto Front (PF) are optimal in the Pareto trade-off sense. For example, for any solution on the PF, no solution exists that is improved in one objective without being degraded on another objective. Often, some solutions on PFs are "unsupported" – these are valid optimal solutions in the Pareto trade-off sense, but for any linear combination of the objectives that might be used in a single-objective simplification of the problem, they would not be optimal.

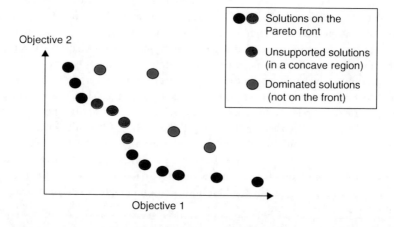

Since solving such a vector optimization problem usually leads to a set of solutions, rather than a single one, there is, in most practical applications, a need for decision making to select one solution from this set. This aspect of multi-objective optimization is important and well studied (Fonseca and Fleming 1998; Miettinen 1999; Branke and Deb 2005), and the various alternative approaches will not be covered here. Suffice to say that in the experiences detailed later, the EMO algorithms were designed to find whole Pareto fronts, with the expectation that a human decision maker would make the final decision using the information incorporated in the output Pareto front.

3.1.3 Example 1: Instrument Optimization in Bioanalytical Chemistry

Modern biotechnology and bioanalytics often involve large-scale experiments which impose heavy demands on sophisticated laboratory instruments. To achieve timely throughput, these experiments often necessitate using configurations of instruments that go beyond the manufacturer's recommended settings. This situation happened in the "HUSERMET" project, which was a collaboration between several UK health authorities, two pharmaceutical companies, and the University of Manchester, undertaken between 2006 and 2009 (www.husermet. org). In this project, human blood samples were collected from around 2,000 people over a 3 year period, with the aim of understanding ovarian cancer and Alzheimer's disease in terms of the variations in metabolites (the chemical products of metabolism) present in patients suffering from, or free from, these diseases. The samples were analyzed with the help of various modern technologies for characterizing complex samples, including laboratory instruments that performed gas chromatography mass spectrometry. The configuration of such instruments is always subject to a degree of optimization in order to ensure that the analytes being detected can be seen, with maximal sharpness and minimal noise. This optimization is usually (though not always) *ad hoc*, subject to much domain knowledge.

In the HUSERMET project, the need for a better instrument configuration optimization process arises from: the unusually large number and diversity of metabolites to be detected (around 2,000), the potential to vary around ten interacting instrument parameters, and the significantly conflicting nature of the optimization objectives. Instrument settings were needed that allowed fast processing of samples (preferably well under an hour), which conflicts with the desire to maximize the detection of the full complement of metabolites at low noise.

Optimizations of two instruments have been reported in detail in O'Hagan et al. (2005) and O'Hagan et al. (2007), respectively. The former study successfully used the evolutionary multi-objective algorithm PESA-II (Corne et al. 2001), but that study also directly inspired the development of the ParEGO algorithm (Knowles 2006), a multi-objective algorithm that is a hybrid of a surrogate modeling approach and an evolutionary algorithm. ParEGO was then used in the second study successfully, and settings derived from the evolutionary algorithms in both cases were subsequently used for the instruments to process the tasks of the HUSERMET project.

The major challenge in that project was the limited number of function evaluations that could be done. A function evaluation ties up an expensive instrument for an appreciable time, when it could otherwise be used more directly furthering the project's needs. This was even more bothersome, given the need to try to optimize three objectives simultaneously (chromatogram peaks, signal/noise ratio, and sample throughput). Only one instrument was available, and a single analysis of a serum sample takes between 15 min and over 1 h. The optimization process used 400 evaluations in total, with the EAs controlling the instrument

◼ **Fig. 8**

Chromatograms indicating detection performance of the instrument optimized in the HUSERMET project: (a) from the initial generation of search; (b) toward the end of the search process. In (b), both the number of peaks and the range of retention times over which peaks are detected have improved, while maintaining noise at low levels.

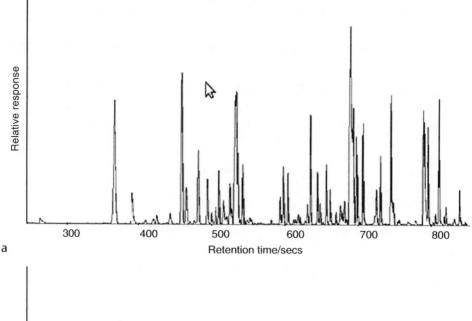

a

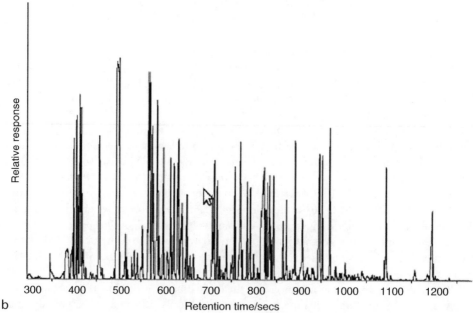

b

settings and loading samples through a robotic interface that was designed especially for the optimization process. In ❷ *Fig. 8* some chromatograms can be seen, which indicate the instrument's performance characteristics before (*upper*) and after (*lower*) optimization. The optimized result achieved approximately threefold increases in the quantity of peaks visible, while at the same time maintaining the signal/noise ratio at low levels, and achieving throughput of samples in around 20 min.

3.1.4 Example 2: Evolving Real DNA on Custom Microarrays

Another example covered in more detail in Knowles (2009) concerns the design of pharmacologically active, highly targeted macromolecules. This is a significant goal in modern medicine, especially in the context of *ab initio* design, where one seeks a molecule with specific properties and activity, but has little or nothing to go on (in the sense of existing molecules with similar properties). In recent research, novel microarray-based technology has been used in the automation of such *ab initio* molecule design. Experimental biotechnology platforms are now available, which can synthesize, and then experimentally test in a variety of ways, any specified DNA sequence. Being able to synthesize any given sequence, and subject it various tests, means there is far less need for computational models which, in the current state of the art, are far from good enough (or fast enough) to support such a process.

The microarray used in the work described next (and more fully in Knowles (2009) and references therein) is the so-called CustomArray technology, available from Combimatrix Corp, which can be used to synthesize up to 90,000 specific bespoke DNA sequences of up to 40 bases long in a single experiment. Once the sequences have been synthesized, they can be tested for a variety of properties, but usually the main property of interest is the ability to bind them to a particular target molecule. In the testing (or assay) process for binding ability, the chip holding the sequences is "washed" with a solution containing the target molecule, and some form of fluorescent tagging is used so that binding can be observed; further automated processes can then estimate the strength of binding.

Short strands of DNA (or RNA) that bind strongly to specific targets are called *aptamers*, and hundreds of these have been developed for a wide variety of applications. Before the recent microarray-based work at Manchester (which is what is discussed here, with full details in Knight et al. 2008), new aptamers were almost always discovered by a method called SELEX (Tuerk and Gold 1990), or in vivo selection, in which the DNA strands are evolved in a test tube by repeated rounds of high-pressure selection and random mutation. As indicated, however, in the microarray approach one knows precisely the sequence information for every sequence tested, and can even exactly specify mutations or other variations to perform. This is not the case in SELEX, and one of the many benefits for the microarray approach is that it allows extremely richer possibilities for borrowing and exploiting algorithms from evolutionary computation, machine learning, and statistics.

Knight et al. (2008) reports the first use of an evolutionary algorithm to produce a DNA aptamer on the B3 Combimatrix platform. This happened after ten generations of evolution, eventually discovering several 30-base long strands that bound very strongly to the target molecule, allophycocyanin. The work in Knight et al. (2008) used a DNA chip that could hold 6,000 strands. With 90,000 strands on a chip now possible in more modern technology, one main challenge (from the optimization perspective) is to determine the best way to exploit such massive population sizes. Wedge et al. (2009) have recently explored such questions in silico simulations using contrived search landscapes, as well as real trials on the DNA landscape, revealing, among other findings, that higher-than-standard mutation rates consistently outperformed a range of other setups. This echoes findings in Corne et al. (2003b), which also explored large population sizes and contrived in silico landscapes, partly to inform the (as then) emerging field of closed-loop protein evolution.

3.1.5   Some Concluding Notes on CL-EMO

For the examples described above, and several more in which CL-EMO has been used, building accurate computational models that could usefully replace real experiments is practically infeasible. The closed-loop alternative offers a more efficient and effective way toward the discovery of innovative solutions, easily making up for the time and expense of tying down the physical kit for the experimental period. One question often worth asking, however, is whether one needs to automate the optimization process at all in such scenarios. There is a processing step in which a computational process (here, e.g., an evolutionary multi-objective optimization algorithm) considers the latest experimental evaluation results, and outputs sample designs for the next sequence of physical evaluations – but this operation could easily be done instead by a domain expert. On the other hand, though, there are several objections to such human involvement: even experts can over-interpret results that are affected by noise or similar factors; similarly, humans are very prone to reason on the basis of simple models, ignoring interactions between parameters. Meanwhile, there is always a very real danger of experts preferring solutions that are (or are close to) known designs.

Problems where accurate computer modeling is infeasible, and for which closed-loop optimization is the efficient solution, are really quite common. For the moment, the main focus of the third author is on problems in modern biology, where there is a growing take-up of multi-objective optimization. Meanwhile, many other substantial areas are able to benefit greatly from CL-EMO; apart from drug discovery and development, large-scale problems such as flood defense design, forest fire control strategies, the location of renewable energy plants, and the task of genetically engineering more pest-resistant food crops and energy crops, can all be seen, to varying degrees, as closed-loop problems.

3.2   Innovization

This section describes a new idea, *innovization*, introduced in Deb and Srinivasan (2005, 2006), which (typically) exploits multi-objective evolutionary computation to find new and innovative design *principles*. Although optimization algorithms are routinely used to find an optimal solution corresponding to an optimization problem, the task of innovization stretches the scope beyond an optimization task and attempts to unveil new and innovative design principles relating to decision variables and objectives, so that a deeper understanding of the problem can be obtained.

Innovation is a common goal for engineers and designers, but there are actually very few (arguably no) systematic procedures for reliably achieving innovations. Goldberg (2002), however, suggests that a "competent" genetic algorithm can be an effective way to achieve an innovative design (and indeed there are numerous examples of innovative designs being discovered by evolutionary computation, including some discussed elsewhere in this chapter). However, the idea of innovization (Deb and Srinivasan 2005, 2006) extends this argument considerably, and gives a systematic procedure that can arrive at a deeper understanding of a given engineering design problem. This systematic procedure may lead to the discovery of new design principles – in particular, principles that are common to the diverse collection of optimal trade-off solutions.

Such common principles may, in many cases, provide a reliable recipe for solving given instances of the problem at hand. In this section, the innovization procedure will be explained and illustrated with two examples in engineering design. The material in this section borrows much from Deb and Srinivasan (2005), which introduces this idea, and contains several more examples. However, before looking at the procedure and examples, it will be helpful to recall some basics about the usually conflicting nature of objectives in the design process.

3.2.1 Multiple Conflicting Objectives in Design

The central idea in innovization involves the presence of at least two conflicting objectives for the design problem at hand. This is far from a limiting constraint – as argued in many places (see in particular Corne et al. 2003a for an introductory account of this argument), almost all realistic problems naturally involve several objectives.

Consider a typical design problem with two or more conflicting goals, such as an engine or generator whose mass needs to be minimized, but whose output needs to be maximized. Such a two-objective optimization task results in a set of Pareto-optimal solutions (see❷ *Fig. 7*). One of the "extreme" solutions will be the best if the interest is only in mass, while the other extreme solution will be ideal for the output consideration, and there will usually be several solutions in between these extremes, also optimal in a sense, all of which share the property that if they are better than another Pareto-optimal solutions in one objective, they will be worse in the other. The intermediate solutions are invariably good compromises within the extremes, and the solution that may eventually be chosen by the designer will often be among these, and its choice will often be helpfully informed by the knowledge that the designer obtains by viewing the shape and the nature of the trade-offs displayed by the entire set of solutions that form the discovered Pareto front. However, what is of particular interest here is that this set of solutions will typically be very diverse, but all sharing the property of Pareto optimality. The idea of innovization arises from the attempt to see whether this property of Pareto optimality, for any given problem, is manifest in concrete features that the diverse solutions share. Another aspect of this is that the process of obtaining such a wide variety of solutions is itself a significant investment in computation time; *innovization* is a way to exploit this significant investment by performing a posterior analysis of the obtained set of trade-off solutions that may result in a set of "innovized" principles relating to the given design problem.

In designing an electrical motor, for example, this posterior analysis might reveal a feature of the diameter of a certain component and the power output that is shared by all of the Pareto-optimal solutions, but not other solutions. Any such relationships discovered would clearly be of great importance to a designer, and perhaps point toward a recipe for future design tasks in the same domain, as well as spark new theoretical insights into the problem. These are just two of a range of benefits that the so-called "innovized principles" could lead to, as discussed further in Deb and Srinivasan (2005), along with a convincing argument that one can often expect such principles to exist.

3.2.2 How to Innovize

The innovization procedure proposed by Deb and Srinivasan (2005) consists of two phases: in the first phase, the idea is to simply try to obtain the Pareto-optimal solutions of the design problem in question. In the second phase, they then analyze the solutions and extract innovized principles. The first phase is not as straightforward as it sounds, since, of course, it can usually never be guaranteed that a true optima has been found for a realistic problem, unless an exhaustive search has been performed. However, the idea of the first phase is to do as well as one can in a reasonable time, since it is expected that the chance of obtaining valid principles of Pareto optimality is improved if one has true (or very close to true) Pareto-optimal solutions. In Deb and Srinivasan's procedure, this centrally involves making use of NSGA-II (Deb et al. 2002) (one of the most prominent and effective evolutionary multi-objective optimization algorithms) as the main engine in finding the Pareto front, but initially informed by a single objective method that has been used to find the extreme points on the Pareto front, and followed by various applications of a local search method and the normal constraint method (NCM) (Messac and Mattson 2004) to locally improve the output solutions from NSGA-II, as far as possible.

The second phase of innovization is then the analysis of the assumed Pareto-optimal solutions that emerge from the first phase. There is no fixed recipe for this process, other than to employ the usual common sense and expertise that underpins a data mining and knowledge discovery task in searching for commonality principles among these solutions that may become plausible innovized relationships. Deb and Srinivasan (2005) also pursue "higher level innovizations" after this phase, which involve returning to the original problem, but investigating different areas of the design space by looking at neighboring problems (e.g., with different boundaries and constraints on the design task); this then enables new principles to be discovered that are likely to be at a higher level than previously, mapping design constraints to design recipes.

Just two from the increasing collection of results of this above innovization procedure in engineering design applications are now described. These were first described in Deb and Srinivasan (2005), among several other examples.

3.2.3   Example 1: Gear Train Design

Deb and Srinivasan (2005) give the example of the design of a compound gear train, in which a specific gear ratio between the driver and driven shafts is desired. The problem is illustrated in ❯ *Fig. 9*, and is a modification to a problem solved elsewhere (Kannan and Kramer 1994;

◘ **Fig. 9**

A gear train with four gears (circles). The task is to achieve as close as possible a gear ratio of 6.931:1 between the driver and follower, while minimizing the sizes of each gear.

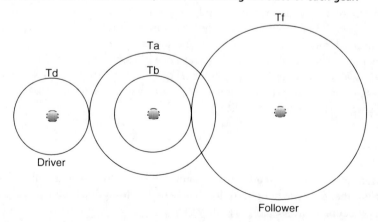

Deb 1997). The objective is to set the number of teeth in each of the four gears in a way that minimizes the error between the obtained gear ratio and a required gear ratio of 6.931:1, while also minimizing the maximum size of the four gears.

The diameter of a gear is proportional to the (integer) number of teeth, so these objectives can be formalized in terms of four integer decision variables: $x = (x_1, x_2, x_3, x_4)$ referring, respectively, to the numbers of teeth in gears T_d (driver), T_b, T_a, and T_f (follower). The problem is then to minimize both f_1 and f_2 below:

$$f_1(x) = \left| 6.931 - \frac{x_3, x_4}{x_1, x_2} \right|$$

$$f_2(x) = \max(x_1, x_2, x_3, x_4)$$

subject to the following constraints:

$$\frac{f_1(x)}{6.931} \leq 0.5,$$

$$12 \leq x_1, x_2, x_3, x_4 \leq 60$$

The constraints ensure that the difference between the designed gear ratio and the desired gear ratio is no more than 50%. After phase one of the innovization procedure, a collection of assumed Pareto-optimal solutions was obtained. ❯ *Table 4* shows the two extreme solutions.

Phase two of the process then revealed several interesting principles relating to the problem, covering the whole set of Pareto-optimal solutions, which is summarized from Deb and Srinivasan's account as follows:

First, in order to minimize the maximal gear size, gears T_d and T_b need to have almost the smallest allowable number of teeth. To get as close as possible to the desired ratio (with error less than 0.1), the T_b and T_d values need to grow somewhat, but still remain close to their lower bounds. Another finding is that the maximum allowed gear size always occurs in a Pareto-optimal solution, either for T_a or for T_f. It is also noted that two distinct types of solutions emerged: (1) gear trains with very low error (very close to the desired gear ratio of 6.931:1), in which there is a great variety of ways in which the numbers on teeth in the four gears combine to almost achieve the 6.931 ratio in the first objective and (2) gear trains with a comparatively large error, with identical first- and second-stage ratios (except the one with the largest error). Although a large error can happen for many different combinations of errors in the two stages, the pressure of the second objective causes both stages of gear-ratios to be identical. Finally, regarding small-error gear trains, half of them have a larger first-stage ratio than second stage, and half have a larger second-stage ratio.

This is a fairly simple and straightforward example, although it brings out several interesting properties of optimal solutions of this type of gear train design problem that are difficult, if not impossible, to infer from the statement of the problem. One implication, for example, concerning recipes for gear train design, is that the process could be guided according to how

◘ Table 4

The extreme solutions obtained for the gear train design problem in Deb and Srinivasan (2005)

Solution	T_d	T_b	T_a	T_f	f_1	f_2
Minimum error	20	13	53	34	~0.00023	53
Minimum maximal gear size	12	12	22	23	3.4171	23

important it is to closely meet the constraint. If a low error is desired, then it is clearly important to examine many possible combinations of gear sizes. If a higher error can be allowed, then solutions with minimal size strongly tend to have equal first-stage and second-stage ratios.

3.2.4 Example 2: Welded Beam Design

This is a much-studied problem in the context of single-objective optimization (Reklaitis et al. 1983), in which a beam needs to be welded onto another beam and must carry a certain load F, as illustrated in ❷ *Fig. 10*.

The problem is to establish the four design parameters (beam thickness and width, respectively, b and t; length and thickness of weld, respectively, l and h) in a way that minimizes the cost of the beam, and also minimizes the vertical deflection at the end of the beam. The overhanging portion of the beam has a fixed length of 14 in., and is subject to a force F of 6,000 lb. Clearly, an ideal design in terms of cost will be less rigid and hence not ideal in terms of deflection, and vice versa. A formulation of this problem (Deb and Kumar 1995; Deb 2000) gives the objectives as follows, where x indicates the vector of design parameters:

$$f_1(x) = 1.10471 \cdot h^2 \cdot l + 0.04811 \cdot t \cdot b(14.0 + l)$$

$$f_2(x) = \frac{2.1952}{t^3 \cdot b}$$

subject to these constraints:

$$\tau(x) \leq 13,600$$
$$\sigma(x) \leq 30,000$$
$$b \geq h$$
$$P_c(x) \geq 6,000$$
$$0.125 \leq h, \ b \leq 5.0$$
$$0.1 \leq l, \ t \leq 10.0$$

The first constraint specifies that the shear stress at the support location is below the allowable shear stress of the material (13,600 psi); the second ensures that the normal stress at the same

◻ **Fig. 10**
The welded beam design problem.

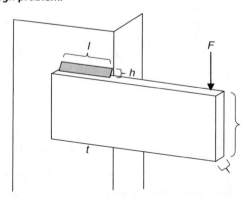

location is below the material's allowable yield strength (30,000 psi); the third ensures the obvious practical consideration that the weld is not thicker than the beam, and the fourth ensures that the applied load F is below the allowable buckling load of the beam. The highly nonlinear stress and buckling terms are as follows (Reklaitis et al. 1983):

$$\tau(x) = \sqrt{(\tau')^2 + (\tau'')^2 + (l\tau'\tau'')/\sqrt{0.25(l^2 + (h+t)^2)}}$$

$$\tau' = \frac{6,000}{\sqrt{2hl}}$$

$$\tau'' = \frac{6,000(14 + 0.5l)\sqrt{0.25(l^2 + (h+t)^2)}}{2(0.707h \cdot l(l^2/12 + 0.25(h+2)^2))}$$

$$\sigma(x) = 504,000/t^2b$$

$$P_c(x) = 64,746.022(1 - 0.282346t)tb^3$$

Following phase one of the innovization procedure, a set of Pareto-optimal solutions was obtained, and ❯ *Table 5* shows the extreme points of this Pareto set, along with an interesting intermediate point which Deb and Srinivasan refer to as *T*, which comes into the innovized principles discussed next.

Deb and Srinivasan's analysis of the many Pareto solutions obtained (spread liberally between those shown in ❯ *Table 5*) revealed the following innovized principles:

First, two distinct behaviors were found: from the intermediate transition solution *T* (shown in ❯ *Table 5*) toward higher-deflection solutions, the objectives behave differently than in the rest of the trade-off region. For small-deflection solutions, the relationship between the objectives was almost polynomial, with f_1 being roughly proportional to $1/f_2^{0.89}$. Next, it was found that, for all Pareto-optimal solutions, the shear stress constraint is active. In the small-deflection (large-cost) cases, the chosen bending strength (30,000 psi) and allowable buckling load (6,000 lb) are quite large compared to the developed stress and applied load. Any Pareto-optimal solution must achieve the maximum shear stress value (13,600 psi). So, to improve the designs in this region without sacrificing deflection, it would be necessary to use a material with a larger shear strength capacity. A third overall principle found was that the transition point (*T*) between two trade-off behaviors is related closely to the buckling constraint. Designs with larger deflection (or smaller cost) reduce the buckling load capacity. When buckling load capacity becomes equal to the allowable limit (6,000 lb), no further reduction is allowed. After this point (toward small-deflection solutions), the beam thickness must reduce in inverse proportion to the deflection objective in order to retain optimality.

◻ **Table 5**
The two extreme solutions and an interesting intermediate solution *T* obtained by Deb and Srinivasan (2005) for the welded beam design problem. The units of the design parameters are inches

Solution	H	l	T	b	f_1	f_2
Minimum cost	0.2443	6.2151	8.2986	0.2443	2.3815	0.0157
Minimum deflection	1.5574	0.5434	10.000	5.000	36.4403	0.00044
Intermediate solution *T*	0.2326	5.3305	10.000	0.2356	2.5094	0.0093

Next, for small-deflection solutions, the beam width remains constant. This indicates that for most Pareto-optimal solutions, the width must be set to its upper limit. Although the beam width has opposite effects on cost and deflection, it is involved in the active shear stress constraint, and since shear stress reduces as beam width increases, it can be argued that fixing beam width to its upper limit would make a design optimal. Thus, if in practice the cost objective is not paramount, solutions which have a fixed width (the maximum 10 in. in this case) may be explored, thereby simplifying the inventory. However, along the Pareto trade-off surface, the weld length increases with increasing deflection, and the weld thickness decreases with increasing deflection. Deb and Srinivasan (2005) noted that these phenomena are counter-intuitive and difficult to explain from the problem formulation. However, the innovized principles for arriving at optimal solutions seem to be as follows: for a reduced cost solution, keep beam width t fixed to its upper limit, increase weld length l, and reduce beam and weld thickness (h and b). This "recipe" is valid while the applied load is strictly smaller than the allowable buckling load.

Beyond that point, any reduction in cost must come from reducing beam width below its upper limit, increasing beam thickness, and adjusting the weld parameters so as to make the buckling and shear stress constraints active. Finally, the minimum cost solution occurs when the bending stress equals the allowable strength (30,000 psi), at which point all four constraints become active.

Finally, to achieve very low-cost solutions, the innovized principles are different: for a reduced cost solution, one needs a smaller beam width, but larger beam thickness and weld parameters.

Deb and Srinivasan report a higher level run of the innovization procedure for this case, in which innovization was redone separately for different values of the three allowable limits in the first, second, and fourth constraints above. It was clear that all three cases produced similar dual behavior (different characteristics on either side of a single transition point) to that observed in the original case. All other innovized principles mentioned above (such as the constant nature of beam width, beam thickness being smaller with increasing deflection, and so on) remained valid. And significant further insights were obtained into the overall design problem, detailed in full in Deb and Srinivasan (2005).

3.2.5 Innovization: Concluding Notes

When facing an optimization problem with at least two conflicting objectives, the set of optimal solutions is very diverse. Having found such a set effectively and efficiently using evolutionary multi-objective optimization (judiciously combined with other methods that help locally optimize), the notion of *innovization* is to analyze this set of solutions to see if there are commonalities and patterns that might translate into general design principles for the problem at hand. It turns out that this is true, and interesting new principles (difficult or impossible to have been obtained otherwise) have emerged from several studies to date. The emerging truth seems to be that solutions along a Pareto front often seem to share similarities that seem to be principles of optimality for the problem at hand, irrespective of location on the Pareto front.

In this section, just two case-study results have been borrowed to illustrate the innovization principle. Deb and Srinivasan (2005) show several more examples, including spring design and multiple-disk clutch brake design, while one more recent study (Datta and Deb 2009) displays an excellent example of the potential impact of innovization (and, hence, indirectly, of evolutionary multi-objective optimization) by finding new innovized principles for the setup parameters of a turning process using a lathe and a cutting tool that are overwhelmingly common in industry workshops. Finally, there is no particular reason to believe that innovization is constrained to engineering design. It will be interesting to see future applications of this idea in other design fields, such as electrical circuits, optical systems, communication networks, and the many other areas in which evolutionary multi-objective optimization is increasingly used.

4 Logistics and Combinatorics Made Easy: Robust Solutions and New Algorithms via Natural Computation

In this section, two areas will be considered that exemplify how natural computing (largely, learning classifier systems and evolutionary computation) has provided highly successful ways to address difficult logistics problems. Logistics usually relates to scheduling and timetabling problems of various kinds, but also included here is the closely related and general field of combinatorial problems in which a discrete collection of items of some kind must be arranged in an optimal way. There are innumerable examples of natural computing applications in this domain, and the first case here is simply a selection of one (of several possibilities) that combines the attributes of: "interesting," "real-world," and "difficult" (looking here at the case of a real-world truck scheduling problem). Next, moving on to perhaps a more profound area that has emerged from the late 1990s, in which, rather than the use of evolutionary computing to solve "one problem at a time," the use of natural computing to discover new *algorithms* is considered, which can then in turn be used on entire classes of problems, solving them efficiently and effectively. This is an area within the emerging field of "hyper-heuristics," but with the particular focus on designing new algorithms which is referred to as "super-heuristics."

4.1 Safe Streets via Robust Route Optimization

This section describes an application of natural computation in a critical seasonal logistical task, covered more fully in Handa et al. (2006). Local authorities in countries such as the UK, with marginal winter climates, are responsible for the precautionary gritting/salting of the road network in order to allow safer travel in icy conditions. This winter road maintenance task is extremely challenging as well as critically important to the locality, with a potentially major impact on both business and day-to-day life.

As Handa et al. (2006) note, in the case of the UK there are around 3,000 precautionary gritting routes that cover about 120,000 km (30% of the entire UK road network). On nights with forecasted snow or ice, these routes need to be treated so as to ensure the safety of road users. This typically costs between £200 and £800 per km of road (Cornford and Thornes 1996). Accurate road surface temperature prediction is required, in order to decide which roads need to be treated, however this decision can often be uncertain. Optimization of the route to be traveled by the gritting/salting trucks also plays a crucial role here. The consequences of a wrong decision – not treating a road that eventually becomes dangerous – are serious, but if grit or salt is spread when it is not actually required, there are obvious financial and environmental drawbacks. The goal of gritting route optimization is to minimize the financial and environmental costs, while ensuring that roads that need treatment will be gritted in time. Further, it is essential that gritting routes are planned in advance, to enable effective use of limited resources (e.g., trucks and salt).

Mostly, the design of gritting routes relies heavily on local knowledge and experience. A "static," often paper-based approach is typically used to optimize gritting routes, staying within constraints imposed by the road network itself, vehicle capacities, the number of vehicles, and the available personnel. This section describes the application of an evolutionary algorithm to this task. Covered in more detail in Handa et al. (2006), discussed here is a Salting Route Optimization (SRO) system that combines evolutionary algorithms with the latest version of the Road Weather Information System (XRWIS) commonly used by local authorities.

4.1.1 The Salting Route Optimization (SRO) System

A very important aspect of the SRO discussed here is its integration with XRWIS, which, recently trialed by the UK Highways Agency, is a high-resolution route-based forecast system that predicts road temperature for a 24-h period. XRWIS models surface temperature and condition at thousands of sites in the road network. Data are collected along each gritting/salting route by conducting a survey of the "sky-view factor" (a measure of the degree of sky obstruction by buildings and trees) (Chapman et al. 2002). This is then combined with other geographic, land use, and updated meteorological data to predict road conditions at typical spatial and temporal resolutions of 20 m and 20 min, respectively. The output is displayed as a color-coded map of forecast road temperatures and conditions that is then disseminated to highway engineers.

In the SRO, XRWIS provides forecast temperature distributions over time that are then input to an evolutionary algorithm module. Each temperature distribution (different distributions for different future timepoints), along with commercially available routing data, is transformed into an instance of a capacitated arc routing problem (CARP) (Lacomme et al. 2004). That is, each temperature distribution suggests a specific set of roads on the networks that need to be treated. The CARP is then defined as the need to find routes that serve this specific set of roads in a reasonable time frame, using no more than the available vehicle numbers and capacities, and ideally minimizing the number of vehicles used. An important point is that each timepoint leads to a different CARP instance, since the set of roads that require treatment may be different. The overall goal of the SRO system is not only to find a suitable series of salting routes, which ensure that the roads that require treatment are treated in time, but also to ensure that the routes do not vary too much, which in turn causes considerable confusion and distraction to the workforce. In this sense, the SRO system finds a *robust* solution, which ensures we can cover the most important sections of the road network.

Given the series of CARP instances, the evolutionary algorithm module finds solutions that are simultaneously good for all or many of these instances. In particular, a specially designed memetic algorithm is used (a combination of evolutionary and local search) as described next. In this approach, the fitness of a solution is calculated according to the entire ensemble of CARP instances. However, at each generation, the operators and local search processes concentrate on a specific instance. The different instances are weighted, and this weighting controls the selection of the instance in each generation, in a way described in the following.

4.1.2 Robust Solutions for Salting Route Optimization

Searching for robust solutions is currently a significant topic in the field of optimization in uncertain environments, since in many problems the decision variables or environmental parameters are subject to noise. In this case, Handa et al. (2006) required that solutions to the different CARP instances be as similar to each other as possible (so that daily changes in the temperature distribution do not lead to significant disturbance in the route to be followed), while at the same time requiring good performance in terms of the costs of the routes. Handa et al. modeled a robust SRO solution as one which optimized the following:

$$F(X) = \int E(X, a)p(a)\,da$$

in which X and a indicate route design variables (routes and possible temperatures), $E(X, a)$ indicates the distance cost of gritting routes X given temperature a, while $p(a)$ indicates the probability of temperature a. Hence, the idea is to find ideal gritting routes for each temperature distribution, but weighted by the prior probabilities of the forecast temperatures.

Although the distribution in temperature will vary daily across a road network, warmer (colder) sections are usually warmer (colder) than the rest of the network. So, even on cold nights, some warmer sections may not require salting, whereas colder sections may need treatment even in relatively warm conditions. The fitness function, as stated above, is impossible to compute exactly since its components are largely unknown; instead it is approximated by using a number of typical temperature distributions. Considering this and other issues, the fitness function used by Handa et al. (2006) was as follows, given a set of temperature distributions A_e:

$$F(X) = \sum_{a \in A_e} w_i \frac{E(X, a_i) - E^*(a_i)}{E^*(a_i)}$$

in which $E^*(a_i)$ represents the difficulty of finding a good route for temperature distribution a_i, and the w_i are weights, summing to 1, which balance the importance of different temperature distributions during the optimization process. The weights are adapted during evolution in a way that maintains a focus on the routes that are proving more costly, while the $E^*(a_i)$ values are lower bounds on the cost of the routes for each temperature distribution a_i, actually predetermined by prior runs of the memetic algorithm for this purpose described in Handa et al. (2005).

Handa et al. (2006) used a permutation-based encoding as follows. An individual solution comprised a permutation of arc IDs (road sections), interspersed with symbols representing individual trucks. For example, the individual:

2 6 s1 5 4 7 1 s2 8 3

indicates a gritting route for two trucks; truck 1's route is road sections 5, 4, 7, and 1 (in that order), and truck 2's route is road sections 8, 3, 2, and 6 (note the wraparound involved in the interpretation).

At each generation of the memetic algorithm, crossover (the EAX operator proposed by Nagata and Kobayashi 1997) and local search methods are applied with regard to only one CARP instance (i.e., one temperature distribution) in every generation. That is, for example, the local search is guided by the fitness according to the selected instance only. The choice of instance is made stochastically according to the current weights of the temperature distributions. However, between generations, the fitness of each solution is calculated according to the

⬛ **Fig. 11**

Left: routes optimized by the SRO system for a cold day in South Gloucestershire; *Right*: existing routes obtained by human experts.

ensemble of instances using the fitness function described, and this then guides the selection of parents for the next generation.

4.1.3 Comparisons and Conclusions

In experiments by Handa et al. (2006) to test and validate this approach, robust solutions were evolved by using ten different temperature distributions, and these were then compared with the routes currently used by South Gloucestershire Council in the UK.

❯ *Figure 11* shows an example of routes found for a cold day, comparing the SRO system's routes (on the *left*) with the existing routes (on the *right*).

In comparison with the routes that were in use at the time, the robust solutions delivered by the SRO were able to provide more than 10% savings in terms of total distance traveled by the available trucks.

The SRO system was developed for finding optimized robust solutions for salting trucks, and as such it is an excellent example of an important real-world combinatorial problem that can be solved effectively via a system with natural computation at the core. In this case, especially given the integration with the XRWIS, the system can be regarded as proof of concept for similar tasks that need careful planning in relation to weather conditions, such as waste collection and parcel delivery.

4.2 Hyper and Super Heuristics

This section briefly considers a fairly new method in search and optimization, variously called hyper-heuristics or, as an emerging term in the community referring to a specific brand of approach, super-heuristics. In the context of selected applications of natural computation, the special aspect of super-heuristics is that they represent the use of a good global optimization or learning method – hence, typically evolutionary computation or a learning classifier system – to discover new *algorithms* that solve problems of a given kind. This is as opposed to, and substantially more general than, using optimization or learning to solve a single problem instance.

In very broad terms, the general notion of hyper-heuristics refers to the idea of using an algorithm that manipulates a set of heuristics in order to solve a given problem. This is indeed a very common activity these days, and can be seen in many published applications of evolutionary algorithms, and of meta-heuristics in general (including, e.g., tabu search and simulated annealing). Typically, such an approach is sometimes called hyper-heuristic in the case that the encoding used involves lower-level heuristics in an integrated way. That is, rather than an encoding of a solution being a direct representation of a solution, the encoding is instead an indirect representation (examples will be looked at later). The interesting point is that, in some cases, an encoded solution for a given problem can actually be interpreted as an algorithm that can be applied to a large collection of instances of that problem, not just the instance currently being solved. In many so-called "hyper-heuristic" applications, this reusability of the encoding of a solution is only a side effect. When the term "super-heuristic" is used herein, this is meant to refer to the idea that evolving new general and reusable algorithms (for classes of instances, rather than a single instance of interest), is the specific goal of the process. However, it is noted that "super-heuristics" seems to have been first used in the literature by Lau and Ho (1999), to denote something more akin to standard hyper-heuristics, in which a set of heuristics are engineered by higher level algorithms in order to solve a specific problem instance. This section describes the ideas, amid some examples and historical notes. For alternative and more detailed accounts, a 2003 book chapter is recommended (Burke et al. 2003), or the current Wikipedia article on hyper-heuristics.

4.2.1 Potential Impact of Super-heuristics

The impact and importance of super-heuristics is partly evidenced by a negative point: despite a large collection of case studies, standard applications of meta-heuristics tend to be "one-off" and resource-intensive. For example, a particle-swarm optimization method developed to solve a small company's daily process scheduling problems may seem successful in its own terms, but its existence does not necessarily accelerate the potential for other companies with similar (but not the same) problems to develop similar solutions. And, typically, the solution itself may be resource intensive, tying up considerable computing resources every morning. Also, the development of this solution will have typically been influenced by a perceived goal to produce solutions as optimal as possible, despite the fact that daily uncertainties and perturbations to the production process underline this optimality – that is, a large number of "reasonably good" solutions will have worked just as well.

In contrast, super-heuristics seem to open up the possibility for producing solutions that, though having an initial cost in development time, are much more flexible. The "solution" in this context would be a fast, constructive algorithm that tends to work well (as well as run much more quickly than a typical meta-heuristic implementation) on the problems typically faced by the company, and may well generalize to similar problems more successfully and easily than the meta-heuristic approach.

4.2.2 Hyper-Heuristics: Further Notions and Examples

Suppose there is an instance of a problem to solve. In particular, it is easier to think in terms of combinatorial and logistics problems, the kind in which a solution might be built step by step by making a series of decisions. For example, if a collection of student examinations have to be timetabled, first a room and a time for the largest exam might be found; then which exam to look at next might be decided, then where to place this next exam might be decided, and so on. For such problem domains, there is usually an available collection of "low-level" heuristics. For example, in timetabling, a common heuristic is to first sort the events that have to be timetabled according to some measure of difficulty. There are several such measures, based on the fact that some events are more difficult to place than others (e.g., can only fit in a small number of rooms, and potentially clash with many of the other events). One way to do timetabling constructively (such algorithms are often called "greedy") is to repeatedly choose an event to timetable based on a difficulty measure, and then timetable it by finding a place and a time that suits. Each potential difficulty measure can be considered a different heuristic. Similarly, deciding where and when to place the event are also activities that can be based on a range of specialized heuristics. Very similar can be said of other, if not all, logistic or combinatorial problems.

With independent roots in the field of automated planning and scheduling systems (Minton 1988; Gratch et al. 1993; Cross and Walker 1994), an early and influential example of the hyper-heuristic approach being used for solving specific problem instances (i.e., one at a time) is concerned with open-shop scheduling problems in Fang et al. (1994). In Fang et al.'s work, several low-level heuristics were considered, all of which were relevant to the problem of "open-shop" scheduling, in which there are, say, j jobs that need to be scheduled, each consisting of a certain number of tasks. Each such task must use a specific resource (usually called a machine) for a specific amount of time, although the tasks that comprise a given job may be done in any order (when the order of tasks within a job is constrained, it is a job-shop problem). For example, PCs may arrive at a processing center with the operating system installed, and need to have a number of applications installed (for which order of installation is unimportant) by a number of experts, each expert in the installation of a particular application. Each "job" is a PC, which may have its own individual specification and subset of applications that need to be installed; this amounts to an open-shop scheduling problem.

Fang et al. (1994) used an evolutionary algorithm which constructed solutions as follows. A chromosome was a series of pairs of integers [t0,h0,t1,h1,...] interpreted from left to right, meaning: for each i, "consider the ti-th uncompleted job (always interpretable, when treating the list of uncompleted jobs as circular) and use heuristic hi to select a task to insert into the growing schedule in the earliest place where it will fit." Examples of the lower-level heuristics used are:

- Choose the task with the largest processing time
- Choose the task with the shortest processing time
- Find the tasks that can start earliest (there may be more than one) and choose the one with largest processing time
- Find the tasks that can be inserted into a gap in the schedule so far, and pick one that best fills this gap

This approach was called "evolving heuristic choice," and led to excellent results on benchmark problems, including some new best results at the time of publication, and it marked the beginning of a wave of interest in what were later termed "hyper-heuristic" approaches.

An example following this work was that of Hart and Ross (1998), who looked at job-shop scheduling problems (where the ordering of tasks within a job is predetermined – for example, in the software installation example, it could well be the case that applications need to be installed in a certain order). Their approach relied on the fact that there is always an optimal schedule which is "active," meaning that to get any task completed sooner, one would need to change the order in which tasks from different jobs get processed on one or more of the machines. Meanwhile, a well-known heuristic algorithm (due to Giffler and Thompson 1960) that generates active schedules was exploited. The explanation by Hart and Ross (1998) and in Burke et al. (2003) is now followed in explaining their approach. Giffler and Thomson's active-schedule generation algorithm is as follows:

1. Let C = the set of all tasks that can be scheduled next
2. Let t = the minimum completion time of tasks in C, and let m = machine on which it would be achieved
3. Let G = the set of tasks in C that are to run on m whose start time is $<$t
4. Choose a member of G, insert it in the schedule
5. Go to step 1

In step 4, there is a choice to be made, which was exploited in Hart and Ross' hyper-heuristic approach. Now consider a simplified version of this algorithm, which only generates so-called "non-delay" schedules.

1. Let C = the set of all tasks that can be scheduled next
2. Let G = the subset of C that can start at the earliest possible time
3. Choose a member of G, insert it in the schedule
4. Go to step 1

This time, there is a choice to be made in step 3. Hart and Ross' approach was to use an encoding of the form [a1,h1,a2,h2,...], again interpreted from left to right, where the *a*is are 0 or 1, indicating whether to use an iteration of the Giffler and Thompson algorithm or an iteration of the non-delay algorithm, in order to decide on the next task to schedule, and the *h*is indicate which of the 12 heuristics to use to make the choice involved in the selected algorithm. This method again produced excellent results on benchmark problems.

Finally, before moving on to two examples of what can be called "super-heuristics" (i.e., where general problem solvers are evolved, rather than algorithms for one instance at a time), an early real-world application of the hyper-heuristic approach is briefly mentioned. Described in Hart et al. (1998), the problem that needed to be solved was to schedule the collection of live chickens from farms in Scotland and Northern England, for delivery to one of two processing factories. A given instance of the problem arises from a set of orders from supermarkets and other retailers, which have to be fulfilled within given time windows. The specific resources that needed scheduling were of two types: the collection of live chickens from farms was done by a set of "catching squads" who moved around the country in minibuses; the delivery of chickens to processing factories was done by a set of lorries. In general, catching squads needed to move from farm to farm collecting chickens, and lorries needed to arrive at farms in time to be loaded with chickens caught by the squads, and then either move to another farm if able to hold more, or proceed to unload at a processing plant (and then perhaps back to a farm). The principal aim was to keep the factories supplied with work, while attempting to ensure that live chickens did not wait too long in the factory yard, for veterinary and legal reasons. There were several constraints. For example, different types of catching squad were distinguished by differences in their contractual arrangements, relating to the amounts of work they would do per day or week (including, e.g., guaranteed minimum amounts of work). Meanwhile, the order in which a given squad could visit farms in 1 day was constrained according to the status of each farm in terms of certain chicken diseases, while lorry schedules also were subject to a range of associated constraints. Overall, the target was to create good schedules satisfying the many constraints, but that were also generally similar to the kinds of work pattern that the staff were already familiar with, and to do so quickly and reliably.

After several approaches which did not work very well, using what were the standard styles of evolutionary algorithm approach at the time (experts in classical scheduling methods had already been consulted by the company, and had tended to retreat in terror once the problem had been described to them), the eventual solution used two evolutionary algorithms in two stages. The first was a hyper-heuristic approach to assign tasks to individual catching squads in a way that was able to cover the current set of customer orders. In detail, a chromosome specified a permutation of customer orders followed by two sequences of heuristic choices. The first sequence of heuristics specified ways to split each order into convenient workloads, and the second sequence of heuristics specified how to assign those workloads to catching squads. The second stage was an evolutionary algorithm that took the set of tasks produced from the first stage, and delivered a schedule of lorry arrivals at each factory. For this real industry problem, a hyper-heuristics approach was central to a solution that worked successfully, whereas no previous approach had met the required standards.

Before moving on to "super-heuristics," we note that the surface of applications that have found hyper-heuristics to be a highly flexible and successful approach has barely been scratched, albeit at the time of writing the application areas tend to be not very diverse, with most either involving timetabling (e.g., Terashima-Marin et al. 1999; Cowling et al. 2000; Burke et al. 2002; Bilgin et al. 2006) or scheduling (e.g., Hart and Ross 1998; Cowling et al. 2002; Ayob and Kendall 2003). For a much more comprehensive discussion of hyper-heuristics, readers may refer again to Burke et al. (2003), as well as Özcan et al. (2008).

4.2.3 Super-heuristics: Evolving and Learning New and Effective Algorithms

In an increasingly influential thread of research, Ross et al. (2002, 2003) extended the notion of hyper-heuristics to see whether new constructive algorithms, which could deal effectively with large sets of problem instances, rather than one instance at a time, could be evolved. In what is termed here as a "super-heuristic" approach, Ross et al. (2002, 2003) used a learning classifier system called XCS (Wilson 1998), and later an evolutionary algorithm, to try to learn an algorithm for solving hard bin-packing problems. The learning was done in Ross et al. (2003) with an evolutionary algorithm aiming to optimize the parameters for a fast constructive bin-packing algorithm, training on a set of test problems (i.e., a collection of different problem instances was involved in the fitness function). When the learned algorithm was then tested on a different set of test problems, its performance was found to be clearly competitive with state-of-the-art human-designed bin-packing constructive algorithms.

In bin-packing (as with many algorithms, and as discussed with scheduling), a typical constructive algorithm will build a solution one step at a time, each step involving the use of some heuristic to choose the next item to pack into a bin, and maybe another heuristic to choose which bin to place it in (or a single heuristic covering the combined decision). The overall goal is to pack a given collection of items of different sizes into a set of fixed capacity bins, using as few bins as possible. In detail, the overall idea in Ross et al. (2002) and their later work (2003) is as follows. At each stage during such a constructive algorithm, there is a particular problem "state" that is characterized by the set of items left to pack, and the current partial packing of items into bins. In this state, it is reasonable to infer that some heuristics will be better than others for deciding on the next item/bin placement. So, Ross et al.'s approach was to define a constructive algorithm as a set of rules. Each rule in the set referred to a particular problem state, and specified what heuristic to use when in that state. Clearly, there are far more potential problem states than one can expect to be represented by the left-hand sides of such a rule; the method gets around this by having the rules essentially refer to points in the space of potential problem states, and the rule that "fires" at any particular time is the one that is closest to the current problem state.

The approach was first tested using 890 benchmark bin-packing problems in Ross et al. (2002), of which 667 were used to train the XCS learning classifier system, and 223 for testing. The single resulting learned constructive algorithm was able to achieve optimal results on 78.1% of the problems in the training set, and 74.6% of the problems in the unseen test set. This compared well with the best single heuristic tested, which achieved optimality 73% of the time. A notable finding in that work was that when the training set was confined to some of the harder problems, the learned algorithm was able to solve seven out of ten of those problems to optimality (compared with zero out of ten for the comparison human-designed heuristics). This approach was improved in Ross et al. (2003), with many interesting findings that showed highly competitive results for evolved algorithms on hard unseen problems.

Finally, we have a brief look at a different style of super-heuristic approach applied to a different domain, specifically the work of Fukunaga (2008), which concerns the satisfiability (SAT) problem. A SAT problem instance is a conjunctive normal form (CNF) expression, such as "(A or B or D) and (B or not(C)) and (D or E)...," involving a number of logical variables (A, B, ...), which may either be true or false, which in turn are the elements of a number of clauses, conjoined into the full statement. The problem is to discover whether or not an assignment of truth values to each of the variables exists, which results in each of the conjuncts, and therefore the entire statement, being true. Fukunaga's work exploited a well-known general local search framework for SAT, as follows:

1. Generate an assignment A of truth values at random (e.g., A = T, B = F, C = F, ...)
2. For a given maximum number of iterations:
    2.1. If A satisfies the formula, return YES
    2.2. Choose a variable V with a *Variable Selection Heuristic*
    2.3. Change A by flipping the value of variable V
3. Return UNKNOWN

The algorithm uses a "Variable Selection Heuristic" in Step 2.2, and this in turn was the focus of Fukunaga's investigations. There are several well-known examples of variable selection heuristics, which are human-designed and typically used within the above algorithm framework. One example is GSAT (Selman et al. 1992), which involves choosing the variable that, if flipped, would cause the highest net gain in satisfied clauses, breaking ties randomly. Another, HSAT (Gent and Walsh 1993), works as GSAT, but breaks ties in favor of *age* – so, the variable that was last flipped longest ago in the overarching local search process is the one chosen to break the tie. Yet another, of several more, is the so-called GWSAT(p) (Selman et al. 1994), in which, with probability p, a random variable from a random unsatisfied clause is selected, else GSAT is used.

Fukunaga (2008) noticed that variable selection heuristics in the SAT literature have certain common building blocks, including

- Scoring variables via a gain metric
- Selecting a variable from a subset of variables
- Ranking variables, and choosing the best (or second best)
- Consideration of a variable's "age"
- Branching (if x do A, else do B)

An insightful comment that Fukunaga makes is that in the history of SAT heuristics, developments typically come from finding new ways to combine these building blocks, rather than entirely novel heuristics. This begs a number of questions, one of which is whether or not automated methods may be able to find better combinations of these building blocks. The latter is in fact exactly what Fukunaga (2008) investigated, by using GP, with a function and terminal set designed in such a way that novel heuristics could be expressed in terms of the above ingredients. As with the previous super-heuristic approach discussed, the GP experiments involved using a large set of different SAT instances in the fitness function, and Fukunaga (2008) evaluates the results by testing the evolved variable selection heuristics on unseen test sets.

On a collection of 1,000 unseen test instances, Fukunaga's evolved variable selection heuristics are very competitive with the state-of-the-art variable selection heuristics, GWSAT, WalkSAT, and Novelty (McAllester et al. 1997). A handful of the new heuristics found in this way dominated the state-of-the-art heuristics in terms of success rate and speed. A further rather interesting finding was that one of the heuristics in a random search of expression trees was almost as good in terms of success rate, but usually faster, than the human-designed state-of-the-art heuristics.

4.2.4 Some Concluding Notes

The super-heuristics concept has the potential to play a major role in optimization over the next few years. One way to view this development is as a thrust toward more "general" optimization systems, which, for a wide variety of application areas, is a significant goal. In just one example of application area, *timetabling*, there has been very extensive research in recent years along the lines of hyper-heuristics and upper-heuristics; this has followed a statement in Ross et al. (1997), which was, ". . . all this naturally suggests a possibly worthwhile direction for timetabling research involving GAs. It is suggested that a GA might be better employed in searching for a good algorithm rather than searching for a specific solution to a specific problem." In agreement with Burke et al. (2003), we emphasize that this suggestion can be generalized to a much wider range of problem areas than has currently been addressed with hyper- and super-heuristic technologies.

5 Design: Art, Engineering, and Software

This penultimate section considers the theme of "Design," and discusses three quite contrasting examples. Design is an area of especial interest when we consider what natural inspiration has to offer to practitioners of various sorts. Today, and for some considerable time still to come, the world is, to most intents and purposes, filled with two kinds of artifacts – those designed by nature, and those designed by human designers. The chief difference between these two kinds of artifacts is the specific design method that was employed. The naturally designed artifacts, as most scientists would agree, were designed by an evolutionary process – essentially an iterated process of randomized generation and test, in which new designs, often failures, sometimes improvements, emerge via slight random changes or randomized recombinations of old designs. With a "survival of the fittest" principle built in to this strategy, the successes are more often chosen than the failures when it comes to being the foundation for (or the parents of) new designs. Over time, this process continues to evolve new designs that are successful in their environment, and the examples seen today include everything from archaea to artichokes, baobabs to brains, *Escherichia coli* to elephants, and from wasps to the sophisticated set of processes that lead to the construction of wasp nests. It is overwhelmingly the case, however, that human-designed artifacts have not adopted this process. Humans prefer to design things in a rational way, which prefers the adoption of designs that have worked before for similar problems, and rejects the notion of any randomized exploration. Humans tend to stick to a battery of accepted design rules for the application in hand, and usually opt for a step-by-step constructive approach, rather than generating and later discarding many different designs at once.

Some criticisms of the human way of designing can be summed up in the following statement: the overreliance on established design rules imposes severe constraints on innovation, and probably limits the effectiveness of the resulting designs. Meanwhile, nature's method for design may well not be perfect – it does indeed seem wasteful – however, it certainly beats the human method for innovation. One cannot yet design, with a rational approach, a biological flying machine as efficient as a mosquito, or an energy transduction system as efficient as photosynthesis. Meanwhile, it is notable that randomization is an integral part of nature's method – undirected perturbations to designs tend to be anathema to the human approach, but are continually tried and tested in nature. Overall, it seems abundantly clear that nature has a lot to teach us about how to design things.

Perhaps unsurprisingly to most of readers, but nevertheless, it is hoped, inspiringly, the documented experiences so far in the arena of natural computation in design show that novel, effective, and unprecedented designs can be found by applying nature's method to design the artifacts that need to be created. The next subsections discuss one of the more prominent and exciting examples in recent years, NASA's use of evolutionary techniques to come up with entirely novel antenna designs that have been deployed on satellite missions. But before that we look at an example of the use of interactive evolutionary computation in artistic design, and this section ends with a brief look at how natural computation is making headway into the design of software.

5.1 Interactive Evolutionary Design of Batik Patterns

Evolutionary Art Systems (EASs) are increasingly popular (Romero and Machado 2008), commonly using evolutionary computation, usually interactively (e.g., Sims 1991; Lutton 2006), to generate aesthetic artworks. In some real-world applications, focusing on particular niches in art and design, EASs have been developed specifically to facilitate a designer's activity. One recent such case, which is described here, is by Li et al. (2009), in which an EAS tool is described for helping designers of Batik patterns, a traditional art in Indonesia and southeast Asia. Batik is a form of painting or writing on cotton cloth, applied with the aid of a tool called a cap (Kerlogue and Zanetini 2004). Nowadays, Batik is used in fashion, furnishing fabrics, and household accessories, as well as paintings and ornamentations in rooms and offices. However, fine-quality handmade Batik is very expensive, so it is potentially valuable to consider ways that would decrease Batik designers' effort and increase production of Batik.

Li et al. (2009) investigated the potential for an EAS-based Batik design system with such goals in mind. In doing so, however, they had to consider the difficulties commonly faced by EAS. First, the evolutionary process is often quite limited by the lack of an explicit correlation between genotypes and phenotypes. Essentially, the common ways in which aesthetic works tend to be encoded by manipulable genes (think of fractal patterns encoded in the typical way by mathematical formulae) are far removed from the works themselves, so that, for example, when human designers select what they think are good parents, they may find that none of the promising features they saw in the parents actually appears in the next generation. Another common difficulty is that the process can be tiresome for a human designer, spending hours sitting at a computer rating generated images. Li et al.'s work attempted to develop a Batik design system with innovations that addressed these issues. In particular, they devised a suitable encoding for various Batik styles, and they devised an "out-breeding" mechanism that provided an additional way to generate new patterns, which seemed to be on the aesthetic path being pursued by the designer. These issues are elaborated in the following subsections, but the reader is referred to Li et al. (2009) for a more complete account.

5.1.1 Encoding Batik Patterns

Li et al. (2009) explored the space of geometrical patterns used in Batik, and classified them into categories. They found that the most common features were repetition, and certain geometric transformations such as rotation, translation, and reflection. This led to a way to encode patterns in genotypes, which specify a number of nonredundant primitives along with transformations. The encoding is therefore based directly on features of Batik patterns, most basic elements of which include: triangle, polygon, circle, dot, star, and flower. Each feature is generated from one gene in the genotype.

A genotype consists of a variable number of genes, each of which represents one feature in the phenotype. Every gene has two evolvable attributes. The first part is a specific basic pattern (e.g., a simple representation of a flower petal, or a circle, or a triangle); the second part, the transformation, is a vector of matrices, which each represent a transformation of the unit set. A matrix is encoded by six numbers, indicating a 2D linear transformation together with a translation. This representation is straightforward and easy to manipulate. The resulting pattern is made up of the union of the patterns induced by the different genes. ❯ *Figure 12* shows some examples of single simple genes in this encoding, with their interpretations above, while ❯ *Fig. 13* shows some patterns produced using the system, contrasted with some human-designed similar Batik patterns.

5.1.2 Boosting the Evolutionary Process

Li et al. (2009) use what they call an "out-breeding" mechanism to invigorate the pool of patterns produced during the interactive evolutionary process. In their EAS, two separate

◨ Fig. 12

Simple examples of Batik pattern genes, and their interpretations.

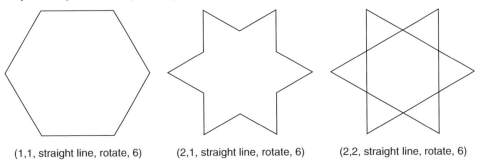

(1,1, straight line, rotate, 6)   (2,1, straight line, rotate, 6)   (2,2, straight line, rotate, 6)

◨ Fig. 13

Above: some real-world Batik patterns; *below*: similar individuals generated by the mathematical model, such as appear in the initial population of the Batik interactive evolutionary system.

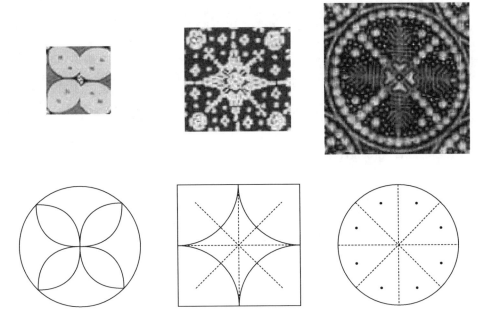

populations of patterns are maintained, displayed to the designer on separate panels. One population evolves in the normal way, based on treating the user's feedback as the fitness function. However, the second population evolves toward individuals that are maximally dissimilar to what seem to be the user's preferences, hence injecting considerable diversity into the displayed patterns. Whenever the first population seems to be stagnating, individuals in the second population will be introduced to the first, contributing diverse input to the gene pool.

The crux of this mechanism is the idea of "dissimilarity," which requires a way to compare patterns. Li et al. (2009) preferred to investigate a measure that was related to the visual difference between patterns, expecting that a method based only on genotypic difference would not be satisfactory. They use a metric based on singular value decomposition (SVD) (Wang et al. 2000). In their approach, a pattern *A* is interpreted as a matrix, and they represent each pattern in terms of the singular values arising from the SVD of *A*, which in turn are likely to capture salient features of the visual perception of *A*. A similarity metric between two patterns is then defined on the basis of a normalized comparison of their vectors of singular values. The outbreeding process then operates as follows: In each generation, while one population continues to regenerate patterns according to the normal process, guided by the user's evaluations, the outbreeding population regenerates in a way guided by using dissimilarity as the fitness measure, measured in terms of dissimilarity from the pattern that the user currently perceives as best. Li et al. (2009) report that the outbreeding mechanism is very effective in aiding the search for innovative patterns, and find that the "outbred" populations tend to be more elaborate and attractive than the "main" population!

Meanwhile, concerning the "standard" interactively evolved population, one notes that the generation of the initial population, and the subsequent evolution based on user-supplied fitnesses, relies on a collection of typical genetic operators as follows. The initial population is informed by using a mathematical model of Batik pattern space based on Li et al.'s preliminary characterization. The model is used to generate collections of genes, and then mutation operators are applied to these: either Gaussian mutation (in which each point in the basic pattern element of each gene is perturbed by the same random amount), or style mutation (in which the elements of a gene reflecting line styles are perturbed, for example from straight-line to curve). During the subsequent interactive evolution process, new patterns are produced by crossover and mutation of patterns deemed good by the user. Standard types of crossover and mutation are used for this in Li et al.'s work so far, for example, including linear combination and gene-swap-based crossover. Also, as explained fully in Li et al. (2009), their system has other features that are meant to aid the user's design process, such as the ability to retrieve patterns that were produced earlier in the evolution.

5.1.3 Empirical Notes

Li et al. (2009) found that some of the traditional Batik designs could be produced by the mathematical model that underpins the generation of the initial population. In ❷ *Fig. 13*, the top three patterns are real-world Batik, while the three underneath were presented in initial populations.

Li et al. (2009) report on five experiments using their system, aimed partly at evaluating the outbreeding technique; each experiment ran the process twice, with and without outbreeding (but starting from the same initial populations). They measured, in particular, the time investment of the user before a satisfactory design was achieved. They found that, with the outbreeding mechanism in place, the design process took on average only 54% of the time taken using the interactive system without outbreeding. Further, the time with outbreeding was roughly 17% of the time it tends to take to design a new Batik pattern by hand. Further experiments confirmed in other ways that the outbreeding mechanism was effective in producing patterns, throughout the process, that tended to be well evaluated by users. ❷ *Figure 14* shows the initial population used for all of these experiments, and

◻ **Fig. 14**

Initial populations used in Li et al.'s experiments.

◻ **Fig. 15**

Tessellations of final-population designs using the Batik pattern interactive evolutionary system (with the outbreeding mechanism).

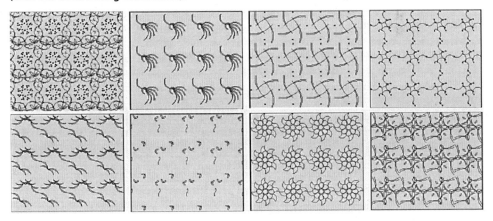

❯ *Fig. 15* shows some final-population designs that satisfied the users (produced with the outbreeding mechanism in operation), converted into tessellations.

5.1.4 Final Points and Notes

The interactive Batik design system described and discussed here is a nice example of how interactive evolutionary computation is beginning to be used in an increasing number of applications that involve creativity. Experience with this system so far shows how it can both speed up and invigorate the process of generating interesting new patterns in the Batik "domain." One of the keys to success in such enterprises is the wise design of the encoding, and we have seen a good example of that in this case. Li et al. (2009), as already seen, also showed an innovative approach to dealing with some of the ever-present problems (and, hence, research issues) in interactive evolution. The outbreeding mechanism was able to enhance diversity in the process, and at the same time reduce the time (and hence fatigue) of human users.

5.2 Novel Antennas for Satellites: Discarding the Rule Book

As elaborated further in Hornby et al. (2006), current practice in antenna design almost invariably involves designing and optimizing them by hand, and this approach is very limited as a way to develop new and better antenna designs. It requires significant time and expertise from human experts in the domain. An ongoing alternative in antenna design (in common with an increasing variety of such specialist areas) is to investigate evolutionary algorithms for this purpose. This has been happening since the early 1990s, with increasing success and take-up as developments in processing power have been seen, and also improvements in the quality of software simulations of antenna performance. To date, many types of antenna have been investigated using evolutionary design approaches. A particularly interesting and useful aspect of this approach is the opportunity to evolve antenna designs specifically for performance in a particular environment, so that the fitness function takes into account the effects of structures surrounding the antenna's intended position. This consideration of the immediate environment is extremely difficult for human expert antenna designers to take into account.

This section summarizes the work reported in Hornby et al. (2006) and other publications from that group, which describe the experience and results of using evolutionary algorithms to evolve antennas for spacecraft associated with a number of NASA missions; in particular, two antennas designed for NASA's Space Technology 5 (ST5) mission, and an antenna for a Tracking and Data Relay Satellite (TDRS) for a mission due to operate after 2010.

5.2.1 Antennas for NASA's Space Technology 5 Mission

NASA's Space Technology 5 (ST5) mission had the goal of launching multiple miniature spacecraft to test various innovative concepts for application in future space missions. Three miniaturized satellites were involved in ST5, called micro-sats, designed to measure the effects of solar activity on the Earth's magnetosphere. These micro-sats were approximately half a meter across and half a meter high, weighing around 25 kg when fully fuelled, and each had two antennas, centered on the top and bottom. They were originally designed to operate in a geosynchronous orbit at approximately 35,000 km above Earth, and had a stringent set of requirements for the communication antennas. Details of the specific requirements are in Hornby et al. (2006), and one need not discuss them here, but (in common with similar antenna design tasks) these requirements were in terms of constraints on the gain patterns, voltage standing wave ratios, and input impedances, at both the transmit and receive frequencies; also the mass of each antenna had to be below 165 g, and the shape had to fit within a cylinder with height and diameter both below 16 cm.

To meet the initial design requirements in this instance, the team decided to constrain their search to a monopole wire antenna with four identical arms, equally spaced around the vertical axis. An evolutionary algorithm was therefore set to work to evolve the shape of a single arm, which in turn defined the entire antenna. Importantly, the encoding used by the team was one that allowed almost arbitrary designs for the arm, with no reference to the limited collection of known standard designs. Essentially it was a GP style approach, in which each node in a tree was an antenna-construction operator. Interpreting the tree top down from the root node, and given an initial "feed-wire" of a given small length and orientation, the operators and leaves of the tree effectively specified three-dimensional movements in the style of "turtle graphics," adding sections of wire of specific lengths and orientations to the current partial design.

Having decoded a tree into an antenna design, the antenna was simulated by means of a sophisticated simulation platform, which yielded estimated performance characteristics which then had to be automatically evaluated against the design requirements. In common with the design requirements themselves, readers are referred to Hornby et al. (2006) for details of the fitness function, but suffice it to say that the requirements themselves and the simulation results are both curves involving performance characteristics at different spatial locations and frequencies, and the fitness function involved such things as estimates of distances between desired and actual curves, weighted in specific ways according to the importance of different requirements.

It so happened that the requirements for the ST5 mission changed while these initial antennas were being designed. New mission requirements effectively forced a single-arm antenna design, and this led to the need to redesign the fitness function for the antenna design process. In the operating environment context of Hornby et al.'s work, it is of particular interest and importance to note that an extremely effective antenna design was produced, for the initial set of requirements, in a short time when compared with the human expert design process. Moreover, with mission requirements altered partway through the process, the evolutionary algorithm approach needed only relatively minor modification and was still able to quickly produce an effective antenna for the new requirements.

To meet the initial mission requirements, the best evolved antenna design that emerged, "ST5-3-10," is shown in ❯ Fig. 16 on the left. This antenna met the initial mission requirements, and was indeed all set to be used on the mission itself, until the mission's orbit (and hence many other aspects) was revised. The new evolved best antenna following the new requirements was the one shown on the right in ❯ Fig. 16, the so-called "ST5-33-142-7." The latter antenna design, which was delivered for prototype fabrication less than a month after the changes to the ST5 mission requirements, was found fully compliant with specifications when the prototype was tested, and on 22 March 2006 the ST5 mission was successfully launched into space using evolved antenna ST5-33-142-7. Hornby et al. (2006) report that this was the first computer-evolved antenna to be deployed for any application and the first evolved hardware in space. (One notes that this is clearly valid, if one confines oneself to hardware produced, by whatever means, in the local solar system; but one does not know about elsewhere.)

Hornby et al. (2006) note that the evolved antenna has a number of advantages over human-designed alternatives. These advantages include reduced power consumption, fabrication time and complexity, and improved performance. The ST5 mission managers had actually hired a contractor to produce antenna designs in addition to awaiting the findings of the evolutionary approach. The contractor used conventional design practices, and came up with a variant of one of the many standard designs. When this design was compared in simulation with the evolved design, it was found that if an ST5 craft used two evolved antennas (recall that

□ **Fig. 16**

Photographs, reproduced with permission, of prototype-fabricated evolved antennas. *Left*: the best obtained antenna for the initial ST5 mission requirements, ST5-3-10; *right*: the best obtained following the revised specifications, ST5-33-142-7.

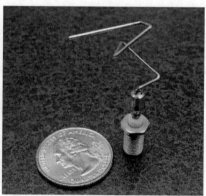

each craft had two antennas), efficiency would be 93% improved over the situation where the craft instead used two of the contractor-designed antennas. Among other explication of the various benefits in Hornby et al. (2006), it is noted that the evolved antenna required approximately 3 person-months to design and fabricate, versus approximately 5 months for the human-designed one.

5.2.2 An Antenna for NASA's TDRS-C Communications Satellite

Later in 2006, the same team evolved an "S-band phased array" antenna element design for NASA's TDRS-C communications satellite, part of a mission that was scheduled for launch sometime between 2010 and 2020. This time the evolutionary algorithm was combined with a hill-climbing algorithm, and the antenna design was somewhat more constrained toward a standard style; nevertheless the resulting design was simpler than the potential competing human designed antennas, consequently reducing testing and integration costs.

As Hornby et al. (2006) reports, the TDRS-C mission will carry several antennas, including among them a 46 element phased array antenna. Readers unfamiliar with the terminology may see ❷ *Fig. 17*, from which it becomes clear what the individual elements are in the phased array. The design and performance specifications for this antenna involved electromagnetic performance issues, as was the case for the ST5 missions, but also certain constraints on the elements and their spacing.

A simpler encoding was used by the team for this case, in which an antenna was represented as a fixed length list of real numbers. Antenna parameters were determined from these simple "genes" in a fairly straightforward way, in which the majority of successive pairs of genes referred to the distance to the next element along the antenna's axis, followed by the size of the next element.

In a similar process used for the ST5 mission antennas, the team set up around 150 separate experiments that each ran an evolutionary algorithm for a total of 50,000 evaluations (antenna simulations) each – the separate evolutionary algorithms each represented a random

◘ Fig. 17
Best evolved TDRS-C antenna.

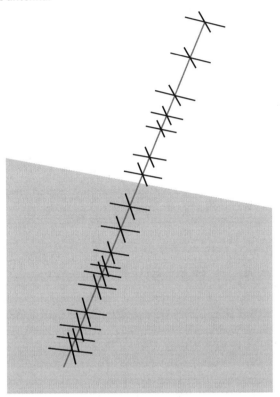

point in parameter space, with different population sizes, mutation rates, and so forth. The best antennas from each of these 150 runs was then subject in a second stage to further improvement via a hill-climbing algorithm for 100,000 evaluations. Finally, the best of these were subject to further hill-climbing.

At the end of this process, most of the evolved antennas were very close to meeting the rather stringent mission specifications, and one of the evolved antennas exceeded the specifications. That one, shown in ❷ *Fig. 17*, was further analyzed by an accurate electromagnetic software (WIPL-D version 5.2), and was subjected to some fine tuning via another evolutionary algorithm, and finally a resulting antenna design was fabricated and tested. The final design, shown in ❷ *Fig. 17*, exceeds the design specifications, and it remains up to the mission leaders whether it is deployed in the TDRS-C mission.

5.2.3 Concluding Points

In this section, the work of Hornby et al. (2006) in evolving antennas for two NASA missions has been described. For both the ST5 mission and the TDRS-C missions it took approximately

3 months to set up the evolutionary algorithms and produce the initial evolved antenna designs. Following the revision in ST5 requirements, it took roughly 1 month for the team to evolve antenna ST5-33-142-7, and the team is indeed very confident (Hornby et al. 2006) that a change in requirements for the TDRS-C mission will result in a similarly fast redesign of an antenna meeting the new requirements.

As well as benefits in relative speed and ease of design, the evolutionary algorithm approach to designing antennas leads to many other advantages over manual design. One such advantage is the potential for performance characteristics that are simply unachievable with conventional design styles. Antenna design is one of several areas in which there is potential for unexplored areas of design space to be examined. These are areas of design space that human experts tend to steer away from, since the current state of theory and understanding is quite limited to the properties of a range of conventional designs. Evolutionary algorithms are far less wary of such ill-understood areas of design space, and, by finding exemplars in such areas that have outstanding performance (such as the ST5 designs discussed in this section), they may lead to more systematic study of such regions of the design space, leading to new design principles and new scientific insights.

5.3 Evolution in Software Design

As noted in Arcuri and Yao (2008), software testing is used to find bugs in computer programs (Myers 1979). Even though successful testing is no guarantee that the software is bug free, testing increases confidence in the software's reliability, and is an integral and extremely important part of modern software engineering. However, testing is very expensive, time consuming, and tedious, amounting to around half the total cost of software development (Beizer 1990). This investment in testing is not begrudged, since releasing bug-ridden software can be immensely more costly in the long run. In fact, it is often argued that far more testing should be done than is usually the case – in the USA, for example, it is estimated that around $20 billion per year could be saved if better testing was done (Tassey 2002). The need for cheaper and faster testing is clear.

This section looks at recent work by Arcuri and Yao (2008), which is part of an area of research called search-based software engineering. In this particular thread, the idea is to investigate the use of evolutionary computation to improve aspects of the testing process. In particular, Arcuri and Yao (2008) are concerned with unit tests (Ellims et al. 2006). This relates to writing small pieces of software code that test as many parts of the project as possible. For example, the test code might call, with specific inputs, a Java method that adds two integers; the returned value is then checked against the expected value. If there is a difference, one can be sure that there is something wrong with the code. However, since testing all possible inputs of a method is usually infeasible, a suitable subset of tests needs to be chosen. Writing code for such "unit tests" requires some way to decide on a good collection of specific input cases, and is a very resource-hungry exercise.

Automated ways to generate unit tests are clearly of interest to the software design process, and this is the topic of Arcuri and Yao's (2008) work. Various approaches have been studied to automatically generate unit tests (McMinn 2004), but there is no known way to generate an optimal set of unit tests for any given program. Also, comparatively little has been done in this area for object-oriented (OO) software. This section describes Arcuri and Yao's (2008) recent work that focuses on a particular type of OO software construct: *containers*. These are data structures (like arrays, lists, vectors, trees, etc.) designed to store arbitrary types of data. What usually distinguishes a container class is the computational cost of operations such as insertion, deletion, and retrieval of data objects. They are used in almost all OO software, so their reliability in commercial code is paramount.

Arcuri and Yao (2007) presented a framework for automatically generating unit tests for container classes, in the context of white box testing. They analyzed a number of different search algorithms, and compared them with more traditional techniques. They used a search space reduction that exploits the characteristics of the containers. Without this reduction, the use of search algorithms would have required too much computational time.

5.3.1 About Testing Java Containers

In each of the many kinds of Java containers (arrays, vectors, lists, trees, and so forth), one usually expects to find methods such as insert, remove, and find. The implementations (and hence computational expense) of these methods can vary much between containers; also, the behavior of such methods is often a function of the container's current contents. This situation considerably complicates the design of unit tests (McMinn and Holcombe 2003, 2005) – just because it is found that a method yields the correct result with certain inputs, it does not mean that it will always give the correct result with those inputs, perhaps depending on the current contents of the container.

The approach to testing containers therefore explicitly considers the sequence, S_i, of function calls. During a testing operation on such a container, the container is referred to here as a "Container under Test" (CuT), and a function call (FC) can be seen as a triple:

<object reference; function name; input list>

This simply refers to calling the given function (method) of the given object (container) with the given list of inputs. In Arcuri and Yao's work, the CuT is subjected to a single sequence S_i of such FCs, rather than a different sequence for each of the functions' branches. Naturally, in the unit test Java code, each FC is embedded in a different try/catch block, so that the paths that throw up exceptions do not forbid the execution of the subsequent FCs in the sequence.

The goal in testing is to achieve a maximal level of "coverage"; broadly speaking, this refers to the amount of code that is tested. All software is replete, for example, with case statements and "if X then Y else…"-style branches, and, without suitable design of test cases, many branches of the code may end up not being followed during the testing process. Given a suitable coverage-related criterion (there are several), it is then important to aim for the short sequence of function calls while achieving excellent coverage. Arcuri and Yao (2008) used *branch coverage* as their coverage criterion, although their approach is easily extensible to other coverage criteria.

In Arcuri and Yao's formulation, they consider a coverage function $cov(S_i)$ which, in relation to a given CuT, returns the number of code branches covered when tested with the sequence of functional calls S_i. Where $len(S_i)$ is simply the number of function calls in the sequence, Arcuri and Yao attempt to optimize both $cov(S_i)$ and $len(S_i)$, preferring a shorter sequence in the case that the coverage of two sequences is the same. To some extent, it is clear that this is a multi-objective problem (see Deb 2001, and Sects. 3.1 and 3.2), however, Arcuri and Yao indicate a definite order of preference in this domain (coverage more important than length), which influences their decision to treat it as a single-objective problem. They therefore attempted to find sequences that optimized $cov(S_i) + 1/(1 + len(S_i))$, although with various modifications and adaptations detailed in Arcuri and Yao (2008).

5.3.2   Smoothing the Test Landscape

Arcuri and Yao (2008) detail several complex factors involved in the enterprise of treating unit testing as an application for evolutionary computation, along with their solutions to these issues. Here, only one such issue will be discussed, of particular pertinence to the "engineering" of problems when considering artificial evolution of solutions. This relates to helping the evolutionary process by making the fitness assessments more informative. The problem in this case is that the number of branches covered (i.e., the number returned by $cov(S_i)$) does not give any indication of how close the sequence S_i is to being able to cover additional branches. Put another way, two sequences S_i and S_j may have the same coverage value, but one may be much "nearer" than the other (i.e., requiring a mutation to just one of its FCs) to a sequence that has higher coverage.

In many branch statements, in which the predicates accessing the branch are quite simple, random sequences of FCs will have little difficulty finding inputs that force coverage of all its branches (this is why a random search, as will be seen later, tends to achieve good coverage). But when the predicate is more complex, it is typically the case that only a very small portion of the space of potential inputs will lead to certain branches being covered. The search is likely to fail.

One approach to this issue is to consider the *Branch Distance* (BD) (Korel 1990). Any particular branch will be entered if a given statement is true (such as $0.2 < x < 0.3$); the BD is a real number that tells how far the relevant predicate is from being true (in the latter case, BD will be low if $x = 0.4$ and high if $x = 10$). Making use of such information in the coverage metric would help the evolutionary search process, by helping to distinguish between pairs of sequences that would otherwise have the same simple coverage value. BD is the topic of much research effort in software testing (e.g., Baresel et al. 2002; Harman et al. 2002; McMinn and Holcombe 2004). This research tends to consider approaches in which different test sequences focus on different branches, without considering the issues involved with the precise sequencing of function calls affecting the results. A difference in Arcuri and Yao's approach is the attempt to evolve a single test sequence that covers all branches.

The technique they adopt is to modify the $cov(S_i)$ defined earlier, incorporating within it a simple measure of branch distance for any uncovered branch, which takes into account how many times the predicate associated with a branch is evaluated. For example, if only one FC in the sequence invokes a predicate with two branches, then only one branch will be covered. However, if in another sequence there are two FCs that test this predicate, both invoking the same branch, then it can be said that this second sequence is closer to covering the second branch, since (if coverage of this other branch is possible at all) this requires only mutation of the input list of one of the FCs in the sequence. There are various ways in which the coverage metric could be modified to take branch distance into account, and it turned out important to use different variations in different circumstances, as is mentioned later, and of course discussed more fully in Arcuri and Yao.

5.3.3 Evaluating Natural Computation for This Task

Arcuri and Yao (2008) tested five approaches: random search (RS), hill-climbing (HC), simulated annealing (SA), a genetic algorithm (GA), and a memetic algorithm (MA). RS is a natural baseline used to understand the effectiveness of other presumably more sophisticated algorithms, and often it brings surprisingly good results. In the current context, one can expect RS to give good results in terms of coverage. The RS worked simply by repeatedly generating random sequences, evaluating them, and returning the best at the end of the process; for pragmatic reasons, it was necessary to specify a maximum length for each sequence.

For the other methods, it was necessary to design neighborhood (and genetic) operators that would operate on sequences to produce variants. In all cases, the encoding of a sequence of FCs was entirely straightforward. The "chromosome" is simply an explicit (variable length) sequence of FCs, each a triple as described above. For neighborhood (mutation) operators, the natural choice was made to use operators of the following type:

- removing an FC from a sequence;
- inserting a new FC into the sequence, in a random position;
- modifying the parameters of a randomly chosen FC in a sequence.

In the GA, single point crossover was also used, in which a child sequence was generated by using the first K (randomly chosen) FCs in one parent, and completing the child with the FCs from position $K + 1$ onward in the second parent. The MA was a simple hybrid of the GA and HC, which repeatedly ran HC on each new individual produced by the GA, until a local optimum was reached. Although the RS, HC, GA, and MC were fairly standard in their SA implementation, various modifications and sophistications were included to control the acceptance of new mutants during the search, in an attempt to balance the coverage and length considerations. These details, and of course other parametric details of all of the algorithms, are explained in Arcuri and Yao (2008).

Arcuri and Yao (2008) performed tests on separate Java containers that implemented Vector, Stack, LinkedList, Hashtable, and TreeMap, respectively, from the Java API 1.4, package java.util, and BinTree and BinomialHeap from the examples in Visser et al. (2006). Described here is only a selection of their results, focusing on the four cases which involved the largest number of public functions under test (PuT). These were: Vector (34 PuT), LinkedList (20 PuT), Hashtable (18 PuT) and TreeMap (17 PuT), respectively, with 1,019, 708, 1,060, and 1,636 lines of code, and achievable coverage of 100, 84, 106, and 191 branches. The latter figures for achievable coverage are based on Arcuri and Yao's experience of around a year's worth of experimentation, with inspection of the container code confirming that non-covered branches seem unreachable.

Each of the five algorithms was tested, to a limit of 100,000 sequence evaluations per trial, using 100 trials per algorithm and container pair; a selection of Arcuri and Yao's results are summarized in ❷ *Table 6*. When one considers the coverage results in the context of the highest achievable coverage mentioned above, it turns out that only TreeMap presents a particularly difficult coverage task. The MA achieves the best mean coverage result on TreeMap, and indeed the MA is reported by Arcuri and Yao (2008) as statistically superior to the other algorithms in all cases except Vector (based on a Mann–Whitney U test). In all container cases except Vector (including the others reported in Arcuri and Yao 2008), the MA shows either the best mean coverage, or it shares first position for coverage while having a better mean length. Not surprisingly, random search tends to have worse performance than the other algorithms,

◻ **Table 6**

Some results from Arcuri and Yao (2008), showing coverage and lengths obtained when evolving sequences of function calls for five separate containers, using five algorithms: random search (RS), hill-climbing (HC), simulated annealing (SA), genetic algorithm (GA), and memetic algorithm (MA)

Container	Algorithm	Mean coverage	Variance in coverage	Mean length	Variance in length
Vector	RS	85.21	1.52	56.99	7.73
	HC	100.00	0.00	47.67	1.05
	SA	99.99	0.01	45.76	1.11
	GA	99.99	0.01	46.87	1.63
	MA	100.00	0.00	47.89	2.64
LinkedList	RS	69.96	1.82	55.27	14.00
	HC	84.00	0.00	38.48	10.27
	SA	82.47	2.25	33.60	5.29
	GA	83.83	0.26	36.66	3.64
	MA	84.00	0.00	36.43	3.58
Hashtable	RS	92.92	1.17	54.45	25.97
	HC	106.00	0.00	35.25	0.19
	SA	105.84	0.74	34.98	0.77
	GA	101.14	6.50	31.10	6.31
	MA	106.00	0.00	35.01	0.01
TreeMap	RS	151.94	5.85	54.11	26.87
	HC	188.76	0.71	51.23	10.08
	SA	184.19	5.75	40.68	5.88
	GA	185.03	3.46	42.14	8.44
	MA	188.86	0.65	50.55	10.31

although it can often achieve reasonable coverage. When coverage from RS is good, however, the length of the sequence of FCs tends to be poor; this is readily understood given the nature of the "difficult" branches in testing, as also indicated above. Finally, it should be pointed out that Arcuri and Yao's system was not able to generate inputs that could cover all branches in the CuT; for example, for pragmatic reasons, branches in private methods were not considered, while around 10% of the public methods were not directly callable.

5.3.4 On Related and Similar Work

Arcuri and Yao (2008) point out the difficulties in comparing their approach with traditional systems in software testing, including the fact that there is no common benchmark scenario, and no reasonable way to replicate the ways that other authors instrumented the software to be tested. However, they point out that traditional techniques (e.g., King 1976; Doong and Frankl 1994; Buy et al. 2000; Marinov and Khurshid 2001; Boyapati et al. 2002; Visser et al. 2004;

Xie et al. 2004, 2005) tend to have considerable challenges with scalability, and invariably rely on considerable prior effort, such as the need to generate algebraic specifications or other formal representations of the functions to be tested; this is particularly tricky when predicates are highly nonlinear, involve loops, nonlinear data types, and so forth. Arcuri and Yao's evolutionary computation-based approach, however, needs no such prior specification effort, and is applicable to any container. Meanwhile, although there are many difficulties with direct comparison with results from traditional techniques reported in the literature, the evolutionary computing approach, especially the MA, seems to have significant benefits in terms of speed. However, much further work is warranted in this field, including hybrids of natural computation and traditional approaches.

As noted by Arcuri and Yao, the use of natural computation in software testing has been gaining a research following in recent years. Other examples include Tonella (2004) who used evolutionary algorithms for generating unit tests of Java programs, while Wappler and Wegener (2006) used strongly typed genetic programming (STGP) also for testing Java programs. Seesing (2006) also investigated STGP for a similar purpose, while Liu et al. (2005) used a hybrid approach, involving ant colony optimization (see the chapter ❷ Swarm Intelligence in this volume) to optimize the sequence of function calls, and a multi-agent evolutionary algorithm to optimize the input parameters of those function calls. Meanwhile, Arcuri and Yao's research described in this section began with presenting new encoding and search operators and a dynamic search space reduction method for testing OO containers (Arcuri and Yao 2007), also testing the Estimation of Distribution Algorithms on this problem (Sagarna et al. 2007).

5.3.5 Summary Thoughts

In this section, we have seen an example of how evolutionary computation is beginning to be used in software engineering. The work focused on showed a comparison with a selection of other methods, and also discussed comparisons with standard techniques in the software engineering industry, and found advantages for the evolutionary computation approach in both scenarios. The empirical tests by Arcuri and Yao (2008) showed that their MA usually performs better than the other algorithms tried. However, there remain clear challenges for improvement (e.g., the performance on the TreeMap container was not completely satisfactory).

Arcuri and Yao conclude, based on their results as well as the related literature, that in testing the OO software, nature-inspired algorithms seem to be better than the standard techniques based on symbolic execution and state matching, since they seem able to solve more complex test problems in less time. Arcuri and Yao's work also included the unusual approach of trying to cover every branch at the same time with a single sequence. This has yet to be compared to the traditional approach of testing each branch separately.

6 Concluding Notes

A *selection* of application areas in which natural computation shows its value in real-world enterprises of various sorts has been discussed. This selection has been quite eclectic. Other authors would have chosen a different set; at another time, the same authors would have chosen a different collection. The main message we mean to convey by such statements is that, for the purposes of demonstrating the significant impact and potential of natural computation in practice, there is certainly no shortage of documented examples that could be selected. We presented here just ten applications, ranging from specific problems to specific domains, and ranging from cases familiar to the authors to highlights known well in the general natural computation community. However, all of them share, one hopes, the property of displaying (each in their own way) a clear indication of the proven promise or great potential for the impact of nature-inspired computation in high-profile and important real-world applications. Similarly, it is hoped that these applications share the property of being inspiring to both students and practitioners; many were selected on the basis of proving particularly popular with students, in the context of getting them interested in the study of natural computation.

When designing a chapter such as this, the first problem one faces is that natural computation is almost too successful in practice. One may ask, for example, why does one not mention more from the thousands of successful real-world applications of neural computation, or fuzzy systems? Well, first of all, they have indeed just been mentioned. But second of all, the positioning of this article in the Broader Perspective part of the Handbook suggests a focus on the novel and unusual; on the less generally known, and on areas whose potential is clear, yet only beginning to be realized in practice.

Naturally, therefore, the center of gravity in this article has turned out to be evolutionary computation. Sandwiched between the more familiar, tried and tested topics of neural and fuzzy systems, and the several emerging areas of natural computation that are currently less "on the map" with application studies, evolutionary computation is a highly flexible child of natural computation that excels in displaying the promise for this field. But, in passing, we saw, in Blondie24, how different areas of nature-inspired computation collaborate to remarkable effect. We have also seen, in the aircraft maneuver study, and in the discussions of super-heuristics, how learning classifier systems – themselves inspired by the adaptive behavior of intelligent organisms – contribute toward natural computing's expanding gallery of successes. Meanwhile, other chapters in this volume cover some of the successes of swarm intelligence, SA, artificial immune systems, and more.

The real-world value of some of the more established natural computing techniques has been proven to be unquestionably immense. It is worth pointing out that this was never anticipated in the "early days" for each individual technique. In the case of neural computation, for example, Minsky and Papert's analysis of the capabilities of two-layer networks led (if not deliberately) to much skepticism and delay in the exploration and take-up of neural networks for pattern recognition. In evolutionary computation's earliest days, the algorithms were usually considered as intellectual curiosities, with the occasional promising application studies considered as one-offs. There seems to be a lesson here for the promise and potential of the several less mature and emerging natural computing ideas – those discussed in this volume, as well as others. In anticipation, one waits and sees.

References

Allen F, Karjalainen R (1999) Using genetic algorithms to find technical trading rules. J Financial Econ 51:245–271

Angeline PJ (1996) Genetic programming's continued evolution. In: Angeline PJ, Kinnear K (eds) Advances in genetic programming, vol 2. MIT Press, Cambridge, pp 89–110

Arcuri A, Yao X (2007) A memetic algorithm for test data generation of object-oriented software. In: IEEE congress on evolutionary computation (CEC), Singapore, pp 2048–2055

Arcuri A, Yao X (2008) Search based software testing of object-oriented containers. Inf Sci 178:3075–3095

Ayob M, Kendall G (2003) A Monte Carlo hyper-heuristic to optimise component placement sequencing for multi head placement machine. In: Proceedings of the international conference on intelligent technologies, Chiang Mai, Thailand, pp 132–141

Banzhaf W, Nordin P, Keller RE, Francone FD (1998) Genetic programming – An introduction: On the automatic evolution of computer programs and its applications. Morgan Kaufmann, San Francisco, CA

Baresel A, Sthamer H, Schmidt M (2002) Fitness function design to improve evolutionary structural testing. In: Genetic and evolutionary computation conference (GECCO). Morgan Kaufmann, San Francisco, New York, CA, pp 1329–1336

Becker LA, Seshadri M (2003a) Comprehensibility and overfitting avoidance in genetic programming for technical trading rules. Computer Science Technical Report WPI-CS-TR-03–09. Worcester Polytechnic Institute, Worcester, Massachusetts, USA

Becker LA, Seshadri M (2003b) Cooperative coevolution of technical trading rules. Computer Science Technical Report WPI-CS-TR-03-15. Worcester Polytechnic Institute, Worcester, Massachusetts, USA

Becker LA, Seshadri M (2003c) GP-evolved technical trading rules can outperform buy and hold. In: Proceedings of sixth international conference on computational intelligence and natural computing, North Carolina, USA, 26–30 September 2003

Beizer B (1990) Software testing techniques. Van Nostrand Rheinhold, New York

Bilgin B, Ozcan E, Korkmaz EE (2006) An experimental study on hyper-heuristics and final exam scheduling. In: Proceedings of the 2006 international conference on the practice and theory of automated timetabling, Brno, Czech Republic, pp 123–140

Box GEP (1957) Evolutionary operation: A method for increasing industrial productivity. Appl Stat 6:81–101

Box G, Hunter W, Hunter J (2005) Statistics for experimenters: design, innovation, and discovery, 2nd edn. Wiley, New York

Boyapati C, Khurshid S, Marinov D (2002) Korat: Automated testing based on java predicates. In: Proceedings of the international symposium on software testing and analysis (ISSTA). ACM, New York

Brabazon A, O'Neill M (2005) Biologically inspired algorithms for financial modelling. Natural computing series. Springer, New York

Branke J, Deb K (2005) Integrating user preferences into evolutionary multi-objective optimization. In: Knowledge incorporation in evolutionary computation. Springer, New York, pp 461–477

Brockhoff D, Zitzler E (2006) Are all objectives necessary? On dimensionality reduction in evolutionary multiobjective optimization. In: Parallel problem solving from nature – PPSN IX. Lecture notes in computer science, vol 4193. Springer, New York, pp 533–542

Burke EK, MacCarthy BL, Petrovic S, Qu R (2002) Knowledge discovery in a hyperheuristic for course timetabling using case based reasoning. In: Proceedings of the fourth international conference on the practice and theory of automated timetabling (PATAT'02) Gent, Belgium. Springer, Berlin

Burke EK, Kendall G, Newall J, Hart E, Ross P, Schulenburg S (2003) Hyper-heuristics an emerging direction in modern search technology. In: Glover F, Kochenberger GA (eds) Handbook of metaheuristics. Springer, New York, pp 457–474

Buy U, Orso A, Pezze M (2000) Automated testing of classes. In: Proceedings of the international symposium on software testing and analysis (ISSTA), pp 39–48

Chapman L, Thornes JE, Bradley AV (2002) Sky-view factor approximation using GPS receivers. Int J Climatol 22(5):615–621

Chellapilla K, Fogel DB (1999a) Evolution, neural networks, games, and intelligence. Proc IEEE 87(9):1471–1496

Chellapilla K, Fogel DB (1999b) Evolving neural networks to play checkers without expert knowledge. IEEE Trans Neural Netw 10(6):1382–1391

Chellapilla K, Fogel DB (2001) Evolving an expert checkers playing program without using human expertise. IEEE Trans Evol Comput 5(4):422–428

Chen SH (2002) Genetic algorithms and genetic programming in computational finance. Kluwer, Boston, MA

Chen SH, Yeh CH (1996) Toward a computable approach to the efficient market hypothesis: an application of genetic programming. J Econ Dyn Cont 21:1043–1063

Cheng SL, Khai YL (2002) GP-based optimisation of technical trading indicators and profitability in FX market. In: Proceeding of the ninth international conference on neural information processing (ICONIP'02), vol 3, Singapore, pp 1159–1163

Chernoff H (1972) Sequential analysis and optimal design. SIAM monograph. SIAM, Philadelphia, PA

Coello C (2000) An updated survey of GA-based multiobjective optimization techniques. ACM Comput Surv (CSUR) 32(2):109–143

Coello C (2006) Twenty years of evolutionary multiobjective optimization: A historical view of the field. IEEE Comput Intell Mag 1(1):28–36

Corne D, Jerram N, Knowles J, Oates M (2001) PESA-II: Region-based selection in evolutionary multiobjective optimization. In: Spector L, Goodman ED, Wu A, Langdon WB, Voigt H-M, Gen M, Sen S,

Dorigo M, Pezeshk S, Garzon MH, Burke E (eds) Proceedings of GECCO-2001: Genetic and evolutionary computation conference. Morgan Kaufmann, San Mateo, San Francisco, CA, pp 283–290

Corne D, Deb K, Fleming P, Knowles J (2003a) The good of the many outweighs the good of the one: Evolutionary multiobjective optimization. IEEE Connections Newsletter 1(1):9–13. ISSN 1543-4281

Corne D, Oates M, Kell D (2003b) Fitness gains and mutation patterns: Deriving mutation rates by exploiting landscape data. In: De Jong K, Poli R, Rowe J (eds) Foundations of genetic algorithms. Morgan Kaufmann, San Francisco, CA, pp 347–364

Cornford D, Thornes JE (1996) A comparison between spatial winter indices and expenditure on winter road maintenance in Scotland. Int J Climatol 16:339–357

Cowling P, Kendall G, Soubeiga E (2000) A hyperheuristic approach to scheduling a sales summit. In: Burke EK, Erben W (eds) Practice and theory of automated timetabling III: Third international conference, PATAT 2000, Konstanz, Germany, August 2000, selected papers. LNCS, vol 2079. Springer, pp 176–190

Cowling P, Kendall G, Soubeiga E (2002) Hyperheuristics: A robust optimisation method applied to nurse scheduling. Technical Report NOTTCS-TR-2002-6. School of Computer Science & IT, University of Nottingham, Nottingham, England

Cross SE, Walker E (1994) Dart: Applying knowledge-based planning and scheduling to crisis action planning. In: Zweben M, Fox MS (eds) Intelligent scheduling. Morgan Kaufmann, San Francisco, CA

Datta R, Deb K (2009) A classical-cum-evolutionary multi-objective optimization for optimal machining parameters. In: Proceedings of NABIC. IEEE CIS Press

Davies ZS, Gilbert RJ, Merry RJ, Kell DB, Theodorou MK, Griffith GW (2000) Efficient improvement of silage additives by using genetic algorithms. Appl Environ Microbiol April:1435–1443

Deb K (1997) Mechanical component design using genetic algorithms. In: Dasgupta D, Michalewicz Z (eds) Evolutionary algorithms in engineering applications. Springer, New York, pp 495–512

Deb K (2000) An efficient constraint handling method for genetic algorithms. Comput Meth Appl Mech Eng 1862(4):311–338

Deb K (2001) Multi-objective optimization using evolutionary algorithms. Wiley, New York

Deb K, Kumar A (1995) Real-coded genetic algorithms with simulated binary crossover: Studies on multimodal and multi-objective problems. Complex Syst 9(6):431–454

Deb K, Srinivasan A (2005) Innovization: Innovation of design principles through optimization. KanGAL Report No. 2005007

Deb K, Srinivasan A (2006) Innovization: Innovating design principles through optimization. In: Proceedings of GECCO. ACM, New York, pp 1629–1636

Deb K, Agrawal S, Pratap A, Meyarivan T (2002) A fast and elitist multi-objective genetic algorithm: NSGA-II. IEEE Trans Evol Comput 6(2):182–197

Doong R, Frankl PG (1994) The ASTOOT approach to testing object-oriented programs. ACM Trans Softw Eng Methodol 3:101–130

Ellims M, Bridges J, Ince DC (2006) The economics of unit testing. Emp Softw Eng 11(1):5–31

Evans JRG, Edirisinghe MJ, Eames PVCJ (2001) Combinatorial searches of inorganic materials using the inkjet printer: Science philosophy and technology. J Eur Ceramic Soc 21:2291–2299

Fang H-L, Ross PM, Corne D (1994) A promising hybrid GA/heuristic approach for open-shop scheduling problems. In: Cohn A (ed) Proceedings of ECAI 94: 11th European conference on artificial intelligence. Wiley, Amsterdam, The Netherlands, pp 590–594

Farnsworth GV, Kelly JA, Othling AS, Pryor RJ (2004) Successful technical trading agents using genetic programming. SANDIA Report SAND2004–4774. SANDIA National Laboratories, California

Fisher R (1971) The design of experiments, 9th edn. Macmillan, New York

Fogel D (1998) Evolutionary computation. The fossil record. Selected readings on the history of evolutionary computation. IEEE Press, Piscataway, New Jersey, USA

Fogel DB (2002) Blondie24: Playing at the edge of AI. Morgan Kaufmann, San Francisco. CA, ISBN 1-55860-783-8

Fogel DB, Hays TJ, Hahn SL, Quon J (2004) A self-learning evolutionary chess program. Proc IEEE 92 (12):1947–1954

Fogel DB, Hays TJ, Hahn SL, Quon J (2006) The Blondie25 chess program competes against Fritz 8.0 and a human chess master. In: Louis S, Kendall G (eds) Proceedings of 2006 IEEE symposium on computational intelligence & games. IEEE, Reno, pp 230–235

Fonseca C, Fleming P (1995) An overview of evolutionary algorithms in multiobjective optimization. Evol Comput 3(1):1–16

Fonseca C, Fleming P (1998) Multiobjective optimization and multiple constraint handling with evolutionary algorithms. I. A unified formulation. IEEE Trans Syst, Man, Cybernetics – Part A 28(1):26–37

Fukunaga A (2008) Automated discovery of local search heuristics for satisfiability testing. Evol Comput 16 (1):31–61

Fyfe C, Marney JP, Tarbert H (1999) Technical trading versus market efficiency: A genetic programming approach. Appl Finan Econ 9:183–191

Gent IP, Walsh T (1993) Towards an understanding of hill-climbing procedures for SAT. In: Proceedings of AAAI'93. AAAI Press/MIT Press, Menlo Park, pp 28–33

Giffler B, Thompson GL (1960) Algorithms for solving production scheduling problems. Oper Res 8 (4):487–503

Goldberg DE (1989) Genetic algorithms in search, optimization, and machine learning. Addison-Wesley, Boston, MA

Goldberg DE (2002) The design of innovation: Lessons from and for competent genetic algorithms. Kluwer, Boston, MA

Gratch J, Chein S, de Jong G (1993) Learning search control knowledge for Deep Space network scheduling. In: Proceedings of the tenth international conference on machine learning, Amherst, MA, pp 135–142

Grefenstette JJ (1988) Credit assignment in rule discovery systems based on genetic algorithms. Mach Learn 3:225–246

Handa H, Chapman L, Yao X (2005) Dynamic salting route optimisation using evolutionary computation. In: Proceedings of the 2005 congress on evolutionary computation, Edinburgh, Scotland, vol 1, pp 158–165

Handa H, Chapman L, Yao X (2006) Robust route optimization for gritting/salting trucks: A CERCIA experience. IEEE Comput Intell Mag February:6–9

Harman M, Hu L, Hierons R, Baresel A, Sthamer H (2002) Improving evolutionary testing by flag removal. In: Genetic and evolutionary computation conference (GECCO). Morgan Kaufmann, San Francisco, New York, CA, pp 1351–1358

Hart E, Ross PM (1998) A heuristic combination method for solving job-shop scheduling problems. In: Eiben AE, Back T, Schoenauer M, Schwefel H-P (eds) Parallel problem solving from nature V. LNCS, vol 1498. Springer, Berlin, pp 845–854

Hart E, Ross PM, Nelson J (1998) Solving a real-world problem using an evolving heuristically driven schedule builder. Evol Comput 6(1):61–80

Holland JH (1992) Adaptation in natural and artificial systems. MIT Press, Cambridge, MA, USA

Holland JH, Holyoak KJ, Nisbett RE, Thagard PR (1986) Induction: Processes of inference, learning, and discovery. MIT Press, Cambridge

Hornby GS, Globus A, Linden DS, Lohn JD (2006) Automated antenna design with evolutionary algorithms. In: AIAA Space, San Jose, CA

Hunter WG, Kittrell JR (1966) Evolutionary operation: A review. Technometrics 8(3):389–397. Available at: http://www.jstor.org/stable/1266686

Kannan BK, Kramer SN (1994) An augmented Lagrange multiplier based method for mixed integer discrete continuous optimization and its applications to mechanical design. ASME J Mech Des 116(2):405–411

Kerlogue F, Zanetini F (2004) Batik: Design, style and history. Thames and Hudson, London

King JC (1976) Symbolic execution and program testing. Commun ACM 19:385–394

Knight CG, Platt M, Rowe W, Wedge DC, Khan F, Day PJ, McShea A, Knowles J, Kell DB (2008) Array-based evolution of DNA aptamers allows modelling of an explicit sequence-fitness landscape. Nucleic Acids Research November:e6

Knowles J (2006) ParEGO: A hybrid algorithm with on-line landscape approximation for expensive multi-objective optimization problems. IEEE Trans Evol Comput 10(1):50–66

Knowles JD (2009) Closed-loop evolutionary multiobjective optimization. IEEE Comput Intell Mag August:77–91

Korel B (1990) Automated software test data generation. IEEE Trans Softw Eng 16:870–879

Koza JR (1992) Genetic programming: On the programming of computers by means of natural selection. MIT Press, Cambridge

Lacomme P, Prins C, Ramdane-Cherif W (2004) Competitive memetic algorithms for arc routing problems. Ann Oper Res 131:159–185

Lau TWE, Ho Y-C (1999) Super-heuristics and their application to combinatorial problems. Asian J Cont 1(1):1–13

Li Y, Hu CJ, Yao X (2009) Innovative Batik design with an interactive evolutionary art system. J Comput Sci Technol 24(6):1035–1047

Liu X, Wang B, Liu H (2005) Evolutionary search in the context of object oriented programs. In: MIC2005: The sixth metaheuristics international conference, Vienna, Austria

Lo AW, Mamaysky H, Wang J (2000) Foundations of technical analysis: Computational algorithms, statistical-inference, and empirical implementation. J Finance 55:1705–1770

Lohpetch D, Corne D (2009) Discovering effective technical trading rules with genetic programming: Towards robustly outperforming buy-and-hold. In: World congress on nature and biologically inspired computing (NABIC). IEEE Press

Lohpetch D, Corne D (2010) Outperforming buy-and-hold with evolved technical trading rules: Daily, weekly and monthly trading. In: EvoApplications. Proceedings of EvoStar 2010. LNCS, vol 6025. Springer pp

Lutton E (2006) Evolution of fractal shapes for artists and designers. Int J Artif Intell Tools 15(4):651–672

Marinov D, Khurshid S (2001) TestEra: A novel framework for testing java programs. In: IEEE international conference on automated software engineering (ASE), San Diego, California, USA. Kluwer, The Netherlands.

Marney JP, Miller D, Fyfe C, Tarbert H (2000) Technical analysis versus market efficiency: A genetic programming approach. Computing in Economics and Finance, Society for Computational Economics, Barcelona, Spain (paper #169)

Marney JP, Fyfe C, Tarbert H, Miller D (2001) Risk adjusted returns to technical trading rules: A genetic programming approach. Computing in Economics and Finance, Society for Computational Economics, Yale University, USA

Marney JP, Tarbert H, Fyfe C (2005) Risk adjusted returns from technical trading: A genetic programming approach. Appl Financial Econ 15: 1073–1077

McAllester D, Selman B, Kautz H (1997) Evidence for invariants in local search. In: Proceedings of the 14th national conference on artificial intelligence. AAAI Press/MIT Press, Menlo Park, Providence, Rhode Island, USA, pp 321–326

McMinn P (2004) Search-based software test data generation: A survey. Softw Test Verif Reliab 14 (2):105–156

McMinn P, Holcombe M (2003) The state problem for evolutionary testing. In: Genetic and evolutionary computation conference (GECCO), Chicago, Illinois, USA, pp 2488–2500

McMinn P, Holcombe M (2004) Hybridizing evolutionary testing with the chaining approach. In: Genetic and evolutionary computation conference (GECCO), Seattle, Washington, USA, pp 1363–1374

McMinn P, Holcombe M (2005) Evolutionary testing of state-based programs. In: Genetic and evolutionary computation conference (GECCO), Washington, DC, USA, pp 1013–1020

Messac A, Mattson CA (2004) Normal constraint method with guarantee of even representation of complete Pareto frontier. AIAA J 42(10):2101–2111

Miettinen K (1999) Nonlinear multiobjective optimization. Springer, New York

Minton S (1988) Learning search control knowledge: An explanation-based approach. Kluwer Academic Publishers Norwell, MA, USA

Murphy JJ (1999) Technical analysis of the financial markets. New York Institute of Finance, New York

Myers G (1979) The art of software testing. Wiley, New York

Myers R, Montgomery D (1995) Response surface methodology: Process and product optimization using designed experiments. Wiley, New York

Nagata Y, Kobayashi S (1997) Edge assembly crossover: A high-power genetic algorithm for the traveling salesman problem. In: Proceedings of the seventh international conference on genetic algorithms, East Lansing, Michigan, USA, pp 450–457

Neely C (2001) Risk-adjusted, ex ante, optimal technical trading rules in equity markets. Working Papers 99-015D, Revised August 2001, Federal Reserve Bank of St. Louis

Nolfi S, Floreano D (2004) Evolutionary robotics: The biology, intelligence, and technology of self-organizing machines. Bradford Book. MIT Press, Cambridge

O'Hagan S, Dunn WB, Brown M, Knowles JD, Kell DB (2005) Closed-loop, multiobjective optimization of analytical instrumentation: Gas chromatography/time-off light mass spectrometry of the metabolomes of human serum and of yeast fermentations. Anal Chem 77(1):290–303

O'Hagan S, Dunn W, Knowles J, Broadhurst D, Williams R, Ashworth J, Cameron M, Kell D (2007) Closed-loop, multiobjective optimization of two-dimensional gas chromatography/mass spectrometry for serum metabolomics. Anal Chem 79(2):464–476

Özcan E, Bilgin B, Korkmaz EE (2008) A comprehensive analysis of hyper-heuristics. Intell Data Anal 12 (1):3–23

Potvin JY, Soriano P, Vallée M (June 2004) Generating trading rules on the stock markets with genetic programming. Comput Oper Res 31(7):1033–1047

Pring MJ (1980) Technical analysis explained. McGraw-Hill, New York

Rechenberg I (1965) Cybernetic solution path of an experimental problem. Royal Aircraft Establishment, Library Translation 1122, Farnborough, Hampshire, UK

Rechenberg I (2000) Case studies in evolutionary experimentation and computation. Comput Meth Appl Mech Eng 186(2–4):125–140

Reklaitis GV, Ravindran A, Ragsdell KM (1983) Engineering optimization methods and applications. Wiley, New York

Romero J, Machado P (2008) The art of artificial evolution: a handbook on evolutionary art and music. Springer, Heidelberg

Ross P, Hart E, Corne D (1997) Some observations about GA-based exam timetabling. In: Burke EK, Carter M (eds) Practice and theory of automated timetabling II: Second international conference, PATAT 1997, Toronto, Canada, August 1997, selected papers. LNCS, vol 1408. Springer, pp 115–129

Ross P, Schulenburg S, Marín-Blázquez JG, Hart E (2002) Hyper-heuristics: Learning to combine simple heuristics in bin-packing problems. In: Genetic and evolutionary computation conference (GECCO 2002), New York

Ross P, Marín-Blázquez JG, Schulenburg S, Hart E (2003) Learning a procedure that can solve hard bin-packing problems: A new GA-based approach to hyper-heuristics. In: Proceedings of the genetic and evolutionary computation conference (GECCO 2003). Lecture notes in computer science, vol 2723. Springer, Chicago, Illinois, USA, pp 1295–1306

Ruggiero MA (1997) Cybernetic trading strategies. Wiley, New York

Russell SJ, Norvig P (2003) Artificial intelligence: A modern approach, 2nd edn. Prentice Hall, Upper Saddle River, NJ, pp 163–171

Sagarna R, Arcuri A, Yao X (2007) Estimation of distribution algorithms for testing object oriented software. In: IEEE congress on evolutionary computation (CEC), Singapore, pp 438–444

Samuel AL (1959) Some studies in machine learning using the game of checkers. IBM J Res Develop 3:210–219

Schaeffer J (1996) One jump ahead: Challenging human supremacy in checkers. Springer, New York, p 97, 447

Schaeffer J, Lake R, Lu P, Bryant M (1996) Chinook: The world man–machine checkers champion. AI Mag 17:21–29

Seesing A (2006) Evotest: Test case generation using genetic programming and software analysis. Master's thesis, Delft University of Technology

Selman B, Levesque HJ, Mitchell DG (1992) A new method for solving hard satisfiability problems. In: Tenth AAAI, San Jose, pp 440–446

Selman B, Kautz HA, Cohen B (1994) Noise strategies for improving local search. In: Proceedings of the 12th national conference on artificial intelligence. AAAI Press/MIT Press, Menlo Park, Seattle, Washington, USA, pp 337–343

Sharpe WF (1966) Mutual fund performance. J Business 39(S1):119–138. doi:10.1086/294846

Shaw RL (1998) Fighter combat: Tactics and maneuvering. United States Naval Institute Press, Annapolis, Maryland, USA

Sims K (1991) Artificial evolution for computer graphics. In: Proceedings of the 18th annual conference on computer graphics and interactive techniques (SIGGRAPH 1991). ACM, New York, Las Vegas, Nevada, USA, pp 319–328

Smith RE, Dike BA (1995) Learning novel fighter combat maneuver rules via genetic algorithms. Int J Expert Syst 8(3):247–276

Smith RE, Dike BA, Mehra RK, Ravichandran B, El-Fallah A (2000) Classifier systems in combat: Two-sided learning of maneuvers for advanced fighter aircraft. Comput Meth Appl Mech Eng 186:431–437

Smith RE, Dike BA, Ravichandran B, El-Fallah A, Mehra RK (2002) Discovering novel fighter combat maneuvers: Simulating test pilot creativity. In: Bentley P, Corne D (eds) Creative evolutionary systems. Morgan Kaufmann, San Francisco, CA, pp 467–486

Tassey G (2002) The economic impacts of inadequate infrastructure for software testing. Final Report. National Institute of Standards and Technology

Terashima-Marín H, Ross PM, Valenzuela-Rendón M (1999) Evolution of constraint satisfaction strategies in examination timetabling. In: Banzhaf W et al. (eds) Proceedings of the GECCO-99 genetic and evolutionary computation conference, Orlando, Florida. Morgan Kaufmann, San Francisco, pp 635–642

Thompson A, Layzell P (1999) Analysis of unconventional evolved electronics. Commun ACM 42(4):71–79

Tonella P (2004) Evolutionary testing of classes. In: Proceedings of the international symposium on software testing and analysis (ISSTA), pp 119–128

Trianni V, Nolfi S, Dorigo M (2006) Cooperative hole avoidance in a swarm-bot. Robot Autonomous Syst 54(2):97–103

Tuerk C, Gold L (1990) Systematic evolution of ligands by exponential enrichment: RNA ligands to bacteriophage T_4 DNA polymerase. Science 249(4968):505

Visser W, Pasareanu CS, Khurshid S (2004) Test input generation with java pathfinder. In: Proceedings of the international symposium on software testing and analysis (ISSTA), Boston, Massachusetts, USA

Visser W, Pasareanu CS, Pelánek R (2006) Test input generation for java containers using state matching. In: Proceedings of the international symposium on software testing and analysis (ISSTA), Portland, Maine, USA, pp 37–48

Wang Y, Tan T, Zhu Y (2000) Face verification based on singular value decomposition and radial basis function neural network. In: Proceedings of fourth Asian conference on computer vision, Taiwan, pp 432–436

Wang SF, Wang S, Takagi H (2006) User fatigue reduction by an absolute rating data-trained predictor in IEC. In: Proceedings of 2006 congress on evolutionary computation, pp 2195–2200

Wappler S, Wegener J (2006) Evolutionary unit testing of object-oriented software using strongly-typed genetic programming. In: Genetic and evolutionary computation conference (GECCO), Seattle, Washington, USA, pp 1925–1932

Wedge D, Rowe W, Kell D, Knowles J (2009) In silico modelling of directed evolution: Implications for experimental design and stepwise evolution. J Theor Biol 257:131–141

Wilson S (1998) Generalisation in the XCS classifier system. In: Koza J (ed) Proceedings of the third genetic programming conference, Madison, Wisconsin. Morgan Kaufmann, San Francisco, CA, USA, pp 665–674

Xie T, Marinov D, Notkin D (2004) Rostra: A framework for detecting redundant object-oriented unit tests. In: IEEE international conference on automated software engineering (ASE), Linz, Austria. IEEE Computer Society, Washington, DC, pp 196–205

Xie T, Marinov D, Schulte W, Notkin D (2005) Symstra: A framework for generating object-oriented unit tests using symbolic execution. In: Proceedings of the 11th international conference on tools and algorithms for the construction and analysis of systems, Edinburgh, UK, pp 365–381

Broader Perspective – Alternative Models of Computation

David W. Corne

53 Artificial Life

Wolfgang Banzhaf[1] · *Barry McMullin*[2]
[1]Department of Computer Science, Memorial University
of Newfoundland, St. John's, NL, Canada
banzhaf@cs.mun.ca
[2]Artificial Life Lab, School of Electronic Engineering, Dublin City
University, Ireland
barry.mcmullin@dcu.ie

G. Rozenberg et al. (eds.), *Handbook of Natural Computing*, DOI 10.1007/978-3-540-92910-9_53,
© Springer-Verlag Berlin Heidelberg 2012

Abstract

Artificial life has now become a mature inter-discipline. In this contribution, its roots are traced, its key questions are raised, its main methodological tools are discussed, and finally its applications are reviewed. As part of the growing body of knowledge at the intersection between the life sciences and computing, artificial life will continue to thrive and benefit from further scientific and technical progress on both sides, the biological and the computational. It is expected to take center stage in natural computing.

1 Introduction and Historical Overview

The phrase *artificial life*, referring to a specific coherent research programme, is normally attributed to Chris Langton. He specifically applied the term as a title for the "interdisciplinary workshop on the synthesis and simulation of living systems" that he organized in September 1987, in Los Alamos, New Mexico (Langton 1989). It was then adopted for a biannual international conference series (now up to *Artificial Life XI*, held in Winchester UK, in August 2008), and alternating with the biannual *European Conference on Artificial Life* (ECAL), first held in Paris in 1991. The *Artificial Life journal* (with Langton as first Editor-in-Chief) was founded in 1993. The journal and the two conference series are now formally coordinated through the International Society for Artificial Life (ISAL) (http://www.alife.org), which was established in 2001.

Langton's original announcement of the 1987 workshop defined the new field as follows:

▶ Artificial life is the study of artificial systems that exhibit behavior characteristic of natural living systems. This includes computer simulations, biological and chemical experiments, and purely theoretical endeavors. Processes occurring on molecular, cellular, neural, social, and evolutionary scales are subject to investigation. The ultimate goal is to extract the logical form of living systems. (Langton 1987)

It is clear that, from the outset, Langton's vision was for a very broad, thoroughly interdisciplinary, endeavor, and indeed that is how the field has largely developed. There has been a consistent focus on using tools of computer science to model and simulate biological systems; but also a specific goal of using *synthesis* to start an exploration of the space of *possible* life. Indeed, Langton's own first use of the phrase "artificial life" was in his paper "*Studying artificial life with cellular automata*" (Langton 1986). This was presented two years before the first artificial life workshop, at the conference *Evolution, Games, & Learning* organized by the Los Alamos Center for Non-Linear Studies (CNLS) in May 1985. In this paper, Langton was already introducing "artificial life" as a direct lineal descendant and inheritor of von Neumann's seminal research programme, begun in the late 1940s, combining "automata theory" with problems of biological organization, self-reproduction, and the evolution of complexity (von Neumann 1949, 1951, 1966).

But while von Neumann's influence was certainly strong, there were many other contributions and precursors to the field. The phrase "artificial life" is, in part, a deliberate play on "Artificial Intelligence" or AI. The original premise of AI, dating from Turing's 1950 paper on "*Computing machinery and intelligence*" (Turing 1950), was that "intelligence" was an intrinsically computational or software phenomenon. It could therefore be divorced from any particular underlying hardware implementation, including all the particular biological details of terrestrial life. By the 1980s however, after 30 years of intensive and well-resourced efforts, AI was still showing little clear progress toward its central goal of realizing human-level intelligence by means of computer programs. Indeed, as early as 1977, Popper was already declaring:

▶ "... computers are totally different from brains, whose function is not primarily to compute but to guide and balance an organism and help it to stay alive. It is for this reason that the first step of nature toward an intelligent mind was the creation of life, and I think that should we artificially create an intelligent mind, we would have to follow the same path. (Popper and Eccles 1977)

In 1988, the mounting challenges and criticisms of GOFAI ("good old fashioned AI" (Haugeland 1989)), were summarized in a special edition of *Daedalus*, the Journal of the American Academy of Arts and Sciences (Graubard 1988). This documented a renaissance of interest in the biological underpinnings of intelligence. Artificial life can therefore be seen as just one of the particular results of this splintering of AI, along with subsumption-architecture robotics (Brooks 1985), parallel distributed processing (Rumelhart et al. 1986) (the "new" connectionism), Neural Darwinism (Reeke and Edelman 1988; Edelman 1987), embodied intelligence (Varela et al. 1992), etc. And indeed, in some ways, artificial life continues to function as an integrative term, bridging between these more specialized and disparate research programmes.

A specific debate that artificial life has inherited from its roots in AI, is that between "weak" and "strong" forms. In the AI case, a distinction is made between the claim that an AI system provides a more or less effective *simulation* of intelligence, and the claim that it is "really" intelligent. The distinction is particularly associated with the philosopher John Searle, whose famous "Chinese Room Argument" claimed to prove that however well digital computers might simulate intelligence, computation could never be sufficient to genuinely *realize* intelligence (Searle 1980). Clearly an analogous distinction can be drawn within artificial life, between merely simulating and actually realizing "real" life (Levy 1993). In that case, it is generally focused on whether "life-like" organization, which exists only in a purely computational virtual world, could ever properly be described as genuinely living. However, Searle's abstract philosophical argument cannot be transferred directly into the artificial life domain as it relies on conscious introspection. Perhaps because of this, the issue has received comparatively less intense attention in this field. In any case, while much artificial life research is carried out in computational/virtual worlds, and thus might be subject to this kind of debate, it is generally considered that work in so-called wet artificial life, working in vitro with real chemical systems (e.g., Luisi and Varela 1989; Rasmussen et al. 2008), is certainly immune to this kind of critique.

Other significant influences and inspirations for artificial life include morphogenesis (Turing 1952), cybernetics (Wiener 1965), general systems theory (Bertalanffy 1976), mathematical/relational biology (Rashevsky 1940), theoretical biology (Waddington 1969), self-organization (Yovits and Cameron 1960), hierarchy theory (Pattee 1973), autopoiesis (Maturana and Varela 1973), and genetic and evolutionary algorithms (Fogel et al. 1966; Holland 1975; Rechenberg 1975).

A specific precursor to the modern artificial life conference series was the conference on *Automata, Languages, Development* (Lindenmayer and Rozenberg 1976) organized by Rozenberg and Lindenmayer, March–April 1975 in Noordwijkerhout, Holland. One of the

participants, Alvy Ray Smith, has subsequently commented that (http://alvyray.com/Papers/
PapersCA.htm):

▶ I like to call this conference ALife0 — for Artificial Life 0 conference — since it was the first attempt
I know to attempt cross-fertilization between biologists and computer scientists. Many of the
players at this conference were present for ALife1, ALife2, etc [...] Other participants at ALife0
were Karel Culik, Pauline Hogeweg, John Holland, Aristid Lindenmayer, and Stanislaw Ulam.

In summary, artificial life is now a mature and well-established discipline (or "inter-
discipline") in its own right. The following sections will present the fundamental questions it is
concerned with, a selection of specific theories and formalisms it relies on, and we review some
applications and conclude with a discussion of the key open problems for future research.

2   Fundamental Questions

2.1   What Is (Artificial) Life?

▶ By extending the horizons of empirical research in biology beyond the territory currently
circumscribed by life-as-we-know-it, the study of Artificial Life gives us access to the domain of
life-as-it-could-be, and it is within this vastly larger domain that we must ground general theories
of biology and in which we will discover practical and useful applications of biology in our
engineering endeavors. (Langton 1995)

The question of "what is life?" is clearly fundamental to a discussion of the field of artificial
life, as the answer (or answers) would be expected to underpin the scope, problems, methods,
and results of its investigations.

The starting point is, of course, natural life, the domain of conventional biology. Historically,
it was conjectured that the very distinctive properties and characteristics of natural living systems
reflected a fundamental *constitutional* distinction: that living systems incorporated (or were
literally "animated by") a distinct and unique "substance" – the so-called *elan vital* that
identified this class of theories as forms of *vitalism*. However, from the pioneering inorganic
synthesis of urea by Wöhler in 1828 onward, vitalism progressively lost credibility. With the
explosive growth of molecular understanding of basic biological mechanisms in the second half
of the twentieth century (traditionally dated from the elucidation of the chemical structure of
DNA in 1953), vitalism in this sense has now all but disappeared from scientific consideration.
There is now a clear scientific consensus that there is no *material* distinction between the
domains of living and nonliving systems; rather the differences (which are, of course, still
objectively real and substantial) arise from differences in material organization.

However, it is still very much an open question as to what exactly these differences in
organization *are*. Conventional molecular biology continues to make rapid progress in teasing
out the detailed molecular basis for particular low-level mechanisms of operation in natural organisms; but by focusing on dismantling organisms into molecular components, these
investigations, though still valuable in themselves, risk eliminating precisely the higher order
organization that is distinctive of natural life. More recently, and partly in reaction to this, there
has been growing interest in what is termed *systems biology*: The explicit study of how low-level
molecular mechanisms are successfully integrated into higher level "systems" (Wolkenhauer
2007); and also, in *synthetic biology* that investigates the synthesis of entirely novel classes
of living systems, either by manipulation of natural organisms and their components

(Smith et al. 2003; Regis 2008) or even by attempting de novo creation of life from entirely nonbiological substrates (Rasmussen 2008).

All these strands of investigation are still essentially premised on the study of existing natural life on Earth; or, as Langton puts it, "life-as-it-is" rather than "life-as-it-could-be" (Langton 1992b) (even synthetic biology, in its current state, largely relies on preexisting biological components, or close analogues of them).

By contrast, the field of artificial life, from its inception, has had a specific goal of investigating the most general possible formulations and instantiations of "living phenomenology." In particular, it has explicitly brought to bear the possibility of creating computer-synthesized universes, or *virtual worlds*, in which even the most basic laws of "physics" or "chemistry" can be arbitrarily manipulated, in whatever ways may prove most conducive to any particular study.

Of course, this lack of hard boundaries makes "artificial life," as a field of study, significantly ill-defined. Unlike the case for natural life, there are, as yet, no clear criteria for what virtual world phenomena should qualify as "living" or sufficiently "life-like" to legitimately count as lying within this field. In large measure, this simply reflects the continuing debate and investigation within conventional biology, of what specific organizational (as opposed to material) system characteristics are critical to properly living systems. The key advantage and innovation in artificial life is precisely that it has this freedom to vary and explore possibilities that are difficult or impossible to investigate in natural living systems. In this context, a precise definition of "life" (natural or artificial) is not a necessary, or even especially desirable condition for progress.

2.2 Hierarchy and Emergence

A specific and pervasive feature of natural life is hierarchical organization – from molecules to cells to tissues to organs to multicellular organisms to societies of organisms. Hierarchical design is also common in artificial engineered systems – for example, from semiconductor materials to logic gates to processors to computers to the Internet. At first sight, the hierarchical structure of natural living systems might be viewed as simply the natural outcome of evolutionary optimization for certain engineering benefits: modularity, standardization, ease of repair, robustness, etc. However, there are significant difficulties in such a simplistic account. In particular, evolution by natural selection must operate without foresight: So the opening up of evolutionary "potential" per se, via a new hierarchical level of organization, cannot be a direct locus for immediate selection to operate. The fundamental question here relates to the emergence and establishment of new levels of Darwinian "actors." This is a process that has evidently occurred repeatedly in natural evolution (the origin of the first protokaryotes; the origin of the eukarotes; multicellularity, sociality, etc.); however, without circular reasoning, this cannot itself simply be assigned to a process of Darwinian selection (not, at least, at the same hierarchical level). In natural biology, this is the focus of investigation into so-called *major transitions in evolution* (Maynard Smith and Szathmáry 1997).

In the case of artificial life, while there have been some examples of model systems demonstrating some degree of emergence of hierarchical organization (e.g., Rasmussen et al. 2001b), the achievement to date has been limited, both in terms of evolutionary potential and in number of hierarchical levels, and is the subject of significant ongoing debate within the community (Gross and McMullin 2001; Rasmussen et al. 2001a). This can be expected to be a particularly active area of continuing research (Baas 1994).

2.3 Constructive and Autopoietic Systems

A problem closely associated with emergence and hierarchy, yet somewhat distinct, is that of "self-constructing" systems. Natural life, at all hierarchical levels, exhibits a characteristic ability to maintain its system organization while simultaneously turning over components at the lower hierarchical levels. Thus, a cell can stably maintain itself (and even grow and divide) while continuously regenerating all its significant molecular components (metabolizing environmental substrates as necessary); similarly, a multicellular organism maintains its organization while constantly replacing its component cells; and a social insect colony may long outlive most, if not all, of its constituent individual organisms. While this is a typical phenomenon of natural life, it is very different from the behavior of conventional engineered systems. The latter may well have complex hierarchical organization, but this organization is typically static: The system is assembled once, and then retains its fixed compositional structure for its functional lifetime. Indeed, it is typical of engineered systems that failure of any component, at any hierarchical level, will at least significantly impair system level function, and will commonly result in full system failure. Although so-called "fault-tolerant" systems may be engineered to incorporate significant redundancy, this often involves static assembly of additional components at manufacture time. Functional lifetime is extended as failing components are automatically removed from service; but once this pre-built redundancy is exhausted, system failure follows from any further component failure. Further, even such redundant systems are generally vulnerable to failures in the "failure-detecting" components that cannot also simply be duplicated without potentially entering into infinite regress.

This is a complex problem to address, but natural life provides a "proof of principle" that such self-constructing organizations are possible. Artificial life certainly offers a potentially very fruitful avenue for further investigation. In particular, artificial life allows "virtual worlds" to be formulated in which problems of self-construction can be simplified, and the core organizational mechanisms can be exposed and subjected both to mathematical analysis and experimental exploration. While a variety of work in artificial life bears on this, some of the most direct contributions might be summarized as follows:

- The Chilean biologists Maturana and Varela pioneered the abstract concept of *autopoiesis* (literally "self-production") as a description of the core mechanism of organizational self-maintenance in biological cells (Varela et al. 1974). Almost 15 years before the modern computational study of artificial life was even named (Langton 1989), they were already using a molecular level, agent-based, abstract chemistry to give a concrete demonstration of this abstract theory of cellular organization. In essence, the proposal is that biological cells are dynamic, self-sustaining, chemical networks (an abstraction of cell metabolism), which also create and maintain a spatial boundary (an abstraction of a cell membrane), where the contained network and the boundary reciprocally rely on each other for stability. While some technical deficiencies were identified in the original presentation of this work (McMullin and Varela 1997), it has given rise to a sustained and continuing programme of active research (McMullin 2004).

- Also in the 1970s, Holland proposed the so-called α-universes (Holland 1976) and, independently, Hofstadter described the *typogenetics* system (Hofstadter 1979). Both involved one-dimensional fragments of computer code, which could interact with each other. In both cases, they were explicitly inspired by aspects of the molecular replication and translation machinery of biology, and were concerned with understanding the reflexive interactions that arise when the "same" class of entities (molecules) can sometimes function as executable "code" (enzyme/protein) and sometimes as literal "data" (nucleic acid), and the possibility of collective self-reproduction and maintenance. These were, however, restricted to theoretical/analytic treatments, and were not subject to empirical investigation until much later (McMullin 1992; Snare 1999; Kvasnicka et al. 2001).

- This same tension between syntax and semantics in materially instantiated dynamic systems is also at the core of Pattee's analysis of what he terms "semantic closure" (Rocha 2000; Cariani 1992).

- A related, but independent line of investigation was pioneered by Kauffman and others, in the form of "collectively autocatalytic sets" (Farmer et al. 1986; Kauffman 1993). Such a set is formally similar (though more strictly defined) to the contained chemical network of an autopoietic system, but without the requirement for a self-generated spatial boundary. The key result here was the demonstration, in completely abstract virtual "chemistries" that such collectively autocatalytic sets can spontaneously arise, quickly and under relatively weak constraints on the underlying chemistry.

- Fontana and Buss self-consciously launched a mathematical and computational investigation of "constructive dynamical systems," based around the so-called *alchemy* system (Fontana and Buss 1994). These systems deliberately diverge from the classical concept of "dynamical systems" by relaxing the normally strict demarcation between "state variable" and "dynamic law." The authors demonstrated and analyzed a number of organizational phenomena, including forms of "self-sustaining closure" where more-or-less complex dynamic aggregates of components successfully sustain themselves, even as all the individual components are continuously diluted or degraded.

- As a comprehensive overview treatment, Dittrich et al. (2001) provide a summary of the general approach of building computational models of arbitrary "artificial chemistries," as a platform for investigating these and other problems. More recently, Dittrich and Speroni have developed a primarily algebraic mathematical analysis in a comprehensive way, in the form of "*Chemical organisation theory*" (Dittrich and di Fenizio 2007). However, while an important advance, the difficulties of incorporating dynamics (chemical kinetics) and the interaction of chemical self-production and spatial demarcation (the self-constructed "boundary" of autopoietic theory) remain substantial.

- A quite different and more radical approach to the problem of biological self-maintenance was proposed by the theoretical biologist Robert Rosen over an extended series of works (Rosen 1959, 1972, 1985, 1991). In brief, Rosen argues that the self-constructing organization characteristic of natural living systems, which he termed "closure under efficient causation" transcends the possibilities of conventional dynamical systems in a fundamental way; with the consequence that it cannot be realized in any "computational" universe. This is a complex and contentious theoretical claim which, not surprisingly, continues to be the focus of considerable critique and criticism (e.g., Chu and Ho 2006). Rosen's work was also a strong influence in Kampis' development of a comprehensive mathematical treatment of what he terms "self-modifying" systems (Kampis 1991).

2.4   Complexity (and Its Growth)

▶ There is a concept which will be quite useful here, of which we have a certain intuitive idea, but which is vague, unscientific, and imperfect ... I know no adequate name for it, but it is best described by calling it "complication." It is effectivity in complication, or the potentiality to do things. I am not thinking about how involved the object is, but how involved its *purposive operations* are. In this sense, an object is of the highest degree of complexity if it can do very difficult and involved things. (von Neumann 1949, p. 78, emphasis added)

Arguably the most fundamental problem in the theory of biology is that of the growth of "complexity"; with "complexity" understood in the sense explained by von Neumann above, of the ability to do complicated things. While this is certainly an informal definition, it suffices to clearly demarcate this notion from purely syntactic, combinatorial, or computational concepts of complexity.

Von Neumann is a key figure in the early development of abstract computational modeling approaches to understanding biological phenomena; indeed, in this respect he can be considered as having instigated the earliest investigations in artificial life, as the term is now understood (McMullin 2000a). In particular, in the late 1940s, (von Neumann 1949) he started to draw attention to an apparent paradox arising from the contrast between any mechanistic (i.e., not vitalistic) understanding of living organisms and common experience of engineering artificial mechanisms or automata. While it is quite generally possible to design machines that construct other machines, this process is normally *degenerative in complexity*: a machine of a given complexity (such as an automated factory) can only construct machines of comparatively lower complexity (such as cars or phones or televisions etc.). Whereas, if the theory of Darwinian evolution is accepted, then biological "machines" (organisms) must be capable of constructing descendant machines of greater complexity. Granted, these increases in complexity may occur only in very small increments, and even then only in some lineages and accumulating over geological periods of time; but nonetheless it must be possible, in principle, for biological organisms to give rise to offspring more complex than themselves.

Von Neumann quickly developed an outline resolution of this paradox, in the abstract form of what he termed a "general constructive automaton." This was inspired by Turing's earlier formulation of a universal computing automaton (Turing 1936), and also by von Neumann's own contemporaneous involvement in designing and building the earliest electronic stored program digital computers (von Neumann 1945). In essence, a general constructive automaton is a *programmable* constructor or assembler, capable of constructing an indefinitely large set of target automata – loosely, a "universal" set comprised of all automata that could be assembled from a given, finite, set of primitive components. Significantly, von Neumann hypothesized that a general constructive automaton, relative to a specific set of components, might itself be realizable as an assembly of these same components. With this conceptual architecture and some relatively minor technical elaboration, von Neumann showed that this would then give rise to an indefinitely large set of automata, spanning an indefinite large range of complexity, all of which would be fully connected by a network of heritable "mutations" (essentially, chance variations in the construction programs). That is, there would exist mutational pathways leading from the simplest to arbitrarily complex members of this set; with all these machines also, incidentally, being capable of self-reproduction. (While terminology in the field is not completely consistent, in this chapter "(self-)*reproduction*" is reserved to mean this von Neumann style process, involving separate actions of copying and decoding; whereas "(self-)*replication*" is used to denote a process of copying only.)

In retrospect, this was already an astonishing achievement. Von Neumann effectively described the abstract architecture of biological self-reproduction, based on separate processes of syntactic copying (replication) and semantic "decoding" (translation) of a passive information carrier; he showed how this architecture supported self-reproduction, heritable mutation, and thus evolutionary growth of complexity; and he presented all these, at least in outline form, in 1948, five years before the chemical structure of DNA was even identified (von Neumann 1949).

However, as yet this was still only a sketch of a solution. To make it properly convincing, he needed to present a concrete set of "primitive parts" and show that with these, it would actually be possible to realize an example of a general constructive automaton. This proved to be a complicated and potentially intractable problem: any plausible "real world" set of components would introduce many ancillary complications – mechanics, thermodynamics, etc. Following a suggestion from Stanislaw Ulam, von Neumann instead formulated an artificial, virtual, world that would simplify away these ancillary complications. He proposed a two-dimensional "tessellation automaton," or, as it is now called, a *cellular automaton* (see ❯ Sect. 3.1.1 for a general introduction to CA). Within this simplified virtual world, von Neumann then successfully developed a fully detailed design for an example of a general constructive automaton. He had planned this as a first step in a much more general and comprehensive "theory of automata." However, he put his unfinished manuscript aside in 1953; and due to his untimely death in 1957, was never to return to this project. The manuscript was subsequently edited by Burks and formally published in 1966 (Burks 1966).

Von Neumann's achievement, both in theoretical insight and technical implementation, was considerable, and it gave rise to an extended program of further research. Many different cellular automata worlds have been formulated and investigated, and many variants on von Neumann's architecture for evolvable self-reproducing automata have been demonstrated (Thatcher 1970; Codd 1968; Langton 1984b). Conway's *Game of Life* has particularly popularized the study of life-like phenomena in cellular automata worlds (Berlekamp et al. 1982). More recently, it has become feasible to actually implement von Neumann's original designs on real computers (Pesavento 1995).

Nonetheless, seminal as this work was, it also explicitly left certain questions still very much unresolved. In particular, as von Neumann himself was clearly aware, the mere existence of mutational pathways from simple to complex automata does not mean that evolution will actually follow such pathways. Indeed, in von Neumann's original cellular automaton world, the embedded self-reproducing machine was extremely fragile. It could only operate successfully at all on the condition that the rest of the world was completely quiescent; and, of course, once it completes one round of reproduction, that could no longer be the case. Parent and offspring would then interfere with each other's structure and operation, and both would rapidly distintegrate. While this problem could be superficially avoided by constraining each complete, successively constructed automaton to restrict its further operation to separate regions of the space, it rules out any possibility of natural selection; and, in any case, cannot be sustained in the long term if there is any possibility of stochastic, component level, malfunction, or failure (which itself is actually *desirable*, as a source of the random variation that is the grist to the mill of Darwinian evolution).

2.5   Coreworlds: Spontaneous Evolution of Computer Programs

▶ Nothing in Biology makes sense except in the light of evolution. (Dobzhansky 1973)

▶ Discovering how to make such self-reproducing patterns more robust so that they can evolve to increasingly more complex states is probably the central problem in the study of artificial life. (Farmer and d'A. Belin 1992)

Although the cellular automata models of von Neumann and his successors have not, to date, demonstrated extended evolutionary dynamics, various other artificial life models have addressed evolution more directly, through attempts to demonstrate evolution among populations of virtual (software) agents in virtual worlds. The most systematically investigated framework is to envisage these as small machine code programs, each executed by a separate, parallel processor, but all sharing a single main memory ("core"). Generically, these approaches can be referred to as *coreworlds*. In principle, the parallelism could be implemented directly with a sufficiently large pool of hardware processing elements, dynamically allocated as agents are created or destroyed; but it is typically just realized with conventional timeslice-based multithreading of a much smaller pool of hardware CPUs (often just one).

The earliest work adopting this methodological approach can be traced to the seminal investigations of Nils Barricelli, carried out between 1953 and 1963 (Barricelli 1957, 1963). In fact, this work was carried out on the original computer designed and built at the Princeton Institute for Advanced Studies (IAS) between 1946 and 1952 under the direction of von Neumann. Barricelli was a visitor to the institute at von Neumann's invitation; and took the opportunity to conduct an investigation of the role of "symbiogenesis" in the origin of life. He programmed the IAS computer to directly model patterns of numbers that could interact with each other within a circular array (effectively the main memory system of the machine), according to some fixed local rules. This is conceptually similar to a one-dimensional cellular automaton. Although these patterns of numbers did not have the general computing ability of Turing machines, Barricelli demonstrated the existence of patterns that could self-replicate, as well as various forms of competitive and cooperative (symbiotic) interaction between patterns. Somewhat strangely, in his published description of this work, Barricelli only obliquely related it to von Neumann's own contemporary work on self-reproducing and evolving automata. In any case, the insight and innovation of this work was not widely recognized at the time, and has only recently been properly reinstated through George Dyson's investigations of the early history of digital computing at the IAS (Dyson 1997, 1998).

The first attempt to study the dynamics of competition among co-resident *general purpose programs* in a shared memory was pioneered by Vyssotsky et al. at Bell Labs in the early 1960s (though not published until 1972), and appears to have been independent of Barricelli's study (Vyssotsky 1972). This was effectively created as a form of computer programming game in which different programmers provided hand-coded programs, which were then instantiated in a common core, and executed to see which would survive longest. Again, self-replication was a key feature. An interesting aspect was that, unlike almost all later work, these programs were implemented in the native (IBM 7090) machine code, rather than in a virtual machine code to be executed by an interpreter. This allowed comparatively much higher execution speed, but with some associated limitations. For example, as programs could not be written to be position independent, successful self-replication required active relocation processing, as opposed to simple self-copying. In this early implementation, a "lethal" program was relatively quickly developed by R.H. Morris, and the original game was put aside.

The α-universes described by Holland in the mid-1970s (Holland 1976) may again be mentioned here. Although formulated independently, and with quite different motivations, this proposal was somewhat similar to the Vyssotsky system. In particular, it again envisaged an essentially linear core memory inhabited by concurrently executing fragments of computer code. However, in other ways this was actually much closer to Barricelli's work in that the specific intention was to investigate the spontaneous emergence of crude collective self-reproduction activity, rather than simple competition between programs precoded by human programmers; but, in any case, as already mentioned in ❷ Sect. 2.3, this system was not experimentally investigated until much later (McMullin 1992).

The more direct descendent of the Vyssotsky system was the *Core War* game, developed by Dewdney and others in the early 1980s. This now relied on an interpreter and offered much more varied gameplay opportunities (Dewdney 1984). Following the establishment of an international tournament (Dewdney 1987), Core War has had a sustained following, and remains active to this day (http://corewars.org/).

Although the Core War framework is predicated on the idea of human programmers coding the competing programs, it does also naturally lead to the question of whether, under conditions of potential mutation during self-replication, there could be an autonomous and sustained evolutionary process, not relying on programmer intervention. This was explicitly studied by Rasmussen et al. in a system they called the *coreworld* (Rasmussen et al. 1990). In this instantiation, however, the system suffered from a similar limitation to that of the von Neumann style cellular automata systems. While the world could be seeded with an initial self-replicating program, as its offspring filled up the core they quickly began overwriting each other and the self-replicating functionality and thus the potential for sustained Darwinian evolution was lost. This issue was addressed by Ray, in the *Tierra* system, developed and investigated in the early 1990s (Ray 1992). Tierra was based on the idea of competing and interacting self-replicating programs in a shared core memory. Programs were now given an ability to employ "memory protection" to limit overwriting by other programs. On its own, this would simply lead to the core filling with programs once, and then no further self-replication (much less evolution) would be possible. To overcome this, Ray added a mechanism for random removal of programs (the so-called "reaper"). This addition allowed for continuing self-replication and the possibility for longer term evolution. He also adopted a significantly different instruction set, incorporating a biologically inspired "template" addressing mode. With these innovations, Ray was able to demonstrate an extended process of evolution, with multiple episodes of Darwinian natural selection. The system produced a surprising array of evolutionary phenomena, including optimization of program length, obligate parasitism, and facultative "hyper-parasitism" (as a defense against obligate parasites).

Tierra has led to a wide variety of related work, such as:

- The *Avida* system of Adami and Brown that introduced a more conventional 2-D spatial world in which programs would colonize nodes, and also be evaluated for their ability to complete extrinsically defined computing tasks (Adami and Brown 1994).
- A proposed extension to Tierra into a multiprocessor networked environment ultimately to be distributed across the wider Internet, under suitable "sandbox" execution controls (Ray 1995).
- *Nanopond* (http://adam.ierymenko.name/nanopond.shtml), a highly simplified implementation of a 2-D spatial "program evolution" system, drawing on elements of both Tierra and Avida, but implemented in less than 1,000 lines of C source code.

- The *Amoeba* system, developed by Pargellis, which specifically demonstrated the possibility of spontaneous emergence of a self-replicating program in coreworld-like systems (Pargellis 2001).

Evolution in coreworlds can also usefully be compared and contrasted with several other related but divergent research fields and investigations:

- There has been a separate line of research into the logic of so-called *Quine* programs. These are defined as computer programs that produce their own source code as output. The original discussion is generally credited to Bratley and Millo (1972), though the term "Quine program" is usually attributed to Hofstadter (1979). The existence of Quine programs in general-purpose programming languages essentially follows from Kleene's recursion theorem. But while the formulation of specific Quine programs in different languages has been a hobby among programmers, these programs have not generally been embedded within an execution environment in which exponential growth, Darwinian selection, and evolution could actually take place. It has not therefore, particularly influenced research approaches in artificial life. That said, through its definitional restriction to programs that must self-reproduce without having direct access to self-inspection, the study of Quine programs draws attention to the logical distinction between proper von Neumann style genetic "self-reproduction" and the "self-replication" by copying normally adopted in coreworlds. It is, in fact, a very open question as to what the effect of this architectural choice is on evolutionary potential (McMullin et al. 2001).
- The replication and propagation of programs in controlled coreworld environments obviously also lead to the idea of self-replicating programs propagating in open, networked, computer environments. This is the realm of computer malware – computer viruses, worms, etc. And indeed, this connection was explicitly made by Dewdney already in the immediate aftermath of the Morris Internet worm incident in 1988 (Dewdney 1989). However, while malware development certainly involves an arms race between the human developers on both sides, there is to date no evidence of effective *autonomous* evolution of "free-living" malware.
- There is significant overlap in inspiration between evolution in coreworld-like systems, and work in evolutionary algorithms, and, especially, *genetic programming* (GP) (Koza 1992; Willis et al. 1997; Banzhaf et al. 1998). This, in turn, has a long history extending back at least to Samuel's late 1950s investigations in machine learning (Samuel 1959). However, the major distinction is that GP is generally driven, directly or indirectly, by an externally provided evaluation function (used as an extrinsic "fitness" to drive an imposed evolutionary algorithm) rather than the coreworld approach, which is to investigate spontaneous autonomous evolutionary dynamics, in which the software agents are responsible for their own replication and relative fitness emerges from their bottom-up ecological interactions.
- Investigations into the evolution of cooperation by Axelrod and others should also be mentioned here (Axelrod 1984, 1987). Although not generally using a coreworld-like framework, and focused on a relatively narrowly defined pattern of interaction (the iterated prisoner's dilemma), using an extrinsically applied genetic algorithm, this examination of the problem of cooperation, and coevolution between "cooperating" and "defecting" strategies has been very influential in the wider fields of political science, ecology, and complexity theory.

To summarize, the general experience of the investigation to date of evolution in core-world-like systems is that the evolutionary potential of these systems is interesting but still strictly limited. Indeed, they can be viewed as formally similar to pure artificial "replicator chemistries," comparable to the systems already mentioned in ❯ Sect. 2.3 above, and exhibiting similar phenomena of collective autocatalytic closure and self-maintenance (McMullin 2000b). Thus, both von Neumann-style genetic self-reproduction in cellular automaton worlds, and replication by simple copying or self-inspection in coreworlds, naturally lead directly into the problems of biological robustness, self-maintenance, and hierarchical organization already discussed above. The integration of self-reproducing or self-replicating programs with self-maintenance and individuation (autopoiesis), and the demonstration of sustained evolutionary growth of complexity in a purely virtual world remains perhaps *the* key "grand challenge" problem in the field of artificial life.

3 Theory of and Formalisms in Artificial Life

A number of formalisms have been effective and widely used in artificial life systems. Among them are cellular automata, rewriting systems, and complex dynamical systems. Other formalisms have been used to analyze artificial life systems, like network analysis. In this section, the most important synthetic and analytic tools will be described.

3.1 Automata

The theory of abstract automata defines finite state automata as behavioral models of machines that possess a finite number of states, a set of transitions between those states, and machine actions (such as "read input" or "write output"). Based on this notion, an entire field in computer science (automata theory) has been formulated, which addresses their mathematical features and the use of these entities. One variant of automata of particular relevance to artificial life are cellular automata.

3.1.1 Cellular Automata

A cellular automaton (CA) is composed of a large interconnected collection of component finite state automata. These possess, besides states and transitions, an additional feature, a location within a neighborhood, often constituted as a grid. The neighboring "cells" (although the word "cell" here gives rise to the term "cellular automaton," note that it is *not* being used with its specifically biological sense; it rather just means a discrete, spatially demarcated, component of a larger, static, locally interconnected, array) or "nodes", which are themselves finite state automata with a different location, receive the state of the particular automaton as input, and also, reciprocally, provide their own states as inputs that inform a particular automaton which state transition to realize. CA are typically defined to be structurally homogenous, that is all cells contain the same finite state automaton and all share the same local neighborhood pattern. Of course, cell *states* vary dynamically in time and across the cells of the automaton; and it is such spatiotemporal patterns of states that are used to represent so-called *embedded* machines or automata. Such a model can thus be used to model spatially extended systems and study their dynamical behavior. CA have also been used as a modeling tool to capture behavior otherwise modeled by partial differential equations (PDEs); and indeed can be regarded as a discrete time, discrete space, analogue of PDEs.

Cellular automata play a particularly useful role, in that they possess the twin features of being applicable for synthesis of system behavior, and for analysis of system behavior. They have been thoroughly examined in one-dimensional and two-dimensional contexts (Wolfram 1986, 1994), and have been considered in the fundamental contexts of computability (Wolfram 1984) and parallel computing (Toffoli and Margolus 1987; Hillis 1989).

As mentioned previously (❷ Sect. 2.4), von Neumann, following a suggestion of Ulam, originally introduced CAs as a basis to formulate his system of self-reproducing and evolvable machines. The von Neumann CA was two dimensional with 29 states per cell. According to Kemeny, von Neumann's core design for a genetically self-reproducing automaton would have occupied a configuration of about 200,000 cells (a size dominated by the "genome" component, which would stretch for a linear distance of about 150,000 cells) (Kemeny 1955, p. 66).

In fact, von Neumann's contributions stand at the cradle of the field of artificial life by showing, for the first time, a mathematical proof that an abstract "machine-like" entity can be conceptualized, which can construct itself, with that copy being able to achieve the same feat in turn, while also having the possibility of undergoing heritable mutations that can support incremental growth in complexity. This was a breakthrough showing that above a certain threshold complexity of an entity, this entity is constructionally powerful enough not only to reproduce itself (and thus maintain that level of complexity) but to seed a process of indefinite complexity growth.

While von Neumann achieved this result in principle, a large number of scientists have since succeeded in pushing down the complexity threshold. In the meantime, much simpler CAs (with smaller number of states per cell) have been shown to support self-reproducing configurations, many of those as 2D loops (Langton 1984a; Sipper 1998). In fact, the smallest currently known example in this form is a cellular automaton with eight states per cell where the self-reproducing configuration is of size five cells (Reggia et al. 1993). However, in many of these cases, the core von Neumann requirement to embody a "general constructive automaton" in the self-reproducing configuration has been relaxed: So the potential for evolutionary growth of complexity is correspondingly impoverished.

Cellular automata also have been used in the design of the Game of Life (Gardner 1970, 1971), which draws inspiration from real life in that replication and transformation of moving entities, again in a 2D CA, were observed (for a comprehensive introduction, see Berlekamp et al. (1982) and Conway's Game of Life, http://en.wikipedia.org/wiki/conway and online resources therein).

The discrete nature of states, space, and time available in cellular automata has led to a number of applications of these tools, many of which could be considered artificial life applications in the widest sense. For example, the patterning of seashells has been modeled using discrete embodiments of nonlinear reaction-diffusion systems (Meinhardt 2003). Scientists in Urban Planning have started to use the tool of cellular automata to model urban spread (White and Engelen 1993).

3.2 Rewriting Systems

Rewriting systems are a very general class of formal systems in mathematics, computer science, and logic, in which rules for replacing formal (sub)structures with others are repeatedly applied. Rewriting systems come in many forms, such as term rewriting, string rewriting, equation rewriting, graph rewriting, among others, and have been at the foundation of programming languages in computer science and automated deduction in mathematics. String rewriting systems, first considered by Axel Thue in the early part of the twentieth century (Dershowitz and Jouannaud 1990; Book and Otto 1993), have a close relationship with Chomsky's formal grammars (Salomaa 1973). Many rewriting systems have the goal of transforming particular expressions into a normalized form, deemed to be stable in some sense. Prior to reaching this stable form from a starting expression, the rewriting process goes through a series of transient expressions. For example, mathematical equations can be formally rewritten until they reach a form in which it is possible to read off solutions.

It can be seen, therefore, that rewriting allows a sort of dynamics of symbolic systems. It is worth noting that rewriting systems provide the basis for the λ-calculus (Barendegt 1984), which has been used to implement artificial chemistries.

In a more general artificial life context, transformations based on rewriting systems can be used to describe growth and developmental processes. For example, besides transforming complex expressions into normal form, another approach uses rewriting to generate complex objects from simple start objects. This has provided the formal basis to studies of biological development through what were later called L-systems (Lindenmayer 1968). The difference between L-systems simulating growth and development and formal grammars lies in the order of rewriting. While formal grammars consider rewriting in a sequential fashion, L-systems consider parallel (simultaneous) rewriting, much closer to the asynchrony of the growth of real cells. But central to all these models is the recursive nature of rewriting systems (Herman et al. 1975). Lindenmayer and, later, Prusinkiewicz and Lindenmayer (1990) have pioneered the use of turtle commands of computer languages like LOGO to visualize plant patterning and growth.

P-systems (Paun 1998; Paun and Rozenberg 2002) are a more complex class of rewriting systems used in artificial life. They are similar to artificial chemistries in that production of symbolic chemicals is prescribed by reaction rules that could also be considered as rewriting rules. In addition to reactions, however, P-systems contain membranes, which limit the availability and direct the flow of material. This provides a higher-level means of structuring environments, something artificial chemistries with well-stirred reactions are unable to do (Dittrich et al. 2001).

In the meantime, a large number of applications of rewriting systems have appeared. For example, such classical artificial life systems as *Tierra* have been reformulated as rewriting systems (Sugiura et al. 2003); Giavitto et al. (2004) describe applications of rewriting systems in modeling biological systems, and virtual cities have been modeled with the help of L-systems (Kato et al. 1998), just to name three examples.

3.3 Dynamical Systems Modeling with ODE

Artificial life systems are often couched in terms of the mathematical language of dynamical systems, captured by ordinary differential equations. This formalism is used for describing the continuous change of state of a system via the change in time of n quantified state variables $\mathbf{q}(t) = \{q_1(t), q_2(t), \ldots, q_n(t)\}$. A very simple and linear example of a dynamical system is a system consisting of two-state variables q_1, q_2, which develop according to the following differential equation

$$\frac{dq_1}{dt} = a_{11}q_1 + a_{12}q_2 \tag{1}$$

$$\frac{dq_2}{dt} = a_{21}q_1 + a_{22}q_2 \tag{2}$$

Coefficients a_{11}, ... can be lumped together into a matrix A, and the stability of the dynamical system can be examined based on a solution to the equation $d\mathbf{q}/dt = A \times \mathbf{q}(t) = \mathbf{0}$.

3.3.1 Nonlinear Dynamical Systems

More interesting than this basic dynamical system are those which provide for nonlinear interactions between components and the environment. This can be achieved by higher order terms in the coupling of state variables. For instance

$$\frac{dq_1}{dt} = a_{11}q_1 + a_{12}q_2^2 \tag{3}$$

$$\frac{dq_2}{dt} = a_{21}q_1^2 + a_{22}q_2 \tag{4}$$

would already provide such an example.

Chaotic Systems

Nonlinear couplings are the hallmark of chaotic systems, as nonlinearities in the coupling of state variables tend to amplify small fluctuations (e.g., noise in state variables) into large global state changes. Of course, all these depend on the size of the fluctuations. It is well known that as long as fluctuations are small relative to the strength of nonlinear interactions, systems act in quasi-linear mode and can be formally linearized. However, once either the fluctuation signal is big enough, or the nonlinear interactions become stronger, systems tend to behave in less predictable ways, for example, start oscillating, and might even transit into a chaotic regime characterized by short-term predictability and long-term unpredictability (Argyris et al. 1994). In the artificial life community, the notion of chaos has played an important role as the extreme case of unpredictability beyond life, much as the predictability of dead matter lies at the other end of the spectrum. This notion was first developed in the context of cellular automata (Langton 1992a), but later extended to other systems.

Constructive Systems

Dynamical systems can be extended by considering a notion of change in the number of state variables. For instance, if each of the variables characterizes the concentration of a chemical substance, and some of the substances are consumed or produced for the first time, this can be captured by changing the number and/or assignment of state variables. Simulation might proceed by alternating between a low-level simulation of, for example, individuals of species, and high-level events that signal the birth or extinction of a particular species (Ratze et al. 2007). While couched here in ecological terms, the same approaches apply to the substances of artificial chemistries. Constructive dynamical systems are an active area of research whose exploration has barely begun.

3.4   Complex Systems

The relation between complex systems and artificial life is complicated. Some might argue that artificial life is a subfield of complex systems. This opinion draws from the argument, valid in both complex systems research and artificial life, that these kinds of systems work by producing emergent effects through the interaction of many entities. As such, living systems (and their artificial counterparts) are complex systems, but not the only ones, since, for instance, societies, economies, or legal and norm systems might legitimately be called complex systems too. On the other hand, one might argue that if systems like societies or economies are behaving as complex systems, making use of emergent phenomena and top-down causation, they really can be considered to be alive. If that argument is followed it could be stated that complex systems are part of the research in artificial life.

In this section, it is preferred not to resolve this tension, but rather to focus on one area of research that one legitimately can consider as at the intersection of the two: networks. Networks and their associated formalisms are peculiar since they provide evidence for a phenomenon frequently seen in artificial life systems, as well as in complex systems: There is no easy way to dissect these systems into cause and effect relationships. Indeed, networks allow us to formalize the concept that causes and associated effects have a many-to-many relationship.

If a simple cause–effect relationship is visualized as an edge between two nodes or vertices in a graph, while the nodes themselves symbolize the entities that might or might not be in these relationships, one can easily see that networks, with nodes of in-degree and out-degree larger than one will be able to capture multiple cause–multiple effect relationships. This is a key characteristic of complex systems, as opposed to simple systems, where simple pathways exist between causes and their associated effects.

3.5   Networks

The recent surge in network research (Albert and Barabasi 2002) can be attributed to the realization that networks are the quintessential structure to capture emergent phenomena. Without network infrastructure, emergent behavior of a system would be transient and very quickly disappear, like fluctuations. It is when a network infrastructure is in place that originally weak and transient signals of emergence can be captured and stabilized into identifiable phenomena.

Network studies have revealed widespread characteristics in complex systems, such as small-world features or scale-free connectivity (Watts 1999; Strogatz 2001) that have been ascribed to the generation or evolution of these networks. The dynamics of network evolution is an active area of research (Dorogovtsev and Mendes 2003).

But even the much simpler formalisms of static networks provide useful insights into complex systems and artificial life systems, because there is a deep relationship between autocatalytic sets (Farmer et al. 1986) and closed paths on networks, which in mathematical terms is the result of a connection between graph theory and algebra. Networks can be formally represented by directed graphs $G = G(N,E)$, defined by the set of N nodes and the set of E edges or connections between those nodes. Edges or links between nodes are often represented by an ordered pair of nodes, called the connectivity matrix. If connections are only present or absent, symbolized by a "0" or "1" entry in the matrix, this is called the adjacency matrix A of a graph G. The adjacency matrix has a set of eigenvalues and eigenvectors, which can be easily computed by solving the characteristic equation of A: $|A - \lambda I| = 0$, where λ are eigenvalues of A and I is the unit matrix. Given that A is a nonnegative matrix, the Perron–Frobenius theorem states that the largest eigenvalue λ_1 of A is real. If this eigenvalue is $\lambda_1 \geq 1$, then there exists a closed path through the graph (Rothblum 1975). Closed paths are deeply related to autocatalytic sets, which were first introduced in chemical systems where each sort of molecules present is catalyzed by at least one other sort of molecules (Eigen 1971). This relation would be considered equivalent to an edge in the graph of all molecular sorts (nodes). Autocatalytic sets correspond therefore to closed paths in that graph.

In the wider context of a theory of organizations (Fontana and Buss 1994; Speroni di Fenizio et al. 2000), autocatalytic sets or closed and self-maintaining sets (organizations) are the key to life-like processes, in that they maintain whatever structure or organization has come into existence. This is the core of the previous statement that emergent phenomena need the stabilizing infrastructure of a network (and its closed paths) to move from the status of a fluctuation in a system to that of an identifiable phenomenon.

The deep connections between algebra and graph theory indicated here would merit a much more detailed discussion, yet space is a constraint here. Suffice it to say that networks are the stage for a dynamics of the chemical (or otherwise) sorts of agents symbolized by the nodes. Thus, it can be said that a dynamical system can be imposed on these networks, and once this dynamics starts to interact with the structure of the network, the most interesting phenomena will be observed (Jain and Krishna 2001).

3.6   Other Formalism

The list of formalisms discussed here is not exhaustive. For instance, the area of statistical mechanics has also contributed valuable insights into artificial life (see, e.g., Adami 1998). Category theory, chemical organisation theory, and other approaches can also be legitimately mentioned as contributing to our understanding of artificial life phenomena.

4   Applications

A huge number of applications can be considered to make use of artificial life techniques in the wider sense (Kim and Cho 2006). It is difficult to even list these applications comprehensively. Instead, a few applications have been selected here to demonstrate the width of the applicability of its concepts.

4.1   Biology

The most obvious application of artificial life is in biology. It was already noted that there is a close relation between artificial life models and synthetic biology (Ray 1994) and that artificial life models in fact are a useful tool for the exploration of questions in biology (Taylor and Jefferson 1994). For a more recent example, see Strand et al. (2002).

4.1.1   Synthetic Biology

The current state of synthetic biology does not acknowledge this connection extensively, despite, historically speaking, emerging from artificial life research. Key questions in this area today are, how to generate biological entities (cells, proteins, genes, organisms) that have some purposefully designed function (Ferber 2004). The question how systems function that can be ascribed to be "living" has been separated off into the field of "systems biology," another newcomer to biology with inheritance from artificial life (Kitano 2002). Artificial life was always concerned with "Life as it could be," that is, alternative designs that can be termed living, designed for a purpose (as engineers would do), or for scientific inquiry, and it was following systems thinking long before systems biology was established (Langton 1997).

4.1.2 Health/Medicine

The modern health-care system is so complex now, that complex systems thinking and ideas from artificial life are ripe for application (Plsek and Greenhalgh 2001). This will entail the understanding of system behavior and response to attempted changes, financial issues such as exploding health-care costs and redesign of the system to become more adaptive.

Medicine has long benefited from its relation to biology, which will result in new applications for artificial life methods as well (Coffey 1998; Hamarneh et al. 2009). Notably, regenerative medicine will be the beneficiary of model systems provided by artificial life (Semple et al. 2005).

4.1.3 Environmental Science and Ecology

In environmental science, the *Gaia* theory (Lovelock 1989) has benefited from simulations provided by artificial life models (Gracias et al. 1997; Downing 2000; Lenton 2002). Artificial life approaches have been used to design sustainable architecture, at least on a local level (Magnoli et al. 2002). In the design of other systems, for example, for recycling, artificial life models have been successfully applied (Okuhara et al. 2003).

Ecosystem research in general has made use of artificial life models, from the setting up of "artificial ecosystems" whose behavior can then be explored in the computer (Lindgren and Nordahl 1994) to the examination of characteristics necessary to provide opportunity for ecosystem formation (Wilkinson 2003). Further, artificial life models have been applied to questions of astrobiology (Centler et al. 2003), another offshoot with some roots in artificial life research.

4.2 Engineering

A number of applications in the engineering subdisciplines have been examined, with autonomous robotics being a very prominent one.

4.2.1 Autonomous Robotics

While it is natural to expect artificial life to contribute to biology, its contribution to the field of engineering is less to be expected. But given the tight connection between behavior and autonomous robotics, it is perhaps less a surprise. In fact, an entire conference series has been termed "from animals to animats" (Meyer and Wilson 1991; Asada et al. 1998), carrying with it the notion of machines that behave adaptively. Recent reviews and summaries of autonomous robotics as the field pertains to artificial life can be found in Eaton and Collins (2009), Harvey et al. (2005), and Pfeifer et al. (2005).

Often, emergence of behavior or functionality is studied in behaving artifacts like robots, using the bottom-up principles of artificial life. Notably, the interaction of software or hardware parts of robots can be assumed to be closely influenced by artificial life thinking. For instance, the subsumption architecture of Brooks (1990, 1991) has been a closely studied subject in the interaction of behavioral modules. Different sorts of swarm-like behavior (Reynolds 1987; Bonabeau et al. 1999) and, more recently, specifically constructed swarmbots on the hardware side (Mondada et al. 2004; Groß et al. 2006) have been studied extensively.

Finally, groups of robots cooperating as social robots, and interacting and cooperating with each other, have been studied (Fong et al. 2003). All these applications emphasize the real-world character of robotics problems, and the embodiment of agents (Steels and Brooks 1995).

4.2.2 Transport

Traffic and our societies' transportation systems are collective systems of an enormous complexity. Their modeling, simulation, and design stand to win substantially from new biological principles (Lucic and Teodorovic 2002; Wang and Tang 2004). Notably, principles gleaned from the organization of insect societies (Deneubourg et al. 1994), and from insect behavior (Theraulaz and Bonabeau 1999), that even give rise to new optimization algorithms (Dorigo and Stützle 2004), are relevant here.

Artificial life applications are not restricted to the movement of physical objects on roads, but are also widespread in other infrastructure networks, such as communication networks (Tanner et al. 2005), notably the Internet (Prokopenko et al. 2005). Early on, it was already proposed to consider computer viruses as a form of artificial life (Dewdney 1989; Spafford 1994) spreading in these communication networks (Bedau 2003).

Already there is a large number of engineering applications for artificial life, but this number is only bound to grow over the coming years (Ronald and Sipper 2000). The reader is referred to (Kim and Cho 2006) for a recent overview of artificial life in engineering.

4.3 Computer Graphics, Virtual Worlds, and Games

Terzopoulos (1999) has summarized a number of applications of artificial life in computer graphics.

4.3.1 Computer Graphics

In particular, Sims' work is worth mentioning (Sims 1991, 1994). He used virtual organisms that coevolved in competition with each other to demonstrate the power of artificial evolutionary systems for developing naturally looking moving behavior. Terzopoulos built animated systems resembling simulated ("artificial") fishes, again studying locomotion (Terzopoulos et al. 1994).

The breve visualization environment (Klein 2002) offers a very natural way to interact with artificial life simulations (Spector et al. 2005).

In general, the aim of these tools is animation of agents equipped with virtual sensors and actors, in a life-like fashion (Miranda et al. 2001; Conde and Thalmann 2004). This development was able to close the gap to games (Maes 1995; Reynolds 1999; Rabin 2002).

4.3.2 Virtual Worlds

Magnenat-Thalmann and Thalmann (1994) provide an early overview of how artificial life ideas could be put to use in creating virtual worlds. One key idea, also found in other application areas, is to make use of unpredictability in the form of random events, as they are also present in the real world of life through mutation and chance encounters. This line of inquiry is continued with further work on multi-sensor integration in virtual environments, see, for example, Conde and Thalmann (2004). The more complex the inner workings of agents are, and the more complex the environment is providing stimuli, the more intelligence is required to allow for realistic simulations. Aylett et al. discuss the connection between an adaptive environment and the intelligence of the agents/avatars populating it (Aylett and Luck 2000).

More recently, artificial life virtual worlds have been put to use in social scenarios, such as urban environments. For instance, Shao and Terzopoulos (2007) integrate motor, perceptual, behavioral, and cognitive components in a model of pedestrian behavior. Building on Reynolds (1987, 1999), this allows much more realistic simulation of multi-individual interactions in complex urban settings than was possible previously. An entire area, crowd simulation has now emerged using these techniques (Lee et al. 2007).

Virtual worlds are bound to expand enormously in the future, and artificial life applications will continue to proliferate.

4.3.3 Games

Artificial life helped to grow the area of (evolutionary) game theory (Alexander 2003; Gintis 2009) and provided valuable insights into more traditional board games (Pollack et al. 1997). But its most notable and unique success is in areas where traditional games have not been able to contribute much: multiuser games, distributed over the Internet.

One now classical example, and one of the earliest games that made headlines and really was a commercial success for some time, was Steve Grand's game "Creatures" (Grand et al. 1996). The idea behind this game was very simple: Evolution through breeding of creatures using genomes of characteristics that could be combined in multiple ways. By exchanging creatures over the Internet, a multiuser sphere was created setting in motion something akin to real evolution, with implicit fitness, not controlled by any central agency (Grand 2001). Prior to this commercial application, Ray's Tierra was distributed over the net in a large experiment (Ray 1998).

4.4 Art

The connection between artificial life and art can be easily seen when considering applications in computer graphics. However, there is a much deeper connection, having to do with creativity and surprise, which is one of the hallmarks of evolutionary systems that, like artificial life, concern themselves with emergent phenomena. A discussion of these connections is provided in Whitelaw (2004). In a 1998 special section of the journal *Leonardo* (Rinaldo 1998), the connection was examined from various viewpoints. Bentley and Corne (2002) and references therein look at creative processes from an evolutionary perspective, which might be argued to be a superset of artificial life approaches. Arts applications range from painting (Todd and Latham 1992, 1994) (even painting by robots (Moura 2004)) via video installations to sculpture (Baljko and Tenhaaf 2006). Creativity is also considered in the area of sound generation and music (Bilotta and Pantano 2002; Berry and Dahlstedt 2003). Again, artificial life work is often put together with evolutionary work in composition (Romero and Machado 2007).

4.5 Artificial Societies and Artificial Economies

Agent-based modeling using artificial life ideas has also been applied to formulate systems best described as artificial societies and artificial economies.

The area of artificial societies was spawned by the seminal work of Axtell and Epstein (Epstein and Axtell 1996), which used the idea of bottom-up effects in social systems to generate models of social systems that could be simulated in a computer. These ideas have been further developed by Epstein (2007). Social simulation in general (Gilbert and Troitzsch 2005) has now moved to center stage in the social sciences, and a new journal has been established to examine artificial societies (http://jasss.soc.surrey.ac.uk/JASSS.html). In the context of social science, emergence (Goldspink and Kay 2007) and learning (Gilbert et al. 2006) have become important topics of discussion.

These very same techniques can also be applied to markets, at which point an artificial society mutates into an artificial economy (Zenobia et al. 2008). The formulation of economies as complex adaptive systems is already somewhat older, and Kaufmann (1993) points out models of economies based on ideas of evolution of complex systems. The power of these ideas lies in their natural ability to treat nonequilibrium phenomena, a topic for the most part carefully avoided by many traditional economists, since it would have to be based on nonlinear systems (Arthur 2006). Recently, another subfield of artificial life, artificial chemistries, have been applied to model economies (Straatman et al. 2008).

5 Conclusion

As we have seen in this chapter, artificial life is a vivid research area, spawning interesting research directions, and providing inspiration for a large number of applications. While the boundaries of the area are fuzzy, there is no doubt that over its 20 years of existence as a named field, artificial life has exerted substantial influence on a whole number of other research areas.

To judge the perspectives of a field, it is perhaps best to focus on its main objectives. The proposal of Bedau (Bedau et al. 2001; Bedau 2003) is followed and slightly changed in dividing the central goals of artificial life into three broad thrusts, all posited here as questions:

(A) What is the origin of life, or how does life arise from the nonliving?
(B) What are the potentials and limits of living systems?
(C) How does life relate to intelligence, culture, society, and human artifacts?

These questions belong to the most challenging and most fascinating in all of science today. From each of these questions derives an entire research programme within artificial life that can be formulated with a non-comprehensive number of tasks:

(A1) Generate a molecular proto-organism in vitro.
(A2) Achieve the transition to life in an artificial chemistry *in silico*.
(A3) Determine the threshold of "minimal" life in the organic world in vivo.
(A4) Determine whether fundamentally novel living organizations can arise from inanimate matter.
(A5) Simulate a unicellular organism over its entire life-cycle.

(B1) Determine what is inevitable in the open-ended evolution of life.
(B2) Establish a set of mechanisms for the generation of novelty and innovation.
(B3) Determine minimal conditions for evolutionary transitions from specific to generic response systems.
(B4) Create a formal framework for synthesizing dynamical hierarchies at all scales.
(B5) Determine the predictability of evolutionary manipulations of organisms, ecosystems, and other living systems.
(B6) Develop a theory of information processing, information flow, and information generation for evolving systems.

(C1) Explain how rules and symbols are generated from physical dynamics in living systems.
(C2) Demonstrate the emergence of intelligence and mind in an artificial living system.
(C3) Evaluate the influence of machines on the next major evolutionary transition of life.
(C4) Provide predictive models of artificial economies and artificial societies.
(C5) Provide a quantitative model of the interplay between cultural and biological evolution.
(C6) Establish ethical principles for artificial life.

Inasmuch as life constitutes a large part of the natural world, artificial life will constitute a large part and exert heavy influence on the area of natural computing discussed in this handbook.

Acknowledgments

WB would like to thank the Canadian Natural Science and Engineering Research Council (NSERC) for providing continuous operating funding since 2003 under current contract RGPIN 283304-07. BMcM acknowledges the support of Dublin City University, and of the European Union FP6 funded project ESIGNET, project number 12789.

References

Adami C (1998) Artificial life, an introduction. Springer, Berlin

Adami C, Brown CT (1994) Evolutionary learning in the 2D artificial life system "Avida." In: Brooks RA, Maes P (eds) Artificial life IV: Proceedings of fourth international workshop on the synthesis and simulation of living systems. MIT Press, Cambridge, MA, pp 377–381, http://citeseer.ist.psu.edu/101182.html

Albert R, Barabasi AL (2002) Statistical mechanics of complex networks. Rev Mod Phys 74:47–97

Alexander J (2003) Evolutionary game theory. In: Stanford encyclopedia of philosophy. http://plato.stanford.edu/

Argyris J, Faust G, Haase M (1994) An exploration of chaos. North-Holland, New York

Arthur W (2006) Out-of-equilibrium economics and agent-based modeling. In: Handbook of computational economics, vol 2. North-Holland, Amsterdam, The Netherlands pp 1551–1564

Asada M, Hallam J, Meyer J, Tani J (eds) (2008) From animals to animats: Proceedings of the tenth international conference on simulation of adaptive behavior, Osaka, Japan, July 2008. Springer, Berlin

Axelrod R (1984) The evolution of cooperation. Basic Books, New York

Axelrod R (1987) The evolution of strategies in the iterated prisoner's dilemma. In: Davis L (ed) Genetic algorithms and simulating annealing. Pitman, London

Aylett R, Luck M (2000) Applying artificial intelligence to virtual reality: intelligent virtual environments. Appl Artif Intell 14(1):3–32

Baas N (1994) Hyperstructures. In: Artificial life III, proceedings of the second workshop on artificial life. Addison-Wesley, Redwood City, CA, Santa Fe, NM, June, 1992

Baljko M, Tenhaaf N (2006) Different experiences, different types of emergence: A-life sculpture designer, interactant, observer. In: Proceedings of AAAI fall 2006 symposium on interaction and emergent phenomenon in societies of agents. Arlington, VA, October 2006, pp 104–110

Banzhaf W, Nordin P, Keller R, Francone F (1998) Genetic programming – An introduction. Morgan Kaufmann, San Francisco, CA

Barendegt H (1984) The lambda calculus, its syntax and semantics. Elsevier-North Holland, Amsterdam, The Netherlands

Barricelli N (1957) Symbiogenetic evolution processes realized by artificial methods. Methodos 9(35–36): 143–182

Barricelli N (1963) Numerical testing of evolution theories. Acta Biotheor 16(3):99–126

Bedau M (2003) Artificial life: organization, adaptation and complexity from the bottom up. Trends Cogn Sci 7(11):505–512

Bedau MA et al. (February 2001) Open problems in artificial life. Artif Life 6(4):363–376

Bentley P, Corne D (eds) (2002) Creative evolutionary systems. Academic Press, San Diego, CA

Berlekamp ER, Conway JH, Guy RK (1982) What is life? In: Winning ways for your mathematical plays, vol 2. Academic Press, London, chap 25, pp 817–850

Berry R, Dahlstedt P (2003) Artificial life: Why should musicians bother? Contemp Music Rev 22(3):57–67

Bertalanffy LV (1976) General system theory: foundations, development, applications. George Braziller, New York

Bilotta E, Pantano P (2002) Synthetic harmonies: Recent results. Leonardo 35:35–42

Bonabeau E, Dorigo M, Theraulaz G (1999) Swarm intelligence: from natural to artificial systems. Oxford University Press, New York

Book R, Otto F (1993) String-rewriting systems. Springer Verlag, London

Bratley P, Millo J (1972) Computer recreations; self-reproducing automata. Software Pract Experience 2:397–400

Brooks R (1990) Elephants don't play chess. Robot Autonomous Syst 6:3–15

Brooks R (1991) Intelligence without representation. Artif Intell 47:139–159

Brooks RA (1985) A robust layered control system for a mobile robot. Technical Report A. I. Memo 864, Massachusetts Institution of Technology, Artificial Intelligence Laboratory, http://people.csail.mit.edu/brooks/papers/AIM-864.pdf

Burks AW (ed) (1966) Theory of self-reproducing automata [by] John von Neumann. University of Illinois Press, Urbana, IL

Cariani P (1992) Some epistemological implications of devices which construct their own sensors and effectors. In: Toward a practice of autonomous systems-proceedings of the first European conference on artificial life Paris, France, December 1991. MIT Press, Cambridge, MA, pp 484–493

Centler F, Dittrich P, Ku L, Matsumaru N, Pfaffmann J, Zauner K (2003) Artificial life as an aid to astrobiology: testing life seeking techniques. In: Banzhaf W, Christaller T, Dittrich P, Kim J, Ziegler J (eds) Advances in artificial life. Proceedings of ECAL 2003, Dortmund, Germany, September 2003. Lecture notes in computer science, vol 2801. Springer, Berlin, pp 31–40

Chu D, Ho WK (2006) A category theoretical argument against the possibility of artificial life: Robert Rosen's central proof revisited. Artif Life 12(1):117–134. doi: 10.1162/106454606775186392, http://www.mitpressjournals.org/doi/abs/10.1162/106454606775186392, http://www.mitpressjournals.org/doi/pdf/10.1162/106454606775186392

Codd EF (1968) Cellular automata. ACM Monograph Series. Academic Press, New York

Coffey D (1998) Self-organization, complexity and chaos: the new biology for medicine. Nat Med 4(8):882–885

Conde T, Thalmann D (2004) An artificial life environment for autonomous virtual agents with multi-sensorial and multi-perceptive features. Comput Anim Virtual Worlds 15:311–318

Deneubourg J, Clip P, Camazine S (1994) Ants, buses and robots-self-organization of transportation systems. In: Proceedings of the From Perception to Action Conference, PerAc-94, IEEE Press, New York, pp 12–23

Dershowitz N, Jouannaud J (1990) Rewrite systems. In: Handbook of theoretical computer science.

Volume B, Formal Methods and Semantics. North-Holland, Amsterdam, the Netherlands, pp 243–320

Dewdney AK (1984) Computer recreations: in a game called Core War hostile programs engage in a battle of bits. Sci Am 250:14–22

Dewdney AK (1987) Computer recreations: a program called mice nibbles its way to victory at the first Core Wars tournament. Sci Am 256(1):8–11

Dewdney AK (1989) Computer recreations: of worms, viruses and Core War. Sci Am 260(3):90–93

Dittrich P, di Fenizio PS (2007) Chemical organization theory. Bull Math Biol 69(4):1199–1231

Dittrich P, Ziegler J, Banzhaf W (2001) Artificial chemistries – a review. Artif Life 7(3):225–275

Dobzhansky T (1973) Nothing in biology makes sense except in the light of evolution. Am Biol Teach 35:125–129. http://www.pbs.org/wgbh/evolution/library/10/2/text_pop/l_102_591.html

Dorigo M, Stützle T (2004) Ant colony optimization. MIT Press, Cambridge, MA

Dorogovtsev S, Mendes J (2003) Evolution of networks: from biological nets to the Internet and WWW. Oxford University Press, Oxford

Downing K (2000) Exploring Gaia theory: artificial life on a planetary scale. In: Artificial life VII: proceedings of the seventh international conference on artificial life. MIT Press, Cambridge, MA, p 90

Dyson G (1997) Darwin among the machines; or, the origins of [artificial] life. Presentation hosted by the Edge Foundation, Inc., 8 July 1997, http://edge.org/3rd_culture/dyson/dyson_p1.html

Dyson GB (1998) Darwin among the machines: the evolution of global intelligence. Basic Books, New York

Eaton M, Collins J (2009) Artificial life and embodied robotics: current issues and future challenges. Artif Life Robot 13(2):406–409

Edelman G (1987) Neural Darwinism: the theory of neuronal group selection. Basic Books, New York

Eigen M (1971) Self-organization of matter and the evolution of biological macro-molecules. Naturwissenschaften 58:465–523

Epstein J (2007) Generative social science: studies in agent-based computational modeling. Princeton University Press, Princeton, NJ

Epstein J, Axtell R (1996) Growing artificial societies: social science from the bottom up. MIT Press, Cambridge, MA

Farmer J, Kauffman S, Packard N (1986) Autocatalytic replication of polymers. Physica D 22:50–67

Farmer JD, d'A Belin A (1992) Artificial life: the coming evolution. In: Langton CG, Taylor C, Farmer JD, Rasmussen S (eds) Artificial Life II. Santa Fe Institute studies in the sciences of complexity, vol X. Addison-Wesley, Redwood City, CA, pp 815–838. Proceedings of the workshop on artificial life, Santa Fe, NM, February 1990

Ferber D (2004) Synthetic biology: microbes made to order. Science 303:158–161

Fogel LJ, Owens AJ, Walsh MJ (1966) Artificial intelligence through simulated evolution. Wiley, New York

Fong T, Nourbakhsh I, Dautenhahn K (2003) A survey of socially interactive robots. Robot Autonomous Syst 42(3–4):143–166

Fontana W, Buss L (1994) The arrival of the fittest: toward a theory of biological organization. Bull Math Biol 56:1–64

Gardner M (1970) Mathematical games: The fantastic combinations of John Conway's new solitaire game "Life." Sci Am 223(4):120–123

Gardner M (1971) Mathematical games: on cellular automata, self-reproduction, the garden of Eden and the game "Life." Sci Am 224(2):112–117

Giavitto J, Malcolm G, Michel O (2004) Rewriting systems and the modeling of biological systems. Comparative Funct Genomic 5:95–99

Gilbert G, Troitzsch K (2005) Simulation for the social scientist. Open University Press, London

Gilbert N, den Besten M, Bontovics A, Craenen B, Divina F et al. (2006) Emerging artificial societies through learning. J Artif Soc Soc Simulation 9(2)

Gintis H (2009) Game theory evolving: a problem-centered introduction to modeling strategic interaction, 2nd edn. Princeton University Press, Princeton, NJ

Goldspink C, Kay R (2007) Social emergence: distinguishing reflexive and non-reflexive modes. In: AAAI fall symposium: emergent agents and socialities: social and organizational aspects of intelligence. Washington, DC, November 2007

Gracias N, Pereira H, Lima J, Rosa A (1997) Gaia: An artificial life environment for ecological systems simulation. In: Artificial life V: proceedings of the fifth international workshop on the synthesis and simulation of living systems. MIT Press, Cambridge, MA, pp 124–134

Grand S (2001) Creation: life and how to make it. Harvard University Press, Cambridge, MA

Grand S, Cliff D, Malhotra A (1996) Creatures: artificial life autonomous software agents for home entertainment. In: Proceedings of the first international conference on autonomous agents, Marina del Rey, CA, February 1997. ACM Press, New York

Graubard SR (ed) (1988) Daedalus, Winter 1988: Artificial intelligence. American Academy of Arts and Sciences, issued as Proc Am Acad Arts Sci 117(1)

Gross D, McMullin B (2001) Is it the right ansatz? Artif Life 7(4):355–365

Groß R, Bonani M, Mondada F, Dorigo M (2006) Autonomous self-assembly in swarm-bots. IEEE Trans Robot 22(6):1115–1130

Hamarneh G, McIntosh C, McInerney T, Terzopoulos D (2009) Deformable organisms: an artificial life framework for automated medical image analysis. Chapman & Hall/CRC, Boca Raton, FL, p 433

Harvey I, Paolo E, Wood R, Quinn M, Tuci E (2005) Evolutionary robotics: a new scientific tool for studying cognition. Artif Life 11(1–2):79–98

Haugeland J (1989) Artificial intelligence: the very idea. MIT Press, Cambridge, MA

Herman G, Lindenmayer A, Rozenberg G (1975) Description of developmental languages using recurrence systems. Math Syst Theory 8:316–341

Hillis W (1989) The connection machine. MIT Press, Cambridge, MA

Hofstadter DR (1979) Gödel, Escher, Bach: An eternal golden braid. Penguin Books, London. First published in Great Britain by The Harvester Press Ltd 1979. Published in Penguin Books 1980

Holland JH (1975) Adaptation in natural and artificial systems. The University of Michigan Press, Ann Arbor MI

Holland JH (1976) Studies of the spontaneous emergence of self-replicating systems using cellular automata and formal grammars. In: Lindenmayer A, Rozenberg G (eds) Automata, languages, development. North-Holland, New York, pp 385–404. Proceedings of a conference held in Noordwijkerhout, Holland, 31 March–6 April 1975

Jain S, Krishna S (2001) A model for the emergence of cooperation, interdependence and structure in evolving networks. PNAS 98:543–547

Kampis G (1991) Self-modifying systems in biology and cognitive science. IFSR international series on systems science and engineering, vol 6. Pergamon Press, Oxford, NY. Editor-in-Chief: George J. Klir

Kato N, Okuno T, Okano A, Kanoh H, Nishihara S (1998) An ALife approach to modelling virtual cities. In: IEEE Conference on Systems, Man and Cybernetics, vol 2, San Diego, CA, October 1998. IEEE Press, New York, vol 2, pp 1168–1173

Kauffman SA (1993) The origins of order: self-organization and selection in evolution. Oxford University Press, Oxford

Kemeny JG (1955) Man viewed as a machine. Sci Am 192(4):58–67

Kim K, Cho S (2006) A comprehensive overview of the applications of artificial life. Artif Life 12(1):153–182

Kitano H (2002) Systems biology: a brief overview. Science 295:1662–1664

Klein J (2002) Breve: a 3D environment for the simulation of decentralized systems and artificial life. In: Proceedings of artificial life VIII, the eighth international conference on the simulation and synthesis of living systems, Sydney, Australia, December 2002. MIT Press, Cambridge, MA, pp 329–334

Koza J (1992) Genetic programming. MIT Press, Cambridge, MA

Kvasnicka V, Pospíchal J, Kaláb T (2001) A study of replicators and hypercycles by typogenetics. In: ECAL 2001: Proceedings of the sixth European conference on advances in artificial life, Prague, Czech Republic, September 2001. Springer-Verlag, Berlin, pp 37–54

Langton CG (1984a) Self-reproduction in cellular automata. Physica D 10:135–144

Langton CG (1992a) Life at the edge of chaos. In: Langton C, Taylor C, Farmer J, Rasmussen S (eds) Artificial life II: Proceedings of the second workshop on artificial life. Addison-Wesley, Redwood, CA

Langton CG (1995) What is artificial life? In: Brown T (ed) comp.ai.alife Frequently Asked Questions (FAQ), November 18th 1995. http://www.faqs.org/faqs/ai-faq/alife/

Langton CG (1997) Artificial life: an overview. MIT Press, Cambridge, MA

Langton CG (1984b) Self-reproduction in cellular automata. Physica 10D:135–144

Langton CG (1986) Studying artificial life with cellular automata. Physica 22D:120–149

Langton CG (1987) Announcement of artificial life workshop. http://groups.google.com/group/news.announce. conferences/browse_thread/thread/776a2af47daf7a73/7ab70ed772c50437, posted to news.announce.conferences, April 20 1987, 7:22 pm

Langton CG (ed) (1989) Artificial Life. Sante Fe Institute studies in the sciences of complexity, vol VI. Addison-Wesley, Redwood City, CA. Proceedings of an interdisciplinary workshop on the synthesis and simulation of living systems, Los Alamos, NM, September 1987

Langton CG (1992b) Preface. In: Langton CG, Taylor C, Farmer JD, Rasmussen S (eds) Artificial life II. Santa Fe Institute Studies in the Sciences of Complexity, vol X. Addison-Wesley, Redwood City, CA. Proceedings of the workshop on artificial life, Sante Fe, NM, February 1990

Langton CG, Taylor C, Farmer JD, Rasmussen S (eds) (1992) Artificial life II. Sante Fe Institute Studies in the sciences of complexity, vol X. Addison-Wesley, Redwood City, CA. Proceedings of the workshop on artificial life Sante Fe, NM, February 1990

Lee K, Choi M, Hong Q, Lee J (2007) Group behavior from video: a data-driven approach to crowd simulation. In: Proceedings of the 2007 ACM SIGGRAPH/Eurographics symposium on computer animation. San Diego, CA, August 2007, pp 109–118

Lenton T (2002) Gaia as a complex adaptive system. Philos Trans R Soc B Biol Sci 357(1421):683–695

Levy S (1993) Artificial life: a report from the frontier where computers meet biology. Vintage, New York

Lindenmayer A (1968) Mathematical models for cellular interaction in development, parts I and II. J Theor Biol 18:280–315

Lindenmayer A, Rozenberg G (eds) (1976) Automata, languages, development. Elsevier, Amsterdam, The Netherlands

Lindgren K, Nordahl M (1994) Cooperation and community structure in artificial ecosystems. Artif Life 1:15–37

Lovelock J (1989) Geophysiology, the science of Gaia. Rev Geophys 27(2):215–222

Lucic P, Teodorovic D (2002) Transportation modeling: an artificial life approach. In: ICTAI 2002: Proceedings of the 14th IEEE international conference on tools with artificial intelligence. Washington, DC, November 2002, pp 216–223

Luisi PL, Varela F (1989) Self replicating micelles: a minimal version of a chemical autopoietic system. Orig Life Evol Biosph 19:633–643

Maes P (1995) Artificial life meets entertainment: lifelike autonomous agents. Commun ACM 38(11):108–114

Magnenat-Thalmann N, Thalmann D (eds) (1994) Artificial life and virtual reality. Wiley, Chichester, UK

Magnoli G, Bonanni L, Khalaf R (2002) Designing a DNA for adaptive architecture: a new built environment for social sustainability. In: Brebbia C, Sucharov L, Pascolo P (eds) Design and nature: comparing design in nature with science and engineering. WIT, Southampton, UK, pp 203–213

Maturana HR, Varela FJ (1973) Autopoiesis: the organization of the living. In: Autopoiesis and cognition: the realization of the living. Boston studies in the philosophy of science, vol 42. D. Reidel Publishing Company, Dordrecht, Holland, pp 59–138. First published 1972 in Chile under the title De Maquinas y Seres Vivos, Editorial Universitaria S.A

Maynard Smith J, Szathmáry E (1997) The major transitions in evolution. Oxford University Press, New York

McMullin B (1992) The Holland α-universes revisited. In: Varela FJ, Bourgine P (eds) Toward a practice of autonomous systems: proceedings of the first European conference on artificial life. Complex Adaptive Systems. MIT Press, Cambridge, MA, pp 317–326. series Advisors: John H. Holland, Christopher Langton and Stewart W. Wilson

McMullin B (2000a) John von Neumann and the evolutionary growth of complexity: looking backwards, looking forwards. Artif Life 6(4):347–361. http://www.eeng.dcu.ie/~alife/bmcm-alj-2000/

McMullin B (2000b) Some remarks on autocatalysis and autopoiesis. Ann N Y Acad Sci 901:163–174. http://www.eeng.dcu.ie/~alife/bmcm9901/

McMullin B (2004) 30 years of computational autopoiesis: a review. Artif Life 10(3):277–296. http://www.eeng.dcu.ie/~alife/bmcm-alj-2004/

McMullin B, Varela FJ (1997) Rediscovering computational autopoiesis. In: Husbands P, Harvey I (eds) ECAL-97: Proceedings of the fourth European conference on artificial life, Brighton, UK, July 1997.

Complex adaptive systems. MIT Press, Cambridge, MA. http://www.eeng.dcu.ie/~alife/bmcm-ecal97/

McMullin B, Taylor T, von Kamp A (2001) Who needs genomes? In: Atlantic symposium on computational biology and genome information systems & technology, Regal University Center, Durham, NC, March 2001. http://www.eeng.dcu.ie/~alife/bmcm-cbgi-2001/

Meinhardt H (2003) The algorithmic beauty of sea shells. Springer, Heidelberg

Meyer J, Wilson S (eds) (1991) From animals to animats: Proceedings of the first international conference on simulation of adaptive behavior, London, 1990. MIT Press Cambridge, MA

Miranda F, Kögler J, Del Moral Hernandez E, Lobo Netto M (2001) An artificial life approach for the animation of cognitive characters. Comput Graph 25(6): 955–964

Mondada F, Pettinaro G, Guignard A, Kwee I, Floreano D, Deneubourg J, Nolfi S, Gambardella L, Dorigo M (2004) SWARM-BOT: a new distributed robotic concept. Autonomous Robot 17(2):193–221

Moura L (2004) Man + robots: symbiotic art. Institut d'Art Contemporain, Villeurbanne, France

Okuhara K, Domoto E, Ueno N, Fujita H (2003) Recycling design using the artificial life technology to optimize evaluation function. In: ECO Design'03: 2003 third international symposium on environmentally conscious design and inverse manufacturing. Tokyo, Japan, December 2003, pp 662–665

Pargellis AN (2001) Digital life behavior in the amoeba world. Artif Life 7(1):63–75. doi: 10.1162/106454601300328025, http://www.mitpressjournals.org/doi/abs/10.1162/106454601300328025, http://www.mitpressjournals.org/doi/pdf/10.1162/106454601300328025

Pattee H (ed) (1973) Hierarchy theory. The challenge of complex systems. Georges Braziller, New York

Paun G (1998) Computing with membranes. Technical Report, Turku Center for Computer Science, TUCS Report 208

Paun G, Rozenberg G (2002) A guide to membrane computing. Theor Comput Sci 287:73–100

Pesavento U (1995) An implementation of von Neumann's self-reproducing machine. Artif Life 2(4):337–354

Pfeifer R, Iida F, Bongard J (2005) New robotics: design principles for intelligent systems. Artif Life 11 (1–2):99–120

Plsek P, Greenhalgh T (2001) The challenge of complexity in health care. Br Med J 323(7313): 625–628

Pollack J, Blair A, Land M (1997) Coevolution of a backgammon player. In: Artificial life V: Proceedings of the fifth international workshop on the synthesis and simulation of living systems. Nara, Japan, May 1996, pp 92–98

Popper KR, Eccles JC (1977) The self and its brain: an argument for interactionism. Routledge & Kegan Paul plc, London. First published 1977, Springer-Verlag, Berlin. This edition first published 1983

Prokopenko M, Wang P, Valencia P, Price D, Foreman M, Farmer A (2005) Self-organizing hierarchies in sensor and communication networks. Artif Life 11 (4):407–426

Prusinkiewicz P, Lindenmayer A (1990) The algorithmic beauty of plants. Springer, New York

Rabin S (2002) AI game programming wisdom. Charles River Media, Hingham, MA

Rashevsky N (1940) Advances and applications of mathematical biology. University of Chicago Press, Chicago, IL

Rasmussen S (2008) Protocells. MIT Press, Cambridge, MA

Rasmussen S, Knudsen C, Feldberg R, Hindsholm M (1990) The Coreworld: emergence and evolution of cooperative structures in a computational chemistry. Physica 42D:111–134

Rasmussen S, Baas NA, Mayer B, Nilsson M (2001a) Defense of the ansatz for dynamical hierarchies. Artif Life 7(4):367–373

Rasmussen S, Baas NA, Mayer B, Nilsson M, Olesen MW (2001b) Ansatz for dynamical hierarchies. Artif Life 7(4):329–353

Rasmussen S, Bedau MA, Chen L, Deamer D, Krakauer DC, Packard NH, Stadler PF (eds) (2008) Protocells: bridging nonliving and living matter. MIT Press, Cambridge, MA

Ratze C, Gillet F, Mueller J, Stoffel K (2007) Simulation modelling of ecological hierarchies in constructive dynamical systems. Ecol Complexity 4:13–25

Ray T (1994) An evolutionary approach to synthetic biology. Artif Life 1(1/2):179–209

Ray T (1998) Selecting naturally for differentiation: preliminary evolutionary results. Complexity 3 (5):25–33

Ray TS (1992) An approach to the synthesis of life. In: Langton CG, Taylor C, Farmer JD, Rasmussen S (eds) Artificial life II. Santa Fe Institute studies in the sciences of complexity, vol X. Addison-Wesley, Redwood City, CA, pp 371–408

Ray TS (1995) A proposal to create two biodiversity reserves: one digital and one organic. http://life.ou.edu/pubs/reserves/, Project proposal for Network Tierra

Rechenberg I (1975) Evolution strategies (in German). Holzmann-Froboog, Stuttgart, Germany

Reeke GN Jr, Edelman GM (1988) Daedalus, Winter 1998: Real brains and artificial intelligence. Proc Am Acad Arts Sci 117(1):143–173

Reggia J, Armentrout S, Chou H, Peng Y (1993) Simple systems that exhibit self-directed replication. Science 259:1282–1287

Regis E (2008) What is life? Investigating the nature of life in the age of synthetic biology. Farrar, Straus & Giroux, New York

Reynolds C (1987) Flocks, herds and schools: a distributed behavioral model. In: Proceedings of the 14th annual conference on computer graphics and interactive techniques, July 1987. ACM, New York, Anaheim, CA, July 1987, pp 25–34

Reynolds C (1999) Steering behaviors for autonomous characters. In: Game developers conference, vol 1999, San Jose, CA, 1999, pp 763–782

Rinaldo K (1998) Artificial life art, introduction to the special section of Leonardo. Leonardo 31:370

Rocha LM (ed) (2000) The physics and evolution of symbols and codes: reflections on the work of Howard Pattee. BioSystems (special issue) 60(1–3) 149–157. Elsevier

Romero J, Machado P (2007) The art of artificial evolution: a handbook of evolutionary art and music. Springer, Berlin

Ronald E, Sipper M (2000) Engineering, emergent engineering, and artificial life: unsurprise, unsurprising surprise, and surprising surprise. In: Artificial life VII: Proceedings of the seventh international conference on artificial life. The MIT Press, Cambridge, MA, pp 523–528

Rosen R (1959) On a logical paradox implicit in the notion of a self-reproducing automaton. Bull Math Biophys 21:387–394

Rosen R (1972) Some relational cell models: the metabolism-repair systems. In: Rosen R (ed) Foundations of mathematical biology, vol II. Academic, New York

Rosen R (1985) Organisms as causal systems which are not mechanisms: an essay into the nature of complexity. In: Rosen R (ed) Theoretical biology and complexity. Academic, Orlando, FL, Chap 3, pp 165–203

Rosen R (1991) Life itself. Columbia University Press, New York

Rothblum U (1975) Algebraic eigenspace of nonnegative matrices. Linear Algebra Appl 12:281–292

Rumelhart DE, McClelland JL, the PDP Research Group (1986) Parallel distributed processing, vol 1: Foundations. MIT Press, Cambridge, MA

Salomaa A (1973) Formal languages. Academic Press, New York

Samuel A (July 1959) Some studies in machine learning using the game of checkers. IBM J 3(3): 210–229, http://ieeexplore.ieee.org/xpl/freeabs_all.jsp?arnumber=5392560

Searle JR (1980) Minds, brains, and programs. Behav Brain Sci 3:417–457, includes peer commentaries. Also reprinted (without commentaries) as Haugeland J (ed) (1981) Mild design. MIT Press, Cambridge, MA, Chap 10, pp 282–305 and Boden MA (ed) (1990) The philosophy of artificial intelligence.

Oxford readings in philosophy. Oxford University Press, Oxford, Chap 3, pp 67–88

Semple J, Woolridge N, Lumsden C (2005) Review: in vitro, in vivo, in silico: computational systems in tissue engineering and regenerative medicine. Tissue Eng 11(3–4):341–356

Shao W, Terzopoulos D (2007) Autonomous pedestrians. Graph Model 69(5–6):246–274

Sims K (1991) Artificial evolution for computer graphics. Comput Graph 25(4):319–328

Sims K (1994) Evolving 3D morphology and behavior by competition. In: Artificial life IV: Proceedings of the fourth international workshop on the synthesis and simulation of living systems. MIT Press, Cambridge, MA, pp 28–39

Sipper M (1998) Fifty years of research on self-replication: an overview. Artif Life 4(3):237–257

Smith HO, Hutchison CA, Pfannkoch C, Venter JC (2003) Generating a synthetic genome by whole genome assembly: ϕx174 bacteriophage from synthetic oligonucleotides. Proc Natl Acad Sci 100(26):15440–15445

Snare A (1999) Typogenetics. http://www.csse.monash.edu.au/hons/projects/1999/Andrew.Snare/thesis.pdf, Thesis for Bachelor of Science (Computer Science) Honours, School of Computer Science and Software Engineering, Monash University

Spafford E (1994) Computer viruses as artif life. Artif Life 1(3):249–265

Spector L, Klein J, Perry C, Feinstein M (2005) Emergence of collective behavior in evolving populations of flying agents. Genet Programming Evolvable Mach 6(1):111–125

Speroni di Fenizio P, Dittrich P, Zeigler J, Banzhaf W (2000) Towards a theory of organizations. In: German workshop on artificial life, vol 5, Bayreuth, Germany, April 2000. http://www.cs.mun.ca/~banzhaf

Steels L, Brooks R (1995) The artificial life route to artificial intelligence. Lawrence Erlbaum Associates, New Haven, CT

Straatman B, White R, Banzhaf W (2008) An artificial chemistry-based model of economies. In: Artificial Life XI: Proceedings of the eleventh international conference on simulation and synthesis of living systems. MIT Press, Cambridge, MA, pp 592–602

Strand E, Huse G, Giske J (2002) Artificial evolution of life history and behaviour. Am Nat 159:624–644

Strogatz S (2001) Exploring complex networks. Nature 410:268–276

Sugiura K, Suzuki H, Shiose T, Kawakami H, Katai O (2003) Evolution of rewriting rule sets using string-based tierra. In: Banzhaf W et al. (eds) Advances in artificial life - Proceedings ECAL 2003, Budapest, Hungary, April 2003. Lecture notes in computer science, vol 2801. Springer, Berlin, pp 69–77

Tanner H, Jadbabaie A, Pappas G (2005) Flocking in fixed and switching networks. In: IEEE conference on decision and control, vol 1, Seville, Spain, December 2005. p 2

Taylor C, Jefferson D (1994) Artificial life as a tool for biological inquiry. Artif Life 1(1–2):1–13

Terzopoulos D (1999) Artificial life for computer graphics. Commu of the ACM 42(8):32–42

Terzopoulos D, Tu X, Grzeszczuk R (1994) Artificial fishes with autonomous locomotion, perception, behavior, and learning in a simulated physical world. In: Artificial life IV: Proceedings of the fourth international workshop on the synthesis and simulation of living systems. MIT Press, Cambridge, MA, pp 17–27

Thatcher JW (1970) Universality in the von Neumann cellular model. In: Burks AW (ed) Essays on cellular automata. University of Illinois Press, Urbana, IL, pp 132–186 (Essay Five)

Theraulaz G, Bonabeau E (1999) A brief history of stigmergy. Artif Life 5(2):97–116

Todd S, Latham W (1992) Artificial life or surreal art. In: Toward a practice of autonomous systems: Proceedings of the first European conference on artificial life. MIT Press, Cambridge, MA, p 504

Todd S, Latham W (1994) Evolutionary art and computers. Academic London

Toffoli T, Margolus N (1987) Cellular automata machines. MIT Press, Cambridge, MA

Turing A (1936) On computable numbers, with an application to the Entscheidungsproblem. Proc London Math Soc Ser 2 42:230–265

Turing AM (1950) Computing machinery and intelligence. Mind LIX(236):433–460, also reprinted as, Boden MA (ed) (1990) The philosophy of artificial intelligence. Oxford readings in philosophy. Oxford University Press, Oxford, Chap 2, pp 40–66

Turing AM (1952) The chemical basis of morphogenesis. Philos Trans R Soc B 237:37–72

Varela FJ, Maturana HR, Uribe R (1974) Autopoiesis: the organization of living systems, its characterization and a model. BioSystems 5:187–196

Varela FJ, Thompson ET, Rosch E (1992) The embodied mind: cognitive science and human experience. MIT Press, Cambridge, MA

von Neumann J (1945) First draft of a report on the EDVAC. (A corrected version was formally published in the IEEE Ann Hist Comput 15(4):27–75 1993.)

von Neumann J (1949) Theory and organization of complicated automata. In: Burks AW (ed) (1966) Theory of self-reproducing automata [by] John von Neumann. University of Illinois Press, Urbana, IL, pp 27– (Part One), based on transcripts of lectures delivered at the University of Illinois, in December 1949. Edited for publication by A.W. Burks von Neumann J (1951) The general and logical theory of automata. In: Taub AH (ed) John von Neumann: collected works. Volume V: Design of computers, theory of automata and numerical analysis. Pergamon Press, Oxford, chap 9, pp 288–328. First published Jeffress LAA (ed) (1951) Cerebral mechanisms in behavior—the Hixon symposium. Wiley, New York, pp 1–41

von Neumann J (1966) The theory of automata: Construction, reproduction, homogeneity. In: Burks AW (ed) (1966) Theory of self reproducing automata [by] John von Neumann. University of Illinois Press, Urbana, IL, pp 89–250. Based on an unfinished manuscript by von Neumann. Edited for publication by A.W. Burks

Vyssotsky VA (1972) Darwin: a game of survival and (hopefully) evolution. Software Pract Experience 2: 91–96. http://www.cs.dartmouth.edu/doug/darwin.pdf, Attachment to letter appearing in Computer Recreations column, signed by M. D. McIlroy, R. Morris, and V. A. Vyssotsky

Waddington CH (ed) (1969) Towards a theoretical biology. 2: Sketches. Edinburgh University Press, Edinburgh, UK

Wang F, Tang S (2004) Artificial societies for integrated and sustainable development of metropolitan systems. IEEE Intell Syst 19(4):82–87

Watts D (1999) Small worlds: the dynamics of networks between order and randomness. Princeton University Press, Princeton, NJ

White R, Engelen G (1993) Cellular automata and fractal urban form: a cellular modelling approach to the evolution of urban land-use patterns. Environ Plan A 25:1175–1199

Whitelaw M (2004) Metacreation: art and artificial life. MIT Press, Cambridge, MA

Wiener N (1965) Cybernetics, 2nd edn: or the control and communication in the animal and the machine. MIT Press, Cambridge, MA

Wilkinson D (2003) The fundamental processes in ecology: a thought experiment on extraterrestrial biospheres. Biol Rev 78:171–179

Willis M, Hiden H, Marenbach P, McKay B, Montague GA (1997) Genetic programming: an introduction and survey of applications. In: Zalzala A (ed) GALESIA'97: Second international conference on genetic algorithms in engineering systems: innovations and applications, Glasgow, UK, September 1997. Institution of Electrical Engineers (IEE), Savoy Place, London, UK, pp 314–319, http://ieeexplore.ieee.org/stamp/stamp.jsp?arnumber=00681044

Wolfram S (1984) Computation theory of cellular automata. Commun Math Phys 96(1):15–57

Wolfram S (1986) Theory and applications of cellular automata. Advanced Series on Complex Systems. World Scientific Publication, Singapore

Wolfram S (1994) Cellular automata and complexity: collected papers. Perseus Books Group, Reading, MA

Wolkenhauer O (2007) Why systems biology is (not) systems biology. BIOforum Europe 2007, vol 4, pp 2–3

Yovits MC, Cameron S (eds) (1960) Self-organizing systems. Pergamon Press, Oxford. Proceedings of an interdisciplinary conference, 5 and 6 May, 1959

Zenobia B, Weber C, Daim T (2008) Artificial markets: a review and assessment of a new venue for innovation research. Technovation. doi:10.1016/j.technovation.2008.09.002

54 Algorithmic Systems Biology — Computer Science Propels Systems Biology

Corrado Priami
Microsoft Research, University of Trento Centre for Computational and
Systems Biology (CoSBi), Trento, Italy
DISI, University of Trento, Trento, Italy
priami@cosbi.eu

G. Rozenberg et al. (eds.), *Handbook of Natural Computing*, DOI 10.1007/978-3-540-92910-9_54,

Abstract

The convergence between computer science and biology occurred in successive waves involving deeper and deeper concepts of computing. The current situation makes computer science a suitable candidate to become a philosophical foundation for systems biology with the same importance as mathematics, chemistry, and physics. Systems biology is a complex and expanding applicative domain that can open completely new avenues of research in computing and eventually help it become a natural, quantitative science. This chapter highlights the benefits of relying on an algorithmic approach to model, simulate, and analyze biological systems. The key techniques surveyed are related to programming languages and concurrency theory as they are the main tools to express in an executable form the key feature of computing: algorithms and the coupling executor/execution of descriptions. The concentration here is on conceptual tools that are also supported by computational platforms, thus making them suitable candidates to tackle real problems.

1   Introduction

Computing and biology have been converging for the past two decades, but with a vision of computing as a service to biology that has propelled bioinformatics. This field addresses structural and static aspects of biology and produced databases, patterns of manipulation and comparison, searching tools and data mining techniques (Spengler 2000; Roos 2001). The most relevant success has been the Human Genome Project that was made possible by the selection of the right language abstraction for representing DNA (a language over a 4-character alphabet) (Searls 2002).

In biology, there is now a heightened interest in system dynamics for interpreting living organisms as information manipulators (Hood and Galas 2003) and moving toward systems biology (Kitano 2002). There is no general agreement on a definition of systems biology, but whatever definition one selects, it must embrace at least four characterizing concepts, as systems biology transitions in the following ways:

1. From qualitative biology toward a quantitative science. Often, biological phenomena are explained through textual descriptions and sometimes cartoons that highlight the interaction of components without any quantitative information. Furthermore, the cartoons are superpositions of different snapshots of the interconnection network of biological elements that vary over time. As a consequence, temporal ordering of events is also not easily retrievable.
2. From reductionism to system-level understanding of biological phenomena. Biology has mainly studied single components of systems following the idea that by putting together the knowledge on the single components the system behavior would also have been elucidated. The emergent behavior due to interaction of subsystems makes the unravelling of the system dynamics impossible.
3. From structural, static descriptions to functional, dynamic properties. The list of components and their three-dimensional structure coupled with the description of their chemical and physical properties does not allow one to infer the way in which those components operate in context. A focus on the interactions (dynamics) is needed from the very beginning of the studies.

4. From descriptive biology to mechanistic/causal biology. Although some attempts to model the dynamics of systems exist (and hence partially move in the right direction with respect to item 3 above), most of the approaches hide within mathematical variables the elementary, mechanistic steps that a system performs to accomplish a goal. As a consequence, it is very difficult to disentangle the causality of events that govern the evolution of a system.

The above features highlight that causality between events, temporal ordering of interactions, and spatial distribution of components, within the reference volume of the system at hand, are becoming essential. This poses new challenges to describe the step-by-step mechanistic behavior that enables phenotypical phenomena, a behavior that both bioinformatics and classical mathematical modeling do not address (Cassman et al. 2005).

The complexity of studying the behavior of whole systems, rather than small isolated parts, calls for a stronger formalization of biological concepts that allows one to automatically manipulate biological knowledge. Therefore, the role of modeling becomes an essential and integral part of the new discipline of systems biology. The metabolic network represented in the right part of ❷ *Fig. 1* is an illuminating example that informal reasoning cannot provide reasonable outcomes. Models can no longer be simple cartoons like the ones found in biology textbooks and cannot be equations, because they hide both the mechanistic behavior of systems and the inherent concurrency of biology in representations that are too abstract.

Rosen's description of modeling (left part of ❷ *Fig. 1*) perfectly fits the situation (Rosen 1998) as well as his statement "I have been, and remain, entirely committed to the idea that modeling is the essence of science..." Nature is governed by causal laws that make systems vary over time and space. The complexity of real systems makes it impossible to study and analyze them directly. Abstractions are mandatory to *encode*, into a formal system, the relevant characteristics needed to study the properties of the real system in which one is interested. One then uses the formal system corresponding to the abstract representation to perform inferences and proofs of properties. In this way one derives some knowledge that is only true in

❏ **Fig. 1**
Left Rosen's representation of the relationships between nature and formal models. Right: a graphical representation of a metabolic network.

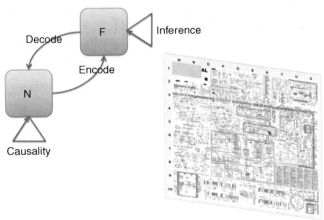

the abstract world; one then needs to *decode* the findings and reinterpret them in the natural world by establishing a precise coupling between inference and causality. However, this is exactly what computer science has been doing since its early days, by using algorithms and semantics of programming languages, to ensure strong and coherent coupling of phenomena at different levels of abstraction. Therefore, algorithms and formal specification languages are suitable candidates to address the modeling challenges of systems biology.

Algorithms expressed by computing languages are a conceptual tool (Wing 2006) that helps elucidate fundamental biological principles by forcing modelers/biologists to think about the mechanisms governing the behavior of the system under question at the right level of abstraction. Since the dynamics of biological systems are mainly driven by quantities such as concentrations, temperatures, gradients, etc., one must clearly focus on quantitative algorithms and languages, that is, the selection mechanism of the next step is determined according to probabilistic/temporal distributions. These distributions allow one to model the kinetics (speed) of interactions that is determined by the chemical and physical quantities listed above. Furthermore, algorithms can help in coherently extracting and organizing general biological principles that underlie the huge amount of data produced by high-throughput technologies, thus producing knowledge from information (data).

Algorithms need a syntax to be described and a semantics to be associated with the descriptions of their intended meaning so that an executor can precisely perform the steps needed to implement the algorithms with no ambiguity. In this way, one is entering the realm of programming languages from both a theoretical and practical perspective.

The use of programming languages to model biological systems is an emerging field that enhances current modeling capabilities (richness of aspects that can be described as well as user-friendliness, composability and reusability of models) (Priami and Quaglia 2004). The metaphor that inspires this (see ❷ *Fig. 2*) is one where biological entities are represented as programs being executed simultaneously and the interaction of two entities is represented by the exchange of a message between the programs representing those entities in the

◘ Fig. 2
The metaphor underlying algorithmic systems biology. The left column represents biological objects and their properties and capabilities, while the right column represents the corresponding computer science concepts.

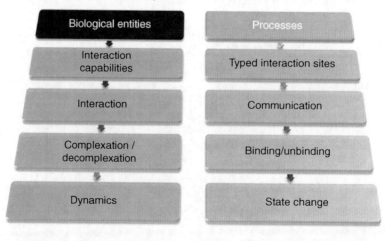

model (Regev and Shapiro 2003). Note that interaction capabilities could also be represented with untyped interfaces or simply through channel names as in most approaches. However, the typing information separates the names used to communicate/interact between entities and the information on the shapes/types of the active domains/receptors that de facto determine the sensitivity or affinity between entities. Therefore, the modeling process is much closer to the natural behavior of systems and helps the modeler think in terms of the fundamental blocks of biological systems. The biological entities involved in the biological process and the corresponding programs in the abstract model are in a 1:1 correspondence, thus coping by construction with the combinatorial explosion of variables needed in the mathematical approach to describe the entire set of states through which a single component can pass.

The simultaneity of the execution of programs requires one to take concurrency into account. Concurrency must not be considered a tool to improve the performance of sequential programming languages and architectures, which is the standard practice in most actual cases. Concurrency is the key concept that permits the execution of algorithms to exhibit the emergent behavior produced at system level by the set of local interactions between components without the need to specify it from the beginning. This aspect is crucial for the predictive power of this approach.

Some programming languages that address concurrency as a core primitive issue and aim at modeling biological systems are emerging, for example, Welch and Barnes (2005) and Dematté et al. (2008b), from the field of process calculi (Bergstra et al. 2001). These concurrent programming languages are suitable candidates to easily and efficiently express the mechanistic rules that propel algorithmic systems biology (Priami 2009). Therefore, it seems natural to check whether the programming and analysis techniques, developed for computer networks and their formal theories, could bring new light to biology when suitably adapted. Note that process calculi are not the only theoretical ground for algorithmic systems biology. Petri nets, logic, rewriting systems, and membrane computing are other relevant examples of formal methods applied to systems biology (for a collection of tutorials see Bernardo et al. (2008) as well as the other chapters of this book). Other approaches that are more closely related to software design principles are the adaptation of UML to biological issues (see www.biouml. org) and statecharts (Harel 2007). Finally, cellular automata (Gutowitz 1990) need to be considered as well with their game of life.

Scalability is an issue for systems biology. Design principles of large software systems can help in developing an algorithmic discipline for systems biology in order to move from toy examples to real case studies. A library of biological components can be easily built and used to derive models of large systems that are ready to be simulated and analyzed just by composing the available modules (Dematté et al. 2008a). Note that here we refer to shallow compositionality (de Alfaro et al. 2001) (i.e., syntactic composition of modules) and hence the well-known problem of composing stochastic semantics does not affect the definition of libraries.

Algorithmic systems biology raises novel issues in computing by stepping away from qualitative descriptions, typical of programming languages, toward a new quantitative computing so that it can fully become an experimental science, as advocated in Denning (2007), that is suitable to support systems biology. This would also foster the move toward a simulation-based science that is needed to address the larger and larger dimensions and complexity of scientific questions and available data. Algorithmic systems biology fully adopts the main assets of computer science: hierarchical, systematic, and algorithmic (computational) thinking and creativity in modeling, programming, and innovating.

1.1 Organization of the Chapter

The organization of the chapter follows a high-level schematic workflow of how models and experiments can be tightly integrated in the algorithmic systems biology vision (see ❷ *Fig. 3*). The next section deals with modeling languages derived from process calculi, which are described in the chapter ❷ Process Calculi, Systems Biology and Artificial Chemistry of this handbook. The modeling language is the core component of the composer in ❷ *Fig. 3*. In order to feed models with quantitative information needed to drive simulations, ❷ Sect. 3 discusses knowledge inference. Then, in ❷ Sect. 4 the simulation of models is addressed and the features of a computational platform are described. These aspects are part of the runner box in the figure that mainly relates to the quantitative implementation of the modeling language. The same section also discusses the post-processing of data obtained from the execution of biological models and shows how to infer new hypotheses to restart the workflow. The key activities of post-processing are collected in the boxes *Abstraction* and *Visualization* in the figure.

◘ Fig. 3

A typical workflow for algorithmic systems biology. All boxes contain examples of the activities that can be performed. (Figure prepared by Ivan Mura.)

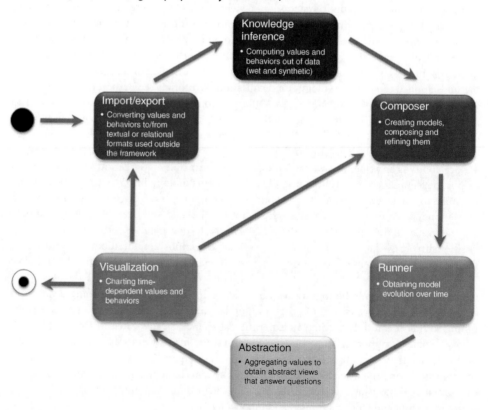

2 Model Specification

The main challenges for building algorithmic models for the system-level understanding of biological processes include the relationship between low-level local interactions and emergent high-level global behavior; the partial knowledge of the systems under investigation; the multilevel and multi-scale representations in time, space, and size; the causal relations between interactions and the context-awareness of the inner components. Therefore, the modeling formalisms that are candidates to propel algorithmic systems biology should be complementary to and interoperable with mathematical modeling; be algorithmic and quantitative; be interaction-driven, compositional, scalable, and modular; address parallelism and complexity; and express causality.

Most of the issues above are naturally addressed by process calculi (see the chapter ❷ Process Calculi, Systems Biology and Artificial Chemistry), and therefore the concentration here is on the conceptual and computational platforms developed from that theoretical foundation. The brief history of this field is recalled by grouping the modeling languages into three different generations. The first two are described in the chapter ❷ Process Calculi, Systems Biology and Artificial Chemistry and the details are not entered into, but their limitations highlighted to model real biological systems in a composable and scalable manner. Guerriero et al. (2008) is referred to for a survey of process calculi applied to biology and here stochastic π-calculus (Priami 1995; Priami et al. 2001) and beta-binders (Priami and Quaglia 2004) are considered as representatives of the two generations of calculi. Then BlenX is examined as an example of the last generation of modeling languages.

2.1 First Generation of Calculi for Biology

For a long time, the only calculus equipped with a quantitative implementation to model and simulate biological systems has been the stochastic π-calculus. This calculus attracted the interest of many researchers and has been used in a variety of case studies that show the feasibility of the programming language approach to model biological systems. Although successful in proving the potential of process calculi, the stochastic π-calculus suffers from some limitations in modeling biology. In fact, it was designed to model and analyze the performance of computer networks and hence it is strictly tuned to computer science. As a matter of fact, many tricks are needed to encode biology, and some natural primitives to describe common biological phenomena are missing.

According to the metaphor, biological entities are represented by π-calculus processes (hereafter for simplicity modeling proteins and their interactions are assumed). Proteins are characterized by functional units called domains. For instance, a protein P with three domains (D_1, D_2, and D_3) can be represented by three processes with the same names as the corresponding domains that are glued together to form a process P representing the protein. The π-calculus has no primitive to model physical attachment of sub-processes into larger processes, and therefore it uses the scope of names to identify boundaries. Intuitively it is assumed that the three domains share a secret (a private name) that states that they are bound together to form a protein. Thus, the protein can be modeled as $P = (\nu x)(D_1 \,|\, D_2 \,|\, D_3)$, where (νx) is the declaration of the new name x with scope of the process P. The behavior of P is determined by the parallel composition of its domains and the shape or physical boundaries of P are determined by the backbone channel x that is shared only by the domains.

The scope of the channel determines the physical boundaries of the biological entities. Real biological systems contain multiple copies of the same protein; since concentrations of species affect the selection mechanism of the next action to be performed in a stochastic simulation, multiplicity is modeled through parallel composition yielding $MultiP = P|..|P$. Finally, a system contains multiple species so that it ends up in $Sys = MultiP|MultiP_1|..|MultiP_n$.

The processes encoding the domains implicitly describe the interaction capabilities of the protein through the names used in their syntactic definition. Recall that the π-calculus communications are determined by complementary actions (send "!" and receive "?") performed on the very same channel name that must be shared between the partners of the interaction. For instance, if $D_3 = a!y.D_3'$, the protein P can interact with all the other entities that are willing to perform an input on the shared channel a, for example, the protein Q made up of the single domain $D_4 = a?w.Q_4'$.

If the interaction of two proteins has to form a complex, the attachment must be programmed through communication of private names so that the scope of the communicated name is changed to include the two proteins. According to the interpretation of the scope of private names as boundaries of physical objects, one has to dynamically manipulate the scope to create and destroy different complexes during the execution. For instance, if one considers the system $((vx)(vy)P)|Q)$ after the interaction of D_3 and D_4, the new system will be $(vy)((vx)(D_1|D_2|D_3')|D_4')$. The residual of protein P and the residual of protein Q after the interaction share the private name y that P sent and made available to Q; therefore, they are interpreted as being physically attached. The detachment of the proteins must be programmed as well by opening the scope of the name y. For instance, if $D_4' = z!y.D_4'$ and it performs the output of the private name with no other process willing to perform the input on the channel z, the resulting system no longer has a restriction on the name y, which becomes available to the whole system and hence the two proteins detach.

Note that both interaction and complexation are based on a key-lock mechanism of affinity. In fact, the partners of a reaction must share the same channel and must perform synchronous complementary operations on it. Real biology does not work in the same manner. Interaction can happen even if the shape of the active domains that are going to react are not exactly complementary to one another. Hence, a given domain can interact with different strengths and probabilities with whole families of partially complementary domains. In order to model this in π-calculus-like formalisms, one must explicitly write the whole set of alternatives relying on the choice operator in the processes. For instance, if the domain D_3 can interact with n different domains, one should specify it within the protein as $P = (vx)(D_1|D_2|a_1!y.D_3' + \cdots + a_n!y.D_3')$. Furthermore, any complementary domain must contain a receive operation on the very same channel a_i representing the compatibility between the two entities.

The main limitations (listed below) of the π-calculus family formalisms stem from the fact that these calculi have been designed for artificial, known computer systems that can interact only when the port and addresses of the partners are perfectly and exactly known. Furthermore, the identity of the programs is not an issue in computer science, in which it is enough to ensure a global behavior as described by the algorithms to be implemented.

Processes. The same syntactical concept is used to model many different biological entities at different levels of abstraction, such as domains, proteins, and whole systems.

Restriction. A primitive intended to declare names and identify their scope is used to represent compartments, membranes, and complexes.

Complex/Decomplex. Creation and destruction of complexes of biological entities is not a primitive operation, but shows up from the interplay of private names and the variation of their scopes through open and close operations. It must be programmed through particular classes of communications. A further confusion that arises in this class of calculi is that channel names are used both to define the boundaries of complexes as well as to manipulate the structure of programs. A characterization of the minimal set of primitives needed to model this phenomena is discussed in Cardelli and Zavattaro (2008).

Low-level programming. The reversibility of reactions as well as the complementarity of complexation and decomplexation must be programmed through communication and name passing. The very same structure of communications determines the network of interaction of the biological entities. In other words, one needs to specify all the arcs of the network through a complementary pair of send and receive on the same channel rather than inferring it from the sensitivity or affinity of interaction of entities as it happens in biology.

Interaction. The interaction of biological entities can occur only if the entities share the very same channel name. This implies a perfect key-lock mechanism of interaction that is not realistic in the biological domain. In fact, the interaction can happen with different strengths or probabilities depending on the complementarity of the active domains and on the concentration of the entities in the reaction volume. The non-key-lock mechanism must be coded; in case of complexation/decomplexation, one also needs to program all the reversible interactions for all the different channels representing the different shapes that can create interactions.

Incremental model building. Most of the information related to the interconnection structure of the entities and on the sensitivity/affinity of interaction is hard-coded into the syntax of the processes through the send/receive pairs. This means that whenever new knowledge, for example, on the capability of interaction of entities, is discovered, a new piece of code must be produced and distributed in various parts of the specification that are not well identified. As a consequence, incremental model building is strongly affected with a negative impact on the scalability of the approach. In fact, the compositional modeling style of all the calculi used so far in the biological context is simply a way of structuring a description of a system by identifying sub-systems that run in parallel rather than a true scalable and compositional methodology.

Identity of entities. Biological entities—for example, proteins—can take on different states during their lifetime depending on the interactions in which they participate. One of the main criticisms of the mathematical modeling of these phenomena is that a new variable is needed for any different state of the same biological entity, thus causing to the well-known combinatorial explosion of the variables and consequently equations. The very reason here is that no identity of the considered entities is maintained in the model. In the π-calculus-like approach, the identity of entities in the initial description can coincide with the processes to which names can be assigned. However, during the execution of the programs, prefixes are consumed and the identity is lost as well.

Implementation. Stochastic simulation selects the action to be performed considering the number of equal entities in the system at a given time (e.g., the number of proteins P). If one assumes that entities are represented by processes, one has to count all the pieces of programs that represent the same object. Due to name passing and the naming of channels in general, it could be that two syntactically different programs represent the very same semantic object, and, hence, the same entity. The technical tool used to address this equivalence is commonly known in the process algebra field as structural congruence. Efficiency in computing is therefore mandatory because it must be done at any step of computation and it is not so immediate for π-calculus. Indeed the entities change their structure after any execution step and their number is then continuously changing. This is due to the fact that processes can be created and updated dynamically. Efficiency in the execution of models is also fundamental due to the large size of the real biological systems. Parallel implementation of algorithms for running models written with process calculi are not available at this stage.

Qualitative versus quantitative descriptions. Biology is mainly driven by quantities, whereas process calculi are usually qualitative descriptions of behavior. Stochastic process calculi handle the kinetics of systems by associating channel names with rates identifying exponential distributions. Most of the time, the kinetic parameters for real biological interactions are not known and must be estimated to fit some phenotypical behavior that is measurable experimentally. Therefore, mixing quantitative and qualitative information imposes the modeler to rewrite the model or to heavily work on it in order to carry out parameter estimation and sensitivity analysis. Furthermore, limiting the description of kinetics to exponential distribution may cause all the high-level abstractions like (such as Hill-functions or even more complex dynamics that are used in mathematical modeling) to explode into elementary steps in order to hide unknown details within models. Unfortunately, most of the time the details to explode these macro-processes are not available and makes it difficult to apply the approach.

Space. Many biological phenomena are highly sensitive to the localization within the reference system volume of the reactants. There is an extremely limited and difficult to manage notion of space, and it must be hard-coded in the language.

Connection with standards. A very limited and preliminary activity is emerging to connect process calculi with SBML (see e.g., Eccher and Priami (2006)). No interoperability between computational tools is available.

Predictive models. Most of the previous items and the introductory description show how any possible dynamic of the modeled system must be coded in the specification. This means that execution can provide the user with different paths due to different selection of alternatives or different temporal ordering of concurrent events, but all the observed action has been created by the modeler and explicitly programmed. Consequently, the predictive power of the models is limited, although it can still provide much more information on the dynamics of systems and their mechanistic behavior than classical ODE-based mathematical modeling.

2.2 Second Generation of Calculi for Biology

The second generation of process calculi directly defined to model biological systems improved the easiness of expressing basic biological principles considerably. The most representative languages are the κ-calculus (Danos and Laneve 2004) and beta-binders (Priami and Quaglia 2004, 2005) because they are equipped with computational platforms to execute specifications and are innovative with respect to many other extensions of process calculi. The main step ahead is the clear 1:1 correspondence between biological entities and objects of these calculi (see ❷ *Fig. 4*). Although an entity can occur in several different states, this

□ **Fig. 4**

κ-calculus (a) and beta-binders (b) maintain the identity of biological entities by relying on boxes with interfaces that describe the interaction capabilities of the modeled entities. Besides the 1:1 correspondence between biological components and boxes specified in the model, there is also a 1:1 correspondence between functional activities of domains and interfaces (binders). The interfaces of the boxes in the κ-calculus are identified by unique names written within the boxes. Interfaces can be either bound (s_2, s_3), visible (s_4) or hidden (s_1). Protein complexes are formed by joining interfaces through arcs. The rule-based manipulation of systems simply creates, deletes, and changes links, thus manipulating complexes. The interfaces of beta-binders boxes have unique names that act as binders for their occurrences within the box (x, y). The names of the interfaces are equipped with types (T, U) over which a compatibility of interaction is defined, thus relaxing the exact complementarity of actions over the same channel name to allow interactions. The boxes of beta-binders contain a π-calculus like description of their internal behavior. Primitives to manipulate the state of the interfaces that can be active/visible or hidden are provided (see the *hide* primitive in the figure that makes a visible interface hidden).

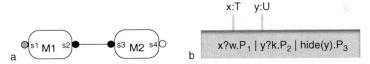

generation of calculi introduces well-defined syntactic boundaries that allow one to group all the states of an entity into single objects. This makes it easier to count and trace the behavior of the modeled entities. Furthermore, the whole set of states through which an entity can pass are implicitly encoded in the program describing the internal behavior of objects, rather than being all listed explicitly and represented with different placeholders, for example, variables in ODEs. Finally, note that other calculi have been defined as well (Cardelli 2005; Regev et al. 2004).

κ-calculus is a formal, rule-based calculus of protein interactions. It was conceived to represent complexation and decomplexation of proteins, using the concept of shared names to represent bonds. The units of κ-calculus are proteins, and operators are meant to represent creation and division of protein complexes. Once the initial system has been specified and the basic reductions have been fixed, the behavior of the system is obtained by rewriting it. This kind of reduction resembles pathway activation. Although this calculus has the merit of being directly inspired by biology, it does not offer a natural support for managing the evolution of compartments. Furthermore, the whole set of possible reactions must be specified since the beginning, so that the predictive power of the models is limited as in the previous generation of calculi.

Beta-binders are strongly inspired by both π-calculus and biological examples. They associate biological entities with boxes. Interaction is still communication-based and can occur either within a box if the same channel name is used or between different boxes if they have compatible interfaces (even if they have different names and hence the interaction is over different channel names). In fact, typed interfaces of boxes enable promiscuity of interaction (sensitivity/affinity versus complementarity). For an example, please see ❱ *Fig. 5*. Note that types here do not play the same role played by names in π-calculus. In fact, the same type can be compatible with many different types, thus allowing interaction over different channels and between different boxes.

◘ **Fig. 5**

The communication between boxes in beta-binders occurs when there are interfaces with compatible types over which one box is willing to output a value and another box is willing to receive a value. For instance, the box A is willing to receive both on its interfaces $x : T$ and $y : U$, while the box B is willing to send a value v over its interface $z : V$. In a separate file, the compatibility between types is specified by inserting a tuple for any pair of types one wants to be compatible. For instance, the tuple $(U, V, r1)$ means that interfaces with types U and V can interact, and the kinetic of the interaction is specified by a rate $r1$. According to the specified compatibilities and the actions enabled over the corresponding interfaces, the two transitions depicted in the figure can occur independently of the channel names used to specify the send and receive operations. Since the compatibility also specifies the rate of the interaction, the actual rate of the transitions is computed taking into account the number of molecules available in the system ($\#A$ and $\#B$).

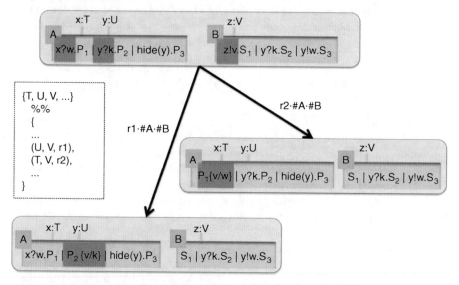

The enclosing surfaces (boxes) of entities cannot be nested and maintain a strict correspondence between processes and biological entities. Primitives to manipulate interfaces can be included in the description of the internal behavior of boxes. Finally, the physical rearrangement of boxes is implemented by defining completely general special split and join functions that operate on the structure of boxes and that are not inserted in the flow of control. This choice enables a higher level of nondeterminism with respect to any calculus defined so far, and better accommodates the specification of systems for which there is a considerable lack of knowledge. In fact, one does not need to completely specify, in the flow of control, the merging or division of boxes as it happens in other approaches, but one only needs to specify general conditions on the structure and on the status of the binders of boxes that enables the corresponding joining and splitting. Furthermore, the higher level of nondeterminism enhances the predictive power of models because the modeler does not have to specify the whole set of reactions from the beginning; they are instead computed by the run-time of the language. For an example, please see ❷ *Fig. 6*.

▢ Fig. 6
The figure shows an example of join of two boxes. The join function returns the list of binders of the new box as well as two renaming functions to change the names of the parallel composition of the original boxes to adapt them to the new binders. The function *comp* defines the affinity between the types T and U and the notation $\beta(x:T)$ denotes an interface (binder) named x with type T.

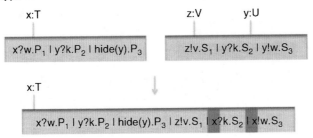

$$f_{join} = \lambda \mathcal{B}_1 \mathcal{B}_2 Q_1 Q_2 . \textbf{if}[\mathcal{B}_1 = \beta(x:T)\mathcal{B}_1^* \text{ and } \mathcal{B}_2 = \beta(y:U)\mathcal{B}_2^* \textbf{ and } comp(T,U)]$$
$$\textbf{then } (\mathcal{B}_1, \sigma_{id}, \{x/y\}) \textbf{ else nothing}$$

Although κ-calculus and beta-binders solve many of the problems identified at the end of the previous section for the first generation of calculi, they still suffer from some drawbacks listed below. (Here one uses the same categories used in the previous section, and those not listed are no longer considered an issue; the ones introduced here refer to issues raised by the second generation of calculi that were not present in the previous generation.)

Complex/Decomplex. κ-calculus provides the best support for manipulating complexes. The main limitation is that the rules for creating complexes and manipulating them must all be specified by the modeler on the basis of the structure of the boxes. No support is provided by the implementation of the language to automatically compute the steps to be performed for this fundamental mechanism of biological interaction.

Low-level programming. Almost nothing is changed on this item with respect to the previous generation of calculi. The only improvement is that the interaction network can be implicitly modeled in beta-binders by adequately defining the compatibility function between types.

Interaction. The key-lock mechanism of interaction is relaxed in beta-binders, while it is hidden in the complexation and decomplexation rules of κ-calculus.

Incremental model building. The identification of entities with boxes in both calculi makes the extension of models easier because the parts of the specifications to be changed are clearly identified.

Implementation. The association between boxes and biological entities simplifies the implementation of the stochastic simulation algorithms. However, the efficiency is still not acceptable and no parallel implementation is available. Furthermore, the generality of split and join functions of beta-binders poses further difficulties to efficiently implement the calculus.

Qualitative versus quantitative descriptions. The most enhanced calculus of the second generation are beta-binders that decouple qualitative descriptions from quantitative ones. In fact, rates are no more merged with the qualitative descriptions, but are externally specified through the compatibility function between the types of the interfaces (see also ❷ *Fig. 5*). However, complex kinetics function cannot yet be incorporated within the models.

◘ **Fig. 7**

The box enclosing Q acts as a sub-compartment of the other box in the figure. In fact, the binder x can only interact with the binder w assuming that the type v does not occur in any other place in the system. Therefore, the box enclosing Q can only interact internally or with its father compartment through x.

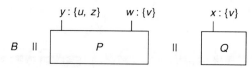

Space. Some preliminary attempt to code compartments and hence implicitly handle localization of molecules is emerging, but very limited computational support is available. For an example, see ❷ *Fig. 7.*

Connection with standards. The situation is still the same as the previous generation of calculi. Only limited activities have been performed (Ciocchetta et al. 2008b).

Predictive models. Something in this direction is obtained by beta-binders as a result of the higher level of nondeterminism within this calculus. However, the predictive capabilities of models specified through calculi are not yet satisfactory. In other words, one is only slightly ahead of ODEs.

Although the second generation of calculi improve the usage and scalability of the models, some relevant issues remain unresolved and no interoperability with mathematical modeling, (for example, ODEs) is available.

2.3 From Theoretical Calculi to Practical Modeling Languages

The very last step in the design of modeling formalisms is moving from theoretical calculi to real programming languages designed for biology. The most relevant platforms emerged from the two calculi discussed in the previous section. The BlenX (Dematté et al. 2008b) language inspired by beta-binders is considered here as an example, because it also incorporates the complex/decomplex features of κ-calculus. BlenX collects all features now available within modeling environments based on process calculi (see ❷ *Fig. 8*). Alternative efforts of BioPEPA (Ciocchetta and Hillston 2008) and BIOCHAM (see http://contraintes.inria.fr/BIOCHAM/) are also mentioned.

The driving principle to move from theoretical descriptions to practical modeling languages is sure to enrich the syntax of the calculi to facilitate the modeling process. This step must be performed by separating concerns as much as possible—that is, by keeping qualitative and quantitative descriptions distinct as well as using different syntactic categories and semantic actions to represent different fundamental biological principles. Finally, the run-time of the language should take into account, as much as possible, the basic dynamic principles of biological systems, like complexation and decomplexation. It would be useful to specify the minimum information that is needed to infer the dynamic interactions between entities.

The dynamic behavior of BlenX models is specified through three classes of actions:

Monomolecular: actions that describe the evolution of single entities (see ❷ *Fig. 9*)

Bimolecular: actions that involve two or more entities (see ❷ *Fig. 5*)

Events: global rewriting rules of the environment that manipulate a set of entities as well as their structure (see ❷ *Fig. 10*)

□ Fig. 8

BlenX is still based on boxes that are associated with biological entities. Boxes have typed and uniquely named interfaces and their internal behavior is specified similarly to beta-binders, although a richer set of primitives to manipulate interfaces is available and the restriction operator is no longer present, thus simplifying the theoretical development. The dynamics of systems are specified through three classes of actions: monomolecular, bimolecular, and events. Monomolecular actions affect a single box and can be either internal communications or internal primitives like *die* to destroy a box or interface manipulations. Bimolecular actions affect two or more boxes and can be either communications between boxes or complex/decomplex operations. Events are actions that are enabled when global conditions on the structure and cardinality of boxes are satisfied.

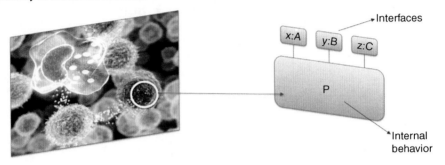

□ Fig. 9

BlenX provides many primitives to manipulate interfaces. Represented in the figure is the *expose* that creates a new interface on a box by ensuring that the name *x* is new (note the renaming to avoid clashes), the complementary actions of hiding and unhiding interfaces. There is also a primitive to let time pass without performing anything significant (*delay*) even if the time causes the choice between the alternatives to be performed. There are other two primitives not represented in the figure: *change(x:W)*, which alters the type of an interface named *x* to *W*, and *die*, which kills a box.

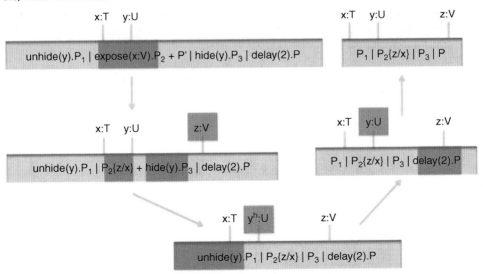

□ **Fig. 10**

Events in BlenX resemble conditional term rewriting systems. A condition on the structure of the system, on time, on the step of the simulation, or on the quantities of the species is tested after the keyword *when*. If the condition is satisfied, the event specified after the condition is enabled and participates in the selection of the next action through the stochastic simulation algorithm. The execution of a *new* or a *delete* event introduces or removes as many copies of boxes equal to the one in the condition as specified by their arguments (see first event in the figure for a *new* example). The execution of a *join* or a *split* event removes the boxes specified in the condition from the system and introduces the ones specified in the event action. For instance, the second event in the figure removes one *Q* and one *R* and introduces one box *P*. Conditions are also equipped with kinetic information. Finally, an *update* event action is provided to assign values to continuous or discrete variables associated with states of the system.

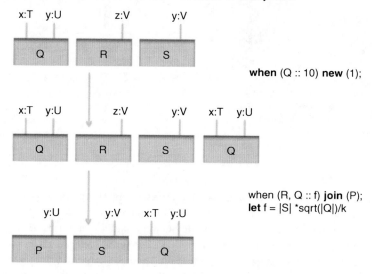

Events are very powerful modeling tools that also enable the execution of perturbative experiments over the models as is usually done in a laboratory. It can be specified that whenever a global condition, possibly time-dependent, is satisfied, a gene can be knocked-out, a drug injected in the system to study its interactions, or some components removed to see how they affect the overall behavior of the systems or, analogously, how robust a system is. Note that any event is associated with a rate and enters the race with all the other actions for selection by the simulation algorithm. Therefore, even if specified differently from the other interaction mechanisms, an event is a normal action of the system that does require modifications of the stochastic engine.

Note that the second event in ❷ *Fig. 10* is specified by a general mathematical function that does not only depend on the concentrations of the boxes affected by the event, but also on the concentrations of other components. Actually BlenX permits the specification of general kinetics by relying on mathematical functions. This is a useful generalization because it easily implements an interoperability with ODEs by translating them into BlenX models automatically. In this case, boxes coincide with variables and the dynamics emerge by simply copying the ODEs into functions associated with a very limited class of special events (see Mura et al. (2009)). As a consequence, the BlenX modeling approach can also handle smoothly hybrid situations in which continuous and discrete aspects are fused. Finally, note that if one uses general distribution functions, the exactness of Gillespie's stochastic simulation algorithm is lost. However, this should not be a concern because most of Gillespie's hypotheses are not satisfied, being, for example, the interior of a cell crowded, not homogeneous and not well-stirred. Actually, the design of a good simulation algorithm for biological systems is an open issue.

Interoperability with mathematical modeling is also ensured in BlenX by the possibility of associating states of the system with discrete and continuous mathematical variables. These variables can be manipulated by events whose conditions are on the values of the state variables and the event action *update* provides the values for the new assignments. This feature can be used, for instance, to easily change the rate of a reaction dynamically depending on some conditions over time, space, or the structure of the system. The same can apply to the reaction volume that can change dynamically (e.g., cell growth).

BlenX includes the management of complexes in a way that specific actions of formation or destruction do not need to be specified (see ❷ *Fig. 11*). Because the formation of complexes and their destruction is a very primitive mechanism of interaction in biological systems, BlenX provides it in its runtime support so that the modeler can think of the basic biological components and their sensitivity without the need to program the interactions in detail. The positive result is an easier modeling approach on one side and a better predictive power of the models on the other. Here, predictive power refers to the set of observable behaviors by executing the systems. In fact, many actions are determined by the runtime of the language according to the compatibility of types and structure of boxes and to binders status. These conditions are dynamically checked and produce potential transitions at runtime that have

◘ **Fig. 11**

In BlenX, it is not necessary to specify the interactions that form or destroy complexes. It is still the affinity between interfaces that determines whether a couple of boxes can complex or not. The parameters for the dynamics of complex formation are specified again in the separate file expressing compatibility. For instance, the tuple (U, V, r_{on}, r_{off}) in the figure means that two boxes having interfaces with types U and V can complex with a rate r_{on} and, once complexed, can detach each other with a rate r_{off}. The stochastic simulation algorithms also consider, at any step, the whole set of possible complexation and decomplexation among the enabled actions. As far as species are concerned, one should note that once two boxes are complexed, they form a new species in which a link between the interfaces complexed is created and visualized (the dashed red line in the figure).

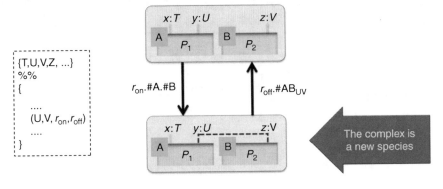

not been coded by the programmer in the flow of control. The degree of freedom added to the runtime of the language increases the potential of observing behaviors, not considered at modeling time.

A BlenX model is made up of a static description of the system obtained by listing the entities of the initial configuration with their amount and specificity for interaction. Then, the execution of the model and the support of the language are in charge of determining, at any step, the possible actions to be performed. This approach enables a drag-and-drop library-based modeling: the modeler simply selects the components that must be considered in the system and then defines the sensitivity of interaction between them. After this step, only a little is needed to equip the boxes with their relevant internal behavior, if particular actions need to be implemented. The BlenX modeling process described above allows modelers to build scalable, modular, and compositional models. Finally, all the combinatorial effects are ruled out at modeling level, because the very same box represents all the possible states through which the corresponding biological entity can pass.

Almost all the concerns expressed in the previous sections for the first two generations of calculi for biology have been greatly reduced by the BlenX language. Some efforts are still ongoing to simplify the management of space and to improve the efficiency of the execution of systems both by parallelizing algorithms (Dematté and Mazza 2008) and by improving the efficiency of them, once contextual information is available (Kuwahara and Mura 2008) (❯ *Fig. 12*).

◻ **Fig. 12**

The figure summarizes the concerns and the characteristics of the calculi of the three generations described in the paper.

	1st generation π- calculus	2nd generation Beta-binders	3rd generation BlenX
Entities	Represented by processes, it is difficult to trace their identity	Encapsulated in boxes and hence easy to trace and count	Encapsulated in boxes and hence easy to trace and count
Complexes	Implicitly represented by scope of names and tricky manipulated through the restriction operator	Explicitly represented through the join/split of boxes or through dedicated links	Improved manipulation of complexes with conditions on their structure and state
Low-level programming	Thought for computing systems, all biological events must be encoded in the available primitives	Thought for biology, primitives closer to biological processes makes model simpler	Enriched and more natural set of primitives. A real programming language
Interaction	Exact complementarity of channel names	Compatibility of the types of the interfaces of boxes	Compatibility of the types of the interfaces of boxes
Incremental model building	Difficult: interconnection structure and affinity of entities hard-coded in the model.	Medium: affinity separated from model behavior and boxes allow easy identification of the connections between models	High: Affinity and complex formation separated from model behavior and handled through types. Boxes and different syntactic components identify connections between models
Implementation	Gillespie algorithm, counting through structural congruence	Gillespie algorithm, counting through structural congruence	Many simulation algorithms, efficient counting due to the design of the language
Qualitative versus quantitative	Exponential distributions with rates associated to channel names	Exponential distributions with rates associated both to typed interfaces and channel names	General distributions, rates as general functions, hybrid models with continuous variables
Space	Extremely limited support	Space can be encoded through tricks	Preliminary notions of space
Standards	Very limited import/export capabilities	Some import/export capabilities	Connection with SBML and ODE
Predictive power	Every interaction must be programmed	Many interactions are inferred by the execution engine	Many interactions are inferred by the execution engine

3 Knowledge Inference

Many efforts have been made toward the refinement of the modeling languages, and a reasonable level of easiness of use and generality to address most of the relevant biological problems has been reached. However, in order to execute these models, kinetic information is needed, which is often difficult to find. In particular, rates of interactions can be inferred from literature or database mining and sometimes directly by manipulating the outcome of wet laboratory experiments (e.g., time-course concentrations from microarrays or outcomes of NMR analyses). Therefore, it is essential, for a conceptual framework that aims at supporting biological research, to cover inference activities, both from a theoretical and computational perspective. Kinetics information inference is referred to here, although network structure inference is an important field as well. The tight integration of inference capabilities with a modeling framework is absolutely mandatory to make the framework practical. ❯ *Figure 13* shows how to integrate the inference of knowledge within a modeling cycle.

Here, one concentrates on model calibration (parameter estimation) since this is the most critical task in building an executable stochastic model following the algorithmic systems biology approach. Interaction networks can be estimated as well, but much more information on this aspect is available through various interactome projects and databases.

◘ Fig. 13

A biological problem is usually modeled by describing the qualitative behavior constrained by quantitative information used to drive the dynamics. Starting from wet laboratory experiments, one needs to infer the parameters governing the kinetics of the reactions as well as the topology of the interaction network of the entities of the system at hand. By varying the parameters and the interaction network, one obtains a class of different models for the same phenomenon and one needs to discriminate between them according to some phenotypical fitness or statistic measures that allows one to rank the relevance of parameters. Once the main parameters are identified, it is possible to provide hints on the design of new experiments to gain the minimum knowledge one needs from the lab, in order to maximize the discrimination between the models. (Figure prepared by Alida Palmisano.)

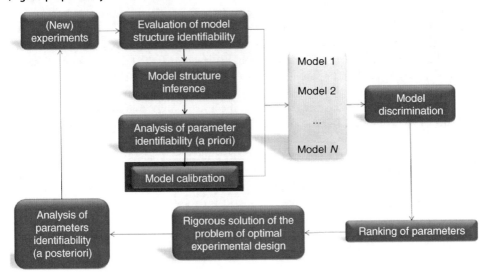

Recent literature reports many examples of methodologies of parameter estimation both in deterministic and stochastic models, including Polisetty et al. (2006), Moles et al. (2003), Tian et al. (2007), and Chou et al. (2006), for example. Tools for parameter fitting, through regression or maximum likelihood methods, can be found as an integral part of simulation tools (e.g., Copasi (Hoops et al. 2006)), but there also exists stand-alone software, exclusively designed for that purpose, such as Splindid (Bashi et al. 2005) and PET (http://mpf.biol.vt.edu/pet/contact.php). Finally, Boys et al. (2008) and Golightly and Wilkinson (2008) developed Bayesian model-based inference techniques. Bayesian schemes offer some advantages over the maximum likelihood methods, such as when the volume of data is limited or the analytic form of the kinetic model makes the maximization of the likelihood difficult.

Most of the current tools for parameter estimation lack robustness to the noise as well as the absence of any estimates of experimental error in their outcome. Experimental uncertainties on parameters propagate from the measurements of the concentrations of the species. Inferring the parameters with an estimate of their uncertainty is essential if one wants to use the tool in the context of optimal experimental design. Moreover, most of the current tools based on optimization techniques suffer from the problem of univocally finding the solution by global optimization, and ask the user to provide a priori the optimization algorithm with the region of parameter space in which to perform the search for the global max/minimum.

A new approach has been recently proposed that enhances the state of the art in parameter estimation by relying on a completely new mathematical method (Lecca et al. 2009). The method is based on a probabilistic, generative model of the variations in reactant concentrations. The time-series concentrations of n reactant species are collected in n state vectors $x_1,\ldots,x_n$. The method discretizes the law of mass action and provides a tool to predict the values of the variables x_i at time t, conditioned by their values at the previous time point. The variations of the concentration of the species, at different time points, are conditionally independent by the Markov nature of the discrete model of the law of mass action. Assuming the observation noise to be Gaussian with variance σ^2, the probability of observing a variation for the concentration of species, between time t_{k-1} and t_k, is a Gaussian with variance depending on σ and the expectation value of the law of mass action under the noise distribution. The discretization of the law of mass action provides a model for the variations of species concentrations, rather than a model for the time-trajectory of species concentrations. This makes the evaluation of the expectation value of the law of mass action function more easily and analytically tractable. The rate coefficients and the level of noise are then obtained by maximizing the likelihood function defined by the observed variations.

This method produces the rate coefficients, the level of noise, and an error range on the estimates of rate constants. Its probabilistic formulation is key to a principled handling of the noise inherent in biological data, and it allows for a number of further extensions, such as a fully Bayesian treatment of the parameter inference and automated model selection strategies based on the comparison between marginal likelihoods of different models.

The implementation of this method in the tool KInfer (see next section) is used as an interface tool, connecting the outcomes of the wet lab activity for the concentration measurements and the software for the simulation of chemical kinetics. Finally, KInfer can also be used as a validation tool for models. If given a model, it is not possible to infer kinetic parameters that allow the model to reproduce the observed behavior in wet labs, it means that something is missing from the model or something is wrong.

4 CoSBi Lab: An Algorithmic Systems Biology Platform

The conceptual framework of algorithmic systems biology can become effective only if a suitable computational support that assists all phases in the life cycle of the modeling, simulation, and analysis of biological systems is implemented. At the time of writing, the algorithmic systems biology platform that covers most of the issues is CoSBi Lab, illustrated in ❯ *Fig. 14*, where the components of the platform and their connections are highlighted. CoSBi Lab is a software platform that implements a new conceptual modeling, analysis, and simulation approach to quantitative, dynamic systems primarily inspired by algorithmic systems biology. CoSBi Lab is intended to become a complete artificial laboratory in which it is possible to replicate, in-silico, all the activities that are usually performed in real wet labs. CoSBi Lab is centered around the new programming language, BlenX, that is the core of the formal modeling facilities. BlenX is directly derived from process calculi and hence has concurrency as a primitive feature that helps simplify models to avoid the combinatorial explosion of variables. CoSBi Lab

◻ **Fig. 14**

CoSBi Lab, the algorithmic systems biology platform in the figure, provides support for quantitative model definition through both textual and graphical formalisms within the BetaWB set of tools. It allows for inference of kinetic rates via KInfer. The platform provides connection to simulators via the BetaWB, Cyto-sim, and Redi. Other kind of analyses on the models are allowed through exporting into Markov chains and SBML descriptions. Finally, Snazer provides support to visualize networks of reactions and annotate them. Snazer also implements a rich set of statistical analyses.

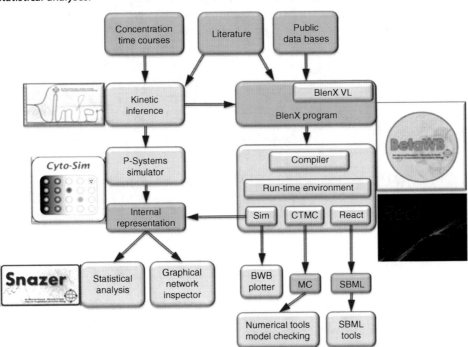

supports Gillespie-based simulations as well as spatial diffusion of entities. Export of models in the SBML format is allowed. CoSBi Lab implements inference procedures for quantitative parameters needed to feed the BlenX models and to drive the simulation engines. The inference starts directly from various sources of data, including the outcome of wet lab experiments that can be mapped (directly or through preprocessing transformations) into time courses of concentrations. CoSBi Lab offers a set of features to analyze the outcome of the experiments (simulations) that include the main statistical techniques as well as the visualization of networks of reactions and plots of concentrations.

The main algorithmic tools populating the platform are now described.

BetaWB (http://www.cosbi.eu/Rpty_Soft_BetaWB.php) (Dematté et al. 2008b) is a collection of tools based on the programming language BlenX, explicitly designed to represent biological entities and their interactions. The BetaWB includes the BetaWB simulator, a stochastic simulator based on an efficient variant of the Gillespie stochastic simulation algorithm (SSA); the BetaWB designer, a graphical editor for developing models; and the BetaWB plotter, a tool to analyze the results of a stochastic simulation run. Generation of Markov chains and export into SBML is allowed.

KInfer (http://www.cosbi.eu/Rpty_Soft_KInfer.php) is a tool for estimating rate constants of biochemical network models from concentration data measured, with error, at discrete time points. A schema of KInfer is reported in ❏ *Fig. 15.* KInfer is inspired by the maximum

❏ **Fig. 15**

KInfer starts from a time series concentration of entities and a list of reactions and automatically builds the mathematical model to be solved in order to estimate the rate constants describing the dynamics. Parameters are selected according to a fitting of the overall model toward some experimental observed data. KInfer also provides the user with a measure of the error performed in the calculation of the parameters, that is, the strength of the noise in the system that can be introduced by the experimental production of the time series. (Figure prepared by Alida Palmisano.)

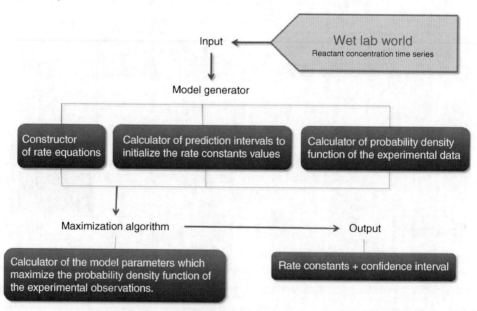

likelihood estimation and assumes a discretized version of the law of mass action as a rate equation. The discretization of the rate equation makes the evaluation of its expectation value analytically tractable. The probabilistic formulation of the KInfer algorithm guarantees the noise-robustness and the possibility of extending it to a Bayesian treatment of the parameters. The principal features of the tools are:

1. Automatic generation of a generalized mass action model
2. Automatic estimation of the initial guesses and bounds for the parameter values
3. Estimation of the experimental errors on the inferred parameters
4. Estimation of the strength of noise in the input data

Redi (http://www.cosbi.eu/Rpty_Soft_Redi.php) (Lecca et al. 2008) is a reaction-diffusion simulator, built to test new diffusion models and algorithms. Redi simulates biochemical systems at the mesoscopic scale of interaction, employing a space discretized variant of the Gillespie SSA. Diffusion coefficients are not fixed, but are computed dynamically in a state-dependent way, inspired by the Maxwell–Stefan model of transport phenomena.

Cyto-Sim (http://www.cosbi.eu/Rpty_Soft_CytoSim.php) (Sedwards and Mazza 2007) is a stochastic simulator of biochemical processes in hierarchical compartments based on P-systems. The compartments may be isolated or may communicate via peripheral and integral membrane proteins. The native syntax is designed to be a compact and intuitive way of describing chemical systems. Arbitrary kinetic rate functions are permitted, allowing seamless import and export to SBML. Export to Matlab is also facilitated.

Snazer (http://www.cosbi.eu/Rpty_Soft_Snazer.php) (Mazza et al. 2010) is a modular tool designed to aid the processes of visualizing and manipulating reactive models, as well as to share and interpret time-course data produced by stochastic simulators or by any other means. Snazer upgrades the viewer of Beta Workbench and interfaces its output format. It loads biological networks encoded in SBML as well, and stores graph layouts in standard GraphML files. Moreover, it loads time-course data exported in CSV, and also compresses and prepares them for remote sharing and statistical processing. Finally, to enhance interoperability, it has been equipped with a public, XML-based schema to allow other tools both to encapsulate and structure their time-course data and to package their interaction networks, if any. A number of features are provided:

1. Import from Beta Workbench output, CSV, SBML, and (proprietary) XML file formats
2. Export to CSV, GraphML, and (proprietary) XML file formats
3. Interacting view of the chemical reactive networks
4. Color auto-tuning for color-blind users
5. Node importance highlighting (degree distribution)
6. Custom and MIRIAM-compliant annotation (see http://www.ebi.ac.uk/miriam/) support
7. Statistical analysis of simulated multi-traces
8. Statistical outcome export in support vector graphics (SVG)

The possible workflow depicted in ❷ *Fig. 16* highlights a relevant aspect of in-silico laboratories based on algorithmic descriptions of the dynamics of systems produced through computing languages. Most biological research is mainly driven by the availability of reactants, by technical capabilities of performing specific perturbation to the modeling organisms, and by the time needed to get the outcome of an experiment. In-silico replica of biological systems can instead exploit the power of algorithmic approaches to speed up experiments and to make thousands of them at a very low cost. The comparison of these in-silico experiments can

◘ **Fig. 16**

A typical workflow supported by the algorithmic systems biology platform starts from the collection of the available biological knowledge and the inference from it of the relevant parameters through KInfer. The outcome of this step is used to feed the quantitative part of the models within BetaWB, that together with the algorithmic description of the behavior allows the simulation of the dynamics of the system. The result of the simulation can be inspected by Snazer and can suggest modifications to the model or new experiments to acquire new knowledge by iterating the cycle. Eventually this process can lead to the discovery of new biological insights. (Figure prepared by Ivan Mura.)

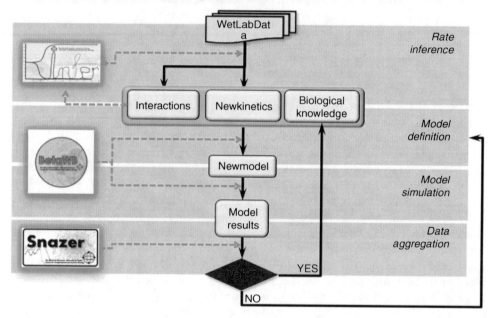

then drive the optimal design of new wet experiments. To make the approach described above practical, it is necessary to interpret the outcome of the execution of a model exactly as the outcome of a wet experiment and to make the same kind of inference on it that is usually done on the output of the bench. This is why the statistical analyses implemented in Snazer, directly on the outcome of multi-run simulations, provide an important added value to the algorithmic systems biology platform described so far.

This section concludes by noting that the described platform is the only integrated environment based on process calculi available. Other implementations are limited to simulators and various graphical tools to visualize the outcome of simulations. The first implementation of a biochemical stochastic π-calculus simulator was BIOSPI (Priami et al. 2001), followed by SPiM (Phillips and Cardelli 2007). A stochastic κ-calculus simulator is described elsewhere (Danos et al. 2007). The Kappa Factory (http://www.lix.polytechnique./fr/krivine/kappaFactory.html) is a graphical platform for the design, analysis, and simulation of biomolecular systems. Different kinds of analysis methods are supported, for example, static dependencies on rules and analysis of traces. Recently, an effort to integrate stochastic simulation with other analyses techniques, like model checking, has been done for BioPEPA (Ciocchetta et al. 2008a).

5   Conclusions

The integration of computer science and systems biology into algorithmic systems biology is a win-win strategy that impacts both disciplines from a scientific and technological perspective (❯ *Fig. 17*).

The scientific impact of accomplishing the vision and the feedback from the biological understanding will be the definition of new quantitative theoretical frameworks that address the challenge of overcoming the increasing concurrency and complexity that is observed in the asynchronous, heterogeneous, and decentralized (natural/artificial) systems in a verifiable, modular, incremental, and composable way. Furthermore, the definition of novel quantitative

◻ **Fig. 17**
Algorithmic systems biology can propel both biological and computer science research. The conceptual and computational framework developed to enhance biological capabilities of understanding basic principles allows biologists to address new and challenging questions (*left arrow*) whose answers provide the requirements for developing new and better computer science frameworks (*right arrow*). At the same time, the environment developed to address the biological challenges can turn computer science into a more quantitative science and can enable simulation-based research in all fields where the dynamics are driven by the interaction of entities (economic markets, social networks, etc.).

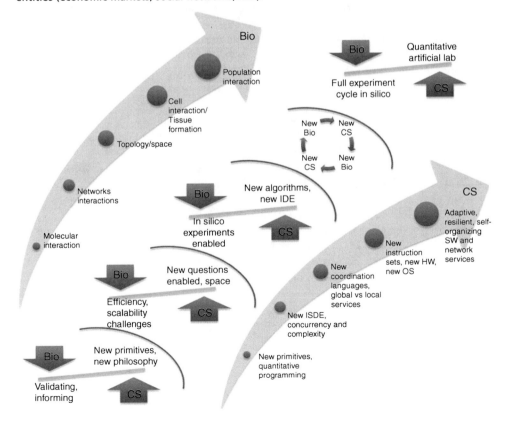

coordination and orchestration mechanisms of the loosely coupled components of the kind of complex systems being studied will produce new conceptual frameworks to cope with the growing paradigm of distributing the logics of applications between local software and global services. The definition of new logical and conceptual schemas to store data related to the dynamics of systems and the new query languages needed to retrieve and examine this new kind of records will create completely new perspectives to build the increasingly valuable data centers that provide added value to global services.

Another major scientific impact will be the definition of a new philosophical foundation of systems biology, which is algorithmic in nature and allows scientists to raise new questions that are out of reach for the current conceptual and computational support. An example of this would be the cross-talks (interaction) between pathways (cascades of interactions of biological entities). Actually the current interpretation of pathways (that do not exist per se in nature) is a reductionist approach (imposed by technological limitations) with respect to the system-level understanding of network behavior (a collection of interwoven pathways working simultaneously and interacting).

The technological impact of accomplishing the full merging of computer science and systems biology will be the design and implementation of artificial biology laboratories that allow many more experiments than the ones currently feasible in wet labs, at lower costs in terms of human and financial resources as well as time. These labs will allow biologists to design, execute, and analyze experiments to generate new hypotheses in a shorter time than needed in wet labs. The new findings in hierarchical networks of biological elements (molecules, cells, organs, and organisms) inform new technology actions coordinated by research requirements to develop novel, massive high-throughput tools. These actions should mainly address advances in experimental design, documentation, and interpretation as well as a deeper integration between dry and wet research. Furthermore, the artificial laboratories will be a main technology to move from single gene diseases to multifactorial diseases that are more than 90% of those affecting society. A deeper look at the causes of multifactorial diseases can positively impact their diagnosis and management. Health is not the only practical application of algorithmic systems biology; the comprehension of the basic mechanisms of life coupled with engineered design environments for synthetic biology will lead one toward the use of ad hoc bacteria to repair environmental damages as well as to produce energy.

Another major technological impact will be allowing computer scientists to properly address and master the challenges originated by the new HW design, which mainly exploit parallelism via many/multi-core architectures rather than the speed of single processors, through new integrated programming environments suitable to address concurrency and complexity.

Unraveling the very basic mechanisms adopted by living organisms to compute and manipulate information leads to the heart of computer science: computability. Life underwent billions of years of tests and was optimized during this time, and thus one can learn new computational paradigms to enhance the field. The same arguments apply to hardware architectures as well. Starting from the very basics, one can further build on top of them to enhance resource management and hence operating systems, primitives to instruct highly parallel systems and consequently (concurrent) programming languages, and also software development environments that ensure higher quality and better properties than current software applications. As a consequence, algorithmic systems biology can contribute to the future of computer science.

Algorithmic systems biology can also contribute to the future of natural and life sciences through connecting models and experiments by means of new conceptual and computational tools integrated in a user-friendly environment equipped with templates of major biological components for drag-and-drop modeling of (artificial) organisms or populations and used by many life scientists to predict the behavior of multi-level, multi-scale biological systems in a modular, compositional, scalable, and executable manner.

Acknowledgments

The author thanks the CoSBi team for many inspiring discussions.

References

Bashi K, Forrest A, Ramanathan M (2005) SPLINDID: a semi-parametric, model-based method for obtaining transcription rates and gene regulation parameters from genomic and proteomic expression profiles. Bioinformatics 21(20):3873–3879

Bergstra J, Ponse A, Smolka S (eds.) (2001) Handbook of process algebras. Elsevier, Amsterdam, The Netherlands

Bernardo M, Degano P, Zavattaro G (eds) (2008) Formal methods for computational systems biology. Lecture notes in computer science, vol 5016. Springer, New York

Boys RJ, Wilkinson DJ, Kirkwood TB (2008) Bayesian inference for a discretely observed stochastic kinetic model. In: Statistics and computing. Springer, The Netherlands

Cardelli L (2005) Brane calculi-interactions of biological membranes. In: CMSB 2004, Paris, France, May 2004. Lecture notes in computer science, vol 3082. Springer, Berlin, pp 257–280

Cardelli L, Zavattaro G (2008) On the computational power of biochemistry. In: AB08, Austria, July–August 2008

Cassman M, Arkin A, Doyle F, Katagiri F, Lauffenburg D, Stokes C (2005) International research and development in systems biology. WTEC Panel on Systems Biology final report

Chou IC, Martens H, Voit EO (2006) Parameter estimation in biochemical systems models with alternating regression. Theor Biol Med Model 3(25)

Ciocchetta F, Hillston J (2008) Calculi for biological systems. In: SFM-08, Bertinoro, Italy, June 2008. Lecture notes in computer science, vol 5016. Springer, pp 265–312

Ciocchetta F, Gilmore S, Guerriero M, Hillston J (2008a) Integrated simulation and model-checking for the analysis of biochemical systems. In: Proceedings of PASM 2008. Electr Notes Theor Comput Sci 238:17–38

Ciocchetta F, Priami C, Quaglia P (2008b) An automatic translation of SBML into beta-binders. IEEE/ACM Trans Comput Biol Bioinform 5(1):80–90

Danos V, Laneve C (2004) Formal molecular biology. TCS 325(1):69–110

Danos V, Feret J, Fontana W, Krivine J (2007) Scalable simulation of cellular signaling networks. In: Proceedings of APLAS'07, Singapore, November–December 2007

de Alfaro L, Henzinger TA, Jhala R (2001) Compositional methods for probabilistic systems. In: CONCUR 2001, Aalborg, Denmark, August 2001. Lecture notes in computer science, vol 2154

Dematté L, Mazza T (2008) On parallel stochastic simulation of diffusive systems. In: Proceedings of CMSB2008, Rostock, Germany, October 2008. Lecture notes in computer science, vol 5307. Springer, pp 191–210

Dematté L, Priami C, Romanel A (2008a) The beta workbench: a tool to study the dynamics of biological systems. Brief Bioinform 9(5):437–449

Dematté L, Priami C, Romanel A (2008b) The BlenX language: a tutorial. In: Formal methods for computational systems biology, SFM 2008, Bertinoro, Italy, June 2008. Lecture notes in computer science, vol 5016. Springer, pp 313–365

Denning P (2007) Computing is a natural science. Commun ACM 50(7):13–18

Eccher C, Priami C (2006) Design and implementation of a tool for translating sbml into the biochemical stochastic π-calculus. Bioinformatics 22(24): 3075–3081

Golightly A, Wilkinson DJ (2008) Bayesian inference for nonlinear multivariate diffusion models observed with error. Comput Stat Data anal 52(3):1674–1693

Guerriero M, Prandi D, Priami C, Quaglia P (2008) Process calculi abstractions for biology. In: Condon A, Harel D, Kok J, Salomaa A, Winfree E (eds) Algorithmic bioprocesses. Springer, Berlin

Gutowitz H (1990) Introduction (to cellular automata). Physica D 45:vii

Harel D (2007) Statecharts in the making: a personal account. In: Proceedings of 3rd ACM SIGPLAN history of programming languages conference (HOPL III), San Diego, CA, June 2007

Hood L, Galas D (2003) The digital code of DNA. Nature 421:444–448

Hoops S, Sahle S, Gauges R, Lee C, Pahle J, Simus N, et al. (2006) COPASI – a COmplex PAthway SImulator. Bioinformatics 22:3067–3074

Kitano H (2002) Systems biology: a brief overview. Science 295:1662–1664

Kuwahara H, Mura I (2008) An efficient and exact stochastic simulation method to analyze rare events in biochemical systems. J Chem Phys 129 (16):165101

Lecca P, Dematté L, Priami C (2008) Modeling and simulating reaction-diffusion systems with state-dependent diffusion coefficients. In: Proceedings of world academy of science, engineering and technology – international conference on bioinformatics and biomedicine, vol 35

Lecca P, Palmisano A, Priami C, Sanguinetti G (2009) A new probabilistic generative model of parameter inference in biochemical networks. In: Proceedings of the 2009 ACM Symposium on Applied Computing, Honolulu, HI, March 2009

Mazza T, Iaccarino G, Priami C (2010) Snazer: the simulations and networks analyzer. BMC Syst Biol 4:1

Moles GC, Mendes P, Banga JR (2003) Parameter estimation in biochemical pathways: a comparison of global optimization methods. Genome Res 13:2467–2474

Mura I, Palmisano A, Priami C (2009) From ODEs to language-based, executable models of biological systems. In: PSB09: Proceedings of the Pacific symposium on biocomputing, Kohala Coast, HI, January 2009

Phillips A, Cardelli L (2007) Efficient, correct simulation of biological processes in stochastic pi-calculus. In: Proceedings of CMSB07, Edinburgh, Scotland, September 2007. Lecture notes in computer science, vol 4695. Springer, pp 184–199

Polisetty PK, Voit EO, Gatzke EP (2006) Identification of metabolic system parameters using global optimization methods. Theor Biol Med Model 3(4):1–15

Priami C (1995) Stochastic π-calclus. Comput J 38 (6):578–589

Priami C (2009) Algorithmic systems biology. Communications of the ACM 52(5):80–88

Priami C, Quaglia P (2004) Modeling the dynamics of biosystems. Brief Bioinform 5:259–269

Priami C, Quaglia P (2005) Beta binders for biological interactions. In: Proceedings of CMSB04, Paris, France, May 2004. Lecture notes in bioinformatics, vol 3082. Springer, pp 21–34

Priami C, Regev A, Shapiro E, Silvermann W (2001) Application of a stochastic name-passing calculus to representation and simulation of molecular processes. Inform Process Lett 80:25–31

Regev A, Shapiro E (2003) Cells as computation. Nature 419:343

Regev A, Panina EM, Silverman W, Cardelli L, Shapiro E (2004) BioAmbients: an abstraction for biological compartments. TCS 325(1):141–167

Roos D (2001) Bioinformatics – trying to swim in a sea of data. Science 291:1260–1261

Rosen R (1998) Essays on life itself. Columbia University Press, New York

Searls D (2002) The language of genes. Nature 420:211–217

Sedwards S, Mazza T (2007) Cyto-sim: a formal language model and stochastic simulator of membrane-enclosed biochemical processes. Bioinformatics 23 (20):2800–2802

Spengler S (2000) Bioinformatics in the information age. Science 287:1221–1223

The Kappa Factory. http://www.lix.polytechnique.fr/ ~krivine/kappaFactory.html

Tian T, Xu S, Burrage K (2007) Simulated maximum likelihood method for estimating kinetic rates in gene expression. Bioinformatics 23(1):84–91

Welch P, Barnes F (2005) Communicating mobile processes: introducing occam-pi. In: Proceedings of CSP25, London, July 2004. Lecture notes in computer science, vol 3525. Springer, pp 175–210

Wing J (2006) Computational thinking. Commun ACM 49(3):33–35

55 Process Calculi, Systems Biology and Artificial Chemistry

Pierpaolo Degano[1] · *Andrea Bracciali*[2]
[1]Dipartimento di Informatica, Università di Pisa, Italy
degano@di.unipi.it
[2]Department of Computing Science and Mathematics,
University of Stirling, UK
braccia@cs.stir.ac.uk

G. Rozenberg et al. (eds.), *Handbook of Natural Computing*, DOI 10.1007/978-3-540-92910-9_55,
© Springer-Verlag Berlin Heidelberg 2012

Abstract

Knowledge about life processes develops through the interplay of theoretical speculation and experimental investigation. Both speculation and experiments present several difficulties that call for the development of faithful and accessible abstract models of the phenomena investigated. Several theories and techniques born in computer science have been proposed for the development of models that rely on solid formal bases and allow virtual experiments to be carried out computationally *in silico*.

This chapter surveys the basics of *process calculi* and their applications to the modeling of biological phenomena at a system level. Process calculi were born within the theory of concurrency for describing and proving properties of distributed interacting systems. Their application to biological phenomena relies on an interpretation of systems as made of interacting components exhibiting a computational kind of behavior, "cells as computation."

The first seminal proposals and the subsequent enhancements for best adapting computer science theories to the domain of biology (with particular reference to chemical, biochemical, and cellular phenomena) are surveyed.

1 Introduction

Knowledge about life processes and ultimately about oneself as a human being develops through the interplay of theoretical speculation and experimental investigation. Experiments have high costs, as materials and techniques are not easily available, may be lengthy, have speed that is determined by the speed of natural processes, pose ethical issues (for example, because they may be invasive), and have available investigation techniques that are not always sufficient to study the phenomena of interest.

A long-standing way of addressing such difficult investigations is to revert to models of the studied problem. As in any other scientific field, a model focuses on the essence of what is studied, abstracting away the unnecessary details so as to reduce complexity. Furthermore, a rigorous, formal description of natural phenomena permits reasoning on them at an abstract level. This is crucial for assessing the adequacy of the proposed model, which can then be used for formally checking aspects and properties of the modeled objects. Particularly relevant for an assessed model is its prescriptive capability, which in the realm of life sciences means that new facts can be discovered and hypotheses tested without resorting to actual experiments. This reduces the costs of the experimental work and drives those tests that remain essential for demonstrating these findings.

The use of formal methods in biology has been advocated for some time. A seminal exploration appeared in Schrödinger's *"What is Life?"* (Schrödinger 1946), which contains *in nuce* a quest for a description of living matter in terms of the mathematics and the techniques used in physics. Biophysics is a fascinating and effective discipline, originally influenced by that book.

Beyond physics and mathematics, computer science also contributes to the development of biology, which becomes more and more a multi-disciplinary research field. On the one hand, computer science has traditionally provided mankind with the information technology infrastructure needed to store the vast amount of data we acquire about biological phenomena and to organize it, as well as with several methods and effective tools to search, retrieve, and reason on these data. On the other hand, more recently, computer science has been fostering the development of *executable* formal models of biology, for which several theories born in computer science are being exploited.

One of the most successful and widely known examples of this is the modeling of DNA as a string and the application of several string-based algorithms in the quest for DNA sequencing. A large body of research has been done on algorithms for mining recurrent patterns for a better understanding of the *structure* of the DNA double chain.

Broadly speaking, these results in representing and searching genomic structures follow from a reductionist approach, which focuses on the single aspects of the components of a biological system, namely the sequential structure of the DNA. However, these efforts seem inadequate for explaining the overall *functioning* of a complex system and the properties emerging from its behavior when considered in its dynamical evolution. Clearly, the complete knowledge of the DNA structure is not enough to describe its metabolic and regulatory functionalities, which rather depend on the interactions among genes, the proteins they code, and also the environment in which these interactions happen.

In the words of Kitano:

▶ To understand biology at the system level, we must examine the structure and dynamics of cellular and organismal function, rather than the characteristics of isolated parts of a cell or organism. . . . many breakthroughs in experimental devices, advanced software, and analytical methods are required. (Kitano 2002)

Such a systemic approach has been embraced by the recently developing *systems biology* research field.

This chapter focuses on the role that theories, models, and analysis techniques born in computer science may have within systems biology. More specifically, theories from the field of *concurrency*—that is, the study and analysis of the interactions happening within computer systems—are considered. This approach relies on interpreting biological systems as interactive computational systems. In both contexts, a potentially vast number of components (for example, biochemical molecules vs. software applications) cooperate together in a similar manner (for example, by chemical reactions vs. communications over a computer network) toward a common goal (for example, sustaining a cell vs. managing a flight control system). This approach was clearly suggested by Regev and Shapiro in their seminal paper (Regev and Shapiro 2002):

▶ We believe that computer science can provide the much-needed abstraction for biomolecular systems. . . .the 'molecule-as-computation' abstraction, in which a system of interacting molecular entities is described and modelled by a system of interacting computational entities. Abstract computer languages, . . . enable simulation of the behaviour of biomolecular systems, . . . supporting qualitative and quantitative reasoning on these systems' properties.

The mentioned abstract computer languages considered here are *process calculi*. In the form of an algebraic representation, they describe the dynamics of an interacting system. This representation is *formal*, that is, mathematically precise, and the modeler can choose the desired *level of abstraction*; it is largely *compositional*, that is, separate subparts of a system can easily be assembled together; it is *executable*, that is, it supports automated simulation; and it offers a repertoire of tools and techniques for a mechanical *verification* of properties.

Process calculi offer their users the possibility of representing both *qualitative* information, (for example, checking whether a metabolite is or is not produced by a cell) and *quantitative* information (for example, the stochastic time-course of specific metabolites along a pathway).

In recent years, various process calculi designed for modeling network distributed applications and their analysis tools have been used to describe the dynamics of biological systems (for example, biochemical pathways and even whole cells). However, it became clear that, as they are, these calculi lack descriptive power under several pragmatic aspects. Then, new process calculi were defined in order to properly take into account biological features, such as the specific structures or the complex interactions that occur within and on membranes and compartmentalized systems (Regev et al. 2004; Cardelli 2004; Priami and Quaglia 2004). These extensions also call for related enhancements in the formal theory that supports them, and so started a new research subfield across concurrency theory and life sciences. In particular, one has to rethink the way the dynamics of systems is specified (that is, their semantics), so as to comprise relevant biological aspects at different levels of detail (for example, the stochastic nature of various phenomena), starting from Hillston (1996) and Priami et al. (2004). This task is not easy at all, and also raises the formal question of the expressivity of the resulting calculi from a theoretical point of view, typically whether they express all the computable functions.

Beyond the mentioned features and despite its early stage, the process calculi for biology often are adequately descriptive and computationally efficient compared with other traditionally adopted modeling techniques, such as those based on the differential calculus. A crucial distinction is that a model in this approach turns out to be, roughly, a computer program that can be executed to *generate* a formal representation of the biological behavior, rather than *being* itself a description of the behavior, as offered, for example, by a system of differential equations. So, it might be easier to build the model, as well as to study the model under different conditions (for example, by varying and tuning parameters, looking for emergent properties, etc.) because each run can be seen as a *virtual* experiment.

Process calculi and the algebraic approach lie in a much bigger effort of defining formal (and executable) models of living matter and to reason on them, ultimately aiming at defining new formal languages for biology. References are only provided for some of these approaches, as completeness would be impossible. Among the various rule-based formalisms presented in the literature, one can find the λ-calculus (Fontana and Buss 1994), Biocham (Calzone et al. 2006), the κ-calculus (Danos et al. 2007), Moleculizer (Larry and Roger 2005), the pathway logic (Eker et al. 2002), the calculus of looping sequences (Milazzo 2008), and closer to formal languages, P-systems (Paun et al. 2007). Formalisms introduced for reactive systems have also been used to model biological entities (Damn and Harel 2001; Sadot et al. 2008). Petri nets offer a graphical framework for formally describing biological networks, and have been widely used (Reddy et al. 1993). The surveys collected in Bernardo et al. (2008) can help to complete the vast picture of the numerous contributions to this field, which cannot be properly summarized in the limits of this chapter.

Synopsis. Process calculi are illustrated in ❷ Sect. 2, where their main features are recapitulated, along with some simple examples of applications taken from biology. In particular, the syntax and the semantics of a simple calculus are introduced, then there is a discussion on how to extend them with quantitative information and illustrate some features that make the modeling of membranes easier. In ❷ Sect. 3, a representation of chemical facts through process calculi is discussed and some expressivity results are surveyed. Classical techniques for analyzing the behavior of processes are in ❷ Sect. 4. Finally, ❷ Sect. 5 surveys a few examples of biological systems that have been modeled and investigated within the approach based on process calculi.

2   Process Calculi for Computational Models of Bio-systems

Process calculi have been defined within concurrency theory to formally represent the behavior of a system made of interacting components, also called *processes*. The represented behavior is an abstraction of the actual behavior of the components at the desired level of abstraction. This abstraction is also referred to as *observable behavior*, suggesting that the aspects of interest that an external observer can perceive are considered, neglecting internal mechanisms or details deemed irrelevant. Process calculi are based on an algebraic notation whose main ingredients are

(i) The basic *actions* that components can perform together with a set of operators for defining *processes* by composing basic actions and simpler processes;

(ii) A *semantics* describing the possible dynamics of a process according to the actions it is enabled to do. Often the semantics is operational and consists of a *labelled transition system (LTS)*. LTS have *states*, roughly processes themselves, and *transitions*, that is, a relation between a state and the next one the process can move to. A transition may have *labels* recording its observable behavior, to keep track of the relevant information about the step. LTS are usually defined compositionally by inductive rules over the structure of the components of a process (SOS rules (Plotkin 2004)). A *trace* is a sequence of transitions, such that the target state of a transition in the sequence is the source of the next one. A trace represents a possible computation of the process in the source of its first transition;

(iii) *Properties, methodologies*, and (mechanical) *tools* to analyze process behavior. Examples of properties can be equivalence relations, such as bisimulations, or reachability properties, such as checking whether a process can reach a given state of interest. Examples of mechanical analysis tools are model checkers, used for verifying whether an LTS can fulfill a logical formula expressing the property of interest, and static analysis tools, which check properties of an LTS by reasoning on an abstraction of it.

Below a simple calculus, called BCCS after basic CCS, will be introduced. It consists of a (restriction- and value passing-free) subset of CCS, an exemplary calculus (Milner 1989). This introduction to process calculi aims at presenting their use in systems biology and is definitely not exhaustive. The interested reader is referred to Milner (1980, 1999) and Bergstra et al. (2001) and citations therein.

Two extensions that have been demonstrated to be relevant for systems biology are then presented. The first one takes into consideration quantitative data. Many biological phenomena need to be understood in terms of the quantities, and their variations, of the elements involved. Often, these are continuous variations derived from deterministic laws. However, at certain scales and levels of abstraction, the phenomena are essentially discrete, as each single element may cause specific behavior to emerge globally. Furthermore, they are better described in a stochastic manner, as the dynamics of the system cannot be fully reduced to deterministic mechanisms (Wilkinson 2006). Process calculi have been equipped with stochastic semantics suitable for studying the performance of computer systems in the presence of events with a probabilistic distribution over time, for example, user requests and queued incoming messages (Hillston 1996; Priami 1995). Stochastic process calculi offer then natural means for formal quantitative modeling of biological phenomena.

The second extension concerns the development of calculi designed to account for a certain form of interaction that peculiarly happens within biological systems. In particular, the focus will be on a few calculi designed to express the notion of membraned components and the behavior that can be expressed by the membranes themselves. The capability of describing intra- and extracellular processes show the versatility of the approach in modeling phenomena at different scales with the desired level of abstraction.

2.1 Process Calculi for Systems Biology

BCCS is illustrated here. Essentially, the processes may perform basic actions that are inputs and outputs over a communication channel. Channels may be seen as the interface of a process with its environment.

Formally, one can assume a set of channels Ch, upon which one can define the set of actions $Act = \{a, \bar{a} : a \in Ch\}$ (one can feel free to use the same letter to denote a channel and an action along it). Actions a and $\bar{a}$ represent complementary input/output actions over channel a (with $\bar{\bar{a}} = a$). Additionally, $\tau \notin Act$ is a distinguished action called *internal action*.

Basic actions can be put in sequence within a process, and their execution will take place in the prescribed order. More interestingly, processes can be built by composing (sub)processes, running in parallel. Two (sub)processes in parallel may proceed independently of each other, or they can synchronize when ready to perform a complementary pair of input/output actions over a channel they share. Furthermore, (sub)processes may also be composed so as to represent mutually exclusive behavior chosen non-deterministically.

More precisely, BCCS is defined by its syntax and its SOS operational semantics, given in ❷ *Fig. 1*. The syntax is in the first line and defines (through the symbol ::=) the various forms a process P may assume, separated by a|. The semantics is specified in a logical style, through inference rules that define the transitions of the calculus, actually its LTS – technically, the minimal transition relation that shall be taken among processes fulfilling the rules in ❷ *Fig. 1*. A transition has the form $P \rightarrow_\alpha Q$, where P is its source process, Q its target, and $\alpha \in Act \cup \{\tau\}$ is its label, recording the activity performed by the transition.

The inference rules are now briefly commented upon. Each of them consists of some premises and a conclusion, written above and below the line, respectively.

The empty process 0 does nothing, so there is no need for any rule governing its behavior.

The rule *pref* specifies the temporal ordering of actions to occur. It has no premise, and directly justifies the transition for any process in the form $\alpha.P$, in words, P prefixed by action α (P plays here the role of process variable, to be instantiated to an actual process). The process $\alpha.P$ becomes P by performing α, which is recorded in the label of the transition.

The parallel composition of two processes is represented by the | operator. Rule *par* says that if a process P can perform an action α, then $P|R$ can also perform the same action

◻ **Fig. 1**

Syntax and operational semantics of BCCS.

$$P, Q ::= 0 \mid \alpha.P \mid P \mid Q \mid P{+}Q \mid N \stackrel{\cdot}{=} P$$

$$\frac{}{\alpha.P \rightarrow_\alpha P} \ (\textit{pref}) \qquad \frac{P \rightarrow_\alpha Q}{P|R \rightarrow_\alpha Q|R} \ (\textit{par}) \qquad \frac{P \rightarrow_\alpha Q}{P{+}R \rightarrow_\alpha Q} \ (\textit{cho})$$

$$\frac{P_1 \rightarrow_\alpha Q_1 \quad P_2 \rightarrow_{\bar{\alpha}} Q_2}{P_1|P_2 \rightarrow_\tau Q_1|Q_2} \ (\textit{comm}) \qquad \frac{A \stackrel{\cdot}{=} P \quad P \rightarrow_\alpha Q}{A \rightarrow_\alpha Q} \ (\textit{def})$$

with Q staying idle. More generally, if the (instantiated) premises of the rule (top row) are themselves justified, possibly by applying other rules, the bottom transition is justified as well. A further comment is in order. As is often the case, processes are considered up to some structural properties of the operators, so the same rule governs the behavior of any process whose structure matches (up to the structural properties) the source of the transition in the conclusion. Here, it is assumed that the parallel operator is associative (i.e., $(P|Q)|R = P|(Q|R)$), commutative ($P|Q = Q|P$), and has the inactive process 0 as the neutral element ($P|0 = P$). Then, the rule *par* also applies to $P|R = R|P$, with R moving and P staying idle.

The second rule governing the parallel operator is *comm*. It states that whenever any two subprocesses P_1 and P_2 in parallel are ready to perform complementary actions, they can synchronize: $P_1|P_2$ becomes $Q_1|Q_2$ (note that this rule has two premises, and both need to be justified to deduce the conclusion). The observable behavior of a synchronization is τ. This use of the internal action is a means of modular information hiding, not disclosing to the environment the details of the synchronization that happened among P_1 and P_2. Technically, it also forces synchronization to be binary, because τ has no complementary action to be used in further synchronizations.

Rule *cho* allows $P + R$ to move as P does, and R is discharged. Also + is associative, commutative, and with 0 as neutral element, and so this rule also prescribes that $R + P$ moves as R and P is discharged. Note that the choice among the two alternatives is purely nondeterministic.

The last syntactic definition permits a process to be denoted by a name, taken from a given set $N = \{A, B, \ldots\}$. A process name can be used to identify a specific fragment of behavior: rule *def* specifies that the process A, standing for P, can move to any process Q to which P moves. Actually, using this feature one can define recursive processes accounting for the repetition of the same behavior. In a recursive definition, the name of the process being defined occurs within the definition itself. For example, the process $N = a.N + 0$ can perform an unbound number of a actions and then become the process 0.

It is worth underlining the *compositional* flavor of these rules. For instance, the behavior of $P|Q$ is defined in terms of the behavior of its components P and Q. This is *abstractly represented* in terms of the mere observable behavior of P and Q respectively, without any reference to their internal structure.

Example 1 (LTS) The simple process $a.0|b.0|\bar{a}.\bar{b}.0$ consists of three parallel sub-processes, the last of which can communicate in a precise order with the other two. This behavior is specified by the following trace:

$$a.0|b.0|\bar{a}.\bar{b}.0 \rightarrow_\tau b.0|\bar{b}.0 \rightarrow_\tau 0.$$

The first transition is justified by recalling that $a.0|b.0|\bar{a}.\bar{b}.0 = a.0|\bar{a}.\bar{b}.0|b.0$; hence rule *comm* can be applied since both its premises hold: $a.0 \rightarrow_a 0$ (by rule *pref*) and $\bar{a}.\bar{b}.0|b.0 \rightarrow_a \bar{b}.0|b.0$ (by rules *par* and *pref*). Being 0 the neutral element for |, it can be elided. The last state 0 has no further possible transition.

Differently, the process $P = a.0|b.0|(\bar{a}.0 + \bar{b}.0)$ has the two traces

$$a.0|b.0|(\bar{a}.0 + \bar{b}.0) \rightarrow_\tau b.0 \rightarrow_b 0 \quad \text{and} \quad a.0|b.0|(\bar{a}.0 + \bar{b}.0) \rightarrow_\tau a.0 \rightarrow_a 0$$

both making available to the environment an action to synchronize with.

Finally, an example of (mutually) recursive processes can be represented by a quite abstract specification of the acquisition/release of energy by the *ATP/ADP* molecules (one can omit here the involved inorganic *ortho*-phosphate). The process representing *ATP* transforms into *ADP* by performing an internal action τ, which needs no synchronization and is executed autonomously. This action represents the mechanisms for the exchange of energy by actually abstracting them away. Analogously, *ADP* performs the inverse process through another τ action.

$$ATP \doteq \tau.ADP \qquad ADP \doteq \tau.ATP$$

A trace of a system made of the simple process *ATP* is

$$ATP \rightarrow_\tau ADP \rightarrow_\tau ATP \rightarrow_\tau \ \ldots$$

The basic calculus introduced comprises several of the ingredients needed to model relevant biological phenomena. Below, another example is considered at the biochemical level and model a fragment of the *Glycolysis* pathway. Recalling the described approach, "cells as computation" metabolites are represented as BCCS processes. The interaction points of metabolites are abstractly represented by channels. Two molecules that may react are then modeled as two processes sharing a channel *a*, and the reaction between them as a synchronization along *a*. Indeed, at this level of abstraction, reactions are often understood as binary phenomena after the consideration that the probability that more than two reactants interact at the very same instant is practically negligible. Hence, calculi based on basic binary interaction operators, such as the one introduced, have been often used. Also, reactions are represented that consider the number of reactants present in the system, rather than the more usual concentrations; this point will be addressed later.

Example 2 Consider the small fraction of glycolysis shown in ❷ *Fig. 2*. We use long names for metabolites (IUPAC convention) and codes (as in Fraser et al. (1995), to which we refer the reader) for enzymes: (i) *mg*023 for fructose-bisphosphate aldolase, (ii) *mg*431 for phosphoglycerate mutase, (iii) *mg*301 for glyceraldehyde 3-phosphate dehydrogenase, and (iv) *mg*050 for 2-deoxyribose-5-phosphate aldolase. We represent the catalytic effects of enzymes in an abstract way: their contribution is simply recorded with a label attached to the τ actions. In this way we do not explicitly represent enzymes as processes, although this would be possible with a specification at a finer level of detail. Note, however, that the labels of transitions are enough to trace the enzymatic activities along the dynamics of the pathway. The freedom

◻ **Fig. 2**

A fragment of the glycolysis pathway.

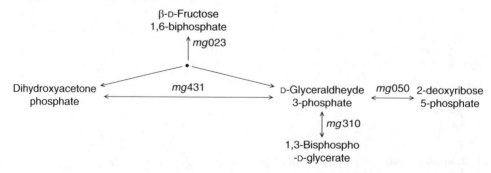

in selecting what will be the focus and what will be represented more abstractly is a feature of process calculi.

The metabolite β-D-fructose-1,6-biphosphate, reacting with the enzyme $mg023$, splits into Dihydroxyacetone phosphate and D-Glyceraldehyde 3-phosphate. In turn, Dihydroxyacetone phosphate, catalyzed by the enzyme $mg431$, becomes D-Glyceraldehyde 3-phosphate. This can become (i) the metabolite 1,3-Bisphospho-D-glycerate via $mg301$, or (ii) Dihydroxyacetone phosphate via $mg431$, or (iii) 2-deoxyribose 5-phosphate via $mg050$, or finally, (iv) it can reconstitute β-D-fructose-1,6-biphosphate via a reverse reaction with respect to the one that has generated it.

The topmost reaction of ❷ *Fig. 2* can be described by defining the behavior of the involved metabolite as follows:

β-D-fructose-1,6-biphosphate$=$
$\qquad \tau_{mg023}.$(Dihydroxyacetone phosphate | D-Glyceraldehyde-3-phosphate)

where the prefix consisting of the internal action represents the biochemical mechanisms driven by the $mg023$ enzyme that triggers the reaction. In this case, the enzyme action is independent of any metabolite other than β-D-fructose-1,6-biphosphate.

The reverse interaction arises from the specification of the interaction capabilities of D-Glyceraldehyde-3-phosphate and of Dihydroxyacetone phosphate. The first offers a synchronization on the channel (representing the) enzyme $mg023$, and then becomes β-D-fructose-1,6-bP; similarly, the second is willing to accept a synchronization on $mg023$, and will then disappear, becoming 0.

$\qquad$ D-Glyceraldehyde-3-phosphate $\doteq \overline{mg023}.$β-D-Fructose-1,6-biphosphate

$\qquad$ Dihydroxyacetone phosphate $\doteq mg023.\ 0$

The *adolase* reaction then spawns from the parallel composition of the two metabolites, modeled by the synchronization "catalyzed" by $mg023$:

$\qquad$ D-Glyceraldehyde-3-phosphate | Dihydroxyacetone phosphate $\longrightarrow_{\tau_{mg023}}$
$\qquad\qquad$ β-D-Fructose-1,6-biphosphate | 0

The one specified above is not the only interaction in which D-Glyceraldehyde-3-phosphate is involved. Its behavior can be incrementally specified by adding the other alternatives to its nondeterministic behavior. Beyond interacting with Dihydroxyacetone-phosphate through the $mg023$ channel/enzyme, it can: (i) execute a τ_{mg301} becoming 1,3-Bisphospho-D-glycerate, or (ii) execute a τ_{mg431} becoming Dihydroxyacetone phosphate, or (iii) execute a τ_{mg050} becoming 2-Deoxyribose 5-phosphate. Similarly, the behavior of Dihydroxyacetone phosphate can be completed by its possibility of becoming D-Glyceraldehyde-3-phosphate

$\qquad$ D-Glyceraldehyde-3-phosphate $\doteq \overline{mg023}.$ β-D-fructose-1,6-biphosphate

$\qquad\qquad + \tau_{mg301}.\ 1,3$-Bisphospho-D-glycerate

$\qquad\qquad + \tau_{mg431}.$ Dihydroxyacetone phosphate

$\qquad\qquad + \tau_{mg050}.\ 2$-Deoxyribose-5-Phosphate

$\qquad$ Dihydroxyacetone phosphate $\doteq \tau_{mg431}.$D-Glyceraldehyde-3-phosphate

$\qquad\qquad + mg023.\ 0$

The following is a trace of a system made of two copies of β-Fructose 1,6-bisphosphate:

$$\beta\text{-}\text{D-fructose\_1,6-bP}|\beta\text{-}\text{D-fructose\_1,6-bP} \rightarrow_{\tau_{mg023}}$$

$$(\text{D-Glyceraldehyde\_3-phosphate}|\text{Dihydroxyacetone\_phosphate})|\beta\text{-}\text{D-fructose\_1,6-bP} \rightarrow_{\tau_{mg301}}$$

$$(1,3\text{-Bisphospho-}\text{D-glycerate}|\text{Dihydroxyacetone\_phosphate})|\beta\text{-}\text{D-fructose\_1,6-bP} \rightarrow_{\tau_{mg023}}$$

$$((1,3\text{-Bisphospho-}\text{D-glycerate}|\text{Dihydroxyacetone\_phosphate})|$$

$$(\text{D-Glyceraldehyde\_3-phosphate}|\text{Dihydroxyacetone\_phosphate})) \rightarrow_{\tau_{mg023}}$$

$$(1,3\text{-Bisphospho-}\text{D-glycerate})|(\beta\text{-}\text{D-fructose\_1,6-bP}|\text{Dihydroxyacetone\_phosphate})$$

2.2 Adding Quantitative Information

The need to embed quantitative information into process calculi has been put forward for studying the performance of systems. Time was the first quantitative dimension that came into play, although it is easy to recognize that other quantities also have to be considered. Indeed, in a distributed system several of the aspects, typically task completion time, that affect performance have a probabilistic nature, as they depend on the probability distribution of the incoming requests, among other factors.

A probabilistic formal model for the study of system performance, depending on the possible occurrence of events, was first defined in Hillston (1996) and Priami (1995) by embedding stochastic aspects into process calculi. This is realized by associating a random variable to each action. The random variable models the delay of action execution, according to the idea that, for each action, it is necessary to consider a completion time that is probabilistic in nature. In other words, the random variable represents the probability, drawn from a given distribution, that the action will be executed within a delay.

Often, the adopted distribution is a negative exponential distribution $1 - e^{-rt}$, meaning that the probability of completing the considered action before time t is $1 - e^{-rt}$. This distribution guarantees that the probability of a system to go from a state to the next via a transition only depends on the current state and not on the way this state has been reached, namely the system enjoys the Markov, or memory-less property. This property has a good correspondence with the modeled reality, because the current state of a distributed system is generally sufficient to determine the next one. Additionally, the memory-less property fosters good semantic definitions.

More formally, the actions Act of the calculus are annotated with a rate r, the parameter of the probability distribution associated with the random variable relative to the action. Basic actions hence consist of pairs (α, r). Compositional operators and process construction work as standard. The stochastic machinery clearly reflects on the semantics. The inference rules label the transitions with actions and rates, for example,

$$\frac{}{(\alpha, r).P \rightarrow_{(\alpha,r)} P} (pref)$$

Noticeably, the meaning of choice is no longer purely nondeterministic. Rather, we have a probabilistic choice between weighted alternatives:

$$\frac{P \rightarrow_{(\alpha,r)} Q}{P + R \rightarrow_{(\alpha,r)} Q} (cho)$$

Furthermore, synchronization also needs to be rethought, as it is clearly related to the probability of the completion time of actions. This point has been heavily debated; see for example, Hillston (1994) and Bradley (1999). A natural choice would be to consider the slowest of the distributions of two synchronizing actions as the distribution of the synchronization, but min/max operations on exponential distributions do not yield an exponential distribution. A suitable solution to compute the rate r is using a function proposed by Hillston (2005), which is written as a side-condition of the rule below:

$$\frac{P_1 \rightarrow_{(\alpha,r_1)} Q_1 \quad P_2 \rightarrow_{(\bar{\alpha},r_2)} Q_2}{P_1|P_2 \rightarrow_{(\tau,r)} Q_1|Q_2} \text{ with } r = \frac{r_1}{r_\alpha(P_1)} \frac{r_2}{r_\alpha(P_2)} min(r_\alpha(P_1), r_\alpha(P_2))$$

Above, it is written $r_\alpha(P_i)$, namely the *apparent rate* of α within P_i, $i = 1, 2$, to denote the sum of all the rates associated with the α's in P_i ready to be performed. The apparent rate takes into account the many possible different ways in which the subprocess P_i offers to its partner a synchronization, which clearly influences the probability of which to occur.

The other rules are straightforward extensions of the usual ones, for example, of (*par*) and (*def*) in ❷ *Fig. 1* for BCCS.

As mentioned above, the use of exponential distributions helps in defining and in exploiting the semantics of a stochastic process calculus. Actually, the operational semantics originates a continuous time Markov chain (CTMC), on which several statistical analyses can be made; for more details, the interested reader should refer to Hillston (2005).

One of the most influential calculi with the stochastic features mentioned above is PEPA (Hillston 1993), which was specifically designed for generating Markov processes that could be solved numerically for performance evaluation. This calculus combines the CCS operations discussed above with others, taken from CSP. As above, PEPA has actions with a duration, represented by a random variable with a negative exponential distribution; in the choice $P + Q$, the first process completing its activity is taken and the other is discarded. Different from the above, processes are composed in parallel through a cooperation operator, indexed by a set of cooperation activities L, much in the CSP style (Hoare 1985). Roughly, the set L determines those activities on which the cooperands are forced to synchronize, and the observable behavior is the activity itself, rather than a τ. This provides one with the possibility of a multi-way synchronization between components: further processes, performing the same actions, can be engaged in the synchronization. Asynchrony occurs when a process performs an action not in L. From CSP, PEPA also inherits the additional operator of hiding, also indexed by a set L, written P/L. When such a process performs an activity belonging to L, it becomes a τ, thus preventing other processes from synchronizing and making the activity internal or private to the process P. Starting from an operational definition, the semantics of PEPA is defined as a CTMC, which can be exploited for extracting performance measures.

Besides being used in the performance evaluation of computer system, BIO-PEPA (Ciocchetta and Hillston 2006) has been recently proposed, which extends PEPA to model various biochemical aspects of living matter, including general kinetic laws.

As a matter of fact, stochastic process calculi have been fruitfully exploited for describing biological phenomena, which generally require quantitative models to be better understood. Consider for instance the biochemical level. System dynamics defines and is affected by the amounts of the reactants produced and consumed. Other measurable quantities, such as temperature and time, play a determinant role in the overall picture. Furthermore, the dynamics of biochemical reactions present a stochastic behavior, just like most natural phenomena.

Although CTMCs exhaustively represent the semantics of interest, they pose computational difficulties, because the interplay of states and transitions over time exponentially increases the size of the model, making the approach computationally intractable. This is particularly important for the models taken into consideration, because they often formalize rather complex biological entities, thus leading to systems with huge dimensions.

A more efficient, computationally successful alternative is using stochastic process calculi for simulation purposes. Instead of considering the whole model of all the possible states, transitions and their probabilities, a single trace, called the *trajectory*, is simulated. This consists of one of the possible time courses of the system whose steps are chosen according to the given probability distribution, obviously obeying the semantics of the calculus.

Pushing further the "cells as computation" analogy, one can see a trajectory as an experiment *in silico*. A simulation starts from the description of the initial state of the system in terms of the number of processes that are initially present (see ❷ Sect. 3 for a discussion on the number of components vs. their concentrations). Recall that processes represent at the desired level of abstraction the components of the biological system specified, and that the models are discrete (and abstract from temperature, pressure, etc.). The initial state above represents then the initial conditions of the biological entity under experimentation.

Trajectory simulation can be repeated, yielding possibly different outcomes: different experiments are carried out, and various behavioral or statistical analysis are possible on the collected "virtual" results.

Technically, trajectories are computed by including Gillespie's stochastic simulation algorithms (SSA) in the semantics of the calculus, which will be outlined in ❷ Sect. 3. Roughly, SSA stochastically selects the pair of complementary actions in the whole process, which are ready to fire. Also, it determines the time at which the resulting synchronization will occur, namely its apparent rate. Note that synchronizations correspond to reactions, so the actions that occur along a channel *a* will have all the same rate, often called the *basal rate*. As will be clear later on, SSA then suitably manipulates basal rates to compute the apparent rates of transitions, taking into account all the synchronizations possible (besides a couple of random numbers).

Several simulation tools have been developed, such as the Stochastic Pi-calculus Machine (Phillips and Cardelli 2007), to cite one. These tools represent a viable computational support for the simulation of phenomena at a sufficiently detailed level of description. See ❷ Sect. 5 for details.

Next, the results of an experiment carried out by using a tool that implements (a variant of) the stochastic version of BCCS are illustrated.

Example 3   Consider again the fragment of Glycolysis in ❷ Example 2, which is taken from the specification of almost all the metabolic pathways of a virtual cell, called VICE (Chiarugi et al. 2007), which is discussed in more detail in ❷ Sect. 5. As said, to obtain a quantitative description of a metabolic pathway, it is necessary to link each transition, and in particular its apparent rate, to a measurable biological parameter. The underlying idea is to associate a channel with a basal rate drawn from the specific (microscopic) rate constant of the described biochemical reaction.

In the original paper (Chiarugi et al. 2004), these basal rates were estimated starting from the constants of Michaelis–Menten kinetics (Fersht 1999; Hammes and Shimmel 1970), and taking care of the control strength of the enzymes involved in a pathway (Fell 1997). These values proved accurate enough, as confirmed also by recent findings (Zhao et al. 2008). The simulation *in silico* of VICE showed, for example, that it reaches its steady state, which is resistant to nonshocking changes in the external environment, for example, in the feeding regimen. This property mimics the homeostatic capability, typical of real cells, which require it in order to regulate their internal medium. Homeostatic biological systems oppose external environment change to maintain their internal equilibrium and succeed in reestablishing their balance, while nonhomeostatic ones eventually stop functioning.

The steady state is manifest when the time course distribution of metabolites reaches a plateau after a certain initial period of time, *in silico* after some transitions. This is shown for VICE in ❷ *Fig. 3* for three selected metabolites, whose behavior is rather informative, because they represent critical nodes in the entire metabolic network. Note that ❷ *Fig. 3* is affected by white Gaussian noise, because the whole model is stochastic, as it embodies Gillespie's SSA.

The first model of a biological system based on a stochastic process calculus, namely the Stochastic Pi-calculus, is in Priami et al. (2004). It extends the Pi-calculus (Milner 1999), associates rates with channels, and enhances its semantics as sketched above (Hillston 1996; Priami 1995).

The Pi-calculus is a nominal calculus that models systems of concurrent communicating processes whose communication network may change in time. Nominal calculi have the distinguished notion of *name*, roughly an identifier that denotes a shared resource, possibly a channel itself. A process that possesses a name can communicate it to another process across the system, but disclosing a name is constrained by *visibility rules*. Typically, a name, say n, is often private to a group of processes, so the resource it denotes can be referred to only by them; for example, when n denotes a channel, it is the mean for each member of the group to communicate with the others. Instead, a process P not belonging to the group cannot use n, unless the group is enlarged and n communicated to P. Technically, this is done by carefully handling the visibility scope of the name n, so to avoid any possible mismatch or ambiguity or leaking of n to other processes. A name n is declared private to a (group of processes) Q by writing $(\nu n)Q$; in words,

◻ **Fig. 3**

Time course distribution of pyruvate (pyr), diacilglycerol (dag), phosphoribosylpyrophosphate (prpp). The concentration of metabolites is plotted vs. the number of transitions.

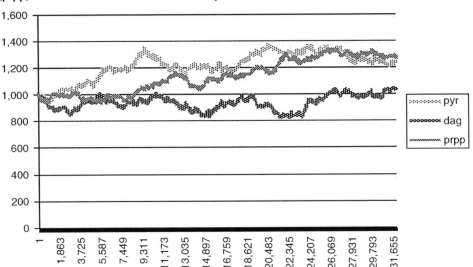

n is restricted within Q (more precisely, ν acts as a binder for n, whose scope is Q, with some *caveat* (Milner et al. 1992), among which α-conversions to avoid name captures).

Clearly, communications are not simple synchronizations, like in BCCS, but also exchange information. As the semantics of the calculus is very rich, instead of giving the details, only a couple of examples are shown.

Notationally, $\bar{n}\langle m\rangle$ is the output action of the name m, through the channel denoted by n; with $n(x)$ one can represent reading a name that replaces the place-holder x. So, a process $\bar{n}\langle m\rangle.Q \mid n(x).P$ will perform a communication that also has the effect of binding within P (suitable occurrences of) the place-holder x to the received name m. If m denotes a channel, the communication additionally alters the interconnection topology of the system. As a matter of fact, now P, rather P with x instantiated to m, is linked to channel m as well, even though m originally was not a channel in the interface of P.

To exemplify the disclosure of a restricted name, consider the process $(\nu m)(\bar{\ell}\langle m\rangle.R \mid m(x).P(x))$ in which the name m is restricted, so both processes in parallel can use it (as a short-hand, one can write $P(x)$ to indicate that the place-holder x occurs within P). Consider now the following trace, starting from the above process, in parallel with two other processes $T = \ell(y).\bar{y}\langle n\rangle.0$ and Q, neither of which can know m. The first transition is a communication of m to T, and it enlarges the scope of the restriction to include $\bar{m}\langle n\rangle.0$ (the *extruded* name is highlighted), but not Q. A further communication on the channel denoted by m is now possible, between processes under the restriction, that is, *within* the same group.

$$(\nu m)(\bar{\ell}\langle m\rangle.R \mid m(x).P(x)) \mid \ell(y).\bar{y}\langle n\rangle.0 \mid Q \rightarrow_\tau (\nu m)(R \mid m(x).P(x) \mid \mathbf{\bar{m}}\langle n\rangle.0) \mid Q \rightarrow_\tau$$
$$(\nu m)(R \mid P(\bar{n})) \mid Q$$

The system described in Priami et al. (2004) consists of a gene regulatory network with positive feedback. It involves two genes, RNA transcription and degradation, and positive feedback. Name machinery plays an important part in the positive feedback of protein A onto protein TF, both coded by the genes of the system. Actually, a complex is formed by the disclosure of the private channels (backbones) of A to TF, basically through the same extrusion mechanism mentioned above. Then, A and TF can interact along newly shared, highly efficient channels, boosting the role of TS in the reciprocal promotion, thus originating the positive feedback. The results of computational simulations clearly show the expected uprise of A in the presence of TF, confirming experimental data.

Sharing restricted channels can be conceived as a natural mechanism for implementing a notion of compounded complex, as is often the case in computer system modeling. However, this seems to be neither completely satisfactory in biological systems modeling, nor expressive enough to faithfully and directly describe compounds, compartmentalized cells, and membranes. In the next subsection, a few calculi will be illustrated with operators explicitly for these features.

2.3 Membrane Systems

Distributed computational paradigms often consider software applications that can move around through the interconnection network, crossing environmental boundaries and interacting at different locations – think of the Internet applications that move along the net and enter/exit from sites where they are executed. The concept of *distributed ambient* has been formalized in the AMBIENT calculus (Cardelli and Gordon 1998). A process P is wrapped within a named environment, called *ambient*, resulting in $n[P]$. Within the ambient n, the process P may contain other (nested) ambients and standard communications actions. Processes can move along the nested ambients by executing the actions *in n* and *out n*, and also dissolve through *open n*. The ambient name referred to by the actions implements a capability discipline: The name of an ambient must be known in order to interact with that ambient. Communication happens only within the same ambient. The premiseless rules in the upper part of ❷ *Fig. 4* are part of the semantics of the calculus.

Quite naturally, ambients may be interpreted as membraned components, for example, cells. Indeed, a system of membraned components has a nested structure and its components often move, interact, and also transform, for instance by dissolving or inglobing membranes.

A first proposal in this direction is the BIOAMBIENT calculus (Regev et al. 2004), which extends the ideas of AMBIENT in several directions to better suit biological interaction; a stochastic version of it is in Brodo et al. (2007), which exploits the SSA algorithm. In BIOAMBIENT, communications can cross ambients in a controlled way, as happens in intra-membrane interaction. The actions for ambient interaction occur synchronously through different channels/modalities, and merging of ambients/membranes has been introduced. Two rules of the operational semantics are in ❷ *Fig. 4*, which may help in understanding its peculiarities. Just as in nominal calculi, communication actions have "directions," for example, $\bar{a}(m)$ represents sending the name m through channel a, while $a(x)$ is the complementary action of receiving, but they are controlled to mimic membrane permeability. See, for example, the parent-to-child communication rule $(p2c)$, where the output with modality $p2c$ matches the input with modality $c2p$. Two ambients can also merge, giving rise to a new ambient, by modeling the analogous activity on membranes. This is specified by the rule (mrg) in ❷ *Fig. 4*, where the name c associated with the actions *merge+ c* and *merge– c* identifies the two membranes that will fuse together. The "direction" $+/-$ (yet another usage of + !) names the resulting membrane m, as *merge + occurs within m*.

The pragmatics of BIOAMBIENT have been illustrated through several proof of concept examples, such as transport layers, protein complexation, and reversible enzyme activity, (Regev et al. 2004). The last one will be briefly surveyed below.

Example 4 Suppose one has the following parallel composition of membraned components corresponding to the enzyme and the substrate molecule:

$$enzyme[E]|\ldots|enzyme[E]|molecule[S]|\ldots|molecule[S]$$

The enzyme E either accepts within its ambient the substrate membrane or an already

❏ Fig. 4

An extract of AMBIENT and BIOAMBIENT semantics.

$$\frac{}{m[in\ n.Q|R]\ |\ n[P] \to_\tau n[m[Q|R]|P]}\ (in)\qquad \frac{}{n[m[out\ n.Q|R]\ |\ P] \to \tau\ m[Q|R]\ |\ n[P]}\ (out)$$

$$\frac{}{n[P]\ |\ open\ n.Q \to_\tau P\ |\ Q}\ (open)\qquad \frac{}{n[p2c\ \bar{a}(b).P\ |\ m[c2p\ a(x).Q(x)]] \to_\tau n[P]\ |\ m[Q(b)]}\ (p2c)$$

$$\frac{}{m[merge+\ c.P|Q]\ |\ n[merge-\ c.R|S] \to_\tau m[P|Q|R|S]}\ (mrg)$$

transformed product molecule. Accordingly, it can then either expel the product, or the original substrate and revert to the original state:

$$E \quad ::= \quad accept\ e\text{-}s\text{-}bind.ES \ + \ accept\ e\text{-}p\text{-}bind.ES$$
$$ES \quad ::= \quad expel\ unbind.E \ + \ expel\ react.E$$

The substrate S can initially only enter the enzyme environment (becoming X), and then, synchronizing with the enzyme, it can either revert to the substrate or exit as the product P. In turn, the product can again be re-processed by the enzyme.

$$S ::= \ enter\ e\text{-}s\text{-}bind.X$$
$$X ::= \ exit\ unbind.S + exit\ react.P$$
$$P ::= \ enter\ e\text{-}p\text{-}bind.X$$

Given a single enzyme and substrate molecule, a possible trajectory (abstracting from the stochastic machinery) follows, in which the substrate transforms into the product:

$$enzyme[E] \mid molecule[S] \rightarrow$$
$$enzyme[\textbf{accept e-s-bind}.ES \ + \ accept\ e\text{-}p\text{-}bind.ES] \mid molecule[\textbf{enter e-s-bind}.X] \rightarrow$$
$$enzyme[(expel\ unbind.E \ + \ \textbf{expel react}.E) \mid molecule[exit\ unbind\ .S + \textbf{exit react}.P]] \rightarrow$$
$$enzyme[E] \mid molecule[P]$$

A further step in the direction of modeling membraned components and the interaction mediated by membranes are the BRANE calculi (Cardelli 2004; Miculan and Bacci 2006). These are a modular set of process calculi expressing different details of membrane activity, and associated logics for describing system properties. An important novelty is that membranes themselves exhibit a behavior.

For brevity, a reduced instance of BRANE, namely sBRANE, is considered below. Its (simpli-fied) syntax is in the topmost part of ❷ *Fig. 5*, which also displays the rules of its qualitative semantics (neither quantities of reactants nor stochastic rates are considered).

A few comments are in order. In sBRANE, membrane complexes (P, Q, . . .) and membrane behavior (σ, ρ, . . .) are kept apart. The complex $\sigma[P]$ consists of an (active) external membrane layer σ and of a complex P contained inside the membrane. The symbol $\diamond$ denotes the null membrane complex.

■ **Fig. 5**

Syntax and operational semantics of sBRANE, a simplified BRANE calculus.

$$P, Q ::= \diamond \mid \sigma[P] \mid P \circ Q \mid r \qquad\qquad \sigma, \rho ::= 0 \mid a.\rho \mid \sigma|\rho \qquad r, s ::= \diamond \mid m \circ r$$
$$a ::= p_n \mid p_n^{\perp}(\sigma), \mid e_n \mid e_n^{\perp}, m_n, \mid m_n^{\perp} \mid r_1(r_2) \xrightarrow{\rightarrow} s_1(s_2)$$

$$\frac{}{m_n.\sigma|\sigma_0[P] \circ m_n^{\perp}.\rho|\rho_0[Q] \rightarrow \sigma|\sigma_0|\rho|\rho_0[P \circ Q]} \ (mate) \qquad \frac{}{e_n^{\perp}.\rho \mid \rho_0\ [e_n.\sigma\mid \sigma_0\ [P] \circ Q] \rightarrow P \circ \sigma\mid \sigma_0\mid \rho\mid \rho_0\ [Q]} \ (exo)$$

$$\frac{}{pn.\sigma \mid \sigma0\ [P] \circ p_n^{\perp}(\rho).\rho1 \mid \rho_0[Q] \rightarrow \rho_1 \mid \rho_0\ [\rho[\sigma\mid \sigma_0\ [P]] \circ Q]} \ (phago) \qquad \frac{P \rightarrow Q}{P \circ R \rightarrow Q \circ R} \ (par)$$

$$\frac{}{r_1 \circ r_1(r_2) \xrightarrow{\rightarrow} s_1(s_2).\sigma \mid \sigma_0[r_2 \circ P] \rightarrow s_1 \circ \sigma \mid \sigma_0[s_2 \circ P]} \ (b\&r) \qquad \frac{P \rightarrow Q}{\sigma[P] \rightarrow \sigma[Q]} \ (mem)$$

The complex $P \circ Q$ is obtained by putting in parallel P and Q, while $m_1 \circ \cdots \circ m_k$ is the complex consisting of a multiset of molecules $(r, s, \ldots)$.

Membrane complexes interact through the membrane layer σ. This can either perform no action if it is 0, the empty one; or perform an action prefixing a layer $a. \sigma$; or behave as two parallel layers $\sigma|\rho$ (other structure can also be added). Here, one can consider the actions $phago_n$ (in brief p_n) and exo_n (in brief e_n), together with the coactions $p_n^{\perp}(\sigma)$ and $e_n^{\perp}$; the name n therein constrains the occurrence of these actions in communications, like in BioAmbients. Moreover, action m_n and coaction $m_n^{\perp}$ model *cell merging* (although this can be expressed by (*phago*) and (*exo*), only (Cardelli 2004)).

A membrane complex $\sigma[P]$ can enter another complex $\rho[Q]$, if σ can execute a phago p_n action and ρ the corresponding coaction $p_n^{\perp}(\rho)$ with the same n (see rule (*phago*)). As a result, $\sigma[P]$ is enclosed within a membrane with part ρ. A nested membrane complex $\rho[\sigma[P] \circ Q]$ allows the subsystem P to leave the $\rho[\ldots]$ membrane complex if σ and ρ are ready to execute e_n and $e_n^{\perp}$, respectively, as dictated by rule (*exo*). Membranes and their contents merge according to rule (*mate*). A membrane complex can also proceed autonomously in a system or within a membrane (rules (*par*) and (*mem*)). As standard, | and $\circ$ are associative and commutative with 0 and $\diamond$ as neutral elements.

Finally, rule (*b&r*) illustrates a kind of membrane permeability. Being molecules r_1 and r_2 outside and inside the complex, respectively, and being the membrane sensitive to them (they appear in the membrane specification as arguments of the $\rightrightarrows$ operator), then the rule can be applied and its effect is to bring inside the complex the molecule s_2 (also an argument of $\rightrightarrows$), while r_1 and r_2 are consumed.

The next example illustrates the use of the chosen calculus in modeling the behavior of systems with membraned components. One can consider (the first steps of) a reference model of viral infection.

Example 5   Via phagocytosis, the virus enters the cell wrapped by a membrane. The external membrane of the virus then merges with a component of the cell, the *endosome*. Finally, through an exocytosis, the viral *nucleocapsid* and the viral RNA it contains are directly released in the *cytosol* of the cell. These first steps on the viral infection can be modeled as follows:

$$virus = p_a.e_a \ [nucap] \qquad nucap = capsid[vRNA] \qquad capsid = p_b \mid bud \mid disasm$$
$$cell = p_a^{\perp}(m_a) \mid e_b^{\perp} \ [cytosol] \quad cytosol = endosome \circ CC \quad endosome = m_a^{\perp} \mid e_a^{\perp} \ [\diamond]$$

The virus membrane is ready to execute a phago p_a and an exo e_a action. Its content *nucap* will take part in later stages, and in turn has a membrane *capsid* that contains the RNA of the virus. The membrane *capsid* is ready to execute a phago action p_b, and two further actions *disasm* and *bud* that are not relevant for the initial steps of the viral infection. The cell membrane is ready for a phago $p_a^{\perp}(m_a)$ and an exo $e_b^{\perp}$ action, with the latter modeling the reproduced virus that eventually leaves the cell. What is inside the membrane, *cytosol*, consists of two parts. The first is the endosome, which is a membrane complex that can merge, via an action $m_a^{\perp}$, with what has been "eaten" by the cell through a phago action. The endosome can also uncoat its content, via $e_a^{\perp}$, as soon as the virus provides a suitable coaction. The second component, *CC*, is not relevant here.

The initial configuration of the system is a complex composed by the virus and the cell. The following trace shows an endocytic pathway, consisting of phagocytosis, merging, and exocytosis. As expected, its last state shows that the genetic material of the virus floats within the cell (recall that *nucap* is defined as *capsid*[*vRNA*]).

$$p_a.e_a[nucap] \quad \circ \quad p_a^\perp(m_a)|e_b^\perp[\ m_a^\perp \mid e_a^\perp\ [\diamond] \circ\ CC\ ] \to$$
$$e_b^\perp[\ m_a[\ e_a[nucap]\ ]\ ] \quad \circ \quad m_a^\perp \mid e_a^\perp\ [\diamond] \circ CC\ ] \to$$
$$e_b^\perp[\ e_a^\perp[e_a[nucap]]\ \circ CC] \to$$
$$e_b^\perp[\ 0[\diamond] \circ\ nucap \circ CC]$$

The last membrane and compartment oriented calculus that is surveyed is BETA BINDERS (Priami and Quaglia 2004), which builds over the Pi-calculus. It has a notion of ambient, called *box*, so a process has the form $B[P]$; boxes cannot be nested. The component B is a set of so-called BETA BINDERS, and represents the possibly modifiable interface of the box. The process P is a Pi-calculus process with standard communication actions, and with some additional actions that manipulate the interfaces. The description of a biological system is then rendered as a set of boxes in parallel.

More technically, a BETA BINDER $\beta(x : \Gamma)$ associates with the name x a set of channels through which the box can communicate with another box, and represents the interaction sites of the modeled membrane. A binder can be either active $\beta^h\ (x : \Gamma)$ or hidden $\beta^h\ (x : \Gamma)$, that is, either ready for or temporarily unavailable for interaction. The interfaces are manipulated through the actions $hide(x)$, $unhide(x)$, and $expose(x, \Gamma)$, the semantics of which is in ❯ *Fig. 6* (the rule for *unhide* and some details are omitted). The first action hides the name x, rather the binder for x in the interface; the second activates a hidden x; the third action makes x available for interaction by creating a binder for it in the interface (name captures are avoided).

A process P is essentially a Pi-process, and communication within a box is standard. Interaction between boxes is instead peculiar: it still occurs on a given channel, but it is not necessary to explicitly say on which. Suppose that one box has in its interface the BETA BINDER $\beta(x : \Gamma)$ and the other has the BETA BINDER $\beta(y : \Delta)$. The two boxes communicate along x and y provided that the sets Γ and Δ are compatible, according to a given notion. The simplest requires that Γ and Δ contain at least one common channel, that is, $\Gamma \cap \Delta \neq \emptyset$. This simple definition of compatibility contains the rule (*inter*) of the operational semantics in ❯ *Fig. 6*; besides this, others conditions can be given.

Here two further actions that can be used to implement a variety of membrane interactions of interest for biological modeling, namely splitting and joining boxes, are not discussed, and the interested reader is directed to refer to the original presentation (Priami and Quaglia 2004) for a discussion on their definition and usage.

The interested reader can find a stochastic semantics for BETA BINDERS in Degano et al. (2006).

Examples of virus infection, similar to the one presented for the sBRANE calculus, have also been modeled in Priami and Quaglia (2004). These illustrate a natural and expressive way of using the loose communication mechanism offered by BETA BINDERS to define which virus can interact with which cells.

◼ **Fig. 6**

Some rules of BETA BINDERS.

$$\frac{}{B\beta(x : \Gamma)[hide(x).P|Q] \to B\beta^h(x : \Gamma)[P|Q]}(hide) \qquad \frac{}{B[expose(x : \Gamma).P|Q] \to B\beta(x : \Gamma)[P|Q]}(exp)$$

$$\frac{}{B\beta(x : \Gamma)[x(w).P|Q] \mid B\beta(y : \Delta)[\overline{y}\langle z\rangle.R(z)|S] \to B\beta(x : \Gamma)[P|Q] \mid B\beta(y : \Delta)[R(w)|S]}(inter) \text{ with } \Gamma \cap \Delta \neq \phi$$

3   Artificial Chemistry

In this section, the focus is on a process calculi-based representation of chemistry. First, some classical notions, in particular those connected with quantitative and stochastic modeling, are briefly discussed.

The dynamics of a chemical system are usually described by transformation rules annotated with quantities, known as Generalized Mass Action Laws (GMA) (Segel 1987). These rules define and are affected by the amounts, traditionally measured as concentrations, of the reactants produced and consumed. Remarkably, GMA deterministically govern the changes in a chemical system. For instance, the reversible interaction of molecules of species A and B generating molecules of species C is described as

$$A + B \underset{k_2}{\overset{k_1}{\rightleftharpoons}} C$$

This stands for the fact that quantities of A and B transform themselves into quantities of C with a rate k_1 (note the different use of + with respect to the nondeterministic operator of the algebra of processes). The rate represents here a proportional constant that determines the amount of quantity transformed in the time unit, in proportion to the joint amount of A and B. Vice versa, C decays back into its constituents, proportionally to the amount of C and the rate k_2. This reaction happens by the spontaneous activity of C. Generally, GMA model systems in terms of variations of continuous *concentrations* of species A, often represented as $[A]$.

Being $[A(t)]$, $[B(t)]$, and $[C(t)]$ the molecular concentrations (numbers of molecules per unit volume) at time t, the following system of ordinary differential equations (ODEs) corresponds to the above kinetic schema:

$$d[C]/dt = k_1[A][B] - k_{-1}[C] \quad d[A]/dt = -k_1[A][B] + k_{-1}[C] \quad d[B]/dt = -k_1[A][B] + k_{-1}[C]$$

Given initial conditions, the solution of the above system provides $[A(t)]$, $[B(t)]$, and $[C(t)]$. In the case of the spontaneous decay $C \overset{k}{\rightarrow} 0$, one can have $d[C]/dt = -k[C(t)]$, hence $[C(t)] = e^{-kt}[C_0]$, which is the well-known exponential law of decay, with $[C_0]$ as the initial concentration.

An extensive treatment of ODEs modeling chemical reactions is dealt with in Voit (2000). Although a valuable and widespread description tool, ODEs present methodological difficulties and rely on assumptions that do not suit all possible scenarios of interest.

Among the first ones, ODEs rarely can be solved analytically; numerical solutions are typically computationally expensive (beyond few tens of variables); and models can hardly be constructed in a compositional manner (the introduction of a new variable/molecule/reaction typically requires the redefinition of the whole model).

As for the necessary assumptions, in many phenomena it happens that extremely low concentrations, up to few molecules in a given volume unit, are responsible for relevant behavior emerging at the whole system level. Moreover, such limited amounts perturb the system so that it behaves in a way that is difficult to interpret deterministically, as is done for concentration variations. In the words of Wilkinson (2006), the deterministic approach fails to capture the nature of chemical kinetics, which at low concentrations is discrete and stochastic. (See also Wolkenhauer for a critical analysis on the adequacy of deterministic ODEs descriptions and the relationships between the continuous deterministic approach and the stochastic discrete one.)

A model accounting for the stochastic aspects of system dynamics can be defined analogously to GMA; its result is particularly suitable for modeling small volumes where only few molecules interact (Gillespie and Petzold 2006; Van Kampen 1992; Gardiner 2001; Wilkinson 2006; Gillespie 1977). The model describes the *stochastic* time-course of *discrete* quantities, that is, the actual number of involved molecules, where rates represent the distribution probability that a selected reaction occurs at a given time-point. This approach leads to the definition of the chemical master equation (CME) (Voit 2000), which describes the interactions of N well-stirred chemical species $S_1, \ldots, S_N$. By naming $x_i(t)$ the number of molecules of S_i at time t, one wants to model the time evolution of the vector $X(t) = (x_1(t), \ldots, x_N(t))$ (sometime written X) from a given initial condition $X(t_0) = X_0$. Species interact by M uni- or bimolecular chemical reactions $R_1, \ldots, R_M$. Each reaction R_i is characterized by two quantities: $a_i(X)$ and v_i, namely the *propensity function* and the *state-change vector*, respectively. The system changes because of the reaction R_i with a probability defined by $a_i(X)dt$. This probability is defined on top of a probability rate constant c_i, such that $c_i dt$ represents the probability that any two, or one, molecules that are randomly chosen among those reacting according to R_i will react in the time interval dt. By knowing c_i and summing over all the possible distinct combinations of R_i reacting molecules (addition law of probability), we can immediatly see that if R_i is a unimolecular reaction $S_j \rightarrow products$, then $a_i(X) = c_i x_j$, while if R_i is a bimolecular reaction $S_j + S_p \rightarrow products$, then $a_i(X) = c_i x_j x_p$. When the reaction R_i occurs, the state vector $X(t)$ becomes $X(t + dt) = X(t) + v_i$.

The CME defines the variations of the probability associated with state changes of the system, which basically corresponds to the probability of reaching the state of interest X from any other possible state minus the probability, being in X, of leaving the state. A possible formalization is the following (see Voit (2000) for further details):

$$\frac{\partial P(X, t | X_0, t_0)}{\partial t} = \sum_{j=1}^{M} [a_j(X - v_j) P(X - v_j | X_0, t_0) - a_j(X) P(X, t | X_0, t_0)]$$

The CME consists of as many coupled differential equations as all the possible molecular reactions within the system, a number that is almost always prohibitively large. Additionally, even for very simple systems, the numerical solution of the CME is computationally hard, given that the transition probability matrix becomes rapidly untractable because of its dimensions (Smith 2005; Ermentrout 2002).

The stochastic formulation of the biochemical intracellular processes of the CME can be simulated by a Monte Carlo method, as proposed by Daniel Gillespie with his stochastic simulation algorithm (SSA) (Gillespie and Petzold 2006; Gillespie 1977). The algorithm determines a numerical realization of $X(t)$ as a function of t, called the *trajectory*, according to the probability values in the system. This realization has been proven to give an exact simulation of the CME. The SSA relies on homogeneous volume hypotheses and models reactions, but not diffusion.

Roughly, the idea behind Gillespie's algorithm can be recapitulated as follows (see provided references for technical details). Assume $P(t, \mu)dt$ is the probability that R_μ will be the next reaction and it will occur in the interval $(t_0 + t, t_0 + t + dt)$, with t_0 the current time. Then, one can get $P(t, \mu) = P_1(t) \times P_2(\mu, t)$, where $P_1(t)dt$ is the probability that the next reaction will occur in the interval $(t_0 + t, t_0 + t + dt)$, and $P_2(\mu, t)$ is the probability that the next reaction will be of type R_μ. Gillespie showed that $P(t, \mu) = a_\mu(X)^{-a_0(X)t}$, where $a_0(X) = \sum_{k=1}^{M} a_k(X)$ represents the sum of the probabilities of all possible reactions. Then the time of the next reaction is $t = (1/a_0(X)) \times \ln(1/r_1)$ and the next occurring reaction R_μ is identified by μ as the smallest integer satisfying the relation $\sum_{k=1}^{\mu} a_k(X) > r_2 a_0(X)$. Here r_1 and r_2 are two random numbers that make the system stochastic. Summing up the application of SSA consists of repeating the steps (2–4) below as many times as needed, after the initialization step (1):

(1) Initialize the initial time t_0 and initial conditions of the system X_0
(2) Evaluate all the $a_k(X)$ and $a_0(X)$
(3) Generate t and μ as indicated above
(4) Replace the time t_0 with $t_0 + t$ and the number of molecules X with $X + \nu_\mu$

Since $P(t, \mu)dt$ follows an exponential distribution, system dynamics is memory-less. Importantly, the next reaction and time determined are *exactly* coherent with the probability distributions considered. Once the time and type of the next reaction are determined, time and quantities are updated as prescribed by the reaction and the algorithmic selection of the next reaction and time can be iterated.

As mentioned in ❯ Sect. 2.2, the stochastic extensions of process calculi are generally based on SSA and its variations. As said, in those settings SSA determines, beyond the elapsed time, the processes that will interact next. The operational semantics then rules the steps of the system. This amounts to saying that SSA and the operational semantics select a trajectory, exactly mimicking the probability distributions described by the rates of the model.

A thorough investigation about the feasibility of a computational modeling of chemistry, and also of biochemistry, is in the work by Cardelli (2009, 2008). A shallow survey follows, in particular considering some examples showing how to model typical chemical reactions, which are often taken literally from the above papers. These examples also describe how some phenomena are best captured by stochastic approaches. Then, some results about the relations between stochastic and deterministic modeling are mentioned. Also, the use of some new features added to the calculi seen so far, which are needed for modeling certain forms of complexation and decomplexation typical of biochemistry, will be illustrated. Finally, the computational expressiveness of the calculi adopted shall be discussed from a theoretical point of view. This allows one to characterize which class of behavior can be expressed by the programs in that formalism and which cannot, and which properties can be mechanically proved about that class.

Cardelli introduces *interacting stochastic automata* to represent species and their capabilities in a graphical notation. For brevity, some of their features will be discarded here, and their *simple* version is expressed as stochastic BCCS processes, each modeling a (state of) a molecule (in their full form interacting stochastic automata exhibit a dynamical interface of communication capabilities, similarly to nominal calculi, mentioned in ❯ Sect. 2.2).

The reactions of kind $A \xrightarrow{r} B$ are *first order*, those like $A + B \xrightarrow{r} C$ are *second order*. Concentrations follow the mass action law $\frac{d[A]}{dt} = -r[A]$ in first-order reactions, and $\frac{d[A]}{dt} = -r[A][B]$ in second-order reactions, under the assumption of working with a fixed interaction volume and a *well-stirred* solution (diffusion effects are then neglected because they are immaterial; refer also to the assumptions of SSA). Once the volume is fixed, the conversion between numbers of molecules and concentration is mediated by Avogadro's number. Three-way reactions are not considered here because they are unlikely to occur.

A molecule performing a first-order reaction waits for a delay, according to a negative exponential distribution, and then reacts. Technically, in this setting, such an action cannot reduce to a synchronization between two processes, because it is performed by a process in isolation. Usually, one assumes there is an internal action annotated with a suitable rate, representing an *ad hoc* stochastic delay construct (note that rates are here associated also with internal actions, and not only with channels). For instance, the decaying process $C \xrightarrow{r} 0$ can be modeled by the process

$$C \stackrel{.}{=} \tau_r.0$$

Second-order reactions, for example, $A + B \xrightarrow{r} C$, follow the above mentioned law of mass action, and can be represented as standard processes, for example, $A \doteq a.0$ and $B \doteq \bar{a}.C$ with the rate r associated to the channel a. Note that the choice of A becoming 0 and B becoming C is arbitrary. A case that requires a bit of care is a second-order reaction within a homogeneous population, for example, $A + A \xrightarrow{r} C$. In this case, each A can interact in two symmetric ways with other A's, for example,

$$A \doteq a.C + \bar{a}.0$$

Since two A's are transformed by each reaction, one can have $\frac{d[A]}{dt} = -2r[A]^2$.

Other examples of reproducible behavior at different levels of abstraction are oscillators, feedbacks, and even operators of the Boolean algebra (see Cardelli (2009) for details). Also zero-order reactions are considered, which obey a law of mass action independent of concentrations, for example, $\frac{d[A]}{dt} = r$. These represent a sort of constant growth of a population, and have no straightforward chemical correspondence, although they show a dynamic similar to that of saturated enzymes.

Some examples facilitate the intuitive understanding of how far one can go with the process-based representation of chemistry. By analyzing the *collective* behavior of large populations of processes/reactants, it is easy to observe stochastically emerging behaviors due to a small number of specific processes spawned in the large populations. One can also observe emerging behavior that can be assimilated to continuous and deterministic variations, as they concern the fluctuations of large populations.

Consider the so-called *celebrity* kind of behavior:

$$A \doteq \bar{a}.A + a.B \qquad B \doteq \bar{b}.B + b.A$$

As soon as any two A's synchronize on a, one of the two decides to become B, snobbishly "differentiating" itself. Analogously, as soon as any two B's synchronize, one of them differentiates by becoming A again. The overall behavior of a system of an equal number of A and B, with equals rates for channels a and b, exhibits a dynamical equilibrium, with the expected stochastic fluctuations. The results are in ❷ *Fig. 7*, and were obtained using SPiM, the Stochastic Pi-calculus Machine (Phillips and Cardelli 2007).

Similarly, *groupies* try to maintain their state as far as they are able to synchronize with others in the same state:

$$A \doteq \bar{a}.A + b.B \qquad B \doteq \bar{b}.B + a.A$$

Under the quantities assumed, rates and number of molecules, the overall behavior tends to uniformity where eventually all the groupies end up in the same state, either A or B, where no further reaction is possible.

It is quite instructive to observe the behavior of the same population of groupies with just one celebrity, ❷ *Fig 8*. The presence of a *single* molecule influences the macroscopic behavior of the system that passes through a deadlock-free random walk: Whenever all groupies get uniform, the celebrity distinguishes itself, followed by some imitating groupie.

In order not to underestimate the differences between stochastic and deterministic models, it is worth comparing ODEs and process calculi-based models on systems with minimal quantities of "destabilizing" molecules. ❷ *Figure 9* compares some ODEs (top row) with processes (bottom row). The leftmost column refers to the example of ❷ *Fig. 8*, which exhibits the most evident differences. The other two cases deal with similar examples, in which the number of molecules increases moving rightward, and the groupies offer growing resistance

▢ Fig. 7
Celebrities (*left*) and groupies (*right*) (from Cardelli 2009).

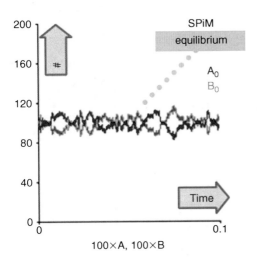
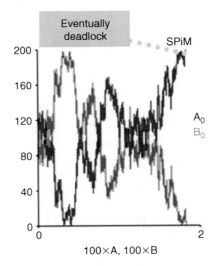

▢ Fig. 8
Groupies and one celebrity (from Cardelli 2009).

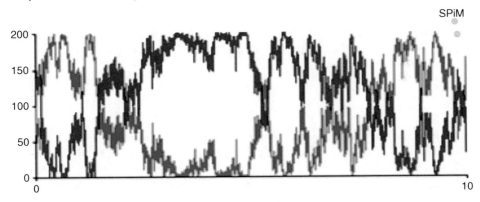

to changes. Consequently, the stochastic behavior tends to assimilate to the continuous case, and the differences between the two approaches blur moving towards the right. As a matter of fact, some phenomena are inherently discrete and stochastic and cannot be faithfully modeled by continuous and deterministic methods.

It is now interesting to link the usual representation of chemical reactions with the process-based representation, through a correspondence between the semantics of the two representations. This involves giving both deterministic (ODEs) and probabilistic (CTMC) semantics to chemistry, as done in Cardelli (2008), starting from a formalization of the language of chemical reactions. Note that while the chemical notation is "reaction centric," that is, the single reactions are listed, the process notation is "reactant centric," as it consists of a list of reactants with an associated behavior. Then, a mapping is needed from processes to the

□ Fig. 9

Stochastic and deterministic models are not the same (from Cardelli 2009).

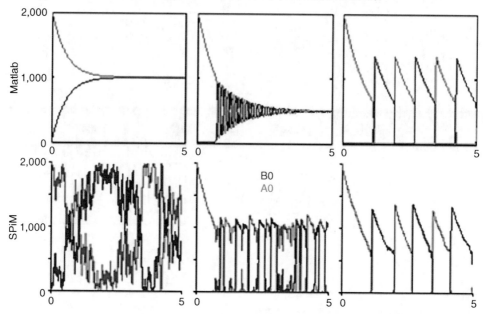

chemical reactions they represent, and backward. We illustrate it through the following example. Consider the processes in the left column below, which will be put in correspondence with the second-order reactions in the right column:

$$A \doteq a.C + \bar{a}.0 \qquad (A.1, A.2) : A + A \xrightarrow{2r_1} C$$
$$B \doteq b.E \quad D \doteq \bar{b}.0 \qquad (B.1, D.1) : B + D \xrightarrow{r_2} E$$

Here r_1 is the rate of channel a and r_2 the rate of channel b (recall that + has a different meaning in reactions and in processes). As part of the mapping, reactions are annotated with information about the reactants/processes that generate them: here the label $(A.1, A.2)$ denotes the two different actions of the same process A. This defines the reaction as occurring among homogeneous reactants, and hence the rate can properly be determined as $2r$. For passing from processes to reactions, these can be labeled with a name that, when translating to processes, may well become the name of the channel representing that specific reaction. This implies that each channel occurs within two processes only (and hence the set of processes is somewhat larger than the set of processes built from reactions).

Having defined a language for reactions and one for processes and a way for passing from one to another, one is obliged to prove a correspondence between them. This requires some technical machinery and will not be detailed here. The relevant steps are only sketched here.

For the discrete interpretation of the models, both reactions and processes can be given a CTMC as semantics. This can be straightforwardly derived once the initial conditions are fixed, that is, the number and the species of reactants/processes initially present in a system. For processes, the CTMC semantics follows from the LTS (which turns out to be a graph) associated with the initial state, that is, the initial set of processes composed in parallel. For reactions, the CTMC semantics follows from a graph representing the possible dynamics of the system of reactions. Relying on a sufficiently analogous structure, the two semantics are then proven equivalent under a suitable equivalence notion. This guarantees that the translation from reactions to processes and vice versa is correct with respect to discrete semantics.

The semantics of the continuous interpretation of the models is the set of ODEs determined according to the Generalized Mass Action Laws. As seen above, once a standard interaction volume is fixed, it is possible to read the number of interacting processes and of reactants as (continuous) concentrations. This requires one to consider the initial conditions as concentrations and to suitably convert rates. Other technical considerations are needed for the hypothesis of continuity in the variation of concentrations. Finally, it has been proven that the continuous semantics for processes is equivalent to the ODEs one.

One can stress again that the often neat correspondence between processes and ODEs models outlined above should not make one believe that the two approaches are freely interchangeable. Indeed, it has been seen that examples of phenomena are intrinsically discrete and stochastic, while others are better modeled by continuous approaches.

We now have a very quick overview of biochemistry and the features required to model it, following Cardelli (2009). Typical biochemical reactions include the creations of complexes, which are best modeled with calculi more expressive than BCCS. A *complex* is obtained by two molecules that interact through complementary reaction sites, a shared channel in these terms, and bind together. As a consequence, those reaction sites are no longer available for further complexations. Furthermore, a complex can split back into the original components, making reaction sites available again. Relevant examples of complexation and decomplexation came from enzymatic activity. Typically, this consists of bounding the enzyme with the substrate, and then splitting the obtained compound either back into enzyme and substrate or, much more likely, into enzyme and product.

Such a notion of composition that preserves the identities of the components cannot be easily represented by BCCS, as so far defined. However, it is sufficient to add a naming mechanism, similar to that of Pi-calculus mentioned in ❷ Sect. 2.2, to obtain the needed expressiveness. Processes of such a calculus are a basic ingredient of the so-called *polyautomata*. These have explicit complexation and decomplexation channels and rates, and they also maintain a state with information about performed complexation actions. When two processes synchronize on a complexation channel, the channel is recorded in a binding list that prevents us reusing it. Moreover, these communications associate a brand new name, say n, with the synchronizing processes, so n exactly identifies the two. Then, only the two processes that have been bound together can further synchronize through the unique name n, by performing a decomplexation action that restores the original components.

Other than complexation and detailed enzymatic activity, processes corresponding to polyautomata have been used to model polymerizations as well.

From the computer science point of view, a natural question to ask now is whether the additional expressive power obtained through the naming machinery is only a matter of pragmatics or if it has an intrinsic computational meaning. More precisely, it is interesting to know whether BCCS is Turing-complete, and if not which extensions are required. Roughly, a Turing-complete, or *universal*, formalism can compute through its programs all the functions that the programs of any other formalism can, often called *the computable* functions (typically, a function is such if it is computed by a Turing machine). Note that the programs are all finite, are made of a fixed finite number of symbols, and can be mechanically executed, with no limitations on the number of steps or on the size of the data they operate on.

This is the case with the programs, viz. the processes. A well-known result of computability theory is that there exist functions that are not computable in this sense. In particular, one cannot mechanically decide, in a finite amount of time, if the execution of a program will never halt, by only looking at the text of the program itself (the so-called *halting problem*).

The question of the formal expressive power of chemistry was first addressed by Magnasco (1997), and then studied by Busi and Gossieri (2006) for the BRANE calculus, by Soloveichik et al. (2008) using the different formalism of Minsky's register machine, and by Cardelli and Zavattaro (2008) for calculi such as those discussed above. Briefly, a first answer is that (a variant of) BCCS is *not* universal, in spite of the intuitions of Magnasco (1997). In chemical terms, this can be rephrased by saying that chemistry can implement, with a finite number of species, no Turing-complete formalism. It cannot thus be used to compute all the computable functions. Intuitively, the reason is that one can decide in finite time whether a certain molecule will be produced, so contradicting the undecidability of the halting problem. (Instead, chemistry with an *unbound* number of species is Turing-complete, but clearly its "programs" are no longer finite, nor do they use a finite number of symbols – nor are they very realistic, either!)

When enhanced with the association and dissociation actions discussed above, BCCS becomes universal (Cardelli and Zavattaro 2008). (Note in passing, these two actions cannot therefore be encoded in BCCS, while they can be encoded in the full CCS, which is known to be Turing-complete.) In other words, biochemistry, which includes operations of complexation and decomplexation formalized as polyautomata, is universal.

4 Formal Analysis of Models

One of the distinguishing features of the presented calculi is that a process can be *executed*, and more generally, *automatically processed*. An example of the first modality is the *simulation* of the modeled reality, of which a few examples have been shown. An example of the second kind is *reasoning* on a process, so as to investigate, prove, or disprove some of its properties, which hopefully are in correspondence with some properties of the modeled reality.

A few techniques that can be fully automated or supported by mechanical tools are briefly reported here. These techniques have been largely investigated in computer science in order to provide means for proving properties of software and hardware systems, and they can usually be smoothly applied also to the formalisms for modeling biological entities. Traditionally, these are techniques based on qualitative models, for example, for checking whether a given condition, a deadlock, or an error condition can ever occur in a system. More recently, and particularly following the development of the quantitative models for systems biology, there are proposals that take into account quantitative models, too.

Theorem proving consists in devising a *proof* for a given fact within a formal system. It often requires human supervision and has been generally applied to well-defined contexts, like restricted topics of mathematics and other sciences. It requires a precise domain knowledge, as well as nontrivial search and heuristic strategies implementing some forms of reasoning, for example, induction. It has not generally been applied to system biology and it will not be discussed any further; for an example, see Paulson (1989).

Model checking is instead a technique that can be carried out in a largely automated way. It was defined independently by Clarke and Emerson (Clarke et al. 1986) and Sifakis (1982) in the early 1980s. In its general formulation, model checking consists of verifying whether a (finite) structure is a *model* for a given property, expressed in a suitable logic. The result can either be positive or negative, in which case a counterexample is returned.

In this case, it consists of checking whether the semantics of a process—that is, its LTS—fulfills a property of interest.

Generally, the logic-expressing properties predicate over LTS; hence, they can also have predicates referring to events that may happen during system evolution, besides the classical Boolean operators such as conjunction, disjunction, and the like. A few operators of the *computation tree logic (CTL*)* (Emerson and Sistla 1983), a well-known logic of this kind, will be considered. A formula f can be an atomic proposition p (which usually depends on the application context), the negation $\neg f$ or the conjunction $f_1 \wedge f_2$ of formulas. A formula holds for a specific state s of the given LTS (the semantics of the logic is a Kripke structure):

$$s \models f$$

Formulas then grow by using modal operators: X is a "next state operator" and is used together with the quantifiers A, always, and E, exists. For example, one can write the formula

$$s \models A X f$$

which is true if and only if f holds in all the states s' reachable after a single transition from s, in symbols if and only if $\forall s'.\ s \rightarrow s'\ s' \models f$.

The method to verify if the above simple formula is true suggests how to automatize the process of checking whether a given structure is a model for the formula at hand. One mechanically builds the relevant LTS, exploiting the operational semantics. The formula is then verified, often by proceeding by structural induction on it and by exploring the LTS, as done for the formula above. The need to visit the LTS (or to incrementally generate pieces of it when needed) is even more evident for formulas with an "eventually"-like operator $\Diamond f$: there will be a state in the future fulfilling f (see below for a formal definition).

Different working hypotheses, like the finiteness of the LTS and the choice of the logical operators available, lead to different decidability results. This kind of logic is generally nonconclusive in its full formulation (there is not a general algorithmic procedure to check every possible formula against every possible structure).

In Miculan and Bacci (2006), a modal logic for reasoning over qualitative BRANE CALCULI models was introduced, and states are direct processes. Beyond classical Boolean operators, it also consists of operators that predicate over the structure and, like *CTL*, over the behavior of BRANE (see ● Sect. 2.3) processes. Skipping a few technical details, there are structural formulae, like $f_1[f_2]$, the intuition of which is that a complex consists of a membrane fulfilling f_1 and of an internal complex fulfilling f_2; the already seen $\Diamond f$, a process can evolve to a state where f is fulfilled; and a modal formula over the structure $\star f$, a process has a sub-term that fulfills f:

$$P \models f_1[f_2] \text{ iff } P \doteq \sigma[Q] \ \wedge \ \sigma \models f_1 \ \wedge \ Q \models f_2$$

$$P \models \Diamond f \text{ iff } \exists Q.\ P \rightarrow {}^*Q \ \wedge \ Q \models f$$

$$P \models \star f \text{ iff } \exists Q.\ P\!\downarrow^*\!Q \ \wedge \ Q \models f$$

with $P \rightarrow^* Q$ indicating a state Q reachable in zero or more transitions, and $\downarrow^*$ defining a suitable notion of sub-term, typically Q can occur anywhere within P.

An example of use of the logic follows. In settings similar to those of ❷ Example 5 about viruses and cells, the following statement can be verified by model checking:

$$Infectable\_cell \models Virus \triangleright (\diamond Infected\_cell)$$

Here *Infectable_cell* is a process modeling a cell that can be infected, *Virus* is a formula fulfilled by processes modeling a given virus (e.g., a formula representing its structure, say), and *Infected_cell* is a formula satisfied by processes modeling infected cells.

Overall, the above formula expresses the fact that an infectable cell of the specified type can eventually become infected if exposed to a virus of the specified type. (Intuitively, the formula $f_1 \triangleright f_2$ is fulfilled by all the processes that, whenever composed with a process fulfilling f_1, form together a complex fulfilling f_2.)

The full logic is not decidable, but subsets of it can be chosen that are decidable, hence allowing for fully automated model checking (under the hypothesis of finiteness of the LTS).

Other approaches involving model checking for biological systems are those based on a combination of qualitative and quantitative properties, where quantities are understood as probability values and time. One example is in Kwiatkowska et al. (2008): the reference LTS has associated probabilities, and properties are expressed in suitable temporal logics with probabilities, for example, Aziz et al. (2000) and Baier et al. (2003). The kind of expressible properties talk about the "*probability* that an event *eventually* occurs" or the "*probability* that an event occurs within *t* time units." Experiments can be done using the PRISM probabilistic model checker (Hinton et al. 2006).

It is worth pointing out that model checking appears to be a mature verification technique, although a full quantitative verification is still an open research topic. Thus, this technique appears adequate to be fruitfully exploited in systems biology. A successful example is shown by the work of Fages and Soliman (2008), which however is carried on a rule-based and reaction-centric approach, rather than the reactant-centric approach surveyed in this paper. A system is given three different semantics, based on ODEs, discrete stochastic processes and Boolean. The last semantics is a sort of "flat" quantitative semantics that only admits the presence or the absence of a component. It is fruitfully used, for example, in checking causal relations among the metabolites of a pathway. Properties can also be expressed by a kind of temporal logic that can express standard qualitative properties (so considering the Boolean semantics) and semiquantitative properties (exploiting discretized ODE-based semantics). The properties are then model checked by verifying the model in hand. Remarkably, the results of model checking permit its user to determine unknown parts of the model, such as reaction rates.

Static techniques refrain from considering the state space a system/process can pass through, as done by model checking. Rather, they analyze "statically" the text of the system, without actually executing it. A typical use of static analysis is in optimizing the code of computer programs, for example, determining from the program code that a variable is dead, that is, it would never be used. In general, these techniques are efficient, being low polynomial in time, but they only give approximate results that err on the safe side. Back to the example above, it may happen that a variable predicted to be used will be never used at execution time (approximation), but it will never happen that a variable is predicted to be dead when it will actually be used (sound).

An example of a static technique used in systems biology is *control flow analysis*, originally proposed for BioAmbient in Nielson et al. (2004a) and further developed in Nielson et al. (2004b), and used for BETA BINDERS in Bodei (2009). For example, properties concerning the nesting of membranes are investigated by considering the set of "possible" interactions as they emerge from process definitions. The approximation comes from the fact that some interactions are statically predicted, but they may be mutually exclusive in the dynamics of the system.

Another problem largely investigated within concurrency is that of establishing the *equivalence* of (the behavior of) processes (Sangiorgi 2004). Behavioral equivalence is at the basis of component interchangeability. Indeed, one can replace a part of a system with an equivalent one, which additionally enjoys a property of interest, or also proceed in incrementally building systems, by replacing (abstract) components with equivalent, more detailed ones. So far, most of the studies have been carried out on qualitative semantics, but proposals for addressing quantitative aspects have been put forward, like notions of stochastic equivalence (see, for example, Doberkat (2007)). Although not yet widely explored, the application of notions of equivalence for biological systems appears of interest, especially for defining equivalences that reflect interchangeability of molecules or of pathways.

The interest in the automated processing of process calculi models has called for the development of a range of computational support, from simulation tools for calculi, like the cited SPiM (Phillips and Cardelli 2007), to extended modeling environments that integrate various analysis tools, both quantitative and qualitative, and support various semantics.

BIOPEPA (Ciocchetta and Hillston 2006) builds upon the theory of PEPA (❷ Sect. 2.2). The essential algebraic representation of calculi is substituted by a richer language. Relevant biological aspects can be explicitly represented, for example, the stoichiometric matrix that summarizes quantitative information about reactions, which is implicitly defined by the overall process interaction in PEPA. Interaction rates are functions that dynamically determine their values according to the current state of the modeled system. Furthermore, the level of detail of the representation can be varied upon according to need, such as by modeling abstract (or non-fully known) interactions involving more than two reactants. Typically, this is done by explicitly associating quantities to reactants, and by assigning a role to actions, such as *activator, product, reactant,* etc. The semantics consists of an LTS enriched with quantitative information, which has a direct correspondence with CTMCs, as usual. A range of diverse kinetic laws can be applied to model the system. Furthermore, the analysis can be carried out through different approaches, like ODE-based descriptions or Gillespie's simulations (with a bit of care depending on the kinetic adopted).

BlenX (Dematté et al. 2008) is explicitly presented as a programming language, with a stochastic execution support based on an efficient variant of Gillespie SSA. It builds upon BETA BINDERS (❷ Sect. 2.3) and consists of a modular architecture, among the modules of which there are the CTMC generator, the one that traces reactions and reagents involved, statistical tools, as well as a graphical user interface and a graphical output module. The programming language includes some mathematics, declarations, binder representations, and its control part relies on (an extended representation of) BETA BINDER processes. Process behavior can also be determined by events, for example, environmental conditions, that may trigger actions, like "no molecules are left of a given species." BlenX supports modeling of a range of phenomena, such as enzymatic reactions, Michaelis–Menten kinetics, oscillatory behaviors, and self-assembly. As far as analysis is concerned, it features the capability of keeping trace of the species that have played an active role during a system simulation and the specific reactions in which they have taken part.

5 Case Studies

In this section, some case studies that show how suitable process calculi help in constructing *in silico* models of biological phenomena will be reported.

The already mentioned virtual cell experiment VICE (Chiarugi et al. 2004) addresses the biological problem of determining a genome as minimally as possible, and checking whether a hypothetical organism possessing it is able to live or not. The starting point was a hypothetical minimal genome (minimal gene set, MGS) (Mushegian and Koonin 1996), obtained by functionally comparing the genomes of two simple bacteria and eliminating duplicated or functionally identical genes.

To check the viability problem of a MGS-based prokaryote, Chiarugi et al. first specified *in silico* its metabolic pathways, using a process calculus very similar to BCCS. Then, they performed dynamic simulations, that is, virtual experiments of cellular metabolic activities, in order to check whether the MGS-prokaryote was able to reach some equilibrium state and to produce the necessary biomass, as these two properties are enjoyed by any living organism. The simulations clearly showed that MGS does not express an organism able to live, as it cannot reach any equilibrium.

Next, some genes were discarded, others added or replaced upon a functional analysis, and the various genomic sets obtained in this way checked for viability. The virtual experiments were made of about 10^8 transitions/reactions and involved up to 10^7 processes. Using a matrix-based interpreter (Chiarugi et al. 2008), a few minutes each were enough on a medium-sized PC.

After several iterations, a genome composed of 187 elements was selected. It expresses a virtual organism that exhibits homeostatic capabilities and produces biomass in the expected manner. Moreover, the steady-state distribution of the concentrations of virtual metabolites that resulted was similar to that experimentally measured in bacteria, as anticipated in ❯ Example 3. At least *in silico*, the virtual cell VICE is able to live.

It might be interesting to compare the above results with those of some authors who recently used a wet-lab approach to characterize the minimal gene set necessary to sustain bacterial life (Glass et al. 2006). Their experimentation consisted of knocking out *Mycoplasma genitalium* genes, which are supposed to form a set close to a minimal one. The difference between the genome proposed after *in silico* experimentations and the one obtained in vitro is about 10% (while the difference is about 30% with MGS).

A process-based model of pharmacodynamics was proposed in Dematté et al. (2007), to study the effects of drugs controlling the hypertension. A typical therapy consists of acting on the vascular tone, the degree of constriction experienced by a blood vessel relative to its maximally dilated state. The vascular tone is primarily dependent on a protein called myosin light chain kinase (MLCK), which is deeply influenced by intracellular calcium concentration.

Dematté et al. fully specified the nitric oxide-cGMP pathway, through which it is possible to modulate several signal transduction mechanisms that control the level of active MLCK in a cell. To study its effects, the considered drug is also represented as a set of processes.

A main contribution of Dematté et al. (2007) is the definition of the *effective index*, which formally gives sensible measures of the effects of a drug, as well as of the dose–response curve. More precisely, an effective index gives the expectation to reach a state, in the transition system of the pathway and the drug, which is safe with respect to an observable property, defined as a logical formula. Remarkably, effective indexes are mechanically computed by relying on Markov process theory (Norris 1970).

Summing up, this case study proves the feasibility of an automatic and system-level decision support in drug discovery and analysis, based on process calculi.

The illustrated framework Bio-PEPA (Ciocchetta and Hillston 2006) has been used to investigate a number of phenomena. A repressilator, that is, a synthetic gene regulatory network with oscillating behavior, is analyzed in Ciocchetta and Hillston (2008). Three genes are connected in a feedback loop, where the protein coded by one inhibits the transcription of another one. The model comprises mRNA transcription, protein synthesis and degradation (the first is regulated by the Hill kinetics and the others by the GMA law). Starting from data in the literature, the biological parameters and initial conditions of the experiments are set. Then, exploiting the facilities of Bio-PEPA, both deterministic and stochastic analysis are carried out. The first consists of ODE-based simulations and reveals the expected oscillatory behavior of the three proteins considered. The second consists of a SSA-based simulation that, when averaged over a number of trials, tends to attenuate the oscillations. This phenomenon is explained by a stochastic difference of phases in the oscillations of each trial, which hence tend to cancel each other. Such a behavior underlines how analysis results may require further interpretations and the utility of frameworks like Bio-PEPA, which smoothly allow a phenomenon to be observed under different viewpoints, especially when there is no privileged viewpoint for all the situations.

Gene regulation networks and their feedback-determined behavior have been long appreciated as test-bed examples of the adequacy of the defined models. It is worth recalling here that the first phenomenon modeled within the process calculus approach is the gene regulatory network described and discussed in ❷ Sect. 2.2.

In cooperation with two neuroscientists, Bracciali et al. (2008a, b) developed the first stochastic and discrete model of the calyx of Held, a big glutamatergic neuronal synapse in the auditorial pathway of the mammalia. The synapses are the places of functional contacts between neurons, where the information is stored and transmitted from one neuron to another. A particular feature of the calyx of Held is its sensitivity to very small quantities of calcium ions, and this makes the continuity assumptions of its ODE-based models not very realistic.

We specified the pre- and the postsynaptic phases in the Stochastic Pi-calculus, starting from existing kinetic models based on ODEs of some sub-components of the synapse, integrating other data from the literature and making some assumptions about non-fully understood processes. A delicate point was determining the stochastic rates of the channels used in processes, starting from the deterministic ones; a similar problem was addressed and solved in Ihekwaba et al. (2007).

In silico experiments have confirmed the coherence of the model with known biological data, also validating the assumptions made. Sensitivity analysis over several parameters of the model has provided results that help to clarify the dynamics of synaptic transmission and explain short-term plasticity mechanisms that are supposed to be at the basis of memory.

It is worth remarking that the compositionality typical of process calculi has permitted us to specify separately, experiment on, and tune the models of the pre- and postsynaptic traits, and then to connect them in a plain and straightforward way.

Acknowledgments

The authors are deeply indebted to Luca Cardelli who kindly gave them permission to reuse here parts of his work, as well as Davide Chiarugi and Roberto Marangoni for joint previous work on VICE. The authors thank Enrico Cataldo for helpful comments and suggestions. The first author wishes to thank all the people at the Microsoft Research—University of Trento Centre for Computational and Systems Biology.

References

Aziz A, Sanwal K, Singhal V, Brayton R (2000) Model checking continuous time Markov chains. ACM Trans Comput Logic 1(1):162–170

Baier C, Haverkort B, Hermanns H, Katoen J-P (2003) Model-checking algorithms for continuous-time Markov chains. IEEE Trans Software Eng 29 (6):524–541

Bergstra JA, Ponse A, and Smolka SA (2001) Handbook of process algebra. North-Holland, Amsterdam, The Netherlands

Bernardo M, Degano P, Zavattaro G (eds) (2008) Formal methods for computational systems biology. In: SFM 2008: 8th international school on formal methods for the design of computer, communication, and software systems, Bertinoro, Italy, June 2008. Lecture notes in computer science, vol 5016. Springer, Berlin

Bodei C (2009) A control flow analysis for beta-binders with and without static compartments. Theor Comput Sci 410(33–34):3110–3127

Bracciali A, Brunelli M, Cataldo E, Degano P (2008a) Stochastic models for the *in silico* simulation of synaptic processes. BMC Bioinform 9(4):S7

Bracciali A, Brunelli M, Cataldo E, Degano P (2008b) Synapses as stochastic concurrent systems. Theor Comput Sci 408(1):66–82, 2008

Bradley J (1999) Towards reliable modelling with stochastic process algebras. PhD thesis, Department of Computer Science, University of Bristol

Brodo L, Degano P, Priami C (2007) A stochastic semantics for BioAmbients. In: Proceedings of PaCT, Pereslarl-Zalessky, Russia, September 2007. Lecture notes in computer science, vol 4671. Springer, Heidelberg

Busi N, Gorrieri R (2006) On the computational power of Brane calculi. In: Transactions on computational systems biology VI. Lecture notes in computer science, vol 4220. Springer, Heidelberg, pp 16–43

Calzone L, Fages F, Soliman S (2006) BIOCHAM: an environment for modeling biological systems and formalizing experimental knowledge. Bioinformatics 22(14):1805–1807

Cardelli L (2009) Artificial biochemistry. In: Condon A, Harel D, Kok JN, Salomaa A, Winfree E (eds) Algorithmic bioprocesses. Springer, New York

Cardelli L (2008) On process rate semantics. Theor Comput Sci 391(3):190–215

Cardelli L (2004) Brane calculi-interactions of biological membranes. In: Danos V, Schachter V (eds) Proceedings of computational methods in systems biology, Paris, France, May 2004. Lecture notes in computer science, vol 3082. Springer, Berlin

Cardelli L, Gordon A (1998) Mobile ambients. In: Nivat M (ed) Proceedings of FoSSaCS'98, Lisbon, Portugal, March–April 1998. Lecture notes in computer science, vol 1378. Springer, Berlin, pp 140–155

Cardelli L, Zavattaro G (2008) On the computational power of biochemistry. In: Proceedings of algebraic biology, Castle of Hagenberg, Austria, July–August 2008. Lecture notes in computer science, vol 5147. Springer, Berlin

Chiarugi D, Curti M, Degano P, Marangoni R (2004) ViCe: a VIrtual CEll. In: Proceedings of 2nd international W/S computational methods in systems biology, Paris, France, May 2004. Lecture notes in computer science, vol 3082. Springer, Berlin

Chiarugi D, Degano P, Marangoni R (2007) A computational approach to the functional screening of genomes. PLoS Comput Biol 3(9):1801–1806

Chiarugi D, Degano P, Bert Van Klinken J, Marangoni R (2008) Cells in silico: a holistic approach. In: Formal methods for computational systems biology, Bertinoro, Italy, June 2008. Lecture notes in computer science, vol 5016. Springer, Berlin, pp 366–386

Ciocchetta F, Hillston J (2006) Bio-PEPA: an extension of the process algebra PEPA for biochemical networks. In: Proceedings of FBTC 2007, Lisbon, Portugal, September 2007. Electr Notes Theor Comput Sci 194(3):101–117

Ciocchetta F, Hillston J (2008) Process algebras in systems biology. In: Bernardo M, Degano P, Zavattaro G (eds) SFM 2008: Formal methods for computational systems biology, Bertinoro, Italy, June 2008. Lecture notes in computer science, vol 5016. Springer, Berlin, pp 265–312

Clarke EM, Emerson EA, Sistla AP (1986) Automatic verification of finite-state concurrent systems using temporal logic specifications. ACM Trans Program Lang Syst 8(2):244–263

Damm W, Harel D (2001) LSCs: breathing life into message sequence charts. Formal Methods Syst Des 19(1):45–80

Danos V, Feret J, Fontana W, Harmer R, Krivine J (2007) Rule-based modelling of cellular signalling. In: Proceedings of CONCUR, Lisbon, Portugal, September 2007. Lecture notes in computer science, vol 4703. Springer, Berlin, pp 17–41

Degano P, Prandi D, Priami C, Quaglia P (2006) Beta-binders for biological quantitative experiments. In: Proceedings of QAPL06, Vienna, Austria, April 2006. Electr Notes Theor Comput Sci 164(3): 101–117

Dematté L, Prandi D, Priami C, Romanel A (2007) Effective Index: A formal measure of drug effects. In: Proceedings of the 2nd Conference Foundations of Systems Biology in Engineering (FOSBE). Stuttgart, Germany, September 2007, pp 485–490

Dematté L, Priami C, Romanel A (2008) The BlenX language: a tutorial. In: Bernardo M, Degano P, Zavattaro G (eds) SFM 2008, Bertinoro, Italy, June 2008. Lecture notes in computer science, vol 5016. Springer, Berlin, pp 313–365

Doberkat E-E (2007) Stochastic relations. Chapman & Hall/CRC, Boca Raton, FL

Eker S, Knapp M, Laderoute K, Lincoln P, Meseguer J, Sönmez MK (2002) Pathway logic: symbolic analysis of biological signaling. In: Altman RB, Dunker AK, Hunter L, Lauderdale K, Klein TE (eds) Pacific symposium on biocomputing. Kauai, HI, 3–7 January 2002, pp 400–412

Emerson EA, Sistla AP (1983) Deciding branching time logic: a triple exponential decision procedure for CTL*. In: Clarke EM, Kozen D (eds) Proceedings logic of programs, Pittsburgh, PA, June 1983. Lecture notes in computer science, vol 164. Springer, Berlin, pp 176–192

Ermentrout B (2002) Simulating, analyzing, and animating dynamical systems. SIAM, Philadelphia, PA

Fages F, Soliman S (2008) Formal cell biology in Biocham. In: Bernardo M, Degano P, Zavattaro G (eds) SFM 2008: Formal methods for computational systems biology, Bertinoro, Italy, June 2008. Lecture notes in computer science, vol 5016. Springer, Berlin, pp 265–312

Fell DA (1997) Understanding the control of metabolism. Portland Press, London

Fersht A (1999) Structure and mechanism in protein science: a guide to enzyme catalysis and protein folding. Freeman, New York

Fontana W, Buss LW (1994) The arrival of the fittest: toward a theory of biological organization. Bull Math Biol 56:1–64

Fraser CM et al. (1995) The minimal gene complement of mycoplasma genitalium. Science 270(1):397–403

Gardiner CW (2001) Handbook of stochastic methods for physics, chemistry and the natural sciences. Springer, Berlin

Gillespie DT (1977) Exact stochastic simulation of coupled chemical reactions. J Phys Chem 81: 2340–2361

Gillespie DT, Petzold LR (2006) Numerical simulation for biochemical kinetics. In: Szallasi Z, Stelling J, Perival V (eds) System modeling in cellular biology, 1st edn. MIT Press, Cambridge, MA, pp 331–354

Glass J, Assad-Garcia N, Alperovich N (2006) Essential genes of a minimal bacterium. PNAS 103:425–430

Hammes GG, Shimmel PR (1970) In: Boyer PD (ed) The enzymes, vol 2. Academic Press, New York

Hillston J (1993) PEPA – performance enhanced process algebra. PhD thesis, University of Edinburgh, Computer Science Department

Hillston J (1994) The nature of synchronisation. In: Herzog U, Rettelbach M (eds) Proceedings of 2nd workshop on Process Algebras and Performance Modelling (PAPM'92). Erlangen, Germany, July 1994, pp 51–70

Hillston J (2005) Process algebras for quantitative analysis. In: LICS 2005: Proceedings of the 20th annual symposium on logic in computer science, Chicago, IL, USA, June 2005. IEEE Computer Society, Washington DC, pp 239–248

Hillston J (1996) A compositional approach to performance modelling. Cambridge University Press, Cambridge

Hinton A, Kwiatkowska M, Norman G, Parker D (2006) PRISM: a tool for automatic verification of probabilistic systems. In: Hermanns H, Palsberg J (eds) Proceedings 12th international conference on tools and algorithms for the construction and analysis of systems, Vienna, Austria. Lecture notes in computer science, vol 3920. Springer, Heidelberg

Hoare CAR (1985) Communicating sequential processes. Prentice-Hall, Englewood Cliffs, NJ

Ihekwaba A, Larcher R, Mardare R, Priami C (2007) BetaWB – a language for modular representation of biological systems. In: Proceedings of ICSB 2007, Long Beach, CA, October 2007

Kitano H (2002) Systems biology: a brief overview. Theor Comput Sci 295(5560):1662–1664

Kwiatkowska MZ, Norman G, Parker D (2008) Using probabilistic model checking for systems biology. SIGMETRICS Performance Evaluation Review 35(4):14–21

Larry L, Roger B (2005) Automatic generation of cellular reaction networks with moleculizer 1.0. Nat Biotechnol 23:131–136

Magnasco MO (1997) Chemical kinetics is Turing universal. Phys Rev Lett 78:1190–1193

Miculan M, Bacci G (2006) Modal logics for Brane calculus. In: Priami C (ed) CMSB06: Computational methods in systems biology, Trento, Italy, October 2006. Lecture notes in computer science, vol 4210. Springer, Heidelberg, pp 1–16

Milazzo P (2008) Formal modeling in systems biology. An approach from theoretical computer Science. VDM - Verlag Dr. Muller, Saarbrücken, Germany

Milner R (1980) A calculus of communicating systems. Lecture notes in computer science, vol 92. Springer, Berlin

Milner R (1989) Communication and concurrency. Prentice-Hall, Englewood Cliffs, NJ

Milner R (1999) Communicating and mobile systems: the π-calculus. Cambridge University Press, Cambridge

Milner R, Parrow J, Walker D (1992) A calculus of mobile processes, I-II. Inform Comput 100(1):1–77

Mushegian AR, Koonin EV (1996) A minimal gene set for cellular life derived by comparison of complete bacterial genome. PNAS 93:10268–10273

Nielson F, Riis Nielson H, Schuch-Da-Rosa D, Priami C (2004a) Static analysis for systems biology. In: Proceedings of workshop on systeomatics - dynamic biological systems informatics, Cancun, Mexico, 2004. Computer Science Press, Trinity College Dublin, pp 1–6

Nielson HR, Nielson F, Pilegaard H (2004b) Spatial analysis of BioAmbient. In: Proceedings of static analysis symposium, Verona, Italy, August 2004. Lecture notes in computer science, vol 3148. Springer, Berlin, pp 69–83

Norris JR (1970) Markov chains. Cambridge University Press, Cambridge, MA

Paulson LC (1989) The foundation of a generic theorem prover. J Automated Reasoning 5(3):363–397

Paun G, Pérez-Jiménez MJ, Salomaa A (2007) Spiking neural P systems: an early survey. Int J Found Comput Sci 18(3):435–455

Phillips A, Cardelli L (2007) Efficient, correct simulation of biological processes in the stochastic pi-calculus. In: Calder M, Gilmore S (eds) Proceedings of computational methods in systems biology, Edinburgh, Scotland, September 2007. Lecture notes in computer science, vol 4695. Springer, Heidelberg, pp 184–199

Plotkin GD (2004) A structural approach to operational semantics. J Log Algebr Program 60–61:17–139

Priami C (1995) Stochastic π-calculus. Comput J 36 (6):578–589

Priami C, Quaglia P (2004) Beta binders for biological interactions. In: Proceedings of CMSB, Paris, France, May 2004. Lecture notes in computer science, vol 3082. Springer, Berlin, pp 20–32

Priami C, Regev A, Shapiro E, Silvermann W (2004) Application of a stochastic name-passing calculus to representation and simulation of molecular processes. Theor Comput Sci 325(1):141–167

Reddy VN, Mavrouvouniotis ML, Liebman MN (1993) Qualitative analysis of biochemical reduction systems. Comput Biol Med 26(1):9–24

Regev A, Shapiro E (2002) Cellular abstractions: cells as computation. Nature 419:343

Regev A, Panina E, Silverman W, Cardelli L, Shapiro E (2004) BioAmbients: an abstraction for biological compartments. Theor Comput Sci 325(1):141–167

Sadot A, Fisher J, Barak D, Admanit Y, Stern MJ, Hubbard EJA, Harel D (2008) Toward verified biological models. IEEE/ACM Trans Comput Biol Bioinform 5(2):223–234

Sangiorgi D (2004) Bisimulation: from the origins to today. In: LICS 2004: Proceeding of 19th IEEE symposium on logic in computer science, Turku, Finland, July 2004. IEEE Computer Society, Washington DC, pp 298–302

Schrödinger E (1946) What is life? Macmillan, New York

Segel LA (1987) Modeling dynamic phenomena in molecular and cellular biology. Cambridge University Press, Cambridge

Sifakis J (1982) A unified approach for studying the properties of transition systems. Theor Comput Sci 18:227–258

Smith GD (2005) Modeling the stochastic gating of ion channels. In: Fall CP, Marland ES, Wagner JM, Tyson JJ (eds) Computational cell biology, 2nd edn. Springer, New York, pp 285–319

Soloveichik D, Cook M, Winfree E, Bruck J (2008) Computation with finite stochastic chemical reaction networks. Nat Comput. doi: 10.1007/s11047-008-9067-y (2008)

Van Kampen NG (1992) Stochastic processes in physics and in chemistry. Elsevier, Amsterdam, The Netherlands

Voit EO (2000) Computational analysis of biochemical systems – a practical guide for biochemists and molecular biologists. Cambridge University Press, Cambridge

Wilkinson DJ (2006) Stochastic modelling for systems biology. Chapman & Hall – CRC Press, London

Wolkenhauer O (2008) Systems biology – Dynamic pathway modelling. Manuscript, available at http://www.sbi.uni-rostock.de/dokumente/t_sb.pdf

Zhao J, Ridgway D, Broderick G, Kovalenko A, Ellison M (2008) Extraction of elementary rate constants from global network analysis of E. Coli central metabolism. BMC Syst Biol 2:41

56 Reaction–Diffusion Computing

Andrew Adamatzky[1] · *Benjamin De Lacy Costello*[2]
[1]Department of Computer Science, University of the West of England, Bristol, UK
andrew.adamatzky@uwe.ac.uk
[2]Centre for Research in Analytical, Material and Sensor Sciences, Faculty of Applied Sciences, University of the West of England, Bristol, UK
ben.delacycostello@uwe.ac.uk

G. Rozenberg et al. (eds.), *Handbook of Natural Computing*, DOI 10.1007/978-3-540-92910-9_56,
© Springer-Verlag Berlin Heidelberg 2012

Abstract

A reaction–diffusion computer is a spatially extended chemical system, which processes information by transforming an input concentration profile to an output concentration profile in a deterministic and controlled manner. In reaction–diffusion computers, the data are represented by concentration profiles of reagents, information is transferred by propagating diffusive and phase waves, computation is implemented via the interaction of these traveling patterns (diffusive and excitation waves), and the results of the computation are recorded as a final concentration profile. Chemical reaction–diffusion computing is among the leaders in providing experimental prototypes in the fields of unconventional and nature-inspired computing. This chapter provides a case-study introduction to the field of reaction–diffusion computing, and shows how selected problems and tasks of computational geometry, robotics, and logics can be solved by encoding data within transient states of a chemical medium and by programming the dynamics and interactions of chemical waves.

1 Introduction

The field of natural computation (sometimes "unconventional computation," "nonclassical computing," and "nonstandard computation" can be used as well, depending on the context and personal preferences) is concerned with the design of computing paradigms, architectures, and experimental implementations of novel computing devices. These devices are based around the principles of information processing in chemical, physical, and biological systems and the implementation of conventional algorithms in nonstandard, non-silicon substrates (Adamatzky and Teuscher 2006; Adamatzky et al. 2007; Akl et al. 2007). The majority of the results published in natural computation (up to 99% of papers related to natural computation are theoretical) deal with algorithms, methodologies, or theories of computing based on paradigms or, occasionally very distant, analogies with biological, physical, and chemical processes. Only a few experimental prototypes of unconventional natural computers have been implemented in laboratories over the last few decades. These prototypes include examples of reaction–diffusion chemical processors (Adamatzky et al. 2005b; Adamatzky and De Lacy Costello 2002b; Tóth and Showalter 1995; Steinbock et al. 1996; Gorecki et al. 2003; 2005; Gorecka and Gorecki 2003; Motoike et al. 2001; Ichino et al. 2003), extended analog computers (Mills 2008), micro-fluidic circuits (Fuerstman et al. 2003), and plasmodium computers (Nakagaki et al. 2001; Shirakawa et al. 2009). This tendency is worrying because it looks like living in a world populated with combustion scientists but no car mechanics, aerodynamic engineers but no pilots. Bearing in mind that there are plenty of "talkers" but on the evidence just a few "doers" in this developing field we decided to focus this chapter entirely on experimental prototypes developed by the authors and coworkers/collaborators (just a couple of illustrations assisted by numerical experiments are included).

The chapter is structured as follows. ❷ Section 2 is a short excursion into the history of reaction–diffusion computing and lists the main achievements in the field. A classification of reaction–diffusion computers is provided in ❷ Sect. 3. In ❷ Sect. 4, chemical recipes are provided for the preparation of the basic types of reaction–diffusion processors: precipitating palladium processors and excitable chemical media in the form of the Belousov–Zhabotinsky (BZ) reaction. Specialized chemical processors for computational geometry and robot control are discussed in ❷ Sect. 5. In ❷ Sect. 6, how to implement basic logical gates in a sub-excitable chemical medium is shown. A few thoughts on the future developments in the field are outlined in ❯ Sect. 7.

This chapter is only able to give a brief introduction to the field of reaction–diffusion computing; for this reason, the existence of working prototypes and some notable practical implementations are mainly stated without going far below the surface. Therefore, the interested reader is strongly encouraged to consult the book by Adamatzky et al. (2005b) for further details on the realization of chemical processors and also the implementation of the reaction–diffusion paradigm in silicon devices.

2 History and Milestones of Reaction–Diffusion Computing

The idea of physics-based computation can be attributed to Joseph Plateau, see references in Courant and Robbins (1941), who thought about methods to devise an experimental solution to the problem of calculating the surface of smallest area bounded by a given closed contour in space. Given a set of planar points, connect the points by a graph with the minimal sum of edge lengths (it is allowed to add more points, however the number of additional points should be minimal). Plateau's solution is simple yet nontrivial: mark the given planar points on a flat surface, insert pins in the points, place another sheet on top of the pins, briefly immerse the device in a soap solution, wait till the soap film dries, then record the topology of the dried soap film. The dried soap film represents a minimal Steiner tree spanning the given planar points (❯ *Fig. 1*).

Due to surface tension, the soap film between the pins, representing points, naturally tries to minimize the total surface area. A length-minimizing curve enclosing a fixed-area region consists of circular arcs of positive outward curvature and line segments (Saltenis 1999). The curvature of the arcs is inversely proportional to the pressure. By gradually increasing the pressure (❯ *Fig. 1*) arcs can be transformed to straight lines, and thus a spanning tree is calculated.

■ **Fig. 1**
Several steps of spanning tree constructions by soap film (Saltenis 1999).

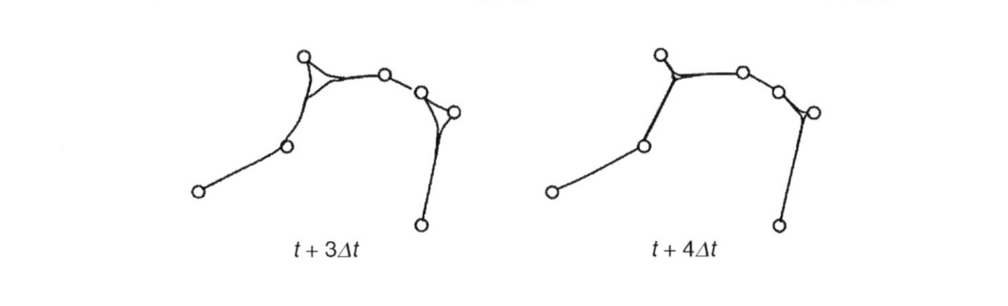

t $t + \Delta t$ $t + 2\Delta t$

$t + 3\Delta t$ $t + 4\Delta t$

The soap film does not involve diffusion or reaction; however, the operation is somewhat analogous to chemical reaction–diffusion computers. Initial data sites are represented by local perturbations of the system's state. Information is transferred between the sites by propagating patterns. The computation is implemented via interaction of the propagating patterns and the result of the computation is represented by a final spatial configuration of the system. After the work of Plateau, it would be another hundred years until the Belousov–Zhabotinsky (BZ) reaction (Zaikin and Zhabotinsky 1970) became widely known and Kuhnert published his pioneering results (Kuhnert 1986a, b; Kuhnert et al. 1989) on implementation of memory devices and basic image processing procedures using a light-sensitive BZ medium. These first BZ processors did not employ the propagation of excitation waves but relied on global switching of the medium between excited and nonexcited states termed phase shifts. In a light sensitive analogue of the BZ reaction, the timing of these phase shifts could be coupled to the projected light intensity due to the light sensitivity of the catalyst used in the reaction. However, the publications encouraged other researchers to experiment with information processing in excitable chemical systems. The image processing capabilities of BZ systems were further explored and the research results were enhanced in Agladze et al. (1995), Rambidi (1997, 1998, 2003), Rambidi and Yakovenchuk (2001), and Rambidi et al. (2002).

In the mid-1990s, the first results concerning the directed, one-way, propagation of excitation waves in geometrically constrained chemical media was announced (Agladze et al. 1996). This led to a series of experimental studies concerning the construction of Boolean (Tóth and Showalter 1995; Steinbock et al. 1996; Motoike and Yoshikawa 2003) and three-valued (Motoike and Adamatzky 2004) logical gates and eventually led to the construction of more advanced circuits (Gorecka and Gorecki 2003; Gorecki et al. 2003; Ichino et al. 2003; Motoike and Yoshikawa 1999) and dynamical memory (Motoike et al. 2001) in the BZ reaction. The BZ medium was also used for one optimization task – shortest path computation (Steinbock et al. 1995; Agladze et al. 1997; Rambidi and Yakovenchuk 2001).

The studies in reaction–diffusion computing were boosted by algorithms of spatial computation in cellular automata, when the first automaton model for Voronoi diagram construction was designed (Adamatzky 1994, 1996). The algorithm was subsequently implemented under experimental laboratory conditions using a precipitating chemical processor (Tolmachiev and Adamatzky 1996; Adamatzky and Tolmachiev 1997), and later a variety of precipitating chemical systems were discovered to be capable of Voronoi diagram approximation (De Lacy Costello 2003; De Lacy Costello and Adamatzky 2003; De Lacy Costello et al. 2004a, b). The precipitating systems also proved to be efficient in the realization of basic logical gates, including many-valued gates (Adamatzky and De Lacy Costello 2002b).

In the early 2000s, the first ever excitable chemical controller mounted on-board a wheeled robot was constructed and tested under experimental laboratory conditions (Adamatzky and De Lacy Costello 2002a; Adamatzky et al. 2003, 2004), and also a robotic hand was interfaced and controlled using a Belousov–Zhabotinsky medium (Yokoi et al. 2004). These preliminary experiments opened the door to the future construction of embedded robotic controllers and other intelligent reaction–diffusion processors.

As for computational universality, sub-excitable analogues of chemical media have been proved by experiment to be powerful collision-based computers, capable of the implementation of universal and arithmetical logical circuits via the collision of propagating wave-fragments (Adamatzky 2004; Adamatzky and De Lacy Costello 2007; Toth et al. 2007).

3   Classification of Reaction–Diffusion Processors

In reaction–diffusion processors, both the data and the results of the computation are encoded as concentration profiles of the reagents. The computation *per se* is performed via the spreading and interaction of wave fronts.

This can be summarized using the following relationships between architecture or operation and implementation:

- Component base, computing substrate $\rightarrow$ Thin layer of reagents
- Data representation $\rightarrow$ Initial concentration profile
- Information transfer, communication $\rightarrow$ Diffusive and phase waves
- Computation $\rightarrow$ Interaction of waves
- Result representation $\rightarrow$ Final concentration profile

Reaction–diffusion computers are parallel because the chemical medium's micro-volumes update their states simultaneously, and molecules diffuse and react in parallel. The architecture and operations of reaction–diffusion computers are based on three principles (Margolus 1984; Adamatzky 2001):

- Computation is dynamical: physical action measures the amount of information
- Computation is local: physical information travels only a finite distance
- Computation is spatial: their nature is governed by waves and spreading patterns

Potentially, a reaction–diffusion computer is a super-computer in a goo. Liquid-phase chemical computing media are characterized by

- Massive parallelism: millions of elementary – 2–4 bit – processors in a small chemical reactor
- Local connectivity: every micro-volume of the medium changes its state depending on the states of its closest neighbors
- Parallel I/O: optical input – control of initial excitation dynamics by illumination masks, output is parallel because the concentration profile representing the results of computation is visualized by indicators
- Fault tolerance and automatic re-configuration: because if some quantity of the liquid phase is removed, the topology is restored almost immediately

As far as reusability is concerned, there are two main classes of reaction–diffusion processors: reusable and disposable. Excitable chemical media are one of a family of reusable processors because given an unlimited supply of reagents, unlimited lifetime of the catalyst and the removal of any by-products, the system can be excited many times and at many simultaneous points on the computing substrate. The excitation waves leave almost no trace a certain time after the propagation. Precipitating chemical systems on the other hand are disposable processors. Once a loci of the substrate is converted to precipitate, it stays in the precipitate state indefinitely and therefore this processor cannot be used again within a realistic timeframe.

With regard to the communication between the domains of the computing medium the chemical media can be classified into either broadcasting or peer-to-peer architectures. Excitable chemical media and most precipitating systems are broadcasting. When a single site of a medium is perturbed the perturbation spreads in all directions, as a classical circular wave. Thus, each perturbed site broadcasts. However, when diffusion or excitation in the medium is limited, for example, the sub-excitable Belousov–Zhabotinsky system, no circular waves are formed but instead self-localized propagating wave fragments are formed (chemical relatives of dissipative solitons). These wave-fragments propagate in a predetermined direction, therefore they can be seen as data packets transferred between any two points of the computing medium.

Broadcasting chemical processors are good for image processing tasks, some kinds of robot navigation and path computation. They are appropriate for all tasks when massive spatially extended data must be analyzed, modified, and processed. The circular-wave-based communication becomes less handy when one tries to build logical circuits: It is preferable to route the signal directly between two gates without affecting or disturbing nearby gates.

When one uses an excitable medium to implement a logical circuit, the propagation of the excitable waves must be restricted, for example, by making conductive channels surrounded by nonconductive barriers, or via the use of localized excitations in a sub-excitable chemical medium. Reaction–diffusion processors where the computing space is geometrically inhomogeneous are called geometrically constrained processors. Sub-excitable chemical media, where wave-fragments propagate freely are called collision-based (because the computation is implemented when solution-like wave-fragments collide) or architecture-less/free-space computers (because the medium is homogeneous, regular, and uniform).

Further, examples of architecture-less chemical processors acting in either broadcasting or peer-to-peer modes are provided.

4   Recipes of the Chemical Processors

In this chapter, only three kinds of chemical reaction–diffusion processors are described: the precipitating palladium processor, and two chemical processors: an excitable Belousov–Zhabotinsky (BZ) medium and a light-sensitive sub-excitable analogue of the BZ reaction. Below we show how to prepare the reaction–diffusion processors in laboratory conditions.

4.1   Palladium Processor

A gel of agar (1.5% by weight, agar select Sigma–Aldrich Company Ltd., Poole, Dorset, UK) containing palladium chloride (Palladium (II) chloride 99%, Sigma–Aldrich Company Ltd.) in the range 0.2–0.9% by weight (0.011–0.051 M) is prepared by mixing the solids in warm deionized water. The mixture is heated with a naked flame and is constantly stirred until it boils to ensure full dissolution of the palladium chloride and production of a uniform gel (on cooling). The boiling liquid is then transferred to Petri dishes or alternatively spread on acetate sheets to a thickness of 1–2 mm and left to set. One can favor the use of Petri dishes because the reaction process is relatively slow and drying of the gel can occur if kept open to the atmosphere. The unreacted gel processors are then kept for 30 min although they remained stable for in excess of 24 h provided drying was controlled. A saturated solution (at 20°C) of potassium iodide (ACS reagent grade, Aldrich Chemical Co.) is used as the outer electrolyte for the reactions. Drops of outer electrolyte are applied to the surface of the gel to initiate the reaction process.

4.2   BZ Processor

A thin layer Belousov–Zhabotinsky (BZ) medium is usually prepared using a recipe adapted from Field and Winfree (1979): an acidic bromate stock solution incorporating potassium bromate and sulfuric acid ($[BrO_3^-] = 0.5$ M and $[H^+] = 0.59$ M) (solution A); solution of malonic acid (solution B) ($[CH_2(CO_2H)_2] = 0.5$ M); and sodium bromide (solution C) ($[Br^-] = 0.97$ M). Ferroin (1,10-phenanthroline iron-II sulfate, 0.025 M) is used as a catalyst and a visual indicator of the excitation activity in the BZ medium. To prepare a thin layer of the BZ medium, one can mix solutions A (7 ml), B (3.5 ml), C (1.2 ml), and finally when the solution becomes colorless, ferroin (1 ml) is added and the mixture is transferred to a Petri dish (layer thickness 1 mm). Excitation waves in the BZ reaction are initiated using a silver colloid solution.

4.3 Light-Sensitive Sub-excitable BZ Processor

The recipe is quoted from Toth et al. (2007). A light-sensitive BZ processor consists of a gel impregnated with catalyst and a catalyst-free solution pumped around the gel. The gel is produced using a sodium silicate solution prepared by mixing 222 ml of the sodium silicate solution with 57 ml of 2 M sulfuric acid and 187 ml of deionized water (Toth et al. 2007). The catalyst $Ru(bpy)_3SO_4$ is recrystallized from the chloride salt with sulfuric acid. Pre-cured solutions for making gels are prepared by mixing 2.5 ml of the acidified silicate solution with 0.6 ml of 0.025 M $Ru(bpy)_3SO_4$ and 0.65 ml of 1.0 M sulfuric acid solution. Using capillary action, portions of this solution were quickly transferred into a custom-designed 25 cm long, 0.3 mm deep Perspex mold covered with microscope slides. The solutions are left for 3 h in the mold to permit complete gelation. After gelation, the adherence to the Perspex mold is negligible leaving a thin gel layer on the glass slide. After 3 h, the slides with the gel on them are carefully removed from the mold and placed into 0.2 M sulfuric acid solution for an hour. Then they are washed in deionized water at least five times to remove by-products. The gels are 26 mm $\times$ 26 mm, with a wet thickness of approximately 300 m. The gels are stored under water and rinsed just before use.

The catalyst-free reaction mixture is freshly prepared in a 30 ml continuously fed stirred tank reactor, which involves the in situ synthesis of stoichiometric bromomalonic acid from malonic acid and bromine generated from the partial reduction of sodium bromate. This reactor is continuously fed with fresh catalyst-free BZ solution in order to maintain a nonequilibrium state. The final composition of the catalyst-free reaction solution in the reactor is the following: 0.42 M sodium bromate, 0.19 M malonic acid, 0.64 M sulfuric acid, and 0.11 M bromide.

A light projector is used to illuminate the computer-controlled image. Images are captured using a digital camera. The open reactor is surrounded by a water jacket thermostated at 22°C. Peristaltic pumps are used to pump the reaction solution into the reactor and remove the effluent.

In some cases, the regular structure of illumination is applied onto the BZ chemical reactor. Four light levels can be used: black (zero light level), at which level the reaction BZ oscillates; dark (0.035 mW cm^{-2}), which represents the excitable medium in which a chemical wave is able to propagate; white (maximum light intensity: 3.5 mW cm^{-2}), the inhibitory level; and finally the light level (1.35 mW cm^{-2}) corresponding to the weakly excitable medium, where excitation just manages to propagate.

Waves were initiated by setting the light intensity to zero within a small square at specific points on the gel surface. A black oscillating square is used to initiate waves periodically. The waves are then directed from source into a weakly excitable area (controlled by projecting a light intensity of 1.35 mW cm^{-2}) such that only small fragments are able to propagate.

5 Specialized Processors

As their name suggests, specialized processors are designed to solve just one possible computational task (or family of very similar tasks). In this section, examples of specialized processors for computational geometry, image processing, and robotics are provided.

5.1 Voronoi Diagram and Skeletonization

Let **P** be a nonempty finite set of planar points. A planar Voronoi diagram of the set **P** is a partition of the plane into such regions, so that for any element of **P** a region corresponding to a unique point p contains all those points of the plane which are closer to p than to any other node of **P**. A unique region $vor(p) = \{z \in \mathbf{R}^2 : d(p, z) < d(p, m) \forall m \in \mathbf{R}^2, \, m \neq z\}$ assigned to point p is called a Voronoi cell of the point p. The boundary of the Voronoi cell of a point p is built of segments of bisectors separating pairs of geographically closest points of the given planar set **P**. A union of all boundaries of the Voronoi cells determines the *planar Voronoi diagram*: $VD(\mathbf{P}) = \cup_{p \in \mathbf{P}} \partial vor(p)$. A variety of Voronoi diagrams and algorithms of their construction can be found in Klein (1990). An example of a Voronoi diagram is shown in ❯ *Fig. 2*.

Voronoi cells of a planar set represent the natural or geographical neighborhood of the set's elements. Therefore, the computation of a Voronoi diagram based on the spreading of some "substance" from the data points is usually the first approach of those trying to design massively parallel algorithms in chemical nonlinear systems.

◘ **Fig. 2**
Voronoi diagram of 50 planar points randomly distributed in the unit disc.

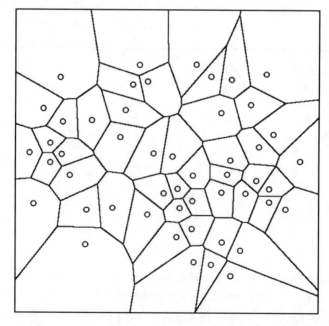

The "spreading substance" can be represented by a potential or an oscillatory field, or diffusing or phase waves, or simply by some traveling inhomogeneities or disturbances in the physical computing medium. This idea was first explored by Blum, Calabi, and Hartnett in their "grass-fire" transformation algorithm (Blum 1967, 1973; Calabi and Hartnett 1968). A fire is started at planar points of a given data set, the fire spreads from the data points on a substrate (that constitutes the computing medium) and the sites where the fire fronts meet represent the edges of the Voronoi diagram of the planar data set.

Quite similarly, to construct the diagram with a potential field one puts the field generators at the data points and then detects sites where two or more fronts of the field waves meet. This technique is employed in the computation of a Voronoi diagram via repulsive potential and oscillatory fields in homogeneous neural networks (Hwang and Ahuja 1992; Lemmon 1991). These approaches are "natural" and intuitive and their simplicity of implementation made them an invaluable tool for massively parallel image processing. The weakness of the approach is that a stationary structure is formed as a result of the computation and the meeting points of the colliding wave fronts must be detected by some "artificial" or extrinsic methods. This disadvantage is eliminated in the reaction–diffusion technique as demonstrated.

To compute a Voronoi diagram of planar points, each point is represented by a unique drop of outer electrolyte. The configuration of the outer electrolyte drops corresponds to the configuration of the data to be subdivided by the Voronoi diagram. Data objects (to be separated) are represented using a clear solution of potassium iodide. The potassium iodide diffuses from the sites of the drops or from the edges of planar shapes into the palladium chloride loaded gel. The potassium iodide reacts with the palladium chloride to form iodo-palladium species. The palladium chloride gel is light yellow colored while iodo-palladium species are dark brown. Thus, during the process of computation, the growth of dark-brown patterns emanating from the initial data sites can be observed. At sites where two or more diffusive fronts meet each other almost no precipitate is formed; these sites therefore remain uncolored, and they represent the bisectors of a Voronoi diagram, generated by the initial geometrical configuration of data objects (❷ Fig. 3a).

Voronoi diagrams of other geometric shapes can be constructed by substituting the drops of the outer electrolyte for pieces of absorbent materials soaked in the electrolyte solution and applying these to the gel surface (❷ Fig. 3b, c).

The processor starts its computation as soon as drops/shapes with potassium iodide are applied; the computation is finished when no more precipitate is formed. Configurations of drops/filter-paper templates of potassium iodide correspond to the input states of the processor; the resulting two-dimensional concentration profile of the precipitate (iodo-palladium species) is the output state of the processor. The processor can be seen as a preprogrammed or hardwired device because the way in which the chemical species react cannot be changed.

It should be noted that the "palladium processor" is just one of a huge number of chemical-based processors capable of Voronoi diagram calculation. Some others, particularly those based on gels containing potassium ferrocyanide and mixed with various metal salts, are discussed in Adamatzky et al. (2005b). Reagents with differing diffusion on the same substrate and differing chemical reactivities to the gel substrate can be used to form a number of generalized weighted Voronoi diagrams.

It is also interesting to note that in these chemical systems, the more data inputted in a given area the faster the computational outcome. The limit of the processors is currently determined by the gel thickness and the integrity and the ability to accurately input data points – but these are just design issues that can realistically be overcome and do not limit the

◻ **Fig. 3**
Voronoi diagram of planar points/discs (**a**), triangles (**b**) and rectangles (**c**) computed in the palladium processor (from Adamatzky et al. 2005b).

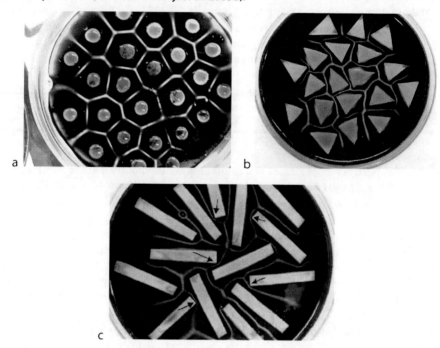

theoretical capabilities of the systems. Also, these systems can be practically used to implement three-dimensional Voronoi tessellations. For example, algorithms of classical computation of Voronoi diagram of planar points and planar objects are very different because when arbitrary-shaped objects are involved, one needs to employ algebraic curves to produce bisectors, which increases complexity. In chemical processors, it does not matter what shape data objects have, the time of computation is determined only by the maximum distance between two geographically neighboring objects.

Skeletonization (❯ *Fig. 4*) of the planar shapes in a reaction–diffusion chemical processor is yet further proof of the correctness of the proposed mechanism, namely, the wave–velocity-dependent bisector formation. During skeletonization, there are no interactions of wave fronts emanating from different sources; there is only one front, generated at the edges of the closed planar shape. Concave parts of the shape produce bisectors; however, the further the bisecting sites are from the given shape the larger the width of the bisectors and the more precipitate they contain (this is because the velocity of the fronts corresponding to the concave parts reduces with time) (Adamatzky and De Lacy Costello 2002c).

5.2 Collision-Free Path Calculation

A BZ parallel processor (Adamatzky and De Lacy Costello 2002a) that computes a collision-free path solves the following problem. Given a space with a set of obstacles and two selected

◘ Fig. 4

Computation of the skeleton of a planar shape in the palladium processor (from Adamatzky and De Lacy Costello 2002c).

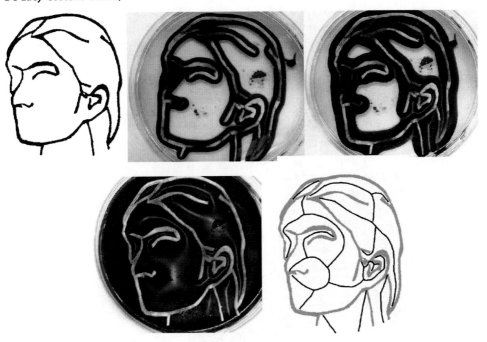

sites, find the shortest path between the source and the destination such that every site of the path is as far from the obstacles as possible. This is not only a classical problem in mathematical optimization but one heavily relied on in robotics, logistics, electronic design, etc.

The problem has already been tackled in a framework of wave-based computing. Three experimental prototypes exist, where BZ processors have been designed to compute the shortest paths:

- extraction of an optimal path in a labyrinth from excitation wave-front dynamics in the labyrinth (Steinbock et al. 1995); the path is extracted from time-lapsed snapshots of the wave-front motion;
- collision-free path using the BZ medium, where obstacles are represented by drops of KCl or strong illumination (Agladze et al. 1997); a path is extracted from the motion of excitation waves;
- approximation of a path on a tree in a light-sensitive BZ-medium (Rambidi and Yakovenchuk 2001); this technique is rather complicated because a data tree is represented by gradients of the medium and the image-processing routines were implemented at every step of the medium's development.

A BZ processor coupled with a two-dimensional cellular automaton has been designed, the hybrid system computes the shortest path by first approximating a distance field generated by the obstacles (BZ medium) and then calculating the shortest path in the field (cellular automaton) (Adamatzky and De Lacy Costello 2003).

A configuration of obstacles (❯ *Fig. 5a*) is represented by an identical configuration of silver wires parallel to each other and perpendicular to the surface of the BZ medium. Each circular obstacle is represented by a unique wire positioned exactly at the center of the obstacle. To start the computation in the BZ processor, the tips of all the wires are briefly immersed into the BZ mixture (in the reduced steady state). Immersing silver wires at specific points of the BZ mixture reversibly removes bromide ions from these sites. Bromide ions act

◻ **Fig. 5**
Converting obstacles to excitations: **(a)** mapping experimental arena (30 m in diameter) to Belousov–Zhabotinsky medium (9 cm in diameter). **(b)** Approximation of a distance field (from Adamatzky and De Lacy Costello 2003).

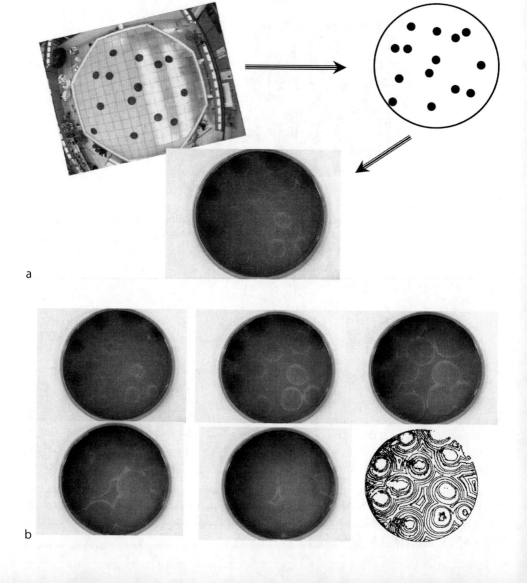

■ Fig. 6
Extraction of the shortest collision-free path (from Adamatzky et al. 2005b).

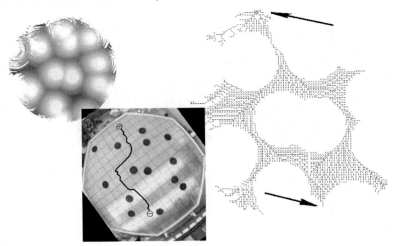

as primary inhibitors of the autocatalytic reaction in the BZ medium, therefore local removal of bromide ions stimulates wave generation. The digital images of the BZ medium were taken at certain intervals until all primary wave fronts had been annihilated (❷ *Fig. 5b*). Then we could transform a series of color snapshots taken to a gray-level matrix, which represents a distance scalar field derived from a configuration of obstacles (❷ *Fig. 6*, on the left).

The distance matrix is mapped to a two-dimensional excitable cellular automaton, which calculates the shortest path between any two points of the experimental arena, labeling the path with the local pointers (❷ *Fig. 6*, on the right). The configuration of the local pointers, representing the shortest collision-free path, can be communicated to a mobile robot that navigates the experimental arena (❷ *Fig. 6*, center).

5.3   Taxis

The Belousov–Zhabotinsky reaction is capable of assisting in the navigation of a mobile robot in a real-life environment (Adamatzky et al. 2004). Given an onboard thin-layer chemical reactor with the liquid-phase Belousov–Zhabotinsky medium, one can apply intermittent stimulation of the medium to sensibly guide the robot.

To make a BZ robotic taxis controller, a thin layer Belousov–Zhabotinsky (BZ) reaction is prepared using a recipe adapted from Field and Winfree (1979). Excitation waves in the BZ reaction are initiated using a silver colloid solution, which is added to the medium by a human operator. The chemical controller is placed onboard a wheeled mobile robot (❷ *Fig. 7a*). The robot is about 23 cm in diameter and able to turn on the spot; wheel motors are controlled by a Motorola 68332 onboard processor. The robot features a horizontal platform, where the Petri dish (9 cm in diameter) is fixed, and a stand with a digital camera Logitech QuickCam (in 120×160 pixels resolution mode), to record excitation dynamics. The robot controller and the camera are connected to a PC via serial port RS-232 and USB 1.0, respectively.

Because vibrations affect processes in BZ reaction systems, the movements of the robot are made as smooth as possible, thus in the experiments the robot moves with a speed

■ **Fig. 7**
Wheeled robot controlled by BZ medium (**a**) and snapshots of excitation dynamics with robot's velocity vector extracted (**b**) (from Adamatzky et al. 2004).

a

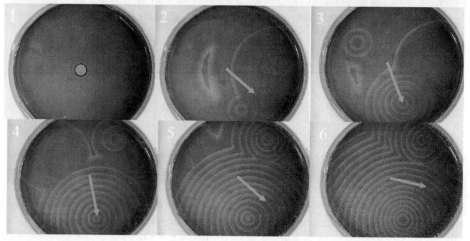

b

of circa 1 cm s^{-1}. It was found that rotational movements were particularly disruptive and caused spreading of the excitation waves from the site of the existing wave fronts faster than would be seen if the process was just diffusion. Therefore, to minimize this effect, the robot was made to rotate with a very low speed of around 1°s^{-1}.

To enhance the images of the excitation dynamics, the Petri dish is illuminated from underneath by a flexible electroluminescent sheet (0.7 mm thick) cut to the shape of the dish. The sheet, powered through an inverter from the robot's batteries, produces a cool uniform blue light not affecting (at least in conditions of the experimental setup) the physicochemical processes in the reactor. It was found preferable to implement experiments in dark areas to avoid interference of daylight and lamps with the light from the luminescent sheet.

When the medium is excited with a silver wire, a local generator of target waves is formed (❷ *Fig. 7b*). The robot extracts the position of the stimulation point, relative to the center of the chemical reactor, from the topology of the excitation target waves. The robot adjusts its velocity vector to match the vector toward the source of stimulation. To guide the robot more precisely one can excite the chemical medium in several points (❷ *Fig. 7b*). It should be noted that eventually it was intended that environmental interaction would become the source of stimulation for the excitation waves. The forced stimulation was simply to test the algorithms on board with real chemical controllers and to learn about any problems coupling chemical reactors and robots. Something that was identified as a problem, physical disruption of chemical waves, could actually turn out to be advantageous as it may feed back both direction, velocity, and even rotational movement of the robot.

5.4 Control of Robotic Hand

In the previous sections, we described the designs of chemical controllers for robots, which can calculate a shortest collision-free path in the robotic arena and guide the robot toward the source of stimulation (taxis). However, the controllers described lacked feedback from the robot. In a set of remarkable experiments undertaken by Hiroshi Yokoi and Ben De Lacy Costello (Yokoi et al. 2004), it was demonstrated that when a closed-loop interface between the chemical controller and the robot is established, the behavior of the hybrid systems becomes even more intriguing.

In the chemical controller of the robotic hand (Yokoi et al. 2004) the excitation waves propagating in the BZ reactor are sensed by photodiodes, which in turn trigger finger motion. When the bending fingers touch the chemical medium glass, "nails" fitted with capillary tubes release small quantities of colloidal silver into the solution and this triggers additional circular waves in the medium (Adamatzky et al. 2005b) (❷ *Fig. 8*). Starting from any initial configuration, the chemical-robotic system always reaches a coherent activity mode, where fingers move in regular, somewhat melodic patterns, and just a few generators of target waves govern the dynamics of the excitation in the reactor (Yokoi et al. 2004).

5.5 Parallel Actuators

How can a reaction–diffusion medium manipulate objects? To find out the answer, a simulated abstract parallel manipulator (Adamatzky et al. 2005a) can be coupled with an experimental Belousov–Zhabotinsky (BZ) chemical medium. The simulated manipulator is a two-dimensional array of actuating units, each unit can apply a force vector of unit length to the manipulated objects. The velocity vector of the object is derived via the integration of the results of all local force vectors acting on the manipulated object.

◫ **Fig. 8**
Robotic hand interacts with Belousov–Zhabotinsky medium (from Yokoi et al. 2004).

Coupling an excitable chemical medium with a manipulator allows one to convert experimental snapshots of the BZ medium to a force vector field and then simulate the motion of manipulated objects in the force field, thus achieving reaction–diffusion medium controlled actuation. To build an interface between the recordings of space–time snapshots of the excitation dynamics in the BZ medium and simulated physical objects, one can calculate the force fields generated by mobile excitation patterns and then simulate the behavior of an object in this force field (Skachek et al. 2005).

The chemical medium used to perform the actuation is prepared following the typical recipe, see Field and Winfree (1979), based on a ferroin catalyzed BZ reaction. A silica gel plate is cut and soaked in a ferroin solution. The gel sheet is placed in a Petri dish and the BZ solution is added. The dynamics of the chemical system is recorded at 30 s intervals using a digital camera.

The cross-section profile of the BZ wave-front recorded on a digital snapshot shows a steep rise of red color values in the pixels at the wave-front's head and a gradual descent in the pixels along the wave-front's tail. Assuming that excitation waves push the object, local force vectors generated at each site – a pixel of the digitized image – of the medium should be oriented along local gradients of the red color values. From the digitized snapshot of the BZ medium, an array of red components is extracted from the snapshot's pixels and then the projection of a virtual vector force at the pixel is calculated.

Force fields generated by the excitation patterns in a BZ system (❷ *Fig. 9*) result in tangential forces being applied to a manipulated object, thus causing translational and rotational motions of the object (Adamatzky et al. 2005a).

It was demonstrated that the BZ medium controlled actuators can be used for sorting and manipulating both small objects, comparable in size to the elementary actuating unit, and larger objects, with lengths of tens or hundreds of actuating units.

Pixel-objects, due to their small size, are subjected to random forces, caused by impurities of the physical medium and imprecision of the actuating units. In this case, no averaging of forces is allowed and the pixel-objects sensitively react to a single force vector. Therefore, one can adopt the following model of manipulating a pixel-object: If all force vectors at the 8-pixel neighborhood of the current site of the pixel-object are nil then the pixel-object jumps to

◼ Fig. 9
Force vector field (**b**) calculated from BZ medium's image (**a**) (Adamatzky et al. 2005a).

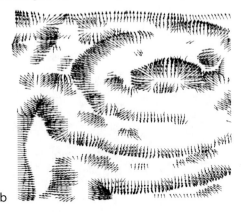

a   b

◼ Fig. 10
Examples of manipulating five pixel-objects using the BZ medium: (**a**) trajectories of pixel-objects. (**b**) Jump-trajectories of pixel-objects recorded every 100th time step. Initial positions of the pixel-objects are shown by circles (Adamatzky et al. 2005a).

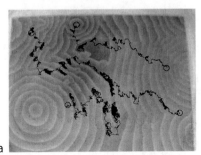
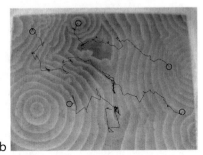

a   b

a randomly chosen neighboring pixel of its neighborhood, otherwise the pixel-object is translated by the maximum force vector in its neighborhood.

When placed on the simulated manipulating surface, pixel-objects move at random in the domains of the resting medium; however, by randomly drifting each pixel-object eventually encounters a domain of co-aligned vectors (representing the excitation wave front in the BZ medium) and is translated along the vectors. An example of several pixel-objects transported on a "frozen" snapshot of the chemical medium is shown in ❷ *Fig. 10*. Trajectories of pixel-objects (❷ *Fig. 10a*) show distinctive intermittent modes of random motion separated by modes of directed "jumps" guided by traveling wave fronts. Smoothed trajectories of pixel-objects (❷ *Fig. 10b*) demonstrate that despite a very strong chaotic component in manipulation, pixel-objects are transported to the sites of the medium where two or more excitation wave fronts meet.

The overall speed of pixel-object transportation depends on the frequency of wave generations by sources of target waves. As a rule, the higher the frequency the faster the objects are transported. This is because in parts of the medium spanned by low-frequency target waves

◘ **Fig. 11**

Manipulating planar object in BZ medium. **(a)** Right-angled triangle moved by fronts of target waves. **(b)** Square object moved by fronts of fragmented waves in sub-excitable BZ medium. Trajectories of center of mass of the square are shown by the dotted line. Exact orientation of the objects is displayed every 20 steps. Initial position of the object is shown by ⊖ and the final position by ⊗ (Adamatzky et al. 2005a).

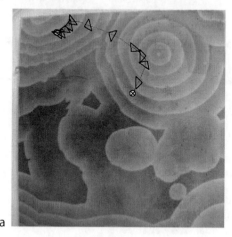
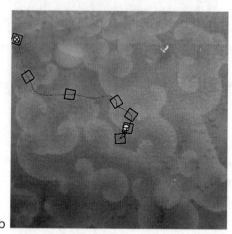

a   b

there are lengthy domains of resting states, where no force vectors are formed. Therefore, a pixel-sized object can wander randomly for a long time until climbing the next wave front (Adamatzky et al. 2005a).

Spatially extended objects follow the general pattern of motion observed for the pixel-sized objects. However, due to the integration of many force vectors, the motion of planar objects is smoother and less sensitive to the orientation of any particular force vector.

The outcome of the manipulation depends on the size of the object, with increasing size of the object – due to larger numbers of local vector forces acting on the object – the objects become more controllable by the excitation wave fronts (❷ *Fig. 11*).

6 Universal Processors

Certain families of thin-layer reaction–diffusion chemical media can implement sensible transformation of the initial (data) spatial distribution of the chemical species concentrations to the final (result) concentration profile (Adamatzky 2001; Sienko et al. 2003). In these reaction–diffusion computers, computation is realized via the spreading and interaction of diffusive or phase waves. Specialized, intended to solve a particular problem, experimental chemical processors implement basic operations of image processing (Kuhnert 1986b; Rambidi 1998; Rambidi et al. 2002; Adamatzky et al. 2005b), computation of optimal paths (Steinbock et al. 1995; Agladze et al. 1997; Adamatzky et al. 2005b), and control of mobile robots (Adamatzky et al. 2005b).

A device is called computationally universal if it implements a functionally complete system of logical gates, for example, a tuple of negation and conjunction, in its space–time dynamics.

A number of computationally universal reaction–diffusion devices were implemented, the findings include logical gates (Tóth and Showalter 1995; Sielewiesiuk and Gorecki 2001) and diodes (Kusumi et al. 1997; Dupont et al. 1998; Motoike and Yoshikawa 1999) in the Belousov–Zhabotinsky (BZ) medium, and an xor gate in a palladium processor (Adamatzky and De Lacy Costello 2002b).

So far, most known experimental prototypes of reaction–diffusion processors exploit the interaction of wave fronts in a geometrically constrained chemical medium. The computation is based on a stationary architecture of the medium's inhomogeneities. Constrained by stationary wires and gates, reaction–diffusion chemical universal processors provide little computational novelty and no dynamical reconfiguration ability because they simply imitate the architectures of conventional silicon computing devices. To appreciate in full the inherent massive-parallelism of thin-layer chemical media and to free the chemical processors from the imposed limitations of fixed computing architectures, an unconventional paradigm of architecture-less, or collision-based, computing has been adopted. An architecture-based, or stationary, computation implies that a logical circuit is embedded into the system in such a manner that all elements of the circuit are represented by the system's stationary states. The architecture is static. If there is any kind of "artificial" or "natural" compartmentalization, the medium is classified as an architecture-based computing device. Personal computers, living neural networks, cells, and networks of chemical reactors are typical examples of architecture-based computers.

A collision-based, or dynamical, computation employs mobile compact finite patterns, mobile self-localized excitations or simply localizations, in active nonlinear medium. The essentials of collision-based computing are as follows.

Truth values of logical variables are given by either the absence or presence of the localizations or other parameters of the localizations. The localizations travel in space and perform computation when they collide with each other. There are no predetermined stationary wires, a trajectory of the traveling pattern is a momentary wire. Almost any part of the medium space can be used as a wire. Localizations can collide anywhere within the overall space of the medium, there are no fixed positions at which specific operations occur, nor location-specified gates with fixed operations. The localizations undergo transformations, form bound states, annihilate, or fuse when they interact with other mobile patterns. The information values of the localizations are transformed as a result of the collisions and thus a computation is implemented (Adamatzky et al. 2005b).

The paradigm of collision-based computing originates from the technique of proving the computational universality of the Game of Life (Berlekamp et al. 1982), conservative logic and the billiard-ball model (Fredkin and Toffoli 1982) and their cellular automaton implementations (Margolus 1984).

Solitons, defects in tubulin microtubules, excitons in Scheibe aggregates, and breathers in polymer chains are most frequently considered candidates for the role of information carrier in nature-inspired collision-based computers, see the overview in Adamatzky (2001). It is experimentally difficult to reproduce all these artifacts in natural systems; therefore, the existence of mobile localizations in an experiment-friendly chemical media opens new horizons for the fabrication of collision-based computers.

The basis for the material implementation of the collision-based universality of reaction–diffusion chemical media was discovered by Sendiña-Nadal et al. (2001). They experimentally proved the existence of localized excitations – traveling wave fragments that behave like quasiparticles – in a photosensitive sub-excitable Belousov–Zhabotinsky medium.

6.1 Basic Gates in Excitable Chemical Medium

Most collisions between wave-fragments in a sub-excitable BZ medium follow the basic rules of the Fredkin–Toffoli billiard ball model but not always conservatively. In most cases, particularly when chemical laboratory experiments are concerned, the wave-fragments do not simply scatter as a result of collisions but rather fuse, split or generate additional new mobile localizations. Two examples of logical gates realized in the BZ medium are shown in ❷ *Fig. 12.*

In the first example (❷ *Fig. 12a, b*) the wave-fragment traveling north-west collides with the wave-fragment traveling west, a new wave-fragment traveling north-west-west is produced as a result of the collision (❷ *Fig. 12a*). This new wave-fragment represents the conjunction of the Boolean variables represented by the colliding wave-fragments (❷ *Fig. 12b*).

In the second example of experimental implementation (❷ *Fig. 12c, d*), one can see two wave-fragments, traveling east and west, colliding "head-on." Two new wave-fragments are produced as a result of the collision. One wave-fragment travels north, another wave-fragment travels south. The corresponding logical gate is shown in ❷ *Fig. 12d.*

◼ **Fig. 12**
Two logical gates implemented in experimental laboratory conditions (Toth et al. 2007). Time lapsed snapshots of propagating wave-fragments are shown in (a) and (c), schemes of the logical gates are shown in (b) and (d).

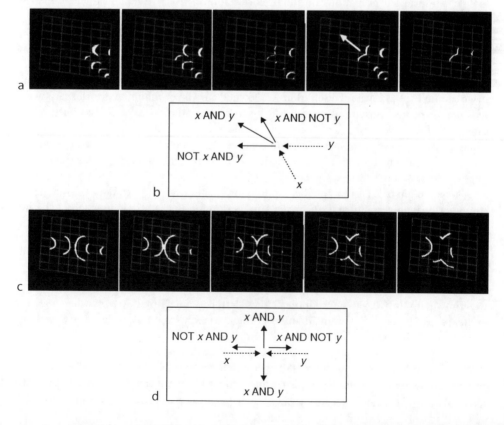

6.2 Constant Truth Generators

In the previous section, the design of logical gates in theoretical and experimental implementations of a sub-excitable BZ medium was demonstrated. The negation operator was the only thing missing in the construction, without negation and just one AND or OR gate a functional completeness could not be achieved. In cellular automata models of collision-based computers, see overview in Adamatzky (2002), negation is implemented with the help of glider guns – an autonomous generator of gliders. Streams of gliders emitted by the glider gun are collided with streams of data gliders to perform the negation operation. Until recently, there was no experimental proof that this missing component of current collision-based circuits could be obtained via experiments with the BZ medium. The gap was filled in Adamatzky et al. (2008), analogues of glider guns were realized in chemical laboratory conditions.

To produce the guns, a BZ light-sensitive sub-excitable processor was used with a slight modification, see Adamatzky et al. (2008) for details. A controllable regular configuration of excitable and non-excitable cells is created by projecting a 10×10 checkerboard pattern comprising low ($0.394 \ \mathrm{mW \ cm^{-2}}$) and high ($9.97 \ \mathrm{mW \ cm^{-2}}$) light intensity cells onto the gel surface using a data projector. In the checkboard pattern, the wave propagation depends on junctions between excitable and non-excitable cells. Junction permeability is affected by the excitability of the cell, the excitability of the neighboring cells, the size and symmetry of the network, the curvature of the wave fragments in the excitable cells, and the previous states of the junctions. When a junction fails, it acts as an impurity in the system, which leads to the

■ **Fig. 13**
Two glider guns where streams of fragments are in collision. (a–c) Snapshots of excitation dynamics in the BZ medium. (d) Scheme of the glider guns: e_1 and e_2 spiral wave fragments generating streams, g_1 and g_2 of localized excitations, a_1 and a_2 auxiliary wave fragments interacting with the wave streams (from De Lacy Costello et al. 2008).

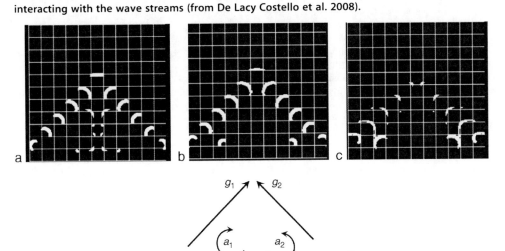

formation of a spiral wave. In certain conditions a tail of the spiral wave becomes periodically severed. Thus, independently moving localized excitations are produced. The spiral generates trains of wave fragments. An example of two glider guns, generating trains of mobile self-localized excitations, is shown in ❷ *Fig. 13*.

7   Conclusion

Major achievements in the field of reaction–diffusion computing were listed, and using selected examples of chemical laboratory prototypes it was shown that chemical reaction–diffusion computers can solve a variety of tasks. They can perform image processing, calculate Voronoi diagrams of arbitrary-shaped planar objects and skeletons of planar shapes, approximate collision-free shortest paths, guide robots toward sources of stimulation, manipulate several small objects in parallel, and implement logical functions. Recipes for the preparation of precipitating and excitable chemical media capable of carrying out such computations were provided. It is hoped that this chapter is sufficient in order for the reader to obtain a basic understanding of reaction–diffusion computing; however, readers inspired to discover and design their own chemical reaction–diffusion computers are advised to consult a more specific and thus more detailed text (Adamatzky et al. 2005b).

References

Adamatzky A (1994) Reaction-diffusion algorithm for constructing discrete generalized Voronoi diagram. Neural Netw World 6:635–643

Adamatzky A (1996) Voronoi-like partition of lattice in cellular automata. Math Comput Modelling 23:51–66

Adamatzky A (2001) Computing in nonlinear media and automata collectives. IoP Publishing, Bristol

Adamatzky A (ed) (2002) Collision-based computing. Springer, London

Adamatzky A (2004) Collision-based computing in Belousov–Zhabotinsky medium. Chaos Solitons Fractals 21:1259–1264

Adamatzky A, De Lacy Costello BPJ (2002a) Collision-free path planning in the Belousov–Zhabotinsky medium assisted by a cellular automaton. Naturwissenschaften 89:474–478

Adamatzky A, De Lacy Costello BPJ (2002b) Experimental logical gates in a reaction-diffusion medium: the XOR gate and beyond. Phys Rev E 66:046112

Adamatzky A, De Lacy Costello BPJ (2002c) Experimental reaction-diffusion pre-processor for shape recognition. Phys Lett A 297:344–352

Adamatzky A, De Lacy Costello BPJ (2007) Binary collisions between wave-fragments in a sub-excitable Belousov-Zhabotinsky medium. Chaos Solitons Fractals 34:307–315

Adamatzky A, Teuscher C (Eds) (2006) From utopian to genuine unconventional computers. Luniver Press, Beckington, UK

Adamatzky A, Tolmachiev D (1997) Chemical processor for computation of skeleton of planar shape. Adv Mater Opt Electron 7:135–139

Adamatzky A, De Lacy Costello B, Melhuish C, Ratcliffe N (2003) Experimental reaction-diffusion chemical processors for robot path planning. J Intell Robot Syst 37:233–249

Adamatzky A, De Lacy Costello B, Melhuish C, Ratcliffe N (2004) Experimental implementation of mobile robot taxis with onboard Belousov–Zhabotinsky chemical medium. Mater Sci Eng C 24:541–548

Adamatzky A, De Lacy Costello B, Skachek S, Melhuish C (2005a) Manipulating objects with chemical waves: open loop case of experimental Belousiv-Zhabotinsky medium. Phys Lett A

Adamatzky A, De Lacy Costello B, Asai T (2005b) Reaction diffusion computers. Elsevier, New York

Adamatzky A, Bull L, De Lacy Costello B, Stepney S, Teuscher C (eds) (2007) Unconventional computing 2007. Luniver, Beckington, UK

Agladze K, Obata S, Yoshikawa K (1995) Phase-shift as a basis of image processing in oscillating chemical medium. Physica D 84:238–245

Agladze K, Aliev RR, Yamaguhi T, Yoshikawa K (1996) Chemical diode. J Phys Chem 100:13895–13897

Agladze K, Magome N, Aliev R, Yamaguchi T, Yoshikawa K (1997) Finding the optimal path with the aid of chemical wave. Physica D 106:247–254

Akl SG, Calude CS, Dinneen MJ, Rozenberg G (2007) In: Unconventional computation: 6th international conference, Kingston, Canada, August 2007. Lecture notes in computer science, vol 4618

Berlekamp ER, Conway JH, Guy RL (1982) Winning ways for your mathematical plays, vol 2. Academic Press, New York

Blum H (1967) A transformation for extracting new descriptors of shape. In: Wathen-Dunn W (ed) Models for the perception of speech and visual form. MIT Press, Cambridge, MA, pp 362–380

Blum H (1973) Biological shape and visual science. J Theor Biol 38:205–287

Calabi L, Hartnett WE (1968) Shape recognition, prairie fires, convex deficiencies and skeletons. Am Math Mon 75:335–342

Courant R, Robbins H (1941) What is mathematics? Oxford University Press, New York

De Lacy Costello BPJ (2003) Constructive chemical processors – experimental evidence that shows that this class of programmable pattern forming reactions exist at the edge of a highly non-linear region. Int J Bifurcat Chaos 13:1561–1564

De Lacy Costello BPJ, Adamatzky A (2003) On multitasking in parallel chemical processors: experimental findings. Int J Bifurcat Chaos 13:521–533

De Lacy Costello BPJ, Hantz P, Ratcliffe NM (2004b) Voronoi diagrams generated by regressing edges of precipitation fronts. J Chem Phys 120 (5):2413–2416

De Lacy Costello BPJ, Adamatzky A, Ratcliffe NM, Zanin A, Purwins HG, Liehr A (2004a) The formation of Voronoi diagrams in chemical and physical systems: experimental findings and theoretical models. Int J Bifurcat Chaos 14(7):2187–2210

De Lacy Costello B, Toth R, Stone C, Adamatzky A, Bull L (2008) Implementation of glider guns in the light-sensitive Belousov–Zhabotinsky medium. Phys Rev E 79:026114

Dupont C, Agladze K, Krinsky V (1998) Excitable medium with left–right symmetry breaking. Physica A 249:47–52

Field RJ, Winfree AT (1979) Travelling waves of chemical activity in the Zaikin–Zhabotinsky–Winfree reagent. J Chem Educ 56:754

Fredkin F, Toffoli T (1982) Conservative logic. Int J Theor Phys 21:219–253

Fuerstman MJ, Deschatelets P, Kane R, Schwartz A, Kenis PJA, Deutch JM, Whitesides GM (2003) Langmuir 19:4714

Hwang YK, Ahuja N (1992) A potential field approach to path planning. IEEE Trans Robot Autom 8:23–32

Gorecka J, Gorecki J (2003) T-shaped coincidence detector as a band filter of chemical signal frequency. Phys Rev E 67:067203

Gorecki J, Yoshikawa K, Igarashi Y (2003) On chemical reactors that can count. J Phys Chem A 107:1664–1669

Gorecki J, Gorecka JN, Yoshikawa K, Igarashi Y, Nagahara H (2005) Phys Rev E 72:046201

Ichino T, Igarashi Y, Motoike IN, Yoshikawa K (2003) Different operations on a single circuit: field computation on an excitable chemical system. J Chem Phys 118:8185–8190

Klein R (1990) Concrete and abstract Voronoi diagrams. Springer, Berlin

Kuhnert L (1986b) Photochemische manipulation von chemischen Wellen. Naturwissenschaften 76:96–97

Kuhnert L (1986a) A new photochemical memory device in a light sensitive active medium. Nature 319:393

Kuhnert L, Agladze KL, Krinsky VI (1989) Image processing using light-sensitive chemical waves. Nature 337:244–247

Kusumi T, Yamaguchi T, Aliev R, Amemiya T, Ohmori T, Hashimoto H, Yoshikawa K (1997) Numerical study on time delay for chemical wave transmission via an inactive gap. Chem Phys Lett 271:355–360

Lemmon MD (1991) 2-degree-of-freedom robot path planning using cooperative neural fields. Neural Comput 3:350–362

Margolus N (1984) Physics-like models of computation. Physica D 10:81–95

Mills J (2008) The nature of the extended analog computer. In: Teuscher C, Nemenman IM, Alexander FJ (eds) Physica D Special issue: Novel Comput Paradigms Quo Vadis. Physica D 237:1235–1256

Motoike IN, Adamatzky A (2004) Three-valued logic gates in reaction-diffusion excitable media. Chaos Solitons Fractals 24:107–114

Motoike IN, Yoshikawa K (1999) Information operations with an excitable field. Phys Rev E 59:5354–5360

Motoike IN, Yoshikawa K (2003) Information operations with multiple pulses on an excitable field. Chaos Solitons Fractals 17:455–461

Motoike IN, Yoshikawa K, Iguchi Y, Nakata S (2001) Real-time memory on an excitable field. Phys Rev E 63:036220

Nakagaki T, Yamada H, Toth A (2001) Biophys Chem 92:47

Rambidi NG (1997) Biomolecular computer: roots and promises. Biosyst 44:1–15

Rambidi NG (1998) Neural network devices based on reaction-diffusion media: an approach to artificial retina. Supramol Sci 5:765–767

Rambidi NG (2003) Chemical-based computing and problems of high computational complexity: the reaction-diffusion paradigm. In: Seinko T, Adamatzky A, Rambidi N, Conrad M (eds) Molecular computing. MIT Press, Cambridge, MA

Rambidi NG, Yakovenchuk D (2001) Chemical reaction-diffusion implementation of finding the shortest paths in a labyrinth. Phys Rev E 63:026607

Rambidi NG, Shamayaev KR, Peshkov G Yu (2002) Image processing using light-sensitive chemical waves. Phys Lett A 298:375–382

Saltenis V (1999) Simulation of wet film evolution and the Euclidean Steiner problem. Informatica 10:457–466

Sendiña-Nadal I, Mihaliuk E, Wang J, Pérez-Muñuzuri V, Showalter K (2001) Wave propagation in subexcitable media with periodically modulated excitability. Phys Rev Lett 86:1646–1649

Shirakawa T, Adamatzky A, Gunji Y-P, Miyake Y (2009) On simultaneous construction of Voronoi diagram and Delaunay triangulation by Physarum polycephalum. Int J Bifurcat Chaos 19(9):3109–3117

Sielewiesiuk J, Gorecki J (2001) Logical functions of a cross junction of excitable chemical media. J Phys Chem A 105:8189–8195

Sienko T, Adamatzky A, Rambidi N, Conrad M (eds) (2003) Molecular computing. MIT Press, Cambridge, MA

Skachek S, Adamatzky A, Melhuish C (2005) Manipulating objects by discrete excitable media coupled with contact-less actuator array: open-loop case. Chaos Solitons Fractals 26:1377–1389

Steinbock O, Tóth A, Showalter K (1995) Navigating complex labyrinths: optimal paths from chemical waves. Science 267:868–871

Steinbock O, Kettunen P, Showalter K (1996) J Phys Chem 100(49):18970

Tolmachiev D, Adamatzky A (1996) Chemical processor for computation of Voronoi diagram. Adv Mater Opt Electron 6:191–196

Toth R, Stone C, Adamatzky A, de Lacy Costello B, Bull L (2009) Experimental validation of binary collisions between wave-fragments in the photosensitive Belousov-Zhabotinsky reaction. Chaos Solitons Fractals 41(4):1605–1615

Tóth A, Showalter K (1995) Logic gates in excitable media. J Chem Phys 103:2058–2066

Yokoi H, Adamatzky A, De Lacy Costello B, Melhuish C (2004) Excitable chemical medium controlled by a robotic hand: closed loop experiments. Int J Bifurcat Chaos 14:3347–3354

Zaikin AN, Zhabotinsky AM (1970) Concentration wave propagation in two-dimensional liquid-phase self-oscillating system. Nature 225:535

57 Rough–Fuzzy Computing

Andrzej Skowron
Institute of Mathematics, Warsaw University, Poland
skowron@mimuw.edu.pl

G. Rozenberg et al. (eds.), *Handbook of Natural Computing*, DOI 10.1007/978-3-540-92910-9_57,
© Springer-Verlag Berlin Heidelberg 2012

Abstract

In recent years, a rapid growth of interest in rough set theory, fuzzy set theory, and their hybridization and applications has been witnessed worldwide. In this chapter, the basic concepts of rough/fuzzy computing are presented. The role of rough/fuzzy computing in the development of Wisdom Technology (Wistech) is also emphasized.

1   Introduction

Gottfried Wilhelm Leibniz, one of the greatest of mathematicians, discussed *calculi of thoughts*. In particular, he wrote

▶ *If controversies were to arise, there would be no more need of disputation between two philosophers than between two accountants. For it would suffice to take their pencils in their hands, and say to each other: "Let us calculate."*

(*Dissertio de Arte Combinatoria* (Leipzig, 1666)).

▶ *...Languages are the best mirror of the human mind, and that a precise analysis of the signification of words would tell us more than anything else about the operations of the understanding.*
(*New Essays on Human Understanding* (*1705*), translated and edited by Peter Remnant and Jonathan Bennett, Cambridge University Press, 1982).

Only much later was it possible to recognize that new tools are necessary for developing such calculi, e.g., due to the necessity of reasoning under uncertainty about objects and vague concepts. Fuzzy set theory (Zadeh 1965) and rough set theory (Pawlak 1982) represent two complementary approaches to vagueness and, in a more general sense, to imperfect knowledge. Both of these approaches aim to approximate vague concepts. Fuzzy set theory addresses the "gradualness" of knowledge and expresses this in terms of fuzzy membership, whereas rough set theory addresses the granularity of knowledge. This granularity is manifest in the classes (knowledge granules) contained in the partition of each set of sample data defined by the indiscernibility relation.

In the rough/fuzzy approach, computations aim to construct granules satisfying a given specification, often vague and expressed in a natural language, to a satisfactory degree (Polkowski and Skowron 1996; Zadeh 2006). Granular computing (GC), introduced by Zadeh (1973, 1998) (Zadeh 2007; Bargiela and Pedrycz 2003; Pedrycz et al. 2008), may now be regarded as a unified framework for theories, methodologies, and techniques for the modeling of calculi of thoughts based on objects called granules. In the generalized theory of uncertainty, introduced by Zadeh (2006), reasoning under uncertainty is treated as a generalized constraint propagation. The generalized constraint language is used to express propositions, commands, and questions expressed in natural language.

One of the issues discussed in connection with the basic notion of a set is *vagueness*. Mathematics requires that all mathematical notions (including that of a set) must be exact, otherwise precise reasoning would be impossible. However, philosophers (Keefe 2000; Read 1994; Russell 1923; Black 1937) and recently computer scientists (see references in Pawlak and Skowron 2007c) as well as other researchers have become interested in *vague* (imprecise) concepts.

In classical set theory, a set is uniquely determined by its elements. In other words, this means that every element must be uniquely classified as belonging to the set or not. That is to say, the notion of a set is a *crisp* (precise) one. For example, the set of odd numbers is crisp because every number is either odd or even. In contrast to odd numbers, the notion of a beautiful painting is vague, because all paintings cannot be classified uniquely into two classes: beautiful and not beautiful. For some paintings, it cannot be decided whether they are beautiful or not, and these paintings thus remain in the "doubtful" area. Hence, *beauty* is not a precise concept, but a vague concept. Almost all concepts in natural language are vague. Therefore, commonsense reasoning based on natural language must be based on vague concepts and not on classical logic. An interesting discussion on this issue can be found in, for example, Read (1994).

The idea of vagueness can be traced back to the ancient Greek philosopher Eubulides of Megara (ca. 400 BC) who first formulated the so-called "sorites" (heap) and "falakros" (bald man) paradoxes (see, e.g., Keefe 2000). The bald man paradox goes as follows: suppose a man has 100,000 hairs on his head; removing one hair from his head surely cannot make him bald. Repeating this step, it can be concluded that a man without any hair is not bald. A similar reasoning can be applied to a heap of stones.

Vagueness is usually associated with the boundary region approach (i.e., the existence of objects that cannot be uniquely classified relative to a set or its complement), which was first formulated in 1893 by the father of modern logic, the German logician, Frege (1848–1925) (Frege 1903).

According to Frege, a concept must have a sharp boundary. To a concept without a sharp boundary there would correspond an area that would not have any sharp boundary – i.e., no line all around the concept that clearly distinguishes the objects to which that concept applies. This means that mathematics must use crisp concepts and not vague ones; otherwise it would be impossible to reason precisely.

However, it should be noted that the modern understanding of the notion of a vague (imprecise) concept has a quite firmly established meaning in context, involving the following issues (Keefe 2000):

1. The presence of borderline cases
2. Boundary regions of vague concepts are not crisp
3. Vague concepts are susceptible to sorites paradoxes

Let it be noted that it is usually assumed that the understanding (approximation) of vague concepts (their semantics is determined by the satisfiability relation) depends, e.g., on the agent's knowledge, which is often changing dynamically. Hence, the approximation of vague concepts by an agent should also be considered changing in time (this is known as concept drift). In the twentieth century, it became obvious that new specialized logic tools needed to be developed to investigate and implement practical problems involving vague concepts. Such tools can be based on rough sets and fuzzy sets. For example, it was shown that the rough set approach makes it possible to deal with vague concepts (see, e.g., Pawlak and Skowron 2007c; Bazan et al. 2006; Skowron 2005).

Vague complex concepts are very often related to one another through a hierarchy induced by the abstraction levels of these concepts. Such hierarchies occur, for instance, when some concepts are components of other concepts. In such a context, there is a great interest in investigating the relation *being a part-of*. This is a different approach from the ontology of modern mathematics (also known as *Cantor ontology*), which is based on the relation *being an element-of*. Such an alternative ontology for mathematics was proposed by Leśniewski (1929) and became the inspiration for an important research area within *rough mereology* (see, e.g., Polkowski and Skowron 1996). Within this approach, the notion of a *rough inclusion relation* plays a central role. The rough inclusion relation describes to what degree some concepts are parts of other concepts. A rough mereological approach is based on the relation "to be a part of to a degree."

Certainly, the mereological approach is not the only attempt at establishing links between vague concepts and rough sets. Among others are links based on treating vague concepts by means of logical values. In this case, one can build an algebra of such vague concepts as an algebra of logical values. Usually, this kind of algebra is a pseudo-Boolean algebra and the relationships between vague concepts can be expressed as relationships of logical values of an intermediate logic (see, e.g., Jankowski and Skowron 2008a).

Summing up, vagueness is (1) not allowed in mathematics, (2) interesting for philosophy, and (3) necessary for natural language, cognitive science, artificial intelligence, machine learning, and computer science.

It is worth mentioning that a large amount of knowledge related to natural computing (Rozenberg 2008; Cooper et al. 2008) is expressed in terms of vague concepts. Hence, natural computing seems to be the domain for future potential applications of rough/fuzzy computing.

The remainder of this chapter is structured as follows. ❷ Section 2 presents the basic concepts of rough computing. ❷ Section 3 reports the basic issues of fuzzy computing. Then, the combination of rough computing and fuzzy computing is surveyed in ❷ Sect. 4. Finally, conclusions together with some comments on future applications of rough/fuzzy computing in Wistech (Jankowski and Skowron 2007, 2008a) are presented.

2   Rough Sets

The first paper on rough sets was published by Professor Pawlak in 1982 (Pawlak 1982, 1991). During recent years, numerous publications on the advances of the foundations of fuzzy sets and applications in many different areas have been published (see, e.g., the survey papers, Pawlak and Skowron 2007a, b, c and the bibliography included in these papers, as well as information about recent publications on rough set theory and applications on Web pages, e.g., http://www.roughsets.org, http://logic.mimuw.edu.pl, http://rsds.wsiz.rzeszow.pl).

The rough set philosophy is founded on the assumption that some information (data, knowledge) is associated with every object in the universe of discourse. For example, if objects are patients suffering from a certain disease, symptoms of the disease form information about patients. Objects characterized by the same information are indiscernible (similar) in view of the available information about them. The indiscernibility relation generated in this way is the mathematical basis of rough set theory. This understanding of indiscernibility is based on the idea of Gottfried Wilhelm Leibniz that objects are indiscernible if and only if all available functionals take identical values on them (Leibnizian indiscernibility).

Any set of all indiscernible (similar) objects is called an elementary set and forms a basic granule (atom) of knowledge about the universe. Any union of some elementary sets is referred to as a crisp (precise) set – otherwise the set is rough (imprecise, vague).

Consequently, each rough set has boundary-line cases, i.e., objects that can neither be classified with certainty as members of the set nor of its complement. Obviously, crisp sets have no boundary-line elements at all. This means that boundary-line cases cannot be properly classified by employing the available knowledge.

Thus, the assumption that objects can be *seen* only through the information available about them leads to the view that knowledge has a granular structure. Due to the granularity of knowledge some objects of interest cannot be discerned and appear as the same (or similar). As a consequence, vague concepts, in contrast to precise concepts, cannot be characterized in terms of information about their elements. Therefore, in the proposed approach, it can be assumed that any vague concept is replaced by a pair of precise concepts – these are called the lower and the upper approximations of the vague concept. The lower approximation consists of all objects that surely belong to the concept and the upper approximation contains all objects that possibly belong to the concept. The difference between the upper and the lower approximation constitutes the boundary region of the vague concept. These upper and lower approximations are two basic operations in rough set theory.

In the following, the basic concepts are presented more formally.

Suppose two finite, nonempty sets U and A are given, where U is the *universe* of *objects*, and A is a set of *attributes*. The pair (U, A) is called an *information table*. With every attribute $a \in A$, a set V_a of its *values* is associated, called the *domain* of a. Any subset B of A determines a binary relation $I(B)$ on U, called an *indiscernibility relation*, defined by

$$xI(B)y \text{ if and only if } a(x) = a(y) \text{ for every } a \in B \tag{1}$$

where $a(x)$ denotes the value of attribute a for object x.

Obviously, $I(B)$ is an equivalence relation. The family of all equivalence classes of $I(B)$, i.e., the partition determined by B, will be denoted by $U/I(B)$, or simply U/B; an equivalence class of $I(B)$, i.e., the block of the partition U/B containing x, will be denoted by $B(x)$ (or $[x]_B$).

If $(x, y) \in I(B)$, it can be said that x and y are B-indiscernible. Equivalence classes of the relation $I(B)$ (or blocks of the partition U/B) are referred to as *B-elementary sets* or *B-elementary granules*. In the rough set approach, the elementary sets are the basic building blocks (concepts) of our knowledge about reality. The unions of *B-elementary sets* are called *B-definable sets*.

Let $\text{Inf}_B(x)$ denote the B-signature of $x \in U$, i.e., the set $\{(a, a(x)) : a \in B\}$. The signature of any object represents incomplete information about this object. The above definition of the indiscernibility relation can be rewritten in the following way:

$$xI(B)y \text{ if and only if } \text{Inf}_B(x) = \text{Inf}_B(y) \tag{2}$$

Hence, one can see that any B-elementary granule is defined by the signature of any of its objects, i.e., $B(x) = \text{Inf}_A^{-1}(x)$, where x is an arbitrary object from the granule (see ❷ *Fig. 1*). Any object from a given elementary granule is labeled by the same signature.

The equality on the right-hand side in the definition in (2) may be substituted by a similarity (tolerance) relation. This leads to the tolerance rough set approach (see, e.g., Skowron and Stepaniuk 1996). It should be noted that the signatures of objects may be incomplete

◘ **Fig. 1**
Elementary granule.

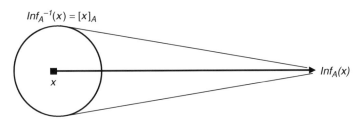

(e.g., when some values of attributes are missing) or can be influenced by noise. Many papers on rough sets are related to missing values (see the bibliography in Pawlak and Skowron 2007a, c). Recently, it was observed that the signature of objects can be extended to the relevant object contexts by hierarchical modeling (Skowron and Szczuka 2010).

The indiscernibility relation will be further used to define basic concepts of rough set theory. The two operations on sets can now be defined as

$$B_*(X) = \{x \in U : B(x) \subseteq X\} \tag{3}$$

$$B^*(X) = \{x \in U : B(x) \cap X \neq \emptyset\} \tag{4}$$

assigning to every subset X of the universe U two sets, $B_*(X)$ and $B^*(X)$, called the *B-lower* and the *B-upper approximation* of X, respectively. The set

$$BN_B(X) = B^*(X) - B_*(X) \tag{5}$$

will be referred to as the *B-boundary region* of X.

If the boundary region of X is the empty set, i.e., $BN_B(X) = \emptyset$, then the set X is *crisp* (*exact*) with respect to B; in the opposite case, i.e., if $BN_B(X) \neq \emptyset$, the set X is referred to as *rough* (*inexact*) with respect to B.

Approximations are illustrated in ❷ *Fig. 2.*

■ **Fig. 2**
A rough set.

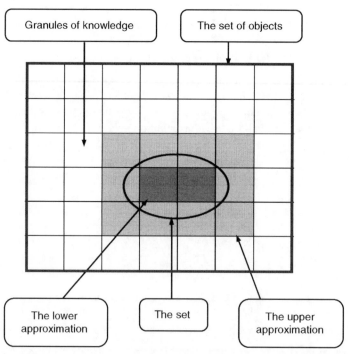

The approximations have the following properties:

$$B_*(X) \subseteq X \subseteq B^*(X)$$
$$B_*(\emptyset) = B^*(\emptyset) = \emptyset, B_*(U) = B^*(U) = U$$
$$B^*(X \cup Y) = B^*(X) \cup B^*(Y)$$
$$B_*(X \cap Y) = B_*(X) \cap B_*(Y)$$
$$X \subseteq Y \text{ implies } B_*(X) \subseteq B_*(Y) \text{ and } B^*(X) \subseteq B^*(Y)$$
$$B_*(X \cup Y) \supseteq B_*(X) \cup B_*(Y) \tag{6}$$
$$B^*(X \cap Y) \subseteq B^*(X) \cap B^*(Y)$$
$$B_*(-X) = -B^*(X)$$
$$B^*(-X) = -B_*(X)$$
$$B_*(B_*(X)) = B^*(B_*(X)) = B_*(X)$$
$$B^*(B^*(X)) = B_*(B^*(X)) = B^*(X)$$

A rough set can also be characterized numerically by the following coefficient:

$$\alpha_B(X) = \frac{|B_*(X)|}{|B^*(X)|} \tag{7}$$

called the *accuracy of approximation*, where $|X|$ denotes the cardinality of $X \neq \emptyset$. Obviously, $0 \leq \alpha_B(X) \leq 1$. If $\alpha_B(X) = 1$, then X is *crisp* with respect to B (X is *precise* with respect to B); otherwise, if $\alpha_B(X) < 1$, then X is *rough* with respect to B (X is *vague* with respect to B).

Several generalizations of the classical rough set approach based on approximation spaces defined by (U, R), where R is an equivalence relation (called indiscernibility relation) in U, have been reported in the literature. Two of the more prominent of these generalizations will now be presented.

A generalized approximation space can be defined by a tuple $AS = (U, I, v)$ where I is the *uncertainty function* defined on U with values in the powerset $P(U)$ of U ($I(x)$ is the *neighborhood* of x) and v is the *rough inclusion function* defined on the Cartesian product $P(U) \times P(U)$ with values in the interval $[0, 1]$ measuring the degree of inclusion of sets. The lower AS_* and upper AS^* approximation operations can be defined in AS by

$$AS_*(X) = \{x \in U : v(I(x), X) = 1\} \tag{8}$$

$$AS^*(X) = \{x \in U : v(I(x), X) = 0\} \tag{9}$$

In the standard case, $I(x)$ is equal to the equivalence class $B(x)$ of the indiscernibility relation $I(B)$; in case of tolerance (similarity) relation $\tau \subseteq U \times U$, $I(x) = \{y \in U : x\tau y\}$, i.e., $I(x)$ is equal to the tolerance class of τ defined by x. The standard rough inclusion relation is defined for $X, Y \subseteq U$ by

$$v(X, Y) = \begin{cases} \frac{|X \cap Y|}{|X|} & \text{if } X \neq \emptyset \\ 1 & \text{otherwise} \end{cases} \tag{10}$$

For applications, it is important to have some constructive definitions of I and v.

One can consider another way to define $I(x)$. Usually together with AS one can consider some set F of formulae describing sets of objects in the universe U of AS defined by semantics $\|\cdot\|_{AS}$, i.e., $\|\alpha\|_{AS} \subseteq U$ for any $\alpha \in F$. Now, one can take the set

$$N_F(x) = \{\alpha \in F : x \in \|\alpha\|_{AS}\} \tag{11}$$

and $I(x) = \{\|\alpha\|_{AS} : \alpha \in N_F(x)\}$. Hence, more general uncertainty functions having values in $P(P(U))$ can be defined. Usually there are considered families of approximation spaces which are labeled by some parameters. By tuning of such parameters one can search for the optimal, under chosen criteria (e.g., the minimal description length), approximation space for a given concept description.

The approach based on rough inclusion functions has been generalized to the *rough mereological approach* (Polkowski and Skowron, 1996). The inclusion relation $x\mu_r y$, with the intended meaning *x is a part of y to a degree at least r*, has been taken as the basic notion of rough mereology, being a generalization of the Leśniewski mereology (Leśniewski, 1929). Research on rough mereology has shown the importance of another notion, namely, the *closeness* of complex objects (e.g., concepts). This can be defined by $xcl_{r,r'} y$ if and only if $x\mu_r y$ and $y\mu_{r'} x$.

Rough mereology offers a methodology for the synthesis and analysis of objects in a distributed environment of intelligent agents, in particular, for the synthesis of objects satisfying a given specification to a satisfactory degree, or for control in such complex environments. Moreover, rough mereology has been used for developing the foundations of the *information granule calculi*, aimed at formalizing the "Computing with Words" paradigm, recently formulated by Zadeh (2001). More complex information granules are defined recursively using already defined information granules and their measures of inclusion and closeness. Information granules can have complex structures like classifiers or approximation spaces. Computations on information granules are performed to discover relevant information granules, e.g., patterns or approximation spaces for complex concept approximations.

It should be noted that rough sets can also be defined by employing the rough membership function instead of approximation. That is, consider

$$\mu_X^B : U \to \langle 0, 1 \rangle$$

defined by

$$\mu_X^B(x) = v(B(x), X) \tag{12}$$

where $x \in X \subseteq U$ and v is the standard rough inclusion (see Eq. 10).

The value $\mu_X^B(x)$ can be interpreted as the degree to which x belongs to X in view of knowledge about x expressed by B, or the degree to which the elementary granule $B(x)$ is included in the set X. This means that the definition reflects subjective knowledge about elements of the universe, in contrast to the classical definition of a set.

The rough membership function can also be interpreted as the conditional probability that x belongs to X given B. It is worth mentioning that set inclusion to a degree has been considered by Łukasiewicz (1970) in studies on assigning fractional truth values to logical formulae.

It can be shown that the rough membership function has the following properties:

1. $\mu_X^B(x) = 1$ iff $x \in B_*(X)$
2. $\mu_X^B(x) = 0$ iff $x \in U - B^*(X)$
3. $0 < \mu_X^B(x) < 1$ iff $x \in BN_B(X)$
4. $\mu_{U-X}^B(x) = 1 - \mu_X^B(x)$ for any $x \in U$
5. $\mu_{X \cup Y}^B(x) \geq \max(\mu_X^B(x), \mu_Y^B(x))$ for any $x \in U$
6. $\mu_{X \cap Y}^B(x) \leq \min(\mu_X^B(x), \mu_Y^B(x))$ for any $x \in U$

From these properties, it follows that the rough membership function differs in essential ways from the fuzzy membership function (Zadeh 1965), since properties (5) and (6) show that membership for the union and intersection of sets, in general, cannot be computed – as in the case of fuzzy sets – from their constituents' membership. Thus, formally, rough membership is more general than fuzzy membership. Moreover, the rough membership function depends on available knowledge (represented by attributes from B). Besides, the rough membership function, in contrast to the fuzzy membership function, has a probabilistic flavor.

Sometimes one can distinguish, in an information table (U, A), a partition of A into two classes $C, D \subseteq A$ of attributes, called *condition* and *decision* (*action*) attributes, respectively. The tuple $\mathcal{A} = (U, C, D)$ is called a *decision table*.

Let $V = \bigcup \{V_a \; a \in C\} \cup V_d$. Atomic formulae over $B \subseteq C \cup D$ and V are expressions $a = v$ called *descriptors* (*selectors*) over B and V, where $a \in B$ and $v \in V_a$. The set $\mathcal{F}(B, V)$ of formulae over B and V is the smallest set containing all atomic formulae over B and V and closed with respect to the propositional connectives $\wedge$ (conjunction), $\vee$ (disjunction), and $\neg$ (negation).

By $\|\varphi\|_{\mathcal{A}}$ we denote the meaning of $\varphi \in \mathcal{F}(B, V)$ in the decision table $\mathcal{A}$, which is the set of all objects in U with the property φ. These sets are defined by $\|a = v\|_{\mathcal{A}} = \{x \in U \mid a(x) = v\}$, $\|\varphi \wedge \varphi'\|_{\mathcal{A}} = \|\varphi\|_{\mathcal{A}} \cap \|\varphi'\|_{\mathcal{A}}$; $\|\varphi \vee \varphi'\|_{\mathcal{A}} = \|\varphi\|_{\mathcal{A}} \cup \|\varphi'\|_{\mathcal{A}}$; $\|\neg\varphi\|_{\mathcal{A}} = U - \|\varphi\|_{\mathcal{A}}$. The formulae from $\mathcal{F}(C, V)$, $\mathcal{F}(D, V)$ are called *condition formulae of $\mathcal{A}$* and *decision formulae of $\mathcal{A}$*, respectively.

Any object $x \in U$ belongs to a *decision class* $\|\bigwedge_{a \in D} a = a(x)\|_{\mathcal{A}}$ of $\mathcal{A}$. All decision classes of $\mathcal{A}$ create a partition of the universe U.

A *decision rule* for $\mathcal{A}$ is any expression of the form $\varphi \Rightarrow \psi$, where $\varphi \in \mathcal{F}(C, V)$, $\psi \in \mathcal{F}(D, V)$, and $\|\varphi\|_{\mathcal{A}} \neq \emptyset$. Formulae φ and ψ are referred to as the *predecessor* and the *successor* of decision rule $\varphi \Rightarrow \psi$. Decision rules are often called "IF ... THEN ..." rules.

The decision rule $\varphi \Rightarrow \psi$ is *true* in $\mathcal{A}$ if and only if $\|\varphi\|_{\mathcal{A}} \subseteq \|\psi\|_{\mathcal{A}}$. Otherwise, one can measure its *truth degree* by introducing some inclusion measure of $\|\varphi\|_{\mathcal{A}}$ in $\|\psi\|_{\mathcal{A}}$. It is important to note that an inclusion measure expressed in terms of the confidence measure, widely used in data mining, was considered by Łukasiewicz (1970) a long time ago in studies on assigning fractional truth values to logical formulae. Given two unary predicate formulae $\alpha(x)$, $\beta(x)$ where x runs over a finite set U, Łukasiewicz proposes to assign to $\alpha(x)$ the value $\|\|\alpha(x)\|\|/|U|$, where $\|\alpha(x)\| = \{x \in U : x \text{ satisfies } \alpha\}$. The fractional value assigned to the implication $\alpha(x) \Rightarrow \beta(x)$ is then $\|\|\alpha(x) \wedge \beta(x)\|\|/\|\|\alpha(x)\|\|$ under the assumption that $\|\alpha(x)\| \neq \emptyset$.

Each object x of a decision table determines a *decision rule* $\bigwedge_{a \in C} a = a(x) \Rightarrow \bigwedge_{a \in D} a = a(x)$.

Decision rules corresponding to some objects can have the same condition parts but different decision parts. Such rules are called *inconsistent* (*nondeterministic, conflicting, possible*); otherwise the rules are referred to as *consistent* (*certain, sure, deterministic, nonconflicting*) rules. Decision tables containing inconsistent decision rules are called *inconsistent* (*nondeterministic, conflicting*); otherwise the table is *consistent* (*deterministic, nonconflicting*).

Numerous methods have been developed within the rough sets literature for generating decision rules. They typically involve searching for decision rules that are (semi) optimal with respect to some optimization criteria describing the quality of decision rules in concept approximations.

In the case of searching for concept approximations in an extension of a given universe of objects (sample), the following steps are typically used. When a set of rules has been induced from a decision table containing a set of training examples, they can be inspected to see if they reveal any novel relationships between attributes that are worth pursuing for further research.

Furthermore, the rules can be applied to a set of unseen cases in order to estimate their classification power. For a systematic overview of rule application methods the reader is referred to the literature (Bazan 1998; Grzymała-Busse 1998; Bazan et al. 2000; Nguyen 2002; Triantaphyllou and Felici 2006) (see also references in Pawlak and Skowron 2007a, b, c).

Another important issue in data analysis is that of discovering dependencies between attributes. Intuitively, a set of attributes D depends totally on a set of attributes C, denoted $C \Rightarrow D$, if the values of attributes from C uniquely determine the values of attributes from D. In other words, D depends totally on C, if there exists a functional dependency between values of C and D. Formally, dependency can be defined in the following way. Let D and C be subsets of A.

We say that D *depends on* C to a *degree* k ($0 \leq k \leq 1$), denoted $C \Rightarrow_k D$, if

$$k = \gamma(C, D) = \frac{|POS_C(D)|}{|U|} \tag{13}$$

where

$$POS_C(D) = \bigcup_{X \in U/D} C_*(X) \tag{14}$$

called a *positive region* of the partition U/D with respect to C, is the set of all elements of U that can be uniquely classified to blocks of the partition U/D, by means of C.

If $k = 1$, we say that D *depends totally* on C, and if $k < 1$, we say that D *depends partially* (to a *degree* k) on C.

The coefficient k expresses the ratio of all elements of the universe, which can be properly classified to blocks of the partition U/D, employing attributes C and will be called the *degree of the dependency*.

It can be easily seen that if D depends totally on C then $I(C) \subseteq I(D)$. This means that the partition generated by C is finer than the partition generated by D. Notice that the concept of dependency discussed above corresponds to that considered in relational databases.

Summing up: D is *totally* (*partially*) dependent on C, if *all* (*some*) elements of the universe U can be uniquely classified to blocks of the partition U/D, employing C.

The question as to whether or not some data from a data-table can be removed while preserving its basic properties, i.e., whether a table contains some superfluous data, is a common one. This idea can be expressed more precisely.

Let $C, D \subseteq A$, be sets of condition and decision attributes, respectively. We say that $C' \subseteq C$ is a *D-reduct* (reduct with *respect* to D) of C, if C' is a minimal subset of C such that

$$\gamma(C, D) = \gamma(C', D) \tag{15}$$

The intersection of all D-reducts is called a *D-core* (core with *respect* to D). Because the core is the intersection of all reducts, it is included in every reduct, i.e., each element of the core belongs to some reduct. Thus, in a sense, the core is the most important subset of attributes, since none of its elements can be removed without affecting of the classification power of attributes.

Many other kinds of reducts and their approximations are discussed in the literature. It turns out that they can all be efficiently computed using heuristics based on a Boolean reasoning approach.

Tasks collected under the labels of data mining, knowledge discovery, decision support, pattern classification, approximate reasoning, and so on, require tools aimed at discovering *templates* (*patterns*) within the data, and classifying them into certain *decision classes*. Templates are in many cases the most frequent sequences of events, the most probable events, regular configurations of objects, the decision rules of highest quality, and so on. Tools for discovering and classifying templates are based on *reasoning schemes* rooted in various paradigms (Hastie et al. 2001). Such patterns can be extracted from data by means of methods based on Boolean reasoning and discernibility.

The discernibility relation is one of the most important relations considered in rough set theory.

The ability to discern between perceived objects is important for constructing many entities like reducts, decision rules, or decision algorithms. In the classical rough set approach, the discernibility relation $DIS(B) \subseteq U \times U$ is defined by $x\, DIS(B)y$ if and only if $non(xI(B)y)$. However, this is in general not the case for the generalized approximation spaces (one can define indiscernibility by $x \in I(y)$ and discernibility by $I(x) \cap I(y) = \emptyset$ for any objects x, y).

The idea of Boolean reasoning is based on the construction of, for a given problem P, a corresponding Boolean function f_P with the following property: the solutions for the problem P can be decoded from the prime implicants of f_P. It should be noted that to solve real-life problems it is necessary to deal with Boolean functions having a large number of variables.

A successful methodology based on the discernibility of objects and Boolean reasoning has been developed for computing many application-critical entities such as reducts and their approximations, decision rules, association rules, discretization of real-value attributes, symbolic value grouping, searching for new features defined by oblique hyperplanes or higher-order surfaces, pattern extraction from data, as well as conflict resolution or negotiation (Nguyen and Skowron 1997; Nguyen and Nguyen 1998; Nguyen 1998, 2006; Skowron 2002; Pawlak and Skowron 2007a).

Most of the problems related to generating the above mentioned entities are NP-complete or NP-hard. However, it has been found possible to develop efficient heuristics yielding suboptimal solutions. The results of experiments on many data sets are very promising (Nguyen and Skowron 1997; Nguyen and Nguyen 1998; Nguyen 1998, 2006; Skowron 2002; Pawlak and Skowron 2007a). They show very good quality of solutions generated by the heuristics in comparison with other methods reported in the literature (e.g., with respect to the classification quality of unseen objects). Moreover, they are very efficient from the point of view of the time necessary for computing the solution.

It is important to note that the methodology makes it possible to construct heuristics having a very important *approximation property*, which can be formulated as follows: expressions generated by heuristics (i.e., implicants) *close* to prime implicants define approximate solutions for the problem (Skowron 2002; Nguyen 2006).

The problem of approximation of concepts over a universe U^∞ (concepts that are subsets of U^∞) is now outlined. It is assumed that the concepts are perceived only through some subsets of U^∞, called samples. This is a typical situation in the machine learning, pattern recognition, or data-mining approaches (Hastie et al. 2001). The rough set approach to induction of concept approximations is explained using the generalized approximation spaces of the form $AS = (U, I, v)$, defined earlier.

Let $U \subseteq U^\infty$ be a finite sample. By Π_U we denote a perception function from $P(U^\infty)$ into $P(U)$ defined by $\Pi_U(C) = C \cap U$ for any concept $C \subseteq U^\infty$. $\Pi_U(C)$ represents partial information relative to the given sample U about the concept C (see ❷ *Fig. 3*).

Let $AS = (U, I, v)$ be an approximation space over the sample U. The problem as to how to extend the approximations of $\Pi_U(C)$ defined by AS to the approximation of C over U^∞ is considered. It is shown that the problem can be described as searching for an extension $AS_C = (U^\infty, I_C, v_C)$ of the approximation space AS, relevant for the approximation of C.

◘ Fig. 3
Concept sample.

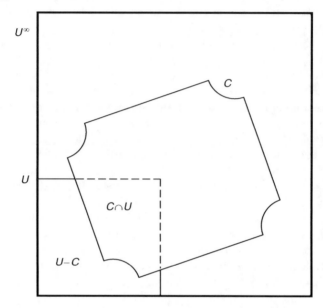

This requires us to show how to extend the inclusion function v from subsets of U to subsets of U^∞ that are relevant for the approximation of C. Observe that for the approximation of C, it is enough to induce the necessary values of the inclusion function v_C without knowing the exact value of $I_C(x) \subseteq U^\infty$ for $x \in U^\infty$.

Let AS be a given approximation space for $\Pi_U(C)$ and let a language L be considered in which the neighborhood $I(x) \subseteq U$ is expressible by a formula $\text{pat}(x)$, for any $x \in U$. This means that $I(x) = \|\text{pat}(x)\|_U \subseteq U$, where $\|\text{pat}(x)\|_U$ denotes the meaning of $\text{pat}(x)$ restricted to the sample U. In the case of rule-based classifiers, patterns of the form $\text{pat}(x)$ are defined by feature value vectors.

It is assumed that for any new object $x \in U^\infty \setminus U$ (e.g., as a result of sensor measurement) a pattern $\text{pat}(x) \in L$ with semantics $\|\text{pat}(x)\|_{U^\infty} \subseteq U^\infty$ can be obtained. However, the relationships between information granules over U^∞ like sets: $\|\text{pat}(x)\|_{U^\infty}$ and $\|\text{pat}(y)\|_{U^\infty}$, for different $x, y \in U^\infty$, are, in general, known only if they can be expressed by relationships between the restrictions of these sets to the sample U, i.e., between sets $\Pi_U(\|\text{pat}(x)\|_{U^\infty})$ and $\Pi_U(\|\text{pat}(y)\|_{U^\infty})$.

The set of patterns $\{\text{pat}(x) : x \in U\}$ is usually not relevant for approximation of the concept $C \subseteq U^\infty$. Such patterns are too specific or not sufficiently general, and can directly be applied only to a very limited number of new objects. However, by using some generalization strategies, one can search, within a family of patterns definable from $\{\text{pat}(x) : x \in U\}$ in L, for such new patterns that are relevant for the approximation of concepts over U^∞. Let a subset $\text{PATTERNS}(AS, L, C) \subseteq L$ chosen as a set of pattern candidates for the relevant approximation of a given concept C be considered. For example, in the case of rule-based classifiers, one can search for such candidate patterns among sets definable by subsequences of feature value vectors corresponding to objects from the sample U. The set $\text{PATTERNS}(AS, L, C)$

can be selected by using some quality measures checked on the meanings (semantics) of its elements restricted to the sample U (such as the number of examples from the concept $\Pi_U(C)$ and its complement that support a given pattern). Then, on the basis of properties of sets definable by these patterns over U, we induce approximate values of the inclusion function v_C on subsets of U^∞ definable by any such pattern and the concept C.

Next, the value of v_C on pairs (X, Y) is induced where $X \subseteq U^\infty$ is definable by a pattern from $\{\text{pat}(x) : x \in U^\infty\}$ and $Y \subseteq U^\infty$ is definable by a pattern from PATTERNS (AS, L, C).

Finally, for any object $x \in U^\infty \setminus U$, the approximation of the degree $v_C(\|\text{pat}(x)\|_{U^\infty}, C)$ is induced applying a conflict resolution strategy *Conflict_res* (a voting strategy, in the case of rule-based classifiers) to two families of degrees:

$$\{v_C(\|\text{pat}(x)\|_{U^\infty}, \|\text{pat}\|_{U^\infty}) : \text{pat} \in \text{PATTERNS}(AS, L, C)\} \tag{16}$$

$$\{v_C(\|\text{pat}\|_{U^\infty}, C) : \text{pat} \in \text{PATTERNS}(AS, L, C)\} \tag{17}$$

Values of the inclusion function for the remaining subsets of U^∞ can be chosen in any way – they do not have any impact on the approximations of C. Moreover, observe that for the approximation of C, the exact values of the uncertainty function I_C need not be known – it is enough to induce the values of the inclusion function v_C. Observe that the defined extension v_C of v to some subsets of U^∞ makes it possible to define an approximation of the concept C in a new approximation space AS_C.

In this way, the rough set approach to induction of concept approximations can be explained as a process of inducing a relevant approximation space.

It is worthwhile mentioning that for any formula α over descriptors one can consider its semantics $\|\alpha\|_U$ over U and semantics $\|\alpha\|_{U^\infty}$ over U^∞, respectively: (1) $\|\alpha\|_U \subseteq U$ and (2) $\|\alpha\|_{U^\infty} \subseteq U^\infty$. In approximate reasoning, hypotheses on relationships of semantics of formulae over U^∞ are induced (estimated) from information on relationships between semantics of formulae over U.

Many tasks in rough computing are based on searching for (semi)optimal approximation spaces (Jankowski et al. 2008). There are numerous significant results and active current research directions on rough set theory and applications (Pawlak and Skowron 2007a, b, c).

3 Fuzzy Sets

Fuzzy set theory has entered its fifth decade of research and development efforts since the first paper on fuzzy sets was published by Zadeh (1965). During these decades, numerous publications on the advances of the foundations of fuzzy sets and applications in many different areas have been published (see, e.g., Dubois and Prade 2000; Hoehle and Rodabaugh 1999; Bezdek et al. 1999a, b; Słowiński 1998; Nguyen and Sugeno 1998; Zimmermann 1999; Klir and Yuan 1995; Klir 2006; Pedrycz and Gomide 2007; Nikravesh et al. 2007, 2008).

In this section, some basic issues that are important for fuzzy computing are discussed.

Fuzzy sets differ from classical sets by rejecting the requirement that for any object it must be possible to determine whether the object is, or is not, a member of the set. More formally, in classical set theory, it can be assumed that for any set $X \subseteq U$, where U is the universe of objects, the following condition holds:

$$\forall x \in U(x \in X \lor x \notin X). \tag{18}$$

Contrary to classical sets, fuzzy sets are not required to have sharp boundaries distinguishing their members from other objects. The membership in a fuzzy set is a matter of degree. Due to their sharp boundaries, classical sets are usually referred to in the fuzzy literature as *crisp* sets.

Any fuzzy set over a given (crisp) universe of objects is defined by a function analogous to the characteristic function of crisp sets. This function is called a (fuzzy) membership function.

Let U be the universe of objects. A *fuzzy set* is defined uniquely by a *fuzzy membership function* of the form $\mu : U \longrightarrow [0, 1]$. For any $x \in U$, the value $\mu(x)$ defines the degree of membership of the object x from U in the fuzzy set identified with the function μ.

The membership of any object from the universe of objects in any fuzzy set is a matter of degree. By using degrees of membership, fuzzy sets make it possible to express gradual transitions from membership to nonmembership. This expressive capability is important for expressing, e.g., the meanings of vague expressions in natural language. Membership degrees in the fuzzy sets define compatibilities of relevant objects with the linguistic expression that the fuzzy sets attempt to approximate. Crisp sets are inadequate for this purpose.

Let an illustrative example of a fuzzy set corresponding to a linguistic hedge *young* be considered. The membership function corresponding to this hedge is presented in ❷ *Fig. 4*.

In the example, the universe U consists of age values. One can see that if the age of a person is not greater than 20, then the membership degree is equal to 1, i.e., this person is classified with certainty as *young*. If the age of a person is greater than 35, then the membership degree is equal to 0, i.e., this person is classified with certainty as not *young*. If the age is between 20 and 35 years the certainty degree is greater than 0 but less than 1, and the higher the age the smaller the degree of membership. Definitely, this membership function is subjective. One can ask how such a function may be acquired from data or from experts, or computed from some other fuzzy sets. One can assume that in the discussed example, the membership function was drawn by an expert. Another solution may be to select the membership function from a parameterized family of functions that are candidates for the membership function so that

■ **Fig. 4**
A fuzzy set corresponding to the linguistic expression: young.

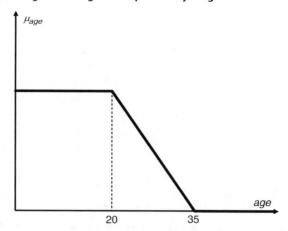

the selected function could match the expert opinions in the best way. One can consider, e.g., the following simple parameterized family of functions:

$$\mu_{a,b}(x) = \begin{cases} 1 & \text{if } x \le a \\ \frac{x-a}{b-a} & \text{if } a < x < b \\ 0 & \text{otherwise} \end{cases} \qquad (19)$$

where a, b are parameters to be tuned to match the expert opinions. In the example, the selected parameters are $a = 20$ and $b = 35$. Certainly, in many applications, the process of construction of the relevant fuzzy membership function may not be so easy. The construction of fuzzy membership functions the most suited for the application under consideration is one of the main tasks in fuzzy computing.

In the above definition of fuzzy sets, membership degrees are expressed by real numbers from the interval $[0, 1]$. In generalizations of the concept of fuzzy sets, which are often referred to as nonstandard fuzzy sets, degrees of membership may be expressed by more complex objects such as intervals of reals or elements of lattices (Goguen 1967). For the sake of simplicity, our considerations are restricted mainly to the case of degrees of membership from the interval $[0, 1]$.

The set of all fuzzy sets over the universe U is denoted by $\mathcal{F}(U)$.

If $\mu, v \in \mathcal{F}(U)$, then the fuzzy set μ is a fuzzy subset of the fuzzy set v, in symbols $\mu \subseteq v$ iff $\mu(x) \le v(x)$ for any $x \in U$.

When U is a finite set, then one can consider a fuzzy subsethood relation, Sub defined for any fuzzy subset μ, v over U by

$$\text{Sub}(\mu, v) = \frac{\sum_{x \in U} \mu(x) - \sum_{x \in U} \max\{0, \mu(x) - v(x)\}}{\sum_{x \in U} \mu(x)} \qquad (20)$$

Many characteristics are defined for any fuzzy set $\mu \in \mathcal{F}(U)$. Some of them are considered here. When U is a finite set, then $|\mu|$ denotes the fuzzy cardinality of μ defined by $|\mu| = \sum_{x \in U} \mu(x)$. Among the most important concepts of fuzzy sets are the concepts of α-cut and strong α-cut, which are defined for any fuzzy set $\mu \in \mathcal{F}(U)$ and any $\alpha \in [0, 1]$ by ${}^{\alpha}\mu = \{x \in U : \mu(x) \ge \alpha\}$ and ${}^{\alpha+}\mu = \{x \in U : \mu(x) > \alpha\}$. The set ${}^{0+}\mu$ is called the *support* and the set ${}^{1+}\mu$ is called the *core* of the fuzzy set μ.

Another task in fuzzy computing is the task of searching for the construction of membership functions (e.g., this is highly relevant for applications) from given fuzzy sets. Some examples of operations that can be used in this process are presented here.

The counterparts of set theoretical operations (i.e., complement, union, and intersection of sets) are not unique for fuzzy sets. These operations for fuzzy sets are defined relative to some functions.

The complement operation for fuzzy sets is defined by *complementation operations*, i.e., functions $n : [0, 1] \longrightarrow [0, 1]$ that satisfy the following conditions: (1) n is order reversing, (2) $n(0) = 1$, (3) $n(1) = 0$, and (4) $n(n(s))$ for all $s \in [0, 1]$ (n is an involution). An example of a complementation function is $n_\lambda(s) = (1 - s^\lambda)^{1/\lambda}$, where $\lambda > 0$. For a given complementation operation and a fuzzy set $\mu \in \mathcal{F}(U)$ the *n-complement* of μ is defined by $\mu'(x) = n(\mu(x))$ for $x \in U$.

The operations of union and intersection on fuzzy sets are defined relative to functions referred to in the research literature as *triangular norms* (or *t-norms*) and *triangular conorms* (or *t-conorms*), defined in terms of functions from the product $[0, 1] \times [0, 1]$ into $[0, 1]$, which are (1) commutative, (2) associative, (3) monotone nondecreasing, and (4) consistent with the characteristic functions of set theoretical intersections and unions, respectively. The standard example of a *t*-norm is *min*, and the standard example of a *t*-conorm is max. There are many other examples of *t*-norms and *t*-conorms such as

$$g(s, w) = 1 - \min\{1, [(1 - s)^\lambda + (1 - w)^\lambda]^{1/\lambda}\} \tag{21}$$

$$h(s, w) = \min\{1, (s^\lambda + w^\lambda)^{1/\lambda}\} \tag{22}$$

where $\lambda > 0$ is a parameter tuned for each application by knowledge acquisition techniques.

For given fuzzy sets $\mu, v \in \mathcal{F}(U)$, *t*-norm *g*, and *t*-conorm *h*, the *g*-union and *h*-intersection of μ and *v* are defined by

$$(\mu \cup v)(x) = g(\mu(x), v(x)) \tag{23}$$

$$(\mu \cap v)(x) = h(\mu(x), v(x)) \tag{24}$$

respectively, where $x \in U$.

There are other operations used for constructing fuzzy sets without counterparts in classical set theory. Among them are modifiers and averaging operations. For example, modifiers are unary operations preserving the order used for modifying fuzzy sets representing linguistic terms. They make it possible to account for linguistic hedges such as *very*, *fairly*, *extreme*, and *more or less*.

Relations are defined in classical set theory as subsets of Cartesian products of some universes of objects. One can also consider fuzzy relations as elements of $\mathcal{F}(U_1 \times \cdots \times U_k)$, where $U_1, \ldots, U_k$ are universes of objects. Fuzzy relations involve additional concepts and operations due to their multidimensionality. Among these operations are projections, cylindric extensions, compositions, joins, and inverses or fuzzy relations.

Let two examples of operations on fuzzy relations, namely, projection and composition be considered. For simplicity, it can be assumed that $\mu \in \mathcal{F}(U_1 \times U_2)$. Then the projection of μ on U_2 is defined by $\mu_{U_2}(x) = \max_{y \in Y} \mu(x, y)$, where $x \in U_1$. In the example of composition of fuzzy relations, let it be assumed that two fuzzy relations $\mu_1 \in \mathcal{F}(U_1 \times U_2)$ and $\mu_2 \in \mathcal{F}(U_2 \times U_3)$ are given. Then the fuzzy composition μ of μ_1 and μ_2 is defined by

$$\mu(x, y) = \max_{z \in U_2} \min\{\mu_1(x, z), \mu_2(z, y)\} \tag{25}$$

where $x \in U_1$ and $y \in U_3$.

An illustrative operation transforming fuzzy sets into fuzzy sets can be defined by a crisp function $f: U_1 \longrightarrow U_2$. Any such function defines an operation $F: \mathcal{F}(U_1) \longrightarrow \mathcal{F}(U_1)$ by the *extension principle*, i.e., by

$$\mu'(y) = \begin{cases} \sup\{\mu(x) : x \in U_1 \text{ and } f(x) = y\} & \text{if } f^{-1}(y) \neq \emptyset \\ 0 & \text{otherwise} \end{cases} \tag{26}$$

Some properties of fuzzy sets can be derived from properties of standard sets. Such properties are called *cutworthy properties*. For example, a fuzzy set over reals is said to be a *fuzzy interval* if and only if all of its α-cuts are closed intervals of reals. The equivalence property of fuzzy binary relations can be considered as another example. A fuzzy relation $\mu \in \mathcal{F}(U \times U)$ is a fuzzy equivalence relation if and only if this relation is reflexive, symmetric, and max–min-transitive, i.e., $\mu(x, y) \geq \max_{z \in U} \min\{\mu(x, z), \mu(z, y)\}$ for all $x \in U \times U$.

The equivalence property of binary relation over U is also a cutworthy property. However, many operations on fuzzy sets are not cutworthy (e.g., when t-norms or t-conorms, different from min and max, are applied in definitions of fuzzy union or intersection).

Several measures of fuzziness of fuzzy sets have been introduced. The basic idea in defining such measures is based on the observation that the less the fuzzy set differs from its fuzzy complement, the fuzzier it is. This idea can be expressed by the following functional $f : \mathcal{F}(U) \longrightarrow \mathcal{R}_+$ defined by

$$f(\mu) = \sum_{x \in U}(1 - |\mu(x) - n(\mu(x))|) \tag{27}$$

where $x \in U$, U is a finite universe of objects, and n is a fuzzy negation.

The degree of fuzzy membership $\mu(x)$, for a given fuzzy set $\mu \in \mathcal{F}(U)$ and $x \in U$ can be interpreted as the degree of truth of the proposition x is a member of μ.

The simplest fuzzy canonical propositional form is defined by

$$f_\mu(\mathbf{x}) : \mathbf{x} \text{ is } \mu \tag{28}$$

where $\mathbf{x}$ is a variable with values in U and $\mu \in \mathcal{F}(U)$, which is called a fuzzy predicate. By substituting into the propositional form instead of the variable $\mathbf{x}$ a particular object $x \in U$, a fuzzy proposition is obtained

$$f_\mu(x) : \mathbf{x} = x \text{ is } \mu \tag{29}$$

where $f_\mu(x)$ also denotes the degree of truth of the proposition x is μ, which is equal to $\mu(x)$ for $x \in U$.

This simple correspondence is used to define the truth of more complex propositions constructed by using negations, conjunctions, disjunctions, or implications of fuzzy propositions. Let the case of implication be considered. The propositions constructed using fuzzy implication are called conditional fuzzy propositions. They are essential for knowledge-based systems (Nguyen and Sugeno 1998). Conditional fuzzy propositions are based on the propositional form

$$f_{\mu'|\mu} : \text{if } \mathbf{x} \text{ is } \mu \text{ then } \mathbf{y} \text{ is } \mu' \tag{30}$$

where $\mathbf{x}$ and $\mathbf{y}$ are variables with values in the universes U_1 and U_2, respectively. This form may also be presented as

$$f_{\mu'|\mu} : (\mathbf{x}, \mathbf{y}) \text{ is } \xi \tag{31}$$

where $\xi \in \mathcal{F}(U_\infty \times U_\in)$ is defined by

$$\xi(x, y) = \imath(\mu(x), \mu'(y)) \tag{32}$$

where $\imath : [0, 1] \times [0, 1] \to [0, 1]$ defines a relevant fuzzy implication in the given application context. The class of such functions defining fuzzy implications is large and the selection of the relevant implication is a challenging task in the given application. One can consider some natural conditions that should satisfy any function $\imath : [0, 1] \times [0, 1] \longrightarrow [0, 1]$ defining fuzzy implications (Klir and Yuan 1995; Pedrycz and Gomide 2007). The basic conditions are the following: (1) (monotonicity) if $b \leq b'$ then $\imath(a, b) \leq \imath(a, b')$, (2) (dominance of falsity) $\imath(0, b) = 1$, and (3) (neutrality of truth) $\imath(1, b) = b$. In addition, also often some other conditions are added (4) if $a \leq a'$ then $\imath(a, b) \geq \imath(a', b)$, (5) $\imath(a, \imath(a', b)) = \imath(a', \imath(a, b))$,

(6) $\iota(a, a) = 1$, (7) $\iota(a, b) = 1$ iff $a \leq b$, (8) $\iota(a, b) = \iota(n(b), n(a))$, where n is a fuzzy negation, and (9) ι is a continuous function. In all conditions (1)–(9) $a, b, a', b' \in [0, 1]$.

In general, one class of implications is defined in terms of negation and disjunction. Another class of fuzzy implications may be defined by the residuation of continuous t-norms. Any t-norm g defines its residuation by

$$\iota(a, b) = \sup_{c \in [0,1]} \{g(a, c) \leq b\} \tag{33}$$

where $a, b \in [0, 1]$.

An important subclass of fuzzy implications known as Łukasiewicz implications is defined by

$$\iota(a, b) = \min\left[1, 1 - a^\lambda + b^\lambda\right]^{1/\lambda} \tag{34}$$

where $\lambda > 0$ is a parameter to be tuned to obtain a relevant implication for the given application and $a, b \in [0, 1]$. There are many other fuzzy implications used in fuzzy set theory and its applications. Among them are the Gödel implication and the Kleene implication defined by

$$\iota(a, b) = \begin{cases} 1 & \text{if } a \leq b \\ b & \text{otherwise} \end{cases} \tag{35}$$

and

$$\iota(a, b) = \max\{1 - a, b\} \tag{36}$$

respectively, where $a, b \in [0, 1]$.

Fuzzy propositions constructed by means of fuzzy connectives may also be quantified by means of fuzzy quantifiers. Fuzzy quantifiers are usually fuzzy intervals.

Reasoning based on fuzzy propositions is related to approximate reasoning. The most fundamental components of approximate reasoning are conditional fuzzy propositions that can be quantified in different ways. Various methods have been developed to deal with different kinds of fuzzy propositions attempting to emulate commonsense reasoning based on natural language. Let one simple example be considered that is related to the *generalized modus ponens* rule, defined by the following scheme:

Fuzzy rule: if **x** is μ then **y** is μ'
Fuzzy fact: **x** is v

Fuzzy conclusion: **y** is v'

where $\mu, v \in \mathcal{F}(U_1)$, $\mu', v' \in \mathcal{F}(U_2)$.

Assuming that the fuzzy rule has been presented in the relational form

$$(\mathbf{x}, \mathbf{y}) \text{ is } \xi \tag{37}$$

where ξ is represented by a fuzzy implication. The fuzzy conclusion v' is obtained by composition v and ξ, i.e., $v' = v \circ \xi$, where $\circ$ denotes the fuzzy composition, i.e.,

$$v'(y) = \max_{x \in U_1}\{\min\{v(x), \xi(x, y)\}\} \tag{38}$$

where $y \in U_2$.

Observe that in any application, it is necessary to choose a relevant fuzzy implication and use it to describe the relation ξ. This requires searching, in particular, for relevant fuzzy implication making it possible to derive conclusions matching the empirical observations or expectations by human experts. There are several methods for the selection of the necessary parameters (Dubois and Prade 2000; Hoehle and Rodabaugh 1999; Bezdek et al. 1999a; Pedrycz and Gomide 2007).

Fuzzy systems are represented by a set of variables and dependencies among values (states) of the variables, which are fuzzy sets. Very often, the values are fuzzy intervals representing linguistic terms such as *small, medium, large, very large,* with the interpretation of these variables depending on the application under consideration. Each linguistic variable is defined by (1) a name of this variable; (2) a base variable with a given set of values, which is often the set of reals; (3) a set of linguistic terms referring to the base variable such as *small, medium, large, very large* and (4) a set of fuzzy intervals defining the semantics of each linguistic term. Fuzzy sets defining semantics of linguistic terms create a *fuzzy partition,* i.e., for any object x from the domain of the base variable the sum of membership degrees in these fuzzy sets is equal to 1.

In knowledge-based systems, relationships between variables are given by a set of *fuzzy inference rules.* These rules are conditional fuzzy propositional forms representing the relevant human expert knowledge, often expressed in natural language. Another class of fuzzy systems involves creating model-based fuzzy systems based on traditional modeling combined with the application of fuzzy sets. By combining knowledge-based systems and model-based systems, such hybrid systems are obtained.

The input–output relationship of such a knowledge-based system can be described by the following scheme of inference:

Rule 1:   if $\mathbf{x}_1$ is μ_{11} and $\mathbf{x}_2$ is μ_{21} …
              then $\mathbf{y}_1$ is μ'_{11} and $\mathbf{y}_2$ is μ'_{21} and …

Rule 2:   if $\mathbf{x}_1$ is μ_{12} and $\mathbf{x}_2$ is μ_{22} …
              then $\mathbf{y}_1$ is μ'_{12} and $\mathbf{y}_2$ is μ'_{22} and …

....

Rule k: if $\mathbf{x}_1$ is μ_{1k} and $\mathbf{x}_2$ is μ_{2k} …
              then $\mathbf{y}_1$ is μ'_{1k} and $\mathbf{y}_2$ is μ'_{2k} and …

Fact: $\mathbf{x}_1$ is v_1 and $\mathbf{x}_2$ is v_2 and …

Conclusion: $\mathbf{y}_1$ is v_1 and $\mathbf{y}_1$ is μ'_2 and …

Different inference methods based on the fuzzy inference rules presented above have been developed. A thorough consideration of all the details of those methods is beyond the scope of this overview. However, it is appropriate to make some general comments. Let it be assumed that (1) each fuzzy inference rule is based on input and output linguistic variables and (2) the values of variables on the left-hand side of rules are given, and the fuzzy intervals (over reals) corresponding to these variables are defined. Then, for each rule r, the inference is performed using the following steps:

1. *Selection of aggregation methods of membership degrees given for fuzzy sets on the left-hand side of the rule r.* This requires, e.g., the selection of some t-norms corresponding to conjunctions occurring on the left-hand side of the rule r. It is worth mentioning that interpretation of all conjunctions by the t-norm *min* may not be satisfactory for a given application and it may then be necessary to search for different t-norms for different conjunctions.

2. *Modification of fuzzy sets on the right-hand side of the rule r using the aggregated membership degree from the previous step.* As a consequence, the result of application of each fuzzy inference rule is a fuzzy interval for each output variable. Next, for each output variable, the fuzzy intervals obtained from different inference rules are aggregated and a new fuzzy interval is obtained. Finally, a defuzzification process is launched, i.e., for each variable, the obtained fuzzy interval is transformed into a single real number representing, in the context of the given application and in the best way, the fuzzy set. One of the basic defuzzification methods, which is called a *centroid method*, can be described for a finite universe U of objects by

$$d(\mu) = \frac{\sum_{x \in U} x \cdot \mu(x)}{\sum x \in U \mu(x)} \tag{39}$$

where $x \in U$.

Methods based on fuzzy logic have been applied to many real-life problems including control problems (Nguyen and Sugeno 1998). Fuzzy logic was inspired by a number of remarkable human capabilities. Among them Zadeh distinguishes (1) the ability to reason and make decisions in cases of imprecision, uncertainty, incompleteness of information and partial truth and (2) the capability to perform many tasks based on perceptions, without any measurements and any computations. In the case of fuzzy logic, we move away from numerical variables and toward linguistic variables or words in natural languages; we sacrifice precision in order to achieve important advantages down the line. This is what is called *the fuzzy gambit*. Let it be noted that in this case, the values of the variables are known precisely. Fuzzy logic is also applied when the values of variables are not known precisely (Zadeh 2007). There are three principal rationales that dictate when words should be used in preference to numbers: (1) when available information is not precise enough to use numbers; (2) when there is a tolerance for imprecision that can be exploited to achieve tractability, robustness, and low solution costs; and (3) when the expressive power of words is higher than the expressive power of numbers.

The problem of construction of fuzzy sets (fuzzy membership functions) is very important for applications. This problem is also related to knowledge acquisition. Methods for constructing fuzzy membership functions can be classified into direct and indirect. In direct methods, experts are expected to define the fuzzy membership function. In indirect methods, experts are asked to compare elements in pairs from the universe of objects relative to the order of the membership degrees. Methods for constructing fuzzy membership functions and relevant operations on fuzzy sets for a given application are crucial for successful applications of fuzzy sets.

There are numerous significant results and active current research directions on fuzzy set theory and applications (Dubois and Prade 2000; Hoehle and Rodabaugh 1999; Bezdek et al. 1999a, b; Słowiński 1998; Nguyen and Sugeno 1998; Zimmermann 1999; Klir and Yuan 1995; Klir 2006; Pedrycz and Gomide 2007; Nikravesh et al. 2007, 2008).

4   Rough/Fuzzy Hybridization

There are numerous theoretical results and real-life applications based on the combination of rough set theory and fuzzy set theory (see, e.g., Dubois and Prade 1987, 1988, 1990; Nanda

1992; Pal and Banerjee 1996; Greco et al. 1998; Pal and Skowron 1999; Greco et al. 1999, 2000, 2006; Pal 2003; Polkowski 2002; Wu et al. 2003; Inuiguchi et al. 2004; Lingras and Jensen 2007; Maji and Pal 2007, 2005; Pawlak and Skowron 2007a, c and references in these papers). These results show the fruitful complementarity of fuzzy set theory and rough set theory, which were both created to deal with imperfect knowledge, in particular, with vague concepts. In fuzzy set theory, any object is characterized by a grade of membership relative to a given concept (*gradualness of membership*), while rough set theory started from the assumption about the possible indiscernibility of perceived objects (*granularity of knowledge*).

Let some examples be considered illustrating the results of the hybridization of fuzzy sets and rough sets in the definition of concept approximation.

Let it be assumed that $\mu \in \mathcal{F}(U)$. The fuzzy set μ can be characterized by a set of α-cuts $\{^{\alpha}\mu\}_{\alpha \in [0, 1]}$. Let it also be assumed that $R \subseteq U \times U$ is an equivalence relation over the universe of objects U. Then, the rough/fuzzy set corresponding to μ is defined by two families of sets, the lower approximation family of cuts and the upper approximation family of cuts defined by

$$\underline{R}(^{\alpha}\mu) = \bigcup_{x \in U} \{[x]_R : [x]_R \subseteq {}^{\alpha}\mu\} \tag{40}$$

and

$$\overline{R}(^{\alpha}\mu) = \bigcup_{x \in U} \{[x]_R : [x]_R \cap {}^{\alpha}\mu \neq \emptyset\} \tag{41}$$

respectively, where $x \in U$, and $[x]_R$ denotes the equivalence class of R defined by x.

Another approach can be used to define the so-called fuzzy/rough sets. In this case, it is assumed that $\mu \in \mathcal{F}(U)$ and $v = \{v_i\}_{i \in I}$ is a fuzzy partition of U. Then two fuzzy sets (over the universe of elements v_i of fuzzy partition v) are defined: the fuzzy/rough lower approximation of μ relative to v and the fuzzy/rough upper approximation of μ relative to v by

$$\underline{v}\mu(v_i) = \inf_{x \in U} \max\{1 - v_i(x), \mu(x)\} \tag{42}$$

and

$$\overline{v}\mu(v_i) = \sup_{x \in U} \min\{v_i(x), \mu(x)\} \tag{43}$$

where $i \in I$.

There are many results on fuzzy/rough sets and rough/fuzzy sets including the generalizations of the above definitions (see, e.g., the references in Lingras and Jensen 2007).

Many methods have also been developed that are based on the combination of rough sets and fuzzy sets for real-life applications. In particular, these methods have been applied in supervised learning, information retrieval, feature selection, clustering, neurocomputing, and evolutionary computing (Lingras and Jensen 2007). There are different strategies for the combination of these two approaches. Let one possible application be illustrated, in which rough sets are used in the approximation of fuzzy sets by searching for *subconcepts* of a given fuzzy set that can be approximated by rough set methods with high quality. The layers defined by the differences of successive cuts are selected as subconcepts. The problem that should be solved is the selection of relevant cuts so that the patterns defined by approximations of layers are relevant for the application under consideration (Skowron and Stepaniuk 2003; Bazan 2008a, b). ❷ *Figure 5* illustrates patterns corresponding to the lower approximations of layers and boundary regions. In applications, it is necessary to select the relevant number of

◘ **Fig. 5**

Learning layers and their approximations.

Patterns defined by the lower approximations of
two successive layers

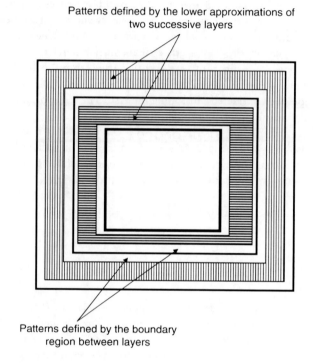

Patterns defined by the boundary
region between layers

layers and regions representing layers before the approximations can be constructed. Optimization strategies supported by domain knowledge are used in searching for layers to obtain the relevant patterns for the given application.

One of the very active research areas in which rough/fuzzy computing plays a very important role is GC (see, e.g., Bargiela and Pedrycz 2003; Pedrycz et al. 2008). In fuzzy logic, everything may be graduated and granulated. The concepts of graduation, granulation, and graduated granulation play a key role in GC. In GC, computations are performed on granules that are clumps of objects drawn together by indistinguishability, similarity, proximity, or functionality. Graduated granulation is inspired by the way in which humans deal with complexity and imprecision (Zadeh 2007). In searching for different relevant granules, rough set methods can be used making it possible to construct, e.g., complex classifiers of high quality. The goal of computations is to construct granules satisfying a given specification (generalized constraints), often expressed in natural language, to a satisfactory degree (Polkowski and Skowron 1996; Zadeh 2006).

5 Wisdom Technology

One of the potential areas for many further applications of rough/fuzzy computing is Wistech discussed in Jankowski and Skowron (2007, 2008a, b) as one of the main paradigms for the development of modern intelligent systems. It will be useful to present some basic ideas of

Wistech because in our opinion Wistech is also very important for developing new methods in natural computing (Rozenberg 2008; Cooper et al. 2008).

There are many indications that we are currently witnessing the onset of an era of radical technological changes. These radical changes depend on the further advancement of technology to acquire, represent, store, process, discover, communicate, and learn "wisdom." We call this technology *wisdom technology* (or Wistech, for short). The term *wisdom* commonly means *rightly judging*. This common notion can be refined. By *wisdom*, we understand an adaptive ability to make judgments correctly to a satisfactory degree (in particular, correct decisions) having in mind real-life constraints. The intuitive nature of wisdom understood in this way can be expressed by the so-called *wisdom equation* (Jankowski and Skowron 2007), as shown metaphorically in (44):

$$\text{Wisdom} = \text{knowledge} + \text{adaptive judgment} + \text{interactions} \qquad (44)$$

Wisdom can be treated as a special type of knowledge processing. To explain the specificity of this type of knowledge processing, let it be assumed that a control system of a given agent Ag consists of a society of agent control components interacting with the other agent Ag's components and with the agent Ag's environments. Moreover, there are special agent components called the agent coordination control components that are responsible for the coordination of control components. Any agent coordination control component mainly searches for answers to the following question: What to do next? Or, more precisely: Which of the agent Ag control components should be activated now? Of course, any agent control component has to process some kind of knowledge representation. In the context of agent perception, the agent Ag itself (by using, e.g., interactions, memory, and coordination among control components) is processing a very special type of knowledge reflecting the agent perception of the hierarchy of needs (objectives, plans, etc.) and the current agent or the environment constraints. This kind of knowledge processing mainly deals with complex vague concepts (such as risk or safety) from the point of view of the selfish agent needs. Usually, this kind of knowledge processing is not necessarily logical reasoning in terms of proving statements (i.e., labeling statements by truth values such as TRUE or FALSE). This knowledge processing is rather analogous to the judgment process in a court aiming at recognition of evidence that could be used as an argument for or against. Arguments for or against are used in order to make the final decision as to which one of the solutions is the best for the agent in the current situation (i.e., arguments are labeling statements by judgment values expressing the action priorities). The evaluation of current needs by agent Ag is realized by using a hierarchy of agent Ag life values/needs). Wisdom style knowledge processing by the agent Ag is characterized by the ability to improve the quality of the judgment process based on the agent Ag's experiences. In order to emphasize the importance of this ability, we use the concept of adaptive judgment in the wisdom equation instead of just judgment. An agent who is able to perform adaptive judgment in the above sense is simply called a judge.

The adaptivity aspects are also crucial from the point of view of interactions (Goldin et al. 2006; Nguyen and Skowron 2008; Skowron 2008; Skowron and Szczuka 2010). The need for adaptation follows, e.g., from the fact that complex vague concepts on the basis of which the judgment is performed by the agent Ag are approximated by classification algorithms (classifiers), which should drift in time following changes in data and represented knowledge.

An important aspect of Wistech is that the complexity and uncertainty of real-life constraints mean that in practice we must reconcile ourselves to the fact that our judgment is based on non-crisp concepts (i.e., concepts with borderline cases) and also does not take into account all the accumulated and available knowledge. This is why our judgments are usually imperfect. But, as a consolation, we also learn to improve the quality of our judgments via observation and analysis of our experience during interaction with the environment. Satisfactory decision-making levels can be achieved as a result of improved judgments.

Thus, wisdom is directly responsible for the focusing of an agent's attention on problems and techniques for their solution that are important in terms of the agent's judgment mechanism. This mechanism is based on the Maslow hierarchy of needs (Jankowski and Skowron 2008a) and agent perception of ongoing interactions with other agents and environments. In particular, the agent's wisdom can be seen as the control at the highest level of hierarchy of the agent's actions and reactions, and is based on concept processing in the metaphoric Aristotle tetrahedron. One can use the following conceptual simplification of agent wisdom: Agent wisdom is an efficient and an online agent judgment mechanism making it possible for agents to answer the following questions: (1) How to construct the most important priority list of problems to be solved? (2) How to solve the top priority problems under real-life constraints? (3) What to do next?

One of the main barriers hindering an acceleration in the development of Wistech applications lies in developing satisfactory computational models implementing the function of *adaptive judgment*. This difficulty primarily consists in overcoming the complexity of integrating the local assimilation and processing of dynamically changing, non-crisp, and incompletely specified concepts necessary to make correct judgments. In other words, we are only able to model tested phenomena using local (subjective) models and interactions between them. In practical applications, usually, we are not able to give perfect global models of analyzed phenomena. However, global models can only be approximated by integrating the various incomplete perspectives of problem perception.

Wistech is based on techniques of reasoning about knowledge, information, and data, which helps apply the current knowledge to problem solving in real-life highly unpredictable environments and autonomous multiagent systems. This includes such methods as identification of the current situation on the basis of interactions or dialogs, extraction of relevant fragments of knowledge from knowledge networks, judgment for prediction for relevant actions or plans in the current situation, or judgment of the current plan reconfiguration.

The concepts surrounding Wistech can be summarized as follows.

Wisdom technology (*Wistech*) is a collection of techniques aimed at the further advancement of technology to acquire, represent, store, process, discover, communicate, and learn *wisdom* in the design and implementation of intelligent systems. These techniques include approximate reasoning by agents or teams of agents about vague concepts concerning real-life dynamically changing, usually distributed, systems in which these agents are operating. Such systems consist of other autonomous agents operating in highly unpredictable environments and interacting with each other. Wistech can be treated as the successor of database technology, information technology, and knowledge management technologies. Wistech is the combination of the technologies represented in (44) and offers an intuitive starting point for a variety of approaches to designing and implementing computational models of Wistech in intelligent systems.

Knowledge technology in Wistech is based on techniques for reasoning about knowledge, information, and data, which helps apply the current knowledge in problem solving. This includes, e.g., extracting relevant fragments of knowledge from knowledge networks for making decisions or reasoning by analogy.

Judgment technology in Wistech covers the representation of agent perception and adaptive judgment strategies based on results of the perception of real-life scenes in environments and their representations in the agent's mind. The role of judgment is crucial, e.g., in adaptive planning relative to the Maslow hierarchy of the agent's needs or goals. Judgment also includes techniques used for perception, analysis of perceived facts, learning, and adaptive improving of approximations of vague complex concepts (from different levels of concept hierarchies in real-life problem solving) applied to modeling interactions in dynamically changing environments (in which cooperating, communicating, and competing agents exist) by using uncertain and insufficient knowledge or resources.

Interaction technology includes techniques for performing and monitoring actions by agents and environments. Techniques for planning and controlling actions are the result of the combination of judgment and interaction technologies.

Clearly, one of the promising possible ways to build Wistech computational models is by using methods of rough/fuzzy computing and other soft computing approaches.

6 Conclusions

Some basic concepts related to rough/fuzzy computing have been outlined. There are numerous research directions based on the foundations and applications of rough sets, fuzzy sets, and their combinations. The interested reader is referred, e.g., to Pawlak and Skowron (2007a, b, c), Dubois and Prade (2000), Hoehle and Rodabaugh (1999), Bezdek et al. (1999a, b), Słowiński (1998), Nguyen and Sugeno (1998), Zimmermann (1999), Klir and Yuan (1995), Klir (2006), Pedrycz and Gomide (2007), and Nikravesh et al. (2007, 2008), and the bibliographies in these papers and books.

A large part of knowledge related to natural computing is expressed by vague concepts. Hence, natural computing seems to be the domain for future potential applications of rough/fuzzy computing, especially in the framework of Wistech.

Acknowledgments

The author would like to express his gratitude to Professor Dave Corne for suggestions and corrections helping to improve this chapter.

This research has been supported by the grant N N516 368334 from the Ministry of Science and Higher Education of the Republic of Poland.

References

Bargiela A, Pedrycz W (2003) Granular computing: an introduction. Kluwer, Dordrecht

Bazan J (2008a) Hierarchical classifiers for complex spatio-temporal concepts. In: Transactions on rough sets IX. Lecture notes in computer science, vol. 5390. Springer, Berlin, pp 470–450

Bazan J (2008b) Rough sets and granular computing in behavioral pattern identification and planning. In:

Pedrycz W, Skowron A, Kreinovich V (eds) Handbook of granular computing. Wiley, New York, pp 777–800

Bazan J, Skowron A, Swiniarski R (2006) Rough sets and vague concept approximation: from sample approximation to adaptive learning. In: Transactions on rough sets V. Lecture notes in computer science, vol 4100. Springer, Berlin, pp 39–62

Bazan JG (1998) A comparison of dynamic and non-dynamic rough set methods for extracting laws from decision tables. In: Polkowski L, Skowron A (eds) Rough sets in knowledge discovery 1: methodology and applications. Studies in fuzziness and soft computing, vol 18. Physica, Heidelberg, pp 321–365

Bazan JG, Nguyen HS, Nguyen SH, Synak P, Wróblewski J (2000) Rough set algorithms in classification problems. In: Polkowski L, Lin TY, Tsumoto S (eds) Rough set methods and applications: new developments in knowledge discovery in information systems. Studies in fuzziness and soft computing, vol 56. Springer/Physica, Heidelberg, pp 49–88

Bezdek J, Dubois D, Prade H (eds) (1999a) Fuzzy sets in approximate reasoning and information systems. In: Dubois D, Prade H (series eds) Handbook of fuzzy sets series, vol 2. Kluwer, Boston/Dordrecht

Bezdek J, Pal N, Keller J, Krishnapuram R (eds) (1999b) Fuzzy set models for pattern recognition and image processing. In: Dubois D, Prade H (series eds) Handbook of fuzzy sets series, vol 2. Kluwer, Boston/Dordrecht

Black M (1937) Vagueness: An exercise in logical analysis. Philos Sci 4(4):427–455

Cooper SB, Löwe B, Sorbi A (eds) (2008) New computational paradigms, changing conceptions of what is computable. Springer, New York

Dubois D, Prade H (1987) Twofold fuzzy sets and rough sets–some issues in knowledge representation. Fuzzy Set Syst 23(1):3–18

Dubois D, Prade H (1988) Fuzzy rough sets. Note on Mult.-Valued Logic in Japan 9(8):1–8

Dubois D, Prade H (1990) Rough fuzzy sets and fuzzy rough sets. Int J Gen Syst 17(2–3):191–209

Dubois D, Prade H (eds) (2000) Fundamentals of fuzzy sets. In: Dubois D, Prade H (series eds) Handbook of fuzzy sets series, vol 1. Kluwer, Boston/Dordrecht

Frege G (1903) Grundgesetze der Arithmetik, 2. Verlag von Hermann Pohle, Jena

Goguen J (1967) L-fuzzy sets. J Math Anal Appl 18:145–174

Goldin D, Smolka S, Wegner P (2006) Interactive computation: the new paradigm. Springer, Heidelberg

Greco S, Matarazzo B, Slowinski R (1998) Fuzzy similarity relation as a basis for rough approximations. In: Polkowski L, Skowron A (eds) Rough sets and current trends in computing. Lecture notes in computer science, vol 1424. Springer, Berlin, pp 283–289

Greco S, Matarazzo B, Słowiński R (1999) The use of rough sets and fuzzy sets in MCDM. In: Gal T, Stewart T, Hanne T (eds) Advances in MCDM models, algorithms, theory, and applications. Kluwer, Dordrecht, pp 14.1–14.59

Greco S, Matarazzo B, Slowinski R (2000) Fuzzy extension of the rough set approach to multicriteria and multiattribute sorting. In: Fodor J, Baets BD, Perny P (eds) Preferences and decisions under incomplete knowledge. Physica, Heidelberg, pp 131–151

Greco S, Inuiguchi M, Slowinski R (2006) Fuzzy rough sets and multiple-premise gradual decision rules. Int J Approx Reason 41(2):179–211

Grzymała-Busse JW (1998) LERS – a knowledge discovery system. In: Polkowski L, Skowron A (eds) Rough sets in knowledge discovery 2. Applications, case studies and software systems. Studies in fuzziness and soft computing. Physica, Heidelberg, pp 562–565

Hastie T, Tibshirani R, Friedman JH (2001) The elements of statistical learning: data mining, inference, and prediction. Springer, Heidelberg

Hoehle U, Rodabaugh S (eds) (1999) Mathematics of fuzzy sets: Logic, topology and measure theory. In: Dubois D, Prade H (series eds) Handbook of fuzzy sets series, vol 2. Kluwer, Boston/Dordrecht

Inuiguchi M, Greco S, Slowinski R (2004) Fuzzy-rough modus ponens and modus tollens as a basis for approximate reasoning. In: Tsumoto S, Slowinski R, Komorowski HJ, Grzymala-Busse JW (eds) Rough sets and current trends in computing. Lecture notes in computer science, vol 3066. Springer, Berlin, pp 84–94

Jankowski A, Skowron A (2007) A Wistech paradigm for intelligent systems. In: Transactions on rough sets VI. Lecture notes in computer science, vol 4374. Springer, Berlin, pp 94–132

Jankowski A, Skowron A (2008a) Logic for artificial intelligence: the Rasiowa-Pawlak school perspective. In: Ehrenfeucht A, Marek V, Srebrny M (eds) Andrzej Mostowski and foundational studies. IOS Press, Amsterdam, pp 106–143

Jankowski A, Skowron A (2008b) Wisdom granular computing. In: Pedrycz W, Skowron A, Kreinovich V (eds) Handbook of granular computing. Wiley, New York, pp 329–346

Jankowski A, Peters J, Skowron A, Stepaniuk J (2008) Optimization in discovery of compound granules. Fund Inform 85(1–4):249–265

Keefe R (2000) Theories of vagueness. Cambridge University Press, Cambridge

Klir G, Yuan B (1995) Fuzzy logic: theory and applications. Prentice-Hall, Englewood Cliffs, NJ

Klir GJ (ed) (2006) Uncertainty and information: foundations of generalized information theory. Wiley, Hoboken, NJ

Leśniewski S (1929) Grundzüge eines neuen Systems der Grundlagen der Mathematik. Fund Math 14:1–81

Lingras P, Jensen R (2007) Survey of rough and fuzzy hybridization. In: Preferences and decisions under incomplete knowledge. FUZZ-IEEE 2007: Proceedings of 2007 IEEE international conference on fuzzy systems. Imperial College, London, 23–26 July, pp 125–130

Łukasiewicz J (1970) Die logischen Grundlagen der Wahrscheinlichkeitsrechnung, Kraków 1913. In: Borkowski L (ed) Jan Łukasiewicz – selected works. North Holland, Amsterdam/London

Maji P, Pal SK (2005) Rough-fuzzy c-medoids algorithm and selection of bio-basis for amino acid sequence analysis. IEEE T Knowl Data Eng 19(6): 859–872

Maji P, Pal SK (2007) RFCM: A hybrid clustering algorithm using rough and fuzzy sets. Fund Inform 80(4):475–496

Nanda S (1992) Fuzzy rough-sets. Fuzzy Set Syst 45:157–160

Nguyen H, Sugeno M (eds) (1998) Fuzzy systems modelling and control. In: Dubois D, Prade H (series eds) Handbook of fuzzy sets series, vol 6. Kluwer, Boston/Dordrecht

Nguyen HS (1998) From optimal hyperplanes to optimal decision trees. Fund Inform 34(1–2):145–174

Nguyen HS (2002) Scalable classification method based on rough sets. In: Alpigini JJ, Peters JF, Skowron A, Zhong N (eds) Rough sets and current trends in computing. Lecture notes in computer science, vol 2475. Springer, Berlin, pp 433–440

Nguyen HS (2006) Approximate Boolean reasoning: Foundations and applications in data mining. In: Transactions on rough sets V. Lecture notes in computer science, vol 4100. Springer, Berlin, pp 334–506

Nguyen HS, Skowron A (1997) Boolean reasoning for feature extraction problems. In: Ras ZW, Skowron A (eds) ISMIS. Lecture notes in computer science, vol 1325. Springer, Berlin, pp 117–126

Nguyen HS, Skowron A (2008) A rough granular computing in discovery of process models from data and domain knowledge. J Chongqing Univ 20(3):341–347

Nguyen SH, Nguyen HS (1998) Pattern extraction from data. Fund Inform 34(1–2):129–144

Nikravesh M, Kacprzyk J, Zadeh LA (eds) (2007) Forging new frontiers: Fuzzy pioneers I. In: Studies in fuzziness and soft computing, vol 217. Springer, Heidelberg

Nikravesh M, Kacprzyk J, Zadeh LA (eds) (2008) Forging new frontiers: Fuzzy pioneers II. In: Studies in fuzziness and soft computing, vol 218. Springer, Heidelberg

Pal S, Banerjee M (1996) Roughness of a fuzzy set. Inform Sci 93(3):235–246

Pal SK (2003) Rough-fuzzy granular computing, case based reasoning and data mining. In: Gesù VD,

Masulli F, Petrosino A (eds) WILF. Lecture notes in computer science, vol 2955. Springer, Berlin, pp 1–10

Pal SK, Skowron A (eds) (1999) Rough fuzzy hybridization: a new trend in decision-making. Springer, Singapore

Pawlak Z (1982) Rough sets. Int J Comput Inf Sci 11:341–356

Pawlak Z (1991) Rough sets: theoretical aspects of reasoning about data. In: System theory, knowledge engineering and problem solving, vol 9. Kluwer, Dordrecht

Pawlak Z, Skowron A (2007a) Rough sets and Boolean reasoning. Inform Sci 177(1):41–73

Pawlak Z, Skowron A (2007b) Rough sets: some extensions. Inform Sci 177(1):28–40

Pawlak Z, Skowron A (2007c) Rudiments of rough sets. Inform Sci 177(1):3–27

Pedrycz W, Gomide F (2007) Fuzzy systems engineering toward human-centric computing. Wiley, Hoboken, NJ

Pedrycz W, Skowron A, Kreinovich V (eds) (2008) Handbook of granular computing. Wiley, New York

Polkowski L (ed) (2002) Rough sets: mathematical foundations. Advances in soft computing. Physica, Heidelberg

Polkowski L, Skowron A (1996) Rough mereology: a new paradigm for approximate reasoning. Int J Approx Reason 51:333–365

Read S (1994) Thinking about logic: an introduction to the philosophy of logic. Oxford University Press, Oxford

Rozenberg G (2008) Computer science, informatics, and natural computing – personal reflections. In: Cooper SB, Löwe B, Sorbi A (eds) New computational paradigms changing conceptions of what is computable. Springer, New York, pp 373–379

Russell B (1923) Vagueness. Austral J Psychol Philos 1:84–92

Skowron A (2002) Rough sets in KDD – plenary talk. In: Shi Z, Faltings B, Musen M (eds) IFIP'00: 16-th world computer congress: IIP'00, Proceedings of conference on intelligent information processing. Publishing House of Electronic Industry, Beijing, pp 1–14

Skowron A (2005) Rough sets and vague concepts. Fund Inform 64(1–4):417–431

Skowron A (2008) Learning complex granules and their interactions. In: Nguyen HS, Huynh VN (eds) SCKT 2008: International workshop on soft computing for knowledge technology at the 10-th Pacific Rim international conference on artificial intelligence, 15–19 May 2008. Hanoi, Vietnam, pp 1–14

Skowron A, Stepaniuk J (1996) Tolerance approximation spaces. Fund Inform 27:245–253

Skowron A, Stepaniuk J (2003) Information granules and rough-neural computing. In: Pal SK, Polkowski L, Skowron A (eds) Rough-neural computing:

techniques for computing with words. Cognitive technologies. Springer, Berlin, pp 43–84

Skowron A, Szczuka M (2010) Toward interactive computations: a rough-granular approach. In: Koronacki J, Ras Z, Wierzchon S, Kacprzyk J (eds) Advances in machine learning II, Dedicated to the memory of Professor Ryszard S. Michalski. Studies in computational intelligence, vol. 263. Springer, Heidelberg, pp 23–42

Słowiński R (ed) (1998) Fuzzy sets in decision analysis, operations research & statistics. In: Dubois D, Prade H (series eds) Handbook of fuzzy sets series, vol 5. Kluwer, Boston/Dordrecht

Triantaphyllou E, Felici G (eds) (2006) Data mining and knowledge discovery approaches based on rule induction techniques. Springer, New York

Wu WZ, Mi JS, Zhang WX (2003) Generalized fuzzy rough sets. Inform Sci 151:263–282

Zadeh L (2007) Granular computing and rough set theory. In: Kryszkiewicz M, Peters JF, Rybiński H, Skowron A (eds) RSEISP 2007: International conference rough sets and intelligent systems paradigms, Warsaw, Poland, 28–30 June 2007. Lecture notes in artificial intelligence, vol 4585. Springer, Heidelberg, pp 1–4

Zadeh LA (1965) Fuzzy sets. Inform Control 8:338–353

Zadeh LA (2001) A new direction in AI – toward a computational theory of perceptions. AI Mag 22(1):73–84

Zadeh LA (2006) Generalized theory of uncertainty (GTU)-principal concepts and ideas. Comput Stat Data Anal 51:15–46

Zimmermann H (ed) (1999) Practical applications of fuzzy technologies. In: Dubois D, Prade H (series eds) Handbook of fuzzy sets series, vol 7. Kluwer, Boston/Dordrecht

58 Collision-Based Computing

Andrew Adamatzky[1] · *Jérôme Durand-Lose*[2]
[1]Department of Computer Science, University of the West of England, Bristol, UK
andrew.adamatzky@uwe.ac.uk
[2]LIFO, Université d'Orléans, France
jerome.durand-lose@univ-orleans.fr

G. Rozenberg et al. (eds.), *Handbook of Natural Computing*, DOI 10.1007/978-3-540-92910-9_58,
© Springer-Verlag Berlin Heidelberg 2012

Abstract

Collision-based computing is an implementation of logical circuits, mathematical machines, or other computing and information processing devices in homogeneous, uniform and unstructured media with traveling mobile localizations. A quanta of information is represented by a compact propagating pattern (gliders in cellular automata, solitons in optical systems, wave fragments in excitable chemical systems). Logical truth corresponds to presence of the localization, logical false to absence of the localization; logical values can also be represented by a particular state of the localization. When two or more traveling localizations collide, they change their velocity vectors and/or states. Post-collision trajectories and/or states of the localizations represent results of logical operations implemented by the collision. One of the principal advantages of the collision-based computing medium – hidden in 1D systems but obvious in 2D and 3D media – is that the medium is architecture-less: nothing is hardwired, there are no stationary wires or gates, a trajectory of a propagating information quanta can be seen as a momentary wire. The basics of collision-based computing are introduced, and the collision-based computing schemes in 1D and 2D cellular automata and continuous excitable media are overviewed. Also a survey of collision-based schemes, where particles/collisions are dimensionless, is provided.

1   Introduction

Edward Fredkin, Tommaso Toffoli, Norman Margolus, Elwyn Berlekamp and John Conway are fully recognized founders of the field of collision-based computing. The term "collision-based computing" was first used in Adamatzky (2002a). There are, however, several equivalent but lesser-used terms such as "signal computing," "ballistic computing," "free space computing," and "billiard ball computing." The idea of collision-based computing is based on studies dealing with collisions of signals traveling along discrete chains, on a one-dimensional cellular automata. The interaction of signals traveling along one-dimensional conductors was among the famous problems in physics, biology, and physiology for centuries, and the problem of interaction was interpreted in terms of finite state machines in the 1960s. The earliest computer science-related results on signal interaction can be attributed to:

- Atrubin (Atrubin 1965), who designed the first-ever multiplier based on a one-dimensional cellular automaton in 1965;
- Fischer (Fischer 1965), who developed a cellular automaton generator of prime numbers in 1965; and
- Waksman (Waksman 1966), who initiated the very popular firing squad synchronization problem and provided an eight-state solution, in 1966.

Banks (1971) showed how to build wires and simple gates in configurations of a two-dimensional binary-state cellular automaton. This was not architecture-free computing, because a wire was represented by a particular stationary configuration of cell states (this was rather a simulation of a "conventional" electrical, or logical, circuit). However, Banks's design was a huge influence on the theory of computing in cellular automata and beyond.

In 1982, Elwyn Berlekamp, John Conway, and Richard Gay proved that Game of Life "can imitate computers" (Berlekamp et al. 1982).

They mimicked electric wires by lines "along which gliders travel" and demonstrated how to do a logical gate by crashing gliders into one another. Chapter 25 of their "Winning Ways" (Berlekamp et al. 1982) demonstrates computing designs that do not simply look fresh 20 years later but are still rediscovered again and again by Game of Life enthusiasts all over the Net.

Berlekamp, Conway, and Gay employed a vanishing reaction of gliders – two crashing gliders annihilate themselves – to build a NOT gate. They adopted Gosper's eater to collect garbage and to destroy glider streams. They used combinations of glider guns and eaters to implement AND and OR gates, and the shifting of a stationary pattern or block, by a mobile pattern, or glider, when designing auxiliary storage of information.

There is even the possibility that space–time itself is granular, composed of discrete units, and that the universe, as Edward Fredkin of M.I.T. and others have suggested, is a cellular automaton run by an enormous computer. If so, what we call motion may be only simulated motion. A moving particle in the ultimate microlevel may be essentially the same as one of our gliders, appearing to move on the macro-level, whereas actually there is only an alteration of states of basic space–time cells in obedience to transition rules that have yet to be discovered. – Berlekamp et al. (1982).

Meanwhile, in 1978, Edward Fredkin and Tommaso Toffoli submitted a 1-year project proposal to DARPA, which got funding and thus started a chain of remarkable events.

Originally, Fredkin and Toffoli aimed to "drastically reduce the fraction of" energy "that is dissipated at each computing step" (Fredkin and Toffoli 2002). To design a non-dissipative computer they constructed a new type of digital logic – conservative logic – that conserves both "the physical quantities in which the digital signals are encoded" and "the information present at any moment in a digital system" (Fredkin and Toffoli 2002).

Fredkin and Toffoli (1982) further developed these ideas in the seminal paper "Conservative Logic," from which a concept of ballistic computers emerged. The Fredkin–Toffoli model of conservative computation – the billiard ball model – explores "elastic collisions involving balls and fixed reflectors." Generally, they proved that "given a container with balls one can do any kind of computation."

The billiard ball model became a masterpiece of cellular automaton theory, thanks to Norman Margolus who invented a cellular automaton (block cellular automata or partitioned cellular automata) implementation of the model. Norman published this result in 1984 (Margolus 1984). "Margolus neighborhood" and "billiard ball model cellular automata" are exploited widely nowadays.

A detailed account of collision-based computing is not provided. There are no excuses to avoid reading original sources (Berlekamp et al. 1982; Fredkin and Toffoli 1982; Margolus 1984). A comprehensive, self-contained report of the modern state of collision-based computing is provided in the book by Adamatzky (2002a). The present chapter rather discusses personal experience designing collision-based computing schemes in one- and two-dimensional cellular automata, and spatially extended nonlinear media. It also provides some hands-on examples of recently discovered collision-based computing devices, with the hope that the examples will help readers to experiment with their own designs.

The chapter is structured as follows. Principles of collision-based computing are outlined in ❷ Sect. 2. ❷ Section 3 shows how basic logical gates can be implemented by colliding localizations in natural systems – simulated reaction–diffusion medium (❷ Sect. 3.1) and light-sensitive Belousov–Zhabotinsky medium (❷ Sect. 3.2). The theoretical foundations of computing with signals in 1D cellular automata are presented in ❷ Sect. 4, including collision-based implementation of 1D Turing machine (❯ Sect. 4.2.1) and cyclic tag systems (CTS) (❯ Sect. 4.2.2). The excursion to architectureless computing is complete with abstract geometrical computation in ❯ Sect. 5, where time and space are continuous and particles/localizations are dimensionless.

2 Principles of Collision-Based Computing

This section attempts to summarize all types of collision-based computers. A collision-based computer is an empty space populated with mobile and stationary localizations. The mobile localizations used for computing so far are:

- Billiard balls (Fredkin and Toffoli 1982)
- Gliders in cellular automata models (Berlekamp et al. 1982; Adamatzky and Wuensche 2007; Delorme and Mazoyer 2002; Rennard 2002; Rendell 2002)
- Solitons in nonlinear media (Jakubowski et al. 1996, 2001; Steiglitz 2001; Anastassiou et al. 2001; Rand et al. 2005; Rand and Steiglitz 2009), and
- Localized wave fragments in excitable chemical media (Adamatzky 2004; Adamatzky and De Lacy Costello 2007)

Examples of stationary localizations are:

- "Still lives," blocks, and eaters in Conway's Game of Life (Berlekamp et al. 1982; Rendell 2002)
- Standing waves in automaton models of reaction–diffusion systems (Adamatzky and Wuensche 2007) and
- Breathers in computing devices based on polymer chains and oscillons in vibrating granular materials (Adamatzky 2002b)

Usually mobile localizations represent signals and stationary localizations are used to route the signals in the space; however, by allowing balls and mirrors to change their states, in addition to velocity vectors, one can, in principle, build multivalued logic circuits.

Classical examples of collision-based gates are shown in ❯ Fig. 1. The interaction gate involves two balls to represent values of variables x and y (❯ Fig. 1a). If two balls are present at the input trajectories, this corresponds to both variables having TRUTH values that collide and deflect as a result of collisions. The deflected trajectories of the balls represent conjunctions of the input variables, xy. If only one ball—say the ball corresponding to the variable x—is present initially, then this ball travels along its original trajectories and does not change its velocity vector. The undisturbed trajectories of the balls represent logical functions $\bar{x}y$ and $x\bar{y}$ (❯ Fig. 1a).

The switch gate (❯ Fig. 1b) is another famous example, which also demonstrates the role of mirrors (stationary localizations). In the switch gate, the signal x is conditionally routed by a control signal c. If the signal is not present, the ball x continues along its original trajectory traveling southeast. The ball x is delayed and its trajectory shifts eastward if the signal c is present in the system. After collision, balls x and c are reflected but their further propagation is restricted by mirrors: the ball c collides with the Northern mirror and the ball x with the Southern mirror (❯ Fig. 1b).

◘ **Fig. 1**
Basics of billiard ball model: Fredkin–Toffoli interaction gate (a) and switch gate (b). (From Fredkin and Toffoli 1982.)

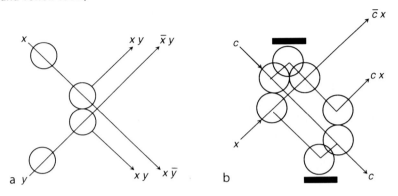

3 Collision-Based Computing in Natural Systems

Those focused on implementation issues may wonder how a scheme of collision-based computing can be implemented in natural, chemical, physical, or biological materials. This section provides two examples of computing with localizations in spatially extended quasi-chemical (reaction–diffusion cellular automata) and excitable chemical (Belousov–Zhabotinsky reaction) systems.

3.1 Computing Schemes in Reaction–Diffusion Cellular Automata: Spiral Rule

The reaction–diffusion cellular automaton Spiral Rule (Adamatzky and Wuensche 2007) exhibits a wide range of mobile and stationary localizations, thus offering unique opportunities to employ both traveling and still patterns in a computation process.

An automaton is designed that emulates nonlinearity of activator (A) and inhibitor (I) interaction for subthreshold concentrations of activator. The following quasi-chemical reaction was used to construct cell-state transition rules (Adamatzky and Wuensche 2007):

$$A + 6S \rightarrow A \quad A + I \rightarrow I \quad A + 3I \rightarrow I$$
$$A + 2I \rightarrow S \quad 2A \rightarrow I$$
$$3A \rightarrow A \quad \beta A \rightarrow I$$
$$I \rightarrow S$$

For subthreshold concentration of the inhibitor and threshold concentrations of activator, the activator is suppressed by the inhibitor. For critical concentrations of the inhibitor, both inhibitor and activator dissociate, producing the substrate.

The quasi-chemical reactions are mapped to cellular automaton rules as follows. Take a totalistic hexagonal cellular automaton (CA), where a cell takes three states – substrate S, activator A, and inhibitor I – and the cell updates its state depending on just the numbers of different cell-states in its neighborhoods. The update rule can be written as follows:

$$x^{t+1} = f(\sigma_I(x)^t, \sigma_A(x)^t, \sigma_S(x)^t)$$

where $\sigma_p(x)^t$ is the number of cell x's neighbors (in seven cells neighborhood) with cell-state $p \in \{I, A, S\}$ at time step t. The rule is compactly represented as a matrix $\mathbf{M} = (M_{ij})$, where $0 \le i \le j \le 7, 0 \le i + j \le 7$, and $M_{ij} \in \{I, A, S\}$ (Adamatzky et al. 2006). The output state of each neighborhood is given by the row-index i (the number of neighbors in cell-state I) and column-index j (the number of neighbors in cell-state A). One does not have to count the number of neighbors in cell-state S, because it is given by $7 - (i + j)$. A cell with a neighborhood represented by indices i and j will update to cell-state M_{ij}, which can be read off the matrix. In terms of the cell-state transition function, this can be presented as follows: $x^{t+1} = M_{\sigma_2(x)^t \sigma_1(x)^t}$. The exact structure of the transition matrix is as follows (Adamatzky and Wuensche 2007):

$$M = \left\{ \begin{array}{llllllll} S & A & I & A & I & I & I & I \\ S & I & I & A & I & I & I \\ S & S & I & A & I & I \\ S & I & I & A & I \\ S & S & I & A \\ S & S & I \\ S & S \\ S \end{array} \right\}$$

The entry $M_{01} = A$ symbolizes the diffusion of activator A, $M_{11} = I$ represents the suppression of activator A by the inhibitor I, and $M_{z2} = I\,(z = 0, \ldots, 5)$ is the self-inhibition of the activator in particular concentrations. $M_{z0} = S\,(z = 1, \ldots, 7)$ means that the inhibitor is dissociated in the absence of the activator, and that the activator does not diffuse in subthreshold concentrations; $M_{zp} = I$, $p \ge 4$ is an upper-threshold self-inhibition.

Starting in a random initial configuration, the automaton will evolve towards a quasi-stationary configuration, with typically two types of stationary localizations, and a spiral generator of mobile localizations, or gliders (❷ Fig. 2). The core of a glider-gun is a discrete analog of a "classical" spiral wave. However, at some distance from the spiral wave tip, the wave front becomes unstable and splits into localized wave fragments. The wave fragments continue traveling along their originally determined trajectories and keep their shape and velocity vector unchanged unless disturbed by other localizations. So, the wave fragments behave as in sub-excitable Belousov–Zhabotinsky systems.

Basic gliders, those with one (activator) head, are found in five types (❷ Fig. 3), which vary by the number of trailing inhibitors. Three types (G_{34}, G_{24}, G_{43}) alternate between two forms. Two types (G_4, G_5) have just one form. The spiral glider-gun in ❷ Fig. 2 emits G_{34} gliders. An alternative, low frequency, spiral glider-gun (Wuensche and Adamatzky 2006) (not shown) releases G_4 gliders. These basic gliders, and also a variety of more complicated gliders including mobile glider-guns, are also generated by many other interactions. Stationary localizations, or eaters (❷ Fig. 3i, j), are another important feature of the CA.

The principal components of any computing device are the input interface, memory, routers, and logical gates. Readers are referred to the paper Adamatzky and Wuensche (2007) to study possible input functions.

◘ Fig. 2

A typical quasi-stable configuration of the CA, which started its development in a random initial configuration (with $1/3$ probability of each cell-state). Cell-state I (inhibitor) is shown by a black disk, cell-state A (activator) by a circle, and cell-state S (substrate) by a dot. One can see there are two types of stationary localizations (glider eaters) and a spiral glider-gun, which emits six streams of gliders, with a frequency of a glider per six time steps in each glider stream. (From Adamatzky and Wuensche 2007.)

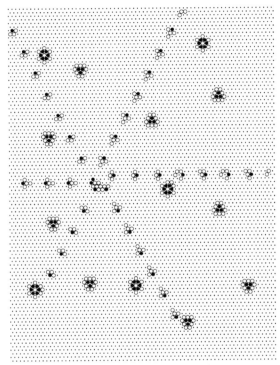

How does one implement a memory in the Spiral Rule CA? The eater E_6 can play the role of a 6-bit flip-flop memory device. The substrate-sites (bit-down) between inhibitor-sites (**◉** *Fig. 3i, j*) can be switched to an inhibitor-state (bit-up) by a colliding glider.

An example of writing one bit of information in E_6 is shown in **◉** *Fig. 4*. Initially E_6 stores no information. The aim is to write one bit in the substrate-site between the northern and northwestern inhibitor-sites (**◉** *Fig. 4a*). A glider G_{34} is generated (**◉** *Fig. 4b, c*) that travels west. G_{34} collides with (or brushes past) the north edge of E_6 resulting in G_{34} being transformed to a different type of glider, G_4 (**◉** *Fig. 4g, h*). There is now a record of the collision – evidence that writing was successful. The structure of E_6 now has one site (between the northern and northwestern inhibitor-sites) changed to an inhibitor-state (**◉** *Fig. 4j*) – a bit was saved.

To read a bit from the E_6 memory device with one bit-up (**◉** *Fig. 5a*), one collides (or brushes past) with glider G_{34} (**◉** *Fig. 5b*). Following the collision, the glider G_{34} is transformed into a different type of basic glider, G_34 (**◉** *Fig. 5g*), and the bit is erased (**◉** *Fig. 5j*).

◼ **Fig. 3**

The basic localizations in a Spiral Rule automaton: gliders (a–g) and eaters (h–i). (a–g) Five types of gliders, shown here traveling west, in the direction of their activator head (cell-state A), with a tail of trailing inhibitors made up of several cell-states I. The glider designator G_{ab} refers to the numbers of trailing inhibitors. (a) and (b) Two forms of glider G_{34}. (c) Glider G_4. (d) Glider G_5. (e) and (f) Two forms of glider G_{24}. (g) and (h) Two forms of glider G_{43}. (i) Eater E_3. (j) Eater E_6. (From Adamatzky and Wuensche 2007.)

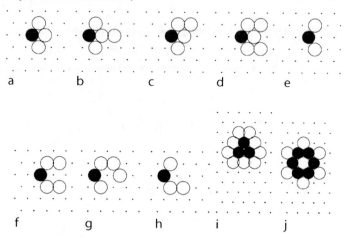

◼ **Fig. 4**

Write bit. (a) t. (b) $t + 1$. (c) $t + 2$. (d) $t + 3$. (e) $t + 4$. (f) $t + 5$. (g) $t + 6$. (h) $t + 7$. (i) $t + 8$. (j) $t + 9$. (From Adamatzky and Wuensche 2007.)

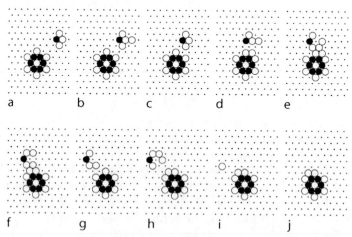

To route signals, one can potentially employ other gliders to act as mobile reflectors. ❯ *Figure 6* shows how a glider traveling northwest collides with a glider traveling west, and is reflected southwest as a result of the collision. However, both gliders are transformed to different types of gliders. This is acceptable on condition that both types of gliders represent the same signal, or signal modality.

◘ Fig. 5
Read and erase bit. (a) t. **(b)** $t + 5$. **(c)** $t + 7$. **(d)** $t + 8$. **(e)** $t + 9$. **(f)** $t + 10$. **(g)** $t + 11$. **(h)** $t + 12$.
(i) $t + 13$. **(j)** $t + 14$. (From Adamatzky and Wuensche 2007.)

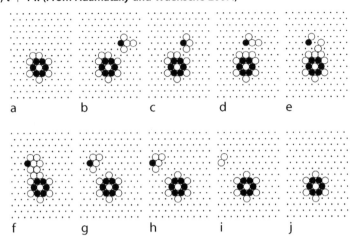

◘ Fig. 6
Glider reflection. (a) t. **(b)** $t + 1$. **(c)** $t + 2$. **(d)** $t + 3$. **(e)** $t + 4$. **(f)** $t + 5$. **(g)** $t + 6$. **(h)** $t + 7$. **(i)** $t + 8$.
(j) $t + 9$. (From Adamatzky and Wuensche 2007.)

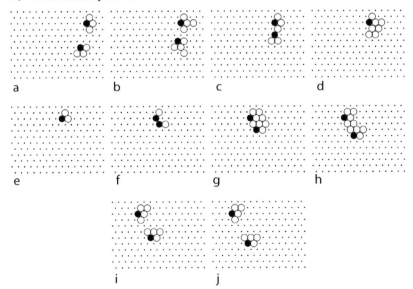

There are two more gates that are useful in designing practical collision-based computational schemes. They are the FANOUT gate and the ERASE gate. The FANOUT gate is based on glider multiplication. There are a few scenarios where one glider can be multiplied by another glider (for details see the original beehive rule [Wuensche 2005]); for example, one can make a FANOUT gate by colliding glider G_{34} with glider G_{24}. The gliders almost annihilate as a result of the collision, but recover into a complicated one, which splits into three G_5 gliders. To annihilate a glider, one can collide it with the central body of an eater, or with another glider; for example, head-on collisions usually lead to annihilation.

The *asynchronous* XOR gate can be constructed from the memory device in ❯ *Figs. 4* and ❯ *5*, employing the eater E_6 and the glider G_{34}. The incoming trajectory of the gliders is an input $x = \langle x, y \rangle$ of the gate, and the state of the cell that is switched to the inhibitor state by the gliders is an output z of the gate (this cell is shown by $\otimes$ in ❯ *Fig. 7a*). As seen in ❯ *Fig. 4*, when glider G_{34} brushes by the eater E_6 it "adds" one inhibitor state to the eater configuration (❯ *Fig. 4, t + 7*), and transforms itself into glider G_{43}. If glider G_{34} brushes by E_6 with an additional inhibitor state (❯ *Fig. 5, t*), it "removes" this additional state and transforms itself into glider G_4 (❯ *Fig. 5, t + 11*).

Assume that the presence of glider G_{34} symbolizes input logical TRUE and its absence – input FALSE, inhibitor state I in cell $\otimes$ – output TRUE and substrate state S – output FALSE. The result of this logical operation can be read directly from the configuration of E_6 or by sending a control glider to brush by E_6 to detect how the glider is transformed. Then the structure implements exclusive disjunction (❯ *Fig. 7b*). The gate constructed is asynchronous, because the output of the operation does not depend on the time interval between the signals but only on the value of signals: when the inhibitor state is added or removed from E_6 the configuration of E_6 remains stable and does not change till another glider collides into it.

The eater E_6 can take four different configurations resulting from the interactions of gliders brushing past, and there are seven types of gliders produced in collisions with the eater (including some basic types flipped). One therefore can envisage (Adamatzky and Wuensche 2007) that a finite state machine can be implemented in the eater-glider system. The internal state of such a machine is represented by the configuration of the eater, the type of the incoming glider symbolizes the input symbol of the machine, and the type of the outgoing glider represents the output state of the machine.

To construct the full state transition table of the eater-glider machine, seven types of gliders are collided into four configurations of the eater and the results of the collisions are recorded. For the sake of compact representation, the configurations of the eater are encoded as shown in ❯ *Fig. 8*. The gliders are denoted as follows: G_{34} as a, G_{43} as b, G_5 as c, G_4 as d, G_{24} as e, G^4 (glider G_4 flipped horizontally) is f, and G^{43} (glider G_{43} flipped horizontally) is g. The state transition table is shown in ❯ *Fig. 9*.

Consider the internal states of the eater-glider machine as unary operators on the set $\{a, b, c, d, e, f, g\}$, that is, the machine's state is reset to its initial state after the collision with the glider. For example, the unary operator α implements the following transformation: $a \to b$, $b \to c$,

□ **Fig. 7**

Asynchronous XOR gate. **(a) Position of output cell is shown by** $\otimes$. **(b) Operation implemented by the gate, input state G_{34} is logical** TRUE, **output state S is** FALSE, **output state I is** TRUE. **(From Adamatzky and Wuensche 2007.)**

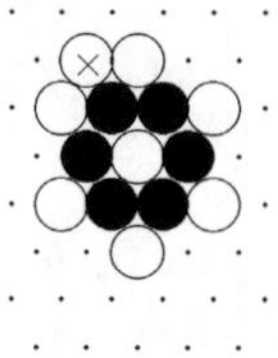

x_1	x_2	y
0	0	S
G_{34}	0	I
0	G_{34}	I
G_{34}	G_{34}	S

a   b

◘ Fig. 8
Encoding the internal states of the glider-eater machine in the configuration of eater E_6. (a) α.
(b) β. (c) χ. (d) δ.

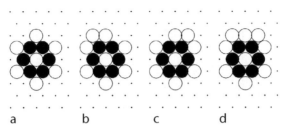

a   b   c   d

◘ Fig. 9
The state transition table of the eater-glider machine. Tuple xy, a pair made up of an eater
state x and glider state y, at the intersection of the ith row and jth column, signifies that being in
state i while receiving input j the machine takes state x and generates output y.

	a	b	c	d	e	f	g
α	βb	δc	αb	αe	δd	αe	δc
β	αd	δe	βc	βc	χg	αa	χe
χ	χd	βe	δf	χa	βb	χa	βe
δ	δb	βc	χg	χe	αf	δe	αa

◘ Fig. 10
Limit sets of unary operators $a, \ldots, g$.

Operator	Limit set
a	$\{\alpha,\beta\}, \{\delta\}$
b	$\{\beta,\delta\}$
c	$\{\alpha\}, \{\beta\}, \{\delta,\chi\}$
d	$\{\alpha\}, \{\beta\}, \{\chi\}$
e	$\{\alpha,\delta\}, \{\beta,\chi\}$
f	$\{\alpha\}, \{\chi\}, \{\delta\}$
g	$\{\alpha,\delta\}, \{\beta,\chi\}$

$c \rightarrow a$, $d \rightarrow a$, $e \rightarrow d$, $f \rightarrow e$, $g \rightarrow e$. The operators have the following limit sets: operator α has the limit set $\{a, b, c\}$, β – set $\{c\}$, χ has two limit sets $\{a, d\}$ and $\{b, c\}$, and operator δ – two limit sets $\{a, b, c, d\}$ and $\{e, f\}$. Considering unary operators $a, \ldots, g$ operating on set $\{\alpha, \beta, \chi, \delta\}$, one obtains the limit sets shown in ❷ *Fig. 10*. Many of the operators have more than two limit sets, which may indicate a significant computational potential of the eater-glider machine.

To characterize the eater-glider machine in more detail, a study was conducted to find out what output strings are generated by the machine when the machine receives the uniform infinite string s^*, $s \in \{a, \ldots, g\}$ on its input. These input string to output string transformations are shown in ❷ *Fig. 11*.

Input string *abcdefg* evokes the following output strings when fed into the machine. The machine starting in state α generates string *begabac*, in state β string *dcgabac*, in state χ string *deccgae*, and in state δ string *bcccgae*.

◻ **Fig. 11**

Input string to output string transformations implemented by the eater-glider machine. String s, at the intersection of the *i*th row and *j*th column, tells one that being initially in state *i* and receiving a uniform string *j*, the machine generates string s.

	a^*	b^*	c^*	d^*	e^*	f^*	g^*
α	$(bd)^*$	$c(ce)^*$	b^*	e^*	$(de)^*$	e^*	$(ca)^*$
β	$(db)^*$	$(ec)^*$	c^*	c^*	$(gb)^*$	ae^*	e^*
χ	d^*	$e(ec)^*$	$(fg)^*$	a^*	$b(gb)^*$	a^*	e^*
δ	b^*	$(ce)^*$	$(gf)^*$	ea^*	$(ed)^*$	e^*	$(ac)^*$

3.2 Collision-Based Computing in Excitable Chemical Media

This section gives a brief introduction to implementation of collision-based circuits in excitable chemical media, the Belousov–Zhabotinsky (BZ) system. Examples discussed here are based on numerical experiments; see the chapter ❯ Reaction–Diffusion Computing of this book for implementation of collision-based circuits in a BZ medium in chemical laboratory conditions. Now, computing is discussed with localized wave fragments in the Oregonator (Field and Noyes 1974; Tyson and Fife 1980) model adapted to a light-sensitive BZ reaction with applied illumination (Beato and Engel 2003; Krug et al. 1990):

$$\frac{\partial u}{\partial t} = \frac{1}{\varepsilon}\left(u - u^2 - (fv + \phi)\frac{u-q}{u+q}\right) + D_u\nabla^2 u, \text{ and}$$

$$\frac{\partial v}{\partial t} = u - v$$

where variables u and v represent local concentrations of bromous acid $HBrO_2$ and the oxidized form of the catalyst ruthenium Ru(III), ε sets up a ratio of time scale of variables u and v, q is a scaling parameter depending on reaction rates, f is a stoichiometric coefficient, ϕ is a light-induced bromide production rate proportional to the intensity of illumination (an excitability parameter – moderate intensity of light will facilitate excitation process, higher intensity will produce excessive quantities of bromide which suppresses the reaction). It is assumed that the catalyst is immobilized in a thin layer of gel; therefore, there is no diffusion term for v. To integrate the system, one uses the Euler method with five-node Laplacian operator, time step $\Delta t = 10^{-3}$ and grid point spacing $\Delta x = 0.15$, with the following parameters: $\phi = \phi_0 + A/2$, $A = 0.0011109$, $\phi_0 = 0.0766$, $\varepsilon = 0.03$, $f = 1.4$, $q = 0.002$.

The chosen parameters correspond to a region of "higher excitability of the sub-excitability regime" outlined in Sendiña-Nadal et al. (2001), which supports propagation of sustained wave fragments (❯ Fig. 12a). These wave fragments are used as quanta of information in the design of CB logical circuits. The waves were initiated by locally disturbing initial concentrations of species; for example, ten grid sites in a chain are given value $u = 1.0$ each. This generates two or more localized wave fragments, similar to counter-propagating waves induced by temporary illumination in experiments. The traveling wave fragments keep their shape for around $4 \cdot 10^3$–10^4 steps of simulation (4–10 time units), then decrease in size and vanish. The wave's life-time is sufficient, however, to implement logical gates; this also allows one not to worry about "garbage collection" in the computational medium.

■ Fig. 12

Basic operations with signals. Overlay of images taken every 0.5 time units. Exciting domains of impurities are shown in black, inhibiting domains of impurities are shown in gray. (a) Wave fragment traveling north. (b) Signal branching without impurities: a wave fragment traveling east splits into two wave fragments (traveling southeast and northeast) when it collides into a smaller wave fragment traveling west. (c) Signal branching with impurity: wave fragment traveling west is split by impurity (shown on the right) into two waves traveling northwest and southwest. (d) Signal routing (U-turn) with impurities: wave fragment traveling east is routed north and then west by two impurities (shown on the right). An impurity-reflector consists of inhibitory (gray) and excitatory (black) chains of grid sites. (From Adamatzky 2004.)

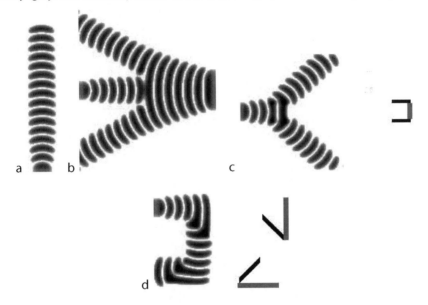

Signals are modeled by traveling wave fragments (Sendiña-Nadal et al. 2001; Beato and Engel 2003): a sustainably propagating wave fragment (❷ *Fig. 12a*) represents the TRUE value of a logical variable corresponding to the wave's trajectory (momentarily wire). To demonstrate that a physical system is logically universal, it is enough to implement negation and conjunction or disjunction in the spatio-temporal dynamics of the system. To realize a fully functional logical circuit, one must also know how to operate input and output signals in the system's dynamics, namely to implement signal branching and routing; delays can be realized via appropriate routing.

One can branch a signal using two techniques. Firstly, one can collide a smaller auxiliary wave to a wave fragment representing the signal, the signal-wave will then split into two signals (these daughter waves shrink slightly down to stable size and then travel with constant shape a further $4 \cdot 10^3$ time steps of the simulation) and the auxiliary wave will annihilate (❷ *Fig. 12b*). Secondly, one can temporarily and locally apply illumination impurities on a signal's way to change properties of the medium and thus cause the signal to split (❷ *Fig. 12c*).

A control impurity, or reflector, consists of a few segments of sites for which the illumination level is slightly above or below the overall illumination level of the medium. Combining excitatory and inhibitory segments, one can precisely control the wave's trajectory, for example, determining a U-turn of a signal (❷ *Fig. 12d*).

A typical billiard ball model interaction gate (Fredkin and Toffoli 1982; Margolus 1984) has two inputs – x and y, and four outputs – $x\bar{y}$ (ball x moves undisturbed in the absence of ball y), $\bar{x}y$ (ball y moves undisturbed in the absence of ball x), and twice xy (balls x and y change their trajectories when they collide into each other). It was not possible to make wave fragments implement exact billiard-ball gates, because the interacting waves either fused or one of the waves was annihilated as a result of the collision with another wave.

However, a BZ (nonconservative) version of a billiard-ball gate with two inputs and three outputs is implemented, which is just one xy output instead of two. This BZ collision gate is shown in ❯ *Fig. 13.*

The rich dynamic of the BZ medium allows one to also implement complicated logical operations just in a single interaction event. An example of a composite gate with three inputs and six outputs is shown in ❯ *Fig. 14.* As one sees in ❯ *Fig. 14,* some outputs, for example $\bar{x}yz$, are represented by gradually vanishing wave fragments. The situation can be dealt with by either using a very compact architecture of the logical gates or by installing temporary amplifiers made from excitatory fragments of illumination impurities.

As known from results of computer simulations and experimental studies, classical excitation waves merge or annihilate when they collide with one another. This may complicate the implementation of nontrivial logical circuits in classical excitable media. Wave fragments, however, behave a bit differently, more like quasi-particles.

In computational experiments with exhaustive analysis of all possible collisions between localized wave fragments (Adamatzky and De Lacy Costello 2007), all the collisions are classified as (❯ *Fig. 15*): quasi-elastic reflection of wave fragments (❯ *Fig. 15a*), pulling and pushing of a wave fragment by another wave fragment (❯ *Fig. 15b, c*), sliding of one wave fragment along the refractory trail of another fragment (❯ *Fig. 15d*), and translation of a wave fragment along one axis by another wave fragment (❯ *Fig. 15e*). Examples of two types of collision are shown in ❯ *Fig. 16.*

4 One-Dimensional Cellular Automata

This section deals with cellular automata in general and discrete signals in particular. It is shown both how they compute in the classical understanding and how they can be used to

◻ **Fig. 13**
Two wave fragments undergo angle collision and implement interaction gate
$\langle x, y \rangle \rightarrow \langle x\bar{y}, xy, \bar{x}y \rangle$. **(a) In this example $x = 1$ and $y = 1$, both wave fragments are present initially. Overlay of images taken every 0.5 time units. (b) Scheme of the gate. In upper left and bottom left corners of (a) one sees domains of wave generation, two echo wave fragments are also generated, they travel outward from the gate area and thus do not interfere with computation. (From Adamatzky 2004.)**

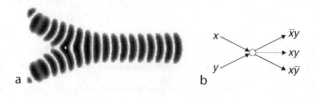

☐ Fig. 14
Implementation of $\langle x,y,z \rangle \rightarrow \langle x\bar{y}, \bar{y}z, xyz, \bar{x}yz, \bar{x}y\bar{z}, xy\bar{z} \rangle$ interaction gate. Overlay of images of
wave fragments taken every 0.5 time units. The following combinations of input configuration are
shown: **(a)** $x = 1$, $y = 1$, $z = 0$, north–south wave collides with east–west wave. **(b)** $x = 1$, $y = 1$, $z = 1$,
north–south wave collides with east–west wave, and with west–east wave. **(c)** $x = 1$, $y = 0$, $z = 1$,
west–east and east–west wave fragments pass near each other without interaction. **(d)** $x = 0$,
$y = 1$, $z = 1$, north–south wave collides with east–west wave. **(e)** Scheme of the gate. (From
Adamatzky 2004.)

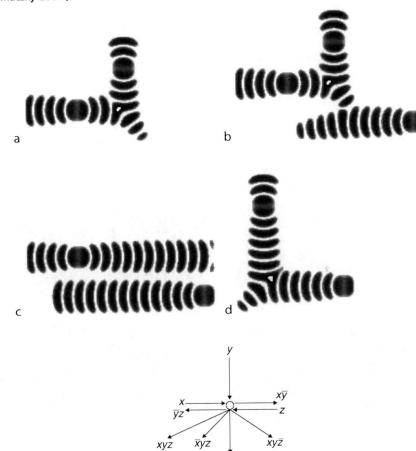

implement some phenomena relevant to massively distributed computing but without any
sequential counterpart.

Cellular automata (CA) were introduced as a discrete model for parallel, local, and
uniform phenomena, whether of engineering, physical, or biological nature. They often
provide a medium where signals naturally appear and are thoroughly used to understand
the global dynamics. On the other hand, a correct handling of such signals is the key to
compute and to design special purpose CA.

A cellular automaton works in the following way. The space is regularly partitioned in cells.
All the cells are identical and are regularly displayed as an infinite array (having as many

☐ **Fig. 15**

Schematic representation of interaction between wave-fragments. (**a**) Reflection. (**b**) Attraction. (**c**) Repulsion. (**d**) Sliding. (**e**) Shifting-sliding. (From Adamatzky and De Lacy Costello 2007.)

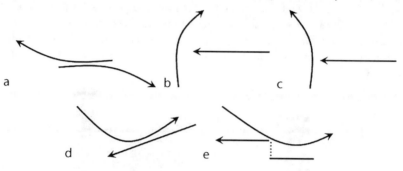

☐ **Fig. 16**

(**a–d**) Sliding: snapshots of wave fragment moving southwest colliding with wave fragment traveling southeast. Center of the initial excitation of the southwest wave fragment is shifted southwards by 22 sites. (From Adamatzky and De Lacy Costello 2007.) (**e–h**) Slide-shift: snapshots of wave fragment moving southeast colliding with wave fragment traveling west.

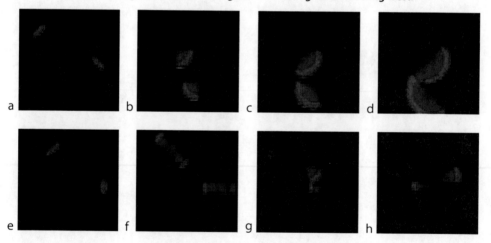

dimensions as the modeled space). Each cell can be in finitely many states and evolves according to its state and the states of the surrounding cells – *locality*. All the cells are identical and behave similarly – *uniformity*. The cells are all updated *synchronously*, like a *massive parallel* process sharing a single clock. In essence, CA form a discrete model: discrete time, discrete space, and discrete values.

Generally, the array is considered to extend infinitely in all directions so that there are no boundaries to deal with. However, borders can be encoded using special states which are never changed and make the two sides independent. Another way to simulate a CA on a computer is to consider that outside of a finite range all cells are in the same stable state (called a *quiescent state*). Finite/periodic configuration can also be generated by displaying the cells to be on a ring (or torus): the last and first are then neighbors.

Previous collision-based systems are CA, or, more accurately, are simulated by CA, some of them work on hexagonal lattices. Computation in dimension 2 and above can be done straightforwardly as already presented by bit encoding and implementation of logic gates.

From now on only one-dimensional space is considered because, on the one hand, in higher dimensions it can – up to some point – be treated alike or correspond to what has been exemplified in previous sections, and, because, on the other hand, dimension 1 is particularly restrictive and needs special focus.

The reader interested in CA might consult Ilachinski (2001), Kari (2005), Sarkar (2000) for various topics not covered here.

4.1 Signals in CA

In space–time diagrams, or orbits, configurations are concatenated as iterations go. They are very important in order to comprehend the dynamics. Since the temporal coordinate is added, this leads to an array with one more dimension than the underlying space.

A signal is anything periodic in the space–time diagram. If the periodicity is in two directions, then it can fill the whole space–time and is referred as a *background*, the substrata upon which signals move. The frontier between two backgrounds is a signal as long as it is periodic. ❷ *Figure 17* provides an example with a complex background (alternatively 0111, 1000, 1101, and 0010 repeating) on the left. On the right, signals are revealed by an ad hoc filtering. This example illustrates the variety in width and pattern of signals. As can be seen, collisions between signals can be pretty complicated.

4.2 Computing in One-Dimensional Systems

In dimension two and above, as soon as it is possible to carry information around (with signal), to make information crossing and duplication, and to process it (with collision) in the proper way to yield a sufficiently large set of logical gates, computation is straightforwardly possible.

❑ Fig. 17
Filtering to exhibit signals and backgrounds on rule 54 (inspired from Boccara et al. 1991, Fig. 7). Time is increasing upwards.

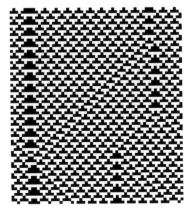 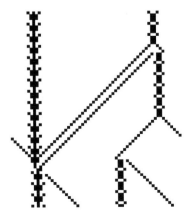

In dimension 1, this is not so easy, because there is no way for a signal to go around another one or some static artifact. This means that signal propagation is really an issue and, for example, handling garbage signals can be cumbersome. One way to cope with it is to have various kinds of signals not identically affected by artifacts (some would pass unaffected while others interact).

This means that to use the bits and gates scheme, one has to be very careful with available signals and have a very clever positioning of the logic circuitry, which is always thought of as two dimensional (time often provides the extra dimension for displaying it). For example, the construction of Ollinger (2002) provides a small CA that is able to compute with circuit implementation where displaying the circuit is not straightforward.

Another way to tackle computation is to use other formalisms. One classical way is to go back to Turing machines, another one is to implement minimal rewriting systems like cyclic tag systems.

4.2.1   Turing Machine

A Turing machine is a finite automaton acting on a potentially infinite array of symbols. The computation starts with the automaton in its initial state and the input written on the tape. The tape is accessed through a read/write head. At each iteration, the automaton reads the symbol under the head, rewrites a symbol, changes its state, and moves the head left or right. A Turing machine is basically as one dimensional as its tape, so that it naturally fits on a one-dimensional CA.

The simulation is rather simple: signals encoding the symbols are motionless and one signal amounting to the state of the automaton is moving round updating the symbols. Collisions are as follows: a moving state meets a stationary symbol, which leaves a new stationary symbol and a moving new state. (❷ *Figure 24* in the next section provides a clear illustration of this in a continuous setting.) This was implemented by, for example, Lindgren and Nordahl (1990).

4.2.2   Cyclic Tag Systems (CTS)

Another way to define computation is to rely on a word that is rewritten until the system halts. The input is given as the starting word and the output is the final word. Connexion with a Turing machine is straightforward when considering the tape as a finite word and the state of the automaton as an extra symbol at the head.

Cyclic tag systems (CTS) is a particular case of many Turing-universal rewriting systems. A CTS considers a binary word together with a circular list of binary words (*appendants*). At each iteration, the first bit is removed from the word; if it is 1, then the first appendant of the list is added at the end of the word, then the list is rotated. The computation stops when the word is empty or a special halting word (denoted h in the example) is activated. An example is provided in ❷ *Fig. 18*.

Cyclic tag systems were proven to be able to perform any computation (Cook 2004) and even to do so in polynomial time (Woods and Neary 2006) and were used to build very small universal Turing machines (Neary and Woods 2009). They have been implemented in one-dimensional CA with collision/signal-based computing by Cook (2004) to produce

◘ Fig. 18
Example of computation of a CTS (given on the first line).

1011	**011**:h:011:01
011**011**	h:**011**:01:011
11011	**011**:01:011:h
1011**011**	**01**:011:h:011
01101**101**	011:**h**:011:01
1101101	**h**:011:01:011
101101	

computation universal CA with only two states (it is impossible to compute with less). The construction relies on signals moving and colliding that are sought and classified in order to work on a more abstract level.

As far as minimal CA are concerned, it is worth mentioning that four states are enough to get a one-dimensional CA, which is able to simulate any other one-dimensional CA (Richard and Ollinger 2008). This is not Turing universality since the simulation encompasses infinite configurations.

4.3 Signal-Specific Issues

Since CA form a model for physical phenomena and also for parallel computing architectures, other issues arise. One is to understand the underlying dynamics when used as a discrete model. The other is to design CA for special purposes.

4.3.1 Signals to Understand Dynamics

Once a simulation is running as a CA, one common way to understand the dynamics is to try to find regularities in the space–time diagram and observe them. They are often the key to the underlying dynamics and predictions. (Although it might happen that what is observed is nothing but an artifact of the model and has no counterpart in reality.) This is made clear by considering two examples.

The first one models sand dripping in a one-dimensional space. The dynamics are quite simple: there is nothing in the initial configuration and at each iteration one grain is added to the first pile/cell. Each time a pile has at least two more grains than the next one, a grain falls.

If only grains dropping at odd time are colored in black, then their position is as in ❷ *Fig. 19a* after 10,000 iterations and only the top grains will ever move. This strange disposal, as well as the two different slopes and precise long-term behavior, are explained by signals (Durand-Lose 1996, 1998). These signals can be identified on ❷ *Fig. 19b* where the successive configurations (iteration 100–150) are set one after the other to form a volume. On this volume, triangles can be seen on the lower (as well as the upper parts), with their frontiers as signals (which can be revealed with an appropriate filtering).

Another example is provided by Das et al. (1995). The aim is to generate a CA with two states and a five-closest-cell neighborhood, such that, on a ring of any size, whatever configuration it is started on it always ends up blinking: all cells are 0 then all cells are 1, alternatively, forever. Instead of trying to build it straightaway, evolutionary programming is used: random

■ **Fig. 19**
One-dimensional sand dripping. (a) Dotting even grains (Durand-Lose 1996, Fig. 6) at iteration
10,000. (b) Successive configurations (Durand-Lose 1996, Fig. 3).

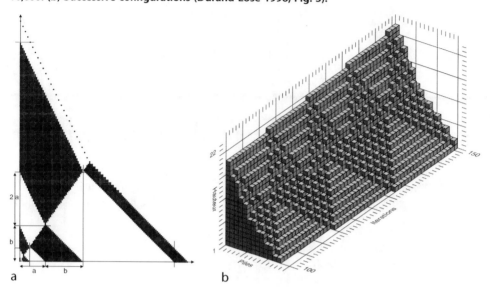

CA are generated; they are ranked according to their "blinking capability"; then the best are
kept, recombined and mutated to form the next generation. It then cycles through ranking and
the next generation until one "fully blinking" CA emerges.

The obtained CA is analyzed in terms of signal (again with some filtering), as illustrated in
❯ *Fig. 20*. (The apparent non-connectivity of some signals comes from the filtering and also
because cells can influence one another up to distance 2.) The evolutionary process goes in
steps where various intermediate levels of "blinking" appear. Analyzing typical members of
each level reveals the progressive apparition of signals and collision rules.

It is very important to notice how, in a context where signals were not asked for by the
evolutionary process, they indeed appear and are the key to both the desired dynamics and the
evolutionary process.

The previous example does not compute in the classical understanding; nevertheless, it
provides a relevant dynamical global property.

4.3.2   Signals to Generate a Particular Behavior

Computing, in the usual understanding, is a special behavior. But CA can also be thought of as
a computing model on its own. Then primitives can be specially defined to take advantage of
the huge parallelism. Signals are the key to managing it.

As already cited, the pioneering work of Fischer (1965) generates prime numbers using a
parallel implementation of the Sieve of Eratosthenes where they are indicated as no signal on
the first cell at the corresponding times. Prime numbers can then be read on the space–time
diagram by observing the state of the first cell.

◻ Fig. 20

Figure and filtering to highlight signals and analyze (from Das et al. 1995, Fig. 1). Time is increasing downward. (a) Space–time diagram. (b) Filtered space–time diagram.

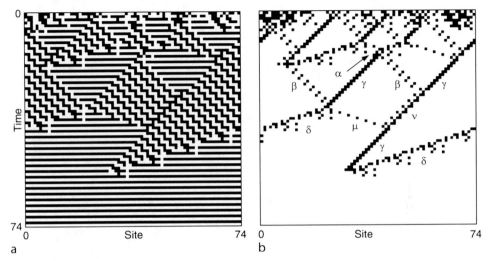

a   b

Another complex behavior is the famous firing squad synchronization problem (FSS). Starting from all but one cell in a quiescent state, one wants all the cells to enter simultaneously the same state – which has not been used before. Like if they would all blink for the very first time synchronously.

Each example of ❯ *Fig. 21* shows the Euclidean conception and then the discrete implementation of a FSS solution. Both constructions rely on recursive cuts in half. In the discrete implementation, at some point, the granularity of space – a cell cannot be divided – is reached. Since CA are synchronous, this point is reached simultaneously everywhere, ensuring the global synchronization of the entire array of cells.

4.3.3 Signals as a Computational Model on Its Own

Mazoyer et al. developed a computing model where multiplication (as in ❯ *Fig. 22a*), composition (as in ❯ *Fig. 22b*), and even iteration are graphically achieved. The programming system is achieved by using a trellis of adaptable size that can be dynamically generated. The computation is carried out over the trellis. Composition is then achieved by having the next computation displayed on a new trellis. Recursion is provided by a scheme that dynamically starts another iteration providing each time an adapted trellis.

5 Abstract Geometrical Computation

Abstract geometrical computation (AGC) is an idealization of collision-based computing where particles/signals are dimensionless (time and space are continuous). Although the number of signals in a bounded space is expected to be finite, it is unbounded. Another way

☐ **Fig. 21**
FSS implementations. **(a)** Divide and conquer in 3n-steps (Yunès 2007a, Fig. 2.3).
(b) Eight-states and 4n-steps (Yunès 2007b, Fig. 1).

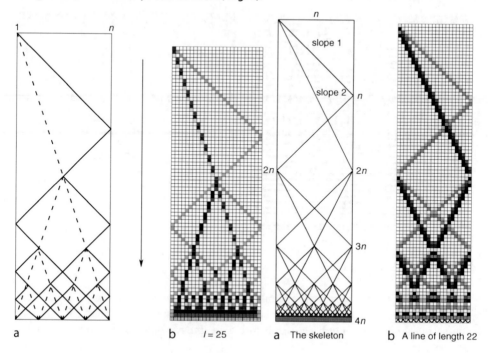

a   b $l = 25$ a   The skeleton   b   A line of length 22

to consider AGC is as a continuous counterpart of CA as a limit when the size of the cells tends to zero. In CA, signals are almost always the key to understanding and designing, and, indeed, in the literature, the discreteness of CA and of their signals is often put aside to reason at a more abstract level and is left to technical details.

In abstract geometrical computation, dimensionless signals move at constant speed. When they meet, they are replaced by others according to some rewriting rules. A *signal machine* (SM) gathers the definition of the nature of available signals (called *meta-signals*), their speeds, and the collision rules. There are finitely many meta-signals and each one is assigned a constant speed (velocity and direction). The constant speed may seem surprising, but each discrete CA signal has indeed a constant speed.

The space–time diagrams generated are continuous in both space and time and the traces of signals form line segments. For a given meta-signal, all these segments are parallel. In this section, only one-dimensional AGC are considered, so that the space–time diagram is always two dimensional with time increasing upwards as illustrated by ❷ *Fig. 23*.

Space–time diagrams can be much more complicated and, as presented below, use continuity to implement the Zeno paradox and emulate the black hole model of computation.

5.1 Computing

Computing, in the classical understanding, is pretty easy. In fact, many constructions for CA directly translate to ACG and are even easier, since discreteness does not have to be taken

⬛ Fig. 22

Mazoyer's system. Time is increasing upwards. **(a)** Multiplication (Mazoyer 1996, Fig. 4).
(b) Composition of multiplications (Mazoyer 1996, Fig. 8).

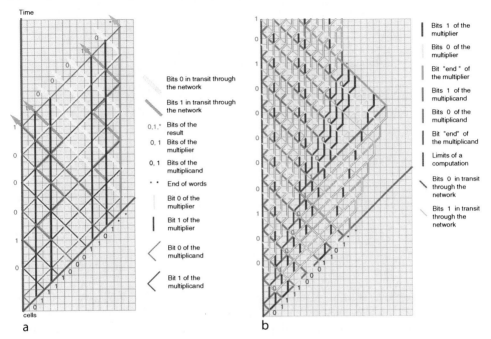

a   b

⬛ Fig. 23

Example of a space–time diagram.

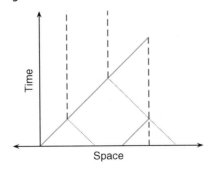

into account. Two ways to achieve computability are presented: Turing machines and cyclic
tag systems.

For Turing machines (as presented in ❷ Sect. 4.2.1), the implementation is quite straightforward: static signals encode the tape – one signal per symbol – and one signal encodes the
state of the automaton and the position of the head. The latter moves forth and back from
symbol to symbol as the read/write head. This is depicted in ❷ *Fig. 24*. The construction is
detailed in Durand-Lose (2009).

◻ **Fig. 24**

Simulating a Turing machine. (a) TM evolution. (b) TM simulation.

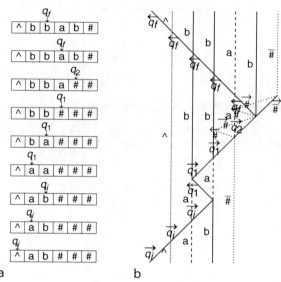

a   b

Cyclic tag systems are presented in ❷ Sect. 4.2.2. The simulation is done by encoding, left to right, the word and then the circular list of appendants by parallel signals encoding the bits as shown in ❷ *Fig. 25a*. At each iteration, the list is rotated to the right, but before that, if the erased bit of the word is 1, a copy is left. The copy is directly added at the right of the word as in ❷ *Fig. 25b*, which presents one full iteration. The full simulation of the example of ❷ *Fig. 18* is given in ❷ *Fig. 25c*. The rightward drifting corresponds to the erasure of the word from the left and to the rightward rotation of the list. Each group of oblique lines corresponds to one rotation of the list.

This simulation is detailed in Durand-Lose (2008a). The signal machine obtained is able to simulate any CTS, and is thus Turing universal. It is the smallest one known (it has only 13 meta-signals).

5.2   Geometric Primitives

As for CA, AGC can also be considered as a model on its own, with particular operators, which might not have any classical computation counterpart (like FSS) or discrete counterpart.

All the following constructions involve adding meta-signals and rules to a given SM, so that the desire capability exists. Signals have to be added to the initial configuration in order to fire the effect. Rules can also be modified in order to dynamically fire it.

Space and time are continuous and scaleless, and since signals have no dimension, there is no limit on the scalability. If all signals are scaled spatially by a given coefficient, the whole computation is also scaled temporally. So, the duration of a computation is meaningless and complexity measures, such as the maximal number of collisions, with a causal link, have to be used.

Spatial rescaling of the initial configuration is a static operation. It is also possible to do it during the computation. The construction relies on the ability to freeze a configuration.

□ Fig. 25

Simulating a cyclic tag system. (a) Initial configuration on 101 and 011 :: 10 :: 10 :: 01. (b) Initial configuration and first iteration on 101 and 011 :: 10 :: 10 :: 01. (c) Full simulation on 101 & 011 :: h :: 0110 :: 01011.

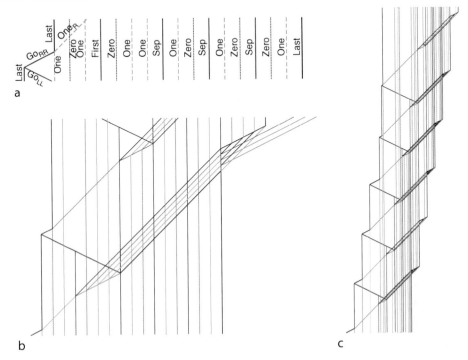

Freezing a computation is quite simple. One signal is added on one side and it crosses the entire configuration. Doing so, each time it meets a signal (or enters a previously existing collision), it replaces it with another signal encoding the meta-signal. All the encoding signals have the same speed: they are parallel and do not interact and, moreover, the distance between them is preserved. The configuration is frozen and shifts. To unfreeze it, a signal comes from the side, crosses the configuration, and replaces each encoding signal by the encoded one (or the result of the collision). Freezing and unfreezing signals must have the same speed so that the configuration is restored exactly as it was, up to a translation. (And indeed they correspond to the same meta-signal toggle.) This is depicted in ❷ Fig. 26.

Meanwhile, when a configuration is frozen into parallel signals, it is possible to act on these signals. One simple trick is to change their direction. In ❷ Fig. 27a, this is done twice so as to restore the original direction. (An extra signal is used on the right to delete the structure.) As a result, since different slopes are used to change direction, the distances between signals are scaled (here by one half). Adding freezing and unfreezing signals (automatically fired in due time), an artifact, to scale down a whole configuration, is generated as can be seen on the ❷ Fig. 27c.

The initial configuration has been modified in order to start straightaway the effect. It is also possible to fire it dynamically during the computation when some special collision happens. This is used to iterate it.

◻ Fig. 26

Freezing and unfreezing. **(a)** Normal computation. **(b)** Translated computation. **(c)** Example.

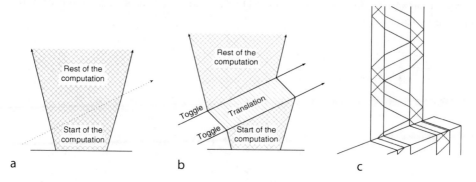

a   b   c

◻ Fig. 27

Scaling. **(a)** Principle. **(b)** Scaled computation. **(c)** Example.

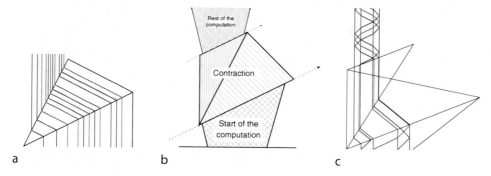

a   b   c

5.3 Infinite Acceleration and Non-recursive Function

Scaling can be automatically restarted ad infinitum; since both space and time are continuous, this is not a problem – at least before the singularity. In ❷ *Fig. 28*, a computation producing an infinite trellis, extending indefinitely in both space and time, is folded into the structure. The right signal and the middle one are very important since the folded computations should be bounded in order to ensure that it is fully scaled.

A *singularity* refers to an accumulation of infinitely many collisions and signals to a given location. ❷ *Figure 28a* shows that the structure alone already creates a singularity. This illustrates the Zeno paradox. The leftmost signals of ❷ *Fig. 28a* form an infinite sequence with infinitely (yet countably) many collisions, although the time elapsed as well as the total distance crossed by signals is finite.

All the signals and collisions that would have existed if there would have been no folding indeed exist, but in a bounded portion of the space and time diagram. (The equivalence of the plane and a bounded part is somehow implemented.) So that up to the presence of the structure and of the frozen parts and the different scales, the computation is the same.

Any computation starting with finitely many signals can be folded. Inside the structure, the rescaling provides speedup (the closer they are, the faster they collide). This speedup is unbounded and goes to infinity. Inside the structure there is a space–time where time-lines are infinitely accelerated compared to the outside. This is the first step to emulate the black-hole model of computation (Hogarth 1994; Etesi and Németi 2002; Lloyd and Ng 2004).

The second step is to find a way for an atomic piece of information to leave the black hole, here the singularity. One meta-signal can be added and rules changed so that the new meta-signal is not affected by the structure but can be generated by the initial computation.

The final step is to add bounding signals on both sides of the folding (called here horizonLe and horizonRi). At their collision point, the folded computation is entirely in the causal past as displayed in ❷ *Fig. 29*. Any signal leaving the folding would have been collected.

This way the black-hole model is emulated: there are two time-lines, one for the machine and one for the observer. The machine one is infinite. On the observer one, after a finite duration, the whole machine one is entirely in the causal past. A single piece of information can be sent by the machine to the observer and it has to be sent after a finite (machine) duration. The ultimate date for the observer to receive anything from the machine is finite and known. To understand the power of this model, just imagine that the machine only sends a signal if its computation ends. The observer receives a signal only if it stops and after a duration the observer is aware that any signal sent would have been received. So, by just

❏ **Fig. 28**
Folding or infinite rescaling. (a) Folding structure. (b) Example.

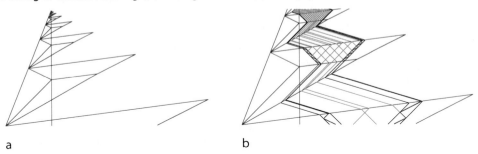

a   b

❏ **Fig. 29**
Framing the folding to collect all signals exiting the folding.

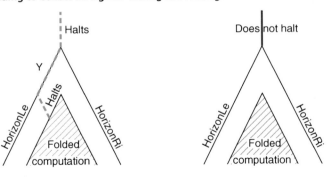

checking its clock, the observer knows that the computation did not halt when it is the case. Altogether, the halting problem (whose undecidability is the cornerstone of Turing computability theory) is decidable!

This model, as well as AGC, clearly computes beyond Turing. In formal term, it can decide Σ_0^1 formulae in the arithmetical hierarchy, which is a recursive/computable predicate prefixed by an existential quantifier. This includes, for example, the consistency of Peano arithmetic and set theory, many conjectures such as the Collatz conjecture and Goldbach's conjecture.

Defining what happens at the singularity point is not easy, especially since the accumulating set can be a line segment, a fractal curve or even a Cantor! (To understand this, take a continuous look at the FSS: what would happen at the bottom of ❷ *Fig. 21*?) For singularity on a single point, a careful continuation has been proposed in Durand-Lose (2009) that allows us to climb the arithmetical hierarchy. There is also another use of isolated singularity as a way to provide limits for analog computation.

5.4 Analog Computation

Unlike collision-based computing and CA, with dimensionless signals in a continuous space, it is possible to encode real numbers as the distance between signals. In the AGC context, this distance is exact—that is, it is a real number in exact precision. As long as signals are parallel, this distance is preserved. This allows one to encode real numbers and to make some computations over them.

Since any real number can be encoded exactly, the model ipso facto falls out of classical computability (because of cardinalities, there is no way to encode all the reals with natural numbers or finite strings). Therefore, an analog model of computation has to be searched for.

With this encoding of real numbers, AGC is equivalent to the linear Blum–Shub–Smale model (BSS) (Blum et al. 1998). In the linear BSS model, variables hold (exact) real numbers and the operations available are addition, multiplication by a constant and branch, according to the sign of a variable. This equivalence is true as long as the BSS machine has an unbounded number of variables (accessed through a context shift like moving a window over an infinite array of variables), there are finite signals and there is no singularity (Durand-Lose 2007).

If singularities are used in a proper way, it becomes possible to multiply two variables. Then the classical BSS model can be implemented in AGC (Durand-Lose 2008b). The simulation is not possible in the other way anymore since, for example, the exact square rooting can also be computed by AGC.

Just as in the discrete case, a proper handling of isolated singularities of any order can be used to decide quantified (over natural but not real numbers) predicates in the BSS model and climb the BSS-arithmetical hierarchy (Durand-Lose 2009).

References

Adamatzky A (ed) (2002a) Collision-based computing. Springer, London

Adamatzky A (ed) (2002b) Novel materials for collision-based computing. Springer, Berlin

Adamatzky A (2004) Collision-based computing in Belousov–Zhabotinsky medium. Chaos Soliton Fract 21:1259–1264

Adamatzky A, De Lacy Costello B (2007) Binary collisions between wave-fragments in sub-excitable Belousov–Zhabotinsky medium. Chaos Soliton Fract 34:307–315

Adamatzky A, Wuensche A (2007) Computing in spiral rule reaction-diffusion hexagonal cellular automaton. Complex Syst 16(4):277–298

Adamatzky A, Wuensche A, De Lacy Costello B (2006) Glider-based computation in reaction-diffusion hexagonal cellular automata. Chaos Soliton Fract 27:287–295

Anastassiou C, Fleischer JW, Carmon T, Segev M, Steiglitz K (2001) Information transfer via cascaded collisions of vector solitons. Optics Lett 26:1498–1500

Atrubin AJ (1965) A one-dimensional real-time iterative multiplier. IEEE Trans Electron Computers EC-14 (1):394–399

Banks E (1971) Information and transmission in cellular automata. Ph.D Dissertation, MIT, cited by Toffoli and Margolus (1987)

Beato V, Engel H (2003) Pulse propagation in a model for the photosensitive Belousov-Zhabotinsky reaction with external noise. In: Schimansky-Geier L, Abbott D, Neiman A, van den Broeck C (eds) Noise in complex systems and stochastic dynamics. Proceedings of SPIE, 2003

Berlekamp ER, Conway JH, Guy RL (1982) Winning ways for your mathematical plays, vol 2 Games in particular. Academic, London

Blum L, Cucker F, Shub M, Smale S (1998) Complexity and real computation. Springer, New York

Boccara N, Nasser J, Roger M (1991) Particle-like structures and interactions in spatio-temporal patterns generated by one-dimensional deterministic cellular automaton rules. Phys Rev A 44(2): 866–875

Cook M (2004) Universality in elementary cellular automata. Complex Syst 15:1–40

Das R, Crutchfield JP, Mitchell M, Hanson JE (1995) Evolving globally synchronized cellular automata. In: Eshelman LJ (ed) International conference on genetic algorithms '95. Morgan Kaufmann, San Mateo, CA, pp 336–343

Delorme M, Mazoyer J (2002) Signals on cellular automata. In: Adamatzky A (ed) Collision-based computing. Springer, Berlin, pp 234–275

Durand-Lose J (1996) Grain sorting in the one dimensional sand pile model. Complex Syst 10(3):195–206

Durand-Lose J (1998) Parallel transient time of one-dimensional sand pile. Theoret Comp Sci 205 (1–2):183–193

Durand-Lose J (2007) Abstract geometrical computation and the linear Blum, Shub and Smale model. In: Cooper S, Löwe B, Sorbi A (eds) Computation and logic in the real world. 3rd Conference Computability in Europe (CiE '07). Springer, no. 4497 in LNCS, pp 238–247

Durand-Lose J (2008a) Abstract geometrical computation: small Turing universal signal machines. In: Neary T, Seda A, Woods D (eds) International workshop on the complexity of simple programs. Cork University Press, Cork, Ireland, December 6–7

Durand-Lose J (2008b) Abstract geometrical computation with accumulations: beyond the Blum, Shub and Smale model. In: Beckmann A, Dimitracopoulos C, Löwe B (eds) Logic and theory of algorithms. CiE 2008 (abstracts and extended abstracts of unpublished papers). University of Athens, Athens, pp 107–116

Durand-Lose J (2009) Abstract geometrical computation 3: Black holes for classical and analog computing. Nat Comput 8(3):455–472

Etesi G, Németi I (2002) Non-Turing computations via Malament-Hogarth space-times. Int J Theor Phys 41 (2):341–370, gr-qc/0104023

Field RJ, Noyes RM (1974) Oscillations in chemical systems. iv. limit cycle behavior in a model of a real chemical reaction. J Chem Phys 60:1877–1884

Fischer PC (1965) Generation of primes by a one-dimensional real-time iterative array. J ACM 12(3):388–394

Fredkin EF, Toffoli T (1982) Conservative logic. Int J Theor Phys 21(3/4)219–253

Fredkin EF, Toffoli T (2002) Design principles for achieving high-performance submicron digital technologies. In: Adamatzky A (ed) Collision-based computing. Springer, Berlin, pp 27–46

Hogarth ML (1994) Non-Turing computers and non-Turing computability. In: Hull D, Forbens M, Burian RM (eds) Biennial meeting of the philosophy of science association. East Lansing, MI, pp 126–138

Ilachinski A (2001) Cellular automata – a discrete universe. World Scientific, Singapore

Jakubowski MH, Steiglitz K, Squier RK (1996) When can solitons compute? Complex Syst 10(1):1–21

Jakubowski MH, Steiglitz K, Squier RK (2001) Computing with solitons: a review and prospectus. Multiple Valued Logic 6(5–6):439–462

Kari J (2005) Theory of cellular automata: a survey. Theoret Comp Sci 334:3–33

Krug HJ, Pohlmann L, Kuhnert L (1990) Analysis of the modified complete oregonator (MCO) accounting for oxygen- and photosensitivity of Belousov-Zhabotinsky systems. J Phys Chem 94:4862–4866

Lindgren K, Nordahl MG (1990) Universal computation in simple one-dimensional cellular automata. Complex Syst 4:299–318

Lloyd S, Ng YJ (2004) Black hole computers. Sci Am 291 (5):31–39

Margolus N (1984) Physics-like models of computation. Phys D 10:81–95

Mazoyer J (1996) Computations on one dimensional cellular automata. Ann Math Artif Intell 16: 285–309

Neary T, Woods D (2009) Four fast universal Turing machines. Fundam Inform 410(4):443–450

Ollinger N (2002) The quest for small universal cellular automata. In: ICALP '02, Springer, Heidelberg, no. 2380 in LNCS, pp 318–329

Rand D, Steiglitz K (2009) Computing with solitons. In: Meyers RA (ed) Encyclopedia of complexity and systems science. Springer, Heidelberg

Rand D, Steiglitz K, Prucnal P (2005) Signal standardization in collision-based soliton computing. Int J Unconvent Comput 1:31–45

Rendell P (2002) Turing universality of the game of life. In: Adamatzky A (ed) Collision-based computing. Springer, Berlin, pp 513–540

Rennard JP (2002) Implementation of logical functions in the game of life. In: Adamatzky A (ed) Collision-based computing. Springer, London, pp 491–512

Richard G, Ollinger N (2008) A particular universal cellular automaton. In: Neary T, Woods D, Seda AK, Murphy N (eds) The complexity of simple programs. National University of Ireland, Cork

Sarkar P (2000) A brief history of cellular automata. ACM Comput Surv 32(1):80–107

Sendiña-Nadal I, Mihaliuk E, Wang J, Pérez-Muñuzuri V, Showalter K (2001) Wave propagation in subexcitable media with periodically modulated excitability. Phys Rev Lett 86:1646–1649

Steiglitz K (2001) Time-gated Manakov spatial solitons are computationally universal. Phys Rev E 63:1660–1668

Toffoli T, Margolus N (1987) Cellular automata machine - a new environment for modeling. MIT Press, Cambridge, MA

Tyson JJ, Fife PC (1980) Target patterns in a realistic model of the Belousov-Zhabotinsky reaction. J Chem Phys 73:2224–2237

Waksman A (1966) An optimum solution to the firing squad synchronization problem. Inform Control 9(1):66–78

Woods D, Neary T (2006) On the time complexity of 2-tag systems and small universal Turing machines. In: 47th Annual IEEE Symposium on Foundations of Computer Science (FOCS '06), IEEE Computer Society, Berkeley, CA, pp 439–448

Wuensche A (2005) Glider dynamics in 3-value hexagonal cellular automata: the beehive rule. Int J Unconventional Comput 1:375–398

Wuensche A, Adamatzky A (2006) On spiral glider-guns in hexagonal cellular automata: activator-inhibitor paradigm. Int J Modern Phys C 17(7): 1009–1026

Yunès JB (2007a) Automates cellulaires; fonctions booléennes. Habilitation à diriger des recherches, Université Paris 7

Yunès JB (2007b) Simple new algorithms which solve the firing squad synchronization problem: a 7-states 4n-steps solution. In: Durand-Lose J, Margenstern M (eds) Machine, Computations and Universality (MCU '07). Springer, Berlin, no. 4664 in LNCS, pp 316–324

59 Nonclassical Computation — A Dynamical Systems Perspective

Susan Stepney
Department of Computer Science, University of York, UK
susan.stepney@cs.york.ac.uk

G. Rozenberg et al. (eds.), *Handbook of Natural Computing*, DOI 10.1007/978-3-540-92910-9_59,
© Springer-Verlag Berlin Heidelberg 2012

Abstract

In this chapter, computation is investigated from a dynamical systems perspective. A dynamical system is described in terms of its abstract *state space*, the system's current state within its state space, and a rule that determines its motion through its state space. In a classical computational system, that rule is given explicitly by the computer program; in a physical system, that rule is the underlying physical law governing the behavior of the system. Therefore, a dynamical systems approach to computation allows one to take a unified view of computation in classical discrete systems and in systems performing nonclassical computation. In particular, it gives a route to a computational interpretation of physical embodied systems exploiting the natural dynamics of their material substrates.

1   Introduction

In this chapter, we investigate computation from a dynamical systems perspective.

A dynamical system is described in terms of its abstract *state space*, the system's current state within its state space, and a rule that determines its motion through its state space. In a classical computational system, that rule is given explicitly by the computer program; in a physical system, that rule is the underlying physical law governing the behavior of the system. So a dynamical systems approach to computation allows us to take a unified view of computation in classical discrete systems and in systems performing nonclassical computation. In particular, it gives a route to a computational interpretation of physical embodied systems exploiting the natural dynamics of their material substrates.

We start with *autonomous* (closed) dynamical systems: those whose dynamics is not an explicit function of time, in particular, those with no inputs from an external environment. We begin with computationally conventional discrete systems examining their computational abilities from a dynamical systems perspective. The aim here is both to introduce the necessary dynamical systems concepts, and to demonstrate how classical computation can be viewed from this perspective. We then move on to continuous dynamical systems, such as those inherent in the complex dynamics of matter, and show how these too can be interpreted computationally, and see how the material embodiment can give such computation "for free," without the need to explicitly implement the dynamics.

We next broaden the outlook to *open* (nonautonomous) dynamical systems, where the dynamics is a function of time, in the form of inputs from an external environment, and which may be in a closely coupled feedback loop with that environment.

We finally look at *constructive*, or developmental, dynamical systems, where the structure of the state space is changing during the computation. This includes various growth processes, again investigated from a computational dynamical systems perspective.

These later sections are less developed than for the autonomous cases, as the theory is less mature (or even nonexistent); however these are the more interesting computational domains, as they move us into the arena of considering biological and other natural systems as computational, open, developmental, dynamical systems.

2   Autonomous Dynamical Systems

Consider a dynamical system with N degrees of freedom; it has an abstract state space $\mathcal{X}^N$. Its state can be defined by N state variables $x_i \in \mathcal{X}$, or, equivalently, by an ND state vector $\mathbf{x} \in \mathcal{X}^N$ (e.g., $\mathbf{x}$ may be a vector of binary bits, or a vector of continuous variables such as position and momentum). The state vector $\mathbf{x}_t$ defines the value of the system state at a given time.

The deterministic dynamics is given by a function $f : \mathcal{X}^N \to \mathcal{X}^N$, which defines how a state vector $\mathbf{x}$ changes with time, that is, it defines the *trajectory* that the system takes through its state space. So the dynamics associates a vector with each point in the state space, defining how that point evolves under the dynamics. (In conventional use, the meaning of this vector is unfortunately different in the discrete and continuous time cases. In the discrete time case (❷ Sects. 2.1 and ❷ 2.2), $\mathbf{x}_{t+1}=f(\mathbf{x}_t)$, and so the vector is the next state; in the continuous time case (❷ Sect. 2.3), $\dot{\mathbf{x}} = f(\mathbf{x})$, and so the vector is the derivative, pointing toward the next state an infinitesimal time later: $\mathbf{x}_{t+dt}=\mathbf{x}_t+f(\mathbf{x}_t)\,dt$. It would be possible to have a uniform meaning, by redefining the discrete case vector to be the difference in states, with $\mathbf{x}_{t+1}=\mathbf{x}_t+f(\mathbf{x}_t)$ (and an implicit $\Delta t=1$). However, here we follow the conventional, and inconsistent, use.) If f is not itself an explicit function of time, then the system is *autonomous*.

In general, dynamical systems theory is not concerned with details of individual trajectories, but rather with the qualitative behaviors of sets of trajectories. For example, consider a set of states occupying some initial volume of the state space: As these states evolve under the dynamics, how does the volume change? We are mostly interested here in *dissipative systems*, where the volume contracts to *attractors* (regions of state space that attract trajectories), and we interpret such attractors from a computational perspective. This contracting behavior is a property of systems that dissipate energy or information. (Closed non-dissipative dynamical systems, on the other hand, have no attractor structure.)

2.1   Discrete Space, Discrete Time Dynamical Systems

We start by considering finite discrete spaces (finite number of finite dimensions), with discrete time dynamics $t \in \mathcal{N}$ (where $\mathcal{N}$ is the set of natural numbers).

We take $\mathcal{X} = \mathcal{S}$, some set with finite cardinality $|\mathcal{S}| \in \mathcal{N}$ (typically $\mathcal{S}$ will be the Boolean set $\mathcal{B}$, but it is not restricted to this). For an N-dimensional system, the state is defined by N state variables $s_i \in \mathcal{S}$, and the state space $\mathcal{S}^N$ comprises $|\mathcal{S}|^N$ distinct discrete states. (When $\mathcal{S} = \mathcal{B}$, these states fall on the vertices of an N-dimensional hypercube.) Let the state vector be $\mathbf{s} \in \mathcal{S}^N$.

The dynamics of a particular system is determined by its particular transition function $f : \mathcal{S}^N \to \mathcal{S}^N$, with $\mathbf{s}_{t+1}=f(\mathbf{s}_t)$. There are $\left(|\mathcal{S}|^N\right)^{|\mathcal{S}|^N}$ such functions f.

Given a particular state $\mathbf{s}_0 \in \mathcal{S}^N$, its trajectory under f is a sequence of states $\mathbf{s}_0, \mathbf{s}_1, \ldots, \mathbf{s}_t, \ldots$. Eventually, because the state space is finite, a state that was met before will be met again: there exists a k such that $\mathbf{s}_k=\mathbf{s}_{k+p}$, for some p. Since the dynamics is deterministic, the trajectory will then recur: for all $i \geq k$, $\mathbf{s}_i=\mathbf{s}_{i+p}$. The system has entered an *attractor*, with cycle length or period p. States not on an attractor are called *transient*.

Given a trajectory $\ldots,\mathbf{s}_t,\mathbf{s}_{t+1},\ldots$, then $\mathbf{s}_t$ is the *pre-image* of $\mathbf{s}_{t+1}$ in this trajectory, and $\mathbf{s}_{t+1}$ is the *successor* of $\mathbf{s}_t$. Every state has precisely one successor (because the dynamics is deterministic). It may have zero, one, or more pre-images (trajectories may merge); if it has zero pre-images, it is a *Garden of Eden* state.

The set of all states $\mathbf{s}_i$ whose trajectories lead to the same attractor forms the *basin of attraction* of that attractor. The total state space is partitioned into these basins: Every state is in precisely one basin. Note that there is no necessary correlation between the volume of the basin (the proportion of state space it occupies, and hence the probability that a state chosen at random will be in it) and the length of the attractor that it leads to.

The *microstate* of the system is which particular $\mathbf{s} \in \mathbf{S}$ it is in. The *macrostate* is which particular attractor (or basin of attraction if the microstate is currently a transient state) the system is in.

For a given dynamics f, there is a minimum of one attractor basin (all states are in the same attractor basin, for example, the zero function), and a maximum of $|\mathbf{S}|^N$ (the identity function where every state forms its own single-state attractor basin). There is a minimum transient length of 0 (all states on some attractor cycle, for example, the increment modulo 2^N function, interpreting the binary encoded state $\mathcal{B}^N$ as a number), and a maximum transient length of $|\mathbf{S}|^N - 1$ (e.g., the decrement and halt on zero function). There is a minimum number of Garden of Eden states of 0 (all states on some attractor cycle), and a maximum number of Garden of Eden states of $|\mathbf{S}|^N - 1$ (e.g., the zero function). Nontrivial dissipative computational systems rarely lie at any of these extremes, however. (Note that reversible, non-dissipative systems have no Garden of Eden states, and no merging trajectories.)

2.1.1 Visualizing the Attractor Field

Visualizing the basins of attraction can help in understanding some aspects of their dynamics. For small systems, the most common approach is to lay out the state transition graph to highlight the separate basins, their attractors, and their symmetries (see ❷ *Fig. 1*). Wolfram (1986b, Fig 9.1) used this approach in early work on cellular automata; Wuensche (Wuensche 2002; Wuensche and Lesser 1992) has developed special purpose layout software, and uses this approach consistently, to highlight aspects of the dynamics.

2.1.2 Computation

Given a finite system in initial microstate $\mathbf{s}_0$ (which may be considered to contain an encoding of any input data), the system follows its dynamics f until it reaches the relevant attractor. If this attractor has a cycle length of one, the system then stays in the single attractor state. For longer cycle lengths, the system perpetually repeats the cycle of states.

In Terms of Attractors

The *computation* performed by the system as it follows its dynamics f can be interpreted as the determination of which attractor basin it is in, by progressing to the attractor from its initial state $\mathbf{s}_0$.

■ **Fig. 1**
Visualization of part of the state transition graph for ECA rule 110 (see ❯ Sect. 2.1.5 on Elementary Cellular Automata), on the periodic lattice $N=12$, showing three of the basins of attraction. Each node corresponds to a state s_i; each edge corresponds to a transition $s_i \rightarrow s_{i+1}$. The leaves of the graphs are Garden of Eden states; the attractor cycles can be seen at the centers of the basins.

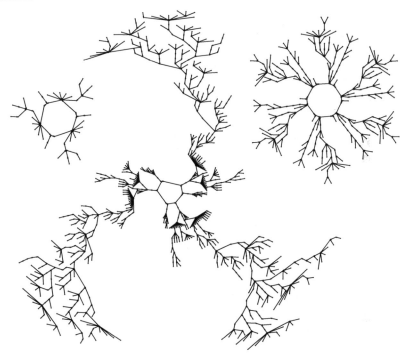

The output of the computation may be the microstates, or some suitable projection thereof, of the discovered attractor cycle. (See, e.g., ❯ Sect. 2.1.5, Example 1: the density classification task.)

In Terms of Trajectories

Alternatively, the *computation* performed by the system as it follows its dynamics f can be considered to be (some projection of) the microstates it passes through along its trajectory, including both transient and attractor cycle states. (See, e.g., ❯ Sect. 2.1.5, Example 2: the Rule 30 PRNG.)

Programming Task

The programming task involves determining a dynamics f that leads to the required trajectories or attractor structure. (See, e.g., ❯ Sect. 2.1.5, Example 1: the density classification task.)

For feasible programs, the discovery of an attractor should be performed in polynomial time (implying polynomial length transients and attractor cycles). At the other complexity extreme, the dynamics should not be defined merely by a $|S|^N$-entry lookup table (which would allow all computations to find the attractor in a single step); it should admit a "compressed" description. (See, e.g., ❷ Sects. 2.1.5 and ❷ 2.1.6 which define the global dynamics f in terms of the composition of local dynamics ϕ_i.)

Implementation

Natural physical systems do not tend to directly implement a discrete dynamics, particularly one that has been designed to perform a specific task. However, any such dynamics can be implemented (or simulated) on a classical digital computer. Hence, there are no implementation constraints imposed on the design of the dynamics f.

Inputs and Outputs

The input is encoded into the initial state; the output is decoded from (a projection of) the resulting attractor state(s). It is important when analyzing the complexity of the computation to take into account any "hidden" computation needed to encode the input, or to decode the output. This is particularly important if the computational interpretation is far removed from the dynamics, for example, if there is some kind of virtual machine present.

2.1.3 Virtual Machine Dynamics

In some cases, a dynamics needs to be accompanied by a very carefully chosen initial condition in order to implement the required computation. For example, when cellular automata are used to implement Turing machines (TMs; see ❷ Sect. 2.1.5 on Universality), they are given a carefully chosen initial configuration that corresponds to the "program" of the TM, plus the "true" input corresponding to its initial tape. This requirement for an exquisitely tuned initial condition constrains the system to traverse only a very small part of its state space (certain basins of attraction are never explored; some transient trajectories are never taken). What is happening in these cases is that the underlying broader dynamics is being used to implement a "virtual machine" with its own dynamics confined to a small subspace of the underlying system; this subspace and its trajectories correspond to the computation of the virtual machine, and may possibly be implemented more directly. In the continuous case, this more direct implementation is what we want: The natural dynamics of the system, with no need for such highly tuned initial conditions, performs the desired computation.

2.1.4 Infinite-Dimensional State Spaces

When N is (countably) infinite, the dynamics of the system can change qualitatively.

In a finite dissipative system (which contains both transients and cycles), there must be Garden of Eden states, but this is no longer true in an infinite system: Transient behavior on the way to the attractor cycle need not have any starting Garden of Eden states (e.g., the decrement and halt on zero function).

An infinite system need not have an attractor cycle: There is no guarantee that the system will reach a previously seen state (e.g., the increment function). Even if there are attractor cycles, there may be states not in their basins (e.g., the function **if** even **then** add 2 **else** halt).

A Turing machine (TM) operates in a state space with a countably infinite (or, more precisely, finite but unbounded) number of dimensions. In dynamical systems terms, halting is reaching one of a number of particular attractor cycles of length 1, which describes the halting states. The output of the TM (the contents of its tape) is simply the microstate of the subspace representing the tape in this halting state. Hence there are potentially many attractors, one for each halting state with different tape contents, corresponding to different initial tape contents (different initial states s_0).

The halting problem means that it is in general undecidable whether a given initial state s_0 is in a halting basin, in some other ("looping") basin, or not in a basin at all.

2.1.5 Cellular Automata

A finite cellular automaton (CA) comprises N *cells* laid out in a regular grid or lattice, usually arranged as an n-dimensional torus (n in common examples is typically 1 or 2). Each cell i at time t has a state value $c_{i,t} \in \mathcal{S}$.

Each cell has a *neighborhood* of k cells, comprising itself and certain nearby cells in the grid. This neighborhood is the same for all cells, in that v_i, the neighborhood of c_i, is v_0, the neighborhood of the origin c_0, translated by i (❷ *Fig. 2a*).

The state of cell c_i's neighborhood v_i at time t is $\chi_{i,t} \in \mathcal{S}^k$, a k-tuple of cell states that is the projection of the full state onto the neighborhood v_i (❷ *Fig. 2b*).

The local state transition rule, or update rule, is $\phi : \mathcal{S}^k \to \mathcal{S}$. (There are $|\mathcal{S}|^{|\mathcal{S}|^k}$ such rules, some of which are related by symmetries. So, since typically $k \ll N$, CA rules capture only a small fraction of all the possible dynamics over an ND space.) These cells form an array of state transition machines. At each timestep, the state of each cell is updated in parallel, $c_{i,t+1} = \phi(\chi_{i,t})$.

The global dynamics f is determined by the local rule ϕ and the shape of the neighborhood. This global behavior from a given initial state is conventionally visualized in the 1D case by drawing the global state at time t as a line of cells (with colors corresponding to the local state), then drawing the state at $t+1$ directly below, and so on (see ❷ *Fig. 3*). Higher dimensional CAs are conventionally visualized as animations.

❷ *Figure 1* shows three basins of attraction of a small CA. There are many "repeated" basins (basins with identical topologies, over different specific states) due to the symmetries in

---

◻ **Fig. 2**

CA neighborhood. (a) A CA's regular neighborhood, *v*, illustrated in a 2D lattice. The neighborhood of cell *i* is the image of the neighborhood of cell 0, translated by *i*. (b) The state of the neighborhood, illustrated in a 1D lattice. χ*i*, the state of the neighborhood of cell *i*, is the projection of the full state s onto the neighborhood *v*i.

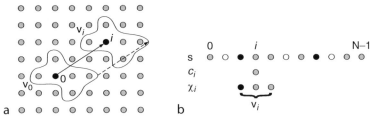

■ **Fig. 3**
Visualization of the time evolution of ECA rule 110 (see ❷ Sect. 2.1.5 on Elementary Cellular Automata), with N=971, with random initial state s₀, over 468 timesteps. Each horizontal line shows the N-bit representation of the state s_t; subsequent lines correspond to subsequent timesteps $s_t, s_{t+1}, \ldots$.

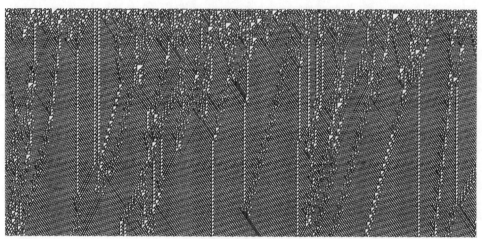

a CA rule (Powley and Stepney 2009a, b). Every rule over a finite space with periodic boundary conditions has "shift" symmetries from shifting the arbitrary origin; additionally, some rules also have reflectional and other symmetries.

Elementary Cellular Automata (ECAs)

ECAs are 2-state ($\mathcal{S} = \mathcal{B}$) CAs, with the cells arranged in a 1D lattice, and with a neighborhood size of 3 (comprising the cell and its immediate left and right neighbors). There are $2^{2^3} = 256$ distinct ECA rules, conventionally referred to by a base 10 number $0 \ldots 255$, representing the base 10 interpretation of the rule table bitstring. After reflection and inversion symmetries have been taken into account, there are 88 essentially distinct rules.

Wolfram's Classification

Wolfram (1984b, a) provides a qualitative characterization of CAs, classifying their long-term evolution into four classes:

1. A unique homogeneous state, independent of the initial state (a single-state attractor cycle); patterns disappear with time
2. A simple periodic pattern of states (short attractor cycles, length $\ll |\mathcal{S}|^N$); patterns become fixed
3. Chaotic aperiodic pattern of states (long attractor cycles, length $O(|\mathcal{S}|^N)$, or no cycles in the infinite case); patterns grow indefinitely
4. Complex localized long-lived structures; patterns grow and contract

Wolfram's classification scheme has been criticized for a variety of reasons, including the fact that even determination of quiescence (long-term fixed patterns) is undecidable (Culik and Yu 1988). This is a recurring problem for any classification scheme: "CA behavior is so complex that almost any question about their long-term behavior is undecidable" (Durand et al. 2003). See also Sutner (2005). Wolfram classes are nevertheless widely used in a qualitative manner to distinguish kinds of behaviors.

Class 3 and class 4 CAs demonstrate *sensitive dependence on initial conditions*: The effect of a minimal (one cell state) change to the initial condition propagates across the system, eventually resulting in a completely different dynamics (see ❷ *Fig. 4*). The 88 essentially distinct ECAs cover all 4 Wolfram classes of behavior.

Universality
Even though CA rules capture only a small fraction of all the possible dynamics over an ND space, there are computationally universal CAs that can emulate a TM. ECA rule 110 (❷ *Fig. 3*) is universal (Cook 2004), as is Conway's Game of Life 2D CA (Berlekamp et al. 1982; Rendell 2002). The proof of universality in these cases involves constructing a virtual machine in the CAs, on which is implemented a TM (or equivalent). In other words, the computation is being performed by a carefully engineered initial condition, and a carefully engineered interpretation of the dynamical behavior. So these systems explore only a small fraction of their full state space: the fraction that corresponds to an interpretation of the state of a TM.

Wolfram (1984b, §8) speculates that "class 4 cellular automata are characterized by the capability for universal computation" (in the case of infinite-dimensional CAs).

Example CA 1: The Density Classification Task
The requirement is to design a local two-state 1D CA rule ϕ, independent of lattice size N (assume N odd for simplicity), such that the global dynamics f has the following properties: (i) the system has two attractors, each of cycle length 1, one being the all zeros state, one being the all ones state; (ii) any initial state s_0 with more zeros than ones is in the basin of the all zeros state, and vice versa; (iii) the maximum transient length is at worst polynomial in n (i.e., the attractor is discovered in polynomial time). The computation determines which attractor basin the initial state is in, and hence, by inference, whether the initial state has more zeros than ones.

No two-state CA with a bounded neighborhood size can solve this problem exactly on arbitrary lattice size N (Land and Belew 1995). The best rules designed or evolved (e.g., Gács et al. 1978; Mitchell et al. 1996; Wolz and de Oliveira 2008) fail to correctly classify about 20%

❏ **Fig. 4**
Sensitive dependence on initial conditions. Each plot overlays the evolutions of two initial states differing in only one bit: the growing central dark region is different; the outer regions are the same. $N=971$, random initial state s_0, over 646 timesteps (a) ECA rule 45; (b) ECA rule 110.

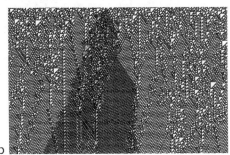

a　　　　　　　　　　　　　　　　　b

of states, that is, they define a dynamics where these states are in the "wrong" basin of attraction. Bossomaier et al. (2000) investigate the problem specifically in terms of the attractor basins.

Example CA 2: The Rule 30 Pseudorandom Number Generator

Wolfram (1986b) discusses ECA rule 30:

$$\phi_{30}(c_{i-1}, c_i, c_{i+1}) = c_{i-1} \text{ XOR } (c_i \text{ OR } c_{i+1}) \tag{1}$$

Starting from an initial state s_0 with a single cell j "on," $c_{i,0} = (\text{if } i = j \text{ then } 1 \text{ else } 0)$, (❷ *Fig. 5*), then the sequence of bits under this single on bit, $\tau = c_{j,0}, c_{j,1}, \ldots, c_{j,t}, \ldots$ forms a (pseudo) random sequence. Wolfram (1986b) presents evidence that the cycle length of the attractor starting from a state with a single nonzero cell grows exponentially with CA lattice size N.

This sequence τ is a projection (onto the single Boolean state variable c_j) of the trajectory from state s_0 under the dynamics defined by rule 30.

2.1.6 Random Boolean Networks

A random Boolean network (RBN) comprises N *nodes*. Each node i at time t has a binary valued state, $c_{i,t} \in \mathcal{B}$. Each node has k inputs assigned randomly from k of the N nodes (an input may be from the node itself); the wiring pattern is fixed throughout the lifetime of the network. This wiring defines the cell's neighborhood, v_i. See ❷ *Fig. 6*.

The state of cell i's neighborhood at time t is $\chi_{i,t} \in \mathcal{B}^k$, a k-tuple of cell states that is the projection of the full state onto the neighborhood v_i.

Each node has its own randomly chosen local state transition rule, or update rule, $\phi_i : \mathcal{B}^k \to \mathcal{B}$. These cells form a network of state transition machines. At each timestep, the state of each cell is updated in parallel, $c_{i,t+1} = \phi_i(\chi_{i,t})$.

◨ **Fig. 5**
ECA rule 30 (❷ Eq. 1), initial state of a single nonzero cell.

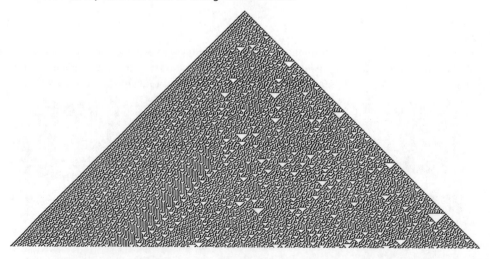

◻ **Fig. 6**
An example random Boolean network (RBN) with $N=6, k=2$. Each node has $k=2$ inputs; it can have any number of outputs. So the neighborhood function is $v_A = (A,F), v_B = (C,F)$, etc. Each node combines its inputs by a random Boolean function ϕ_i: that function might ignore one (or both) of the inputs.

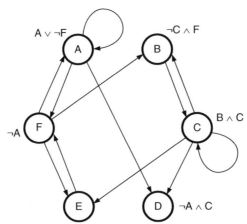

The global dynamics f is determined by the local rules ϕ_i and the connectivity pattern of the nodes v_i. In contrast to the regularity of a CA, in an RBN each node has its own random neighborhood connection pattern and its own random update rule. Like CAs, RBNs capture only a small fraction of all the possible dynamics over an ND space.

Kauffman (1990, 1993) investigates the properties of RBNs as a function of connectivity k. (The wiring conditions given above are not stated explicitly in these references. However, in the $k=N$ case, Kauffman (1993, p.192) states that "Since each element receives an input from all other elements, there is only one possible wiring diagram." This implies that multiple connections from a single node are not allowed (otherwise more wiring diagrams would be possible) whereas self connections are allowed (otherwise k would be restricted to a maximum value of $N-1$). Subsequent definitions (e.g., Drossel 2008) explicitly use the same conditions as given here.) Unlike CAs, RBNs tend not to have repeated basins, because of the random nature of the connections, and hence the relative lack of symmetries. Despite all the randomness, however, "such networks can exhibit powerfully ordered dynamics" (Kauffman 1990), particularly when $k=2$ (❯ *Fig. 7*; ❯ *Table 1*). Drossel (2008) notes that subsequent computer simulation of much larger networks shows that "for larger N the apparent square-root law does not hold any more, but that the increase with system size is faster."

Kauffman identifies $k=1$ as the "ordered" regime, with a very large number of short period attractors. Large k is the chaotic regime, with very long period attractors (compare Wolfram's class 3 behavior). $k=2$ occurs at a "phase transition" (Kauffman 1990), separating the ordered and chaotic regimes; it exhibits a moderate number of moderate period attractors.

Kauffman investigates RBNs as simplified models of gene regulatory networks (GRNs). He notes that "cell types are constrained and apparently stable recurrent patterns of gene expression," and interprets his RBN results as demonstrating that a "cell type corresponds to a state cycle attractor" (Kauffman 1993, p. 467) (in a $k=2$ network).

◘ Fig. 7

Visualization of the time evolution of three typical $k=2$ RBNs, with $N=400$, and initial condition all nodes "off"; after 150 timesteps all nodes are set to "on," then all nodes are randomized (50% "on," 50% "off") every further 150 timesteps, to explore other attractors. They exhibit ordered behavior: short transients and low period attractors. The visualization scheme used here (Stepney 2009) orders the nodes to expose the *frozen core* (Kauffman 1993, p. 203) of nodes that do not change state on the attractor; this frozen core is well preserved on different attractors.

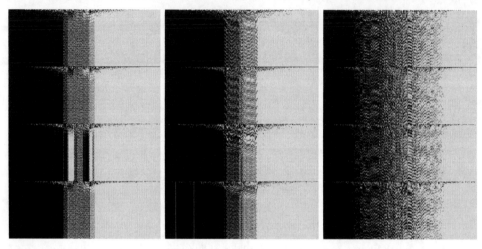

◘ Table 1

Dynamics of random Boolean networks (RBNs) for different k (Adapted from Kauffman (1993, Table 5.1))

k	Attractor cycle length	# Attractors
1	$O(\sqrt{N})$	$O(2^N)$
2	$O(\sqrt{N})$	$O(\sqrt{N})$
>5	$O(2^N)$	$O(N)$

Emergent macrostates of the dynamics, in addition to the attractor cycle length, are the number of nodes whose states change during a cycle, compared to the number that form the static *frozen core*. In the GRN interpretation, the frozen core would correspond to genes whose regulatory state was constant in a particular cell type, and the changing nodes to those genes whose regulatory state was cycling.

2.2 Continuous Space, Discrete Time

We next consider continuous spaces, with $\mathcal{X} = \mathcal{R}$ (where $\mathcal{R}$ is the set of real numbers), with discrete time dynamics $t \in \mathcal{N}$. Let the state vector be $\mathbf{r} \in \mathcal{R}^N$. The dynamics of a particular system is determined by its particular transition function $f : \mathcal{R}^N \rightarrow \mathcal{R}^N$, with $\mathbf{r}_{t+1} = f(\mathbf{r}_t)$.

These systems are called *difference equations* or *iterated maps*. A discrete-time trajectory $\mathbf{r}_t, \mathbf{r}_{t+1}, \mathbf{r}_{t+2}, \ldots$ is also called an *orbit*.

The trajectories can display a range of behaviors, corresponding to a range of types of attractor, depending on the system. Trajectories may diverge ($|\mathbf{r}_t| \to \infty$); they may converge to a fixed point attractor ($\mathbf{r}_t \to \mathbf{r}^*$, where $f(\mathbf{r}^*) = \mathbf{r}^*$); they may converge to a periodic attractor; they may be chaotic, never repeating but still confined to a particular subregion of the state space. So, continuous space systems have chaotic behavior that differs qualitatively from the "chaotic" behavior of discrete space systems, since the finite discrete systems must eventually repeat and hence be periodic (although with exponentially long periods).

2.2.1 Parameterized Families of Systems

It is often convenient to consider a family of dynamics related by some parameter $p \in \mathcal{P}$, that is, $f(\mathbf{r}, p)$, and investigate how the dynamics of a system vary as a function of this parameter. The trajectories of such parameterized systems can display the whole range of behaviors corresponding to the range of types of attractor, depending on the value of the parameter. A small change to the parameter can make a fixed point move, or become unstable, or periodic, or disappear. It can move a system from period P to period $2P$ (*period doubling*), or from periodic to chaotic behavior. The parameter values where these qualitative changes in the dynamics occur are called *bifurcation points*. As the parameter crosses the bifurcation point, the change in the dynamics can be continuous (smooth), or discontinuous (*catastrophic*).

Many systems exhibit a sequence of period doublings as the parameter changes. Subsequent doublings happen ever more rapidly (requiring ever smaller changes to the parameter), then the system moves into a chaotic regime. This is known as the *period doubling route to chaos*. Appearance of a period doubling cascade in a parameterized system is indication that chaos will ensue if the parameter changes further.

Another typical behavior is the *intermittency route to chaos*. Here the parameter starts in a region with periodic dynamics; as it is changed, the periodic behavior is broken by intermittent bursts of irregular behavior. As the parameter changes further, there are more and more of these bursts, until the behavior becomes completely chaotic. Hence a fully deterministic system may be apparently periodic, interrupted by what look like bursts of noise, where these bursts are simply part of the same overall dynamics of the system, and in need of no external explanation.

2.2.2 Logistic Map

The logistic map, a parameterized family of 1D iterated maps,

$$r_{t+1} = \lambda \, r_t (1 - r_t) \tag{2}$$

is a well-studied example of such a system (see ❯ *Figs. 8* and ❯ *9*). It is usually studied for $\lambda \in [0, 4]$, since in this region its dynamics is confined to the unit interval: $r \in [0, 1]$. It does have complex dynamics for other values of λ, but these are not constrained to the unit interval.

Its unintuitively rich, complex properties have been researched in detail, and May (1976) advised:

▶ Not only in research, but also in the everyday world of politics and economics, we would all be better off if more people realised that simple nonlinear systems do not necessarily possess simple dynamical properties.

☑ **Fig. 8**

Orbit diagram of the logistic map, (a) $2 \leq \lambda \leq 4$, showing the period doubling route to chaos; (b) $3.5 \leq \lambda \leq 4$, zooming in on the window of period 3.

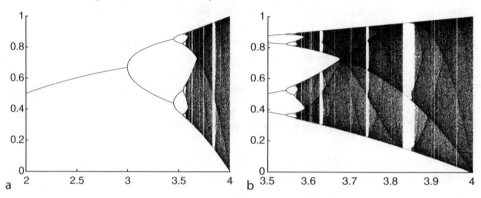

The attractor structure, visible in ❷ *Fig. 8*, is summarized in ❷ *Table 2* as a function of parameter λ. It has period doubling cascades to chaos as λ is increased. The first cascade leads to the onset of chaos at $\lambda = 3.56994\,56718.\ldots$ Within this chaotic region, there are windows of periodicity (such as the window of period 3), which then also period-double back to chaos. Given the existence of a period 3 cycle in a map, then every possible period can also be found in that map (Li and Yorke 1975). In the logistic map, there are windows of every period for some $3 < \lambda < 4$. Each of these periodic windows then period-doubles back to chaos as λ increases. The order in which these cascades occur itself has a complex structure, which can be calculated iteratively, in terms of symbolic sequences (Metropolis et al. 1973); the lowest order sequences are shown in ❷ *Table 2*.

Cycles of the same period can nevertheless have different kinds of behaviors: see, for example, ❷ *Fig. 10*.

The logistic map also exhibits the intermittency route to chaos. If the parameter λ falls in a window of periodic behavior, and is then slowly *reduced*, intermittent behavior is seen (e.g., ❷ *Fig. 9e*), until fully chaotic behavior is reached.

Binary Shift Map

Consider the fully chaotic case, $\lambda = 4$. Changing variables, to $x = \frac{1}{\pi} \cos^{-1}(1 - 2r)$, yields the equation for the binary shift (Bernoulli) map:

$$x_{t+1} = 2x_t \bmod 1 \tag{3}$$

If x is expressed in base 2, each iteration of the map results in a left shift of the number (the multiplication by 2), and dropping any resulting leading 1 in front of the binary point (the mod 1). For example, if $x_t = 0.1110001\ldots$, then $x_{t+1} = 0.110001\ldots$, $x_{t+2} = 0.10001\ldots$, etc. If x_0 is rational, its binary expansion is periodic, and hence the iteration will be periodic; if x_0 is irrational, its binary expansion is nonperiodic, and hence the iteration will be nonperiodic. Separate values of x_0 that are the same up to their nth bit will initially have similar iterations, but will eventually diverge, until they are completely different at the nth iteration: This is a manifestation of sensitive dependence on initial conditions.

◘ **Fig. 9**

Timeseries of the logistic map. 100 iterations, with $r_0=0.1$, for various λ. Top row: (a) $\lambda=2.8$, period 1; (b) $\lambda=3.3$, period 2. Second row: (c) $\lambda=3.5$, period 4; (d) $\lambda=3.74$, period 5. Third row: (e) $\lambda=3.828$, chaos before period 3: period 3 behavior is interleaved with intermittent bursts of chaotic behavior; (f) $\lambda=3.829$, period 3. Bottom row: (g) $\lambda=4$, chaos; (h) $\lambda=4$, with $r_0=0.10001$, demonstrating sensitive dependence on initial conditions.

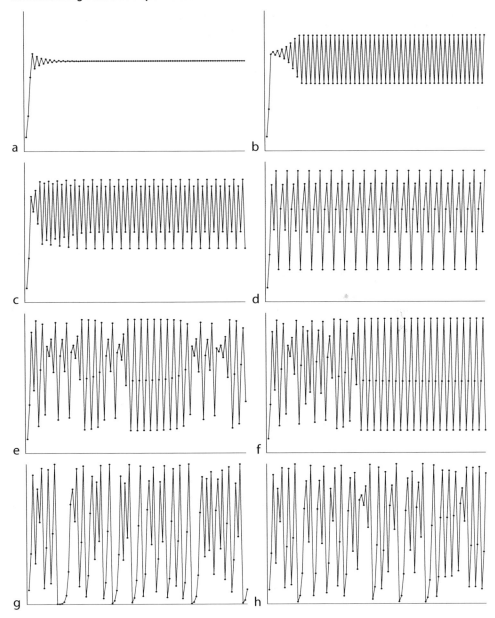

☐ **Table 2**

Dynamical structure of the logistic map as a function of parameter λ (some λ values taken from Sloane (2008)); including order of occurrence of period doubling cascades, for all initial periods up to 7 (Adapted from Hao and Zheng (1998, Table 2.2)). Higher initial periods cascades are interleaved with these

Lambda	Dynamics	Sloane number (Sloane 2008)
$(0,1)$	Fixed point $r^*=0$	
$(1,3)$	Fixed point, function of λ (e.g., $\lambda=2, r^*=0.5$	
3	1st period doubling, to period 2	
$1+\sqrt{6}=3.449\ldots$	2nd period doubling, to period 4	
$3.54409\,03595\ldots$	3rd period doubling, to period 8	A086181
$3.56440\,72660\ldots$	4th period doubling, to period 16	A091517
$3.56994\,56718\ldots$	End of first cascade; first onset of chaos	A098587
$3.62655\,31616\ldots$	Appearance of period 6; period 6 cascade	A118453
$3.70164\,07641\ldots$	1st period 7 cascade	A118746
$3.73817\,23752\ldots$	Appearance of period 5; 1st period 5 cascade	A118452
$3.7741\ldots$	2nd period 7 cascade	
$1+2\sqrt{2}=3.828427\ldots$	Appearance of period 3; sole period 3 cascade	
$3.841499\ldots$	Period doubling, to period 6	
$3.8860\ldots$	3rd period 7 cascade	
$3.9055\ldots$	2nd period 5 cascade	
	4th period 7 cascade	
$3.9375\ldots$	Period 6 cascade	
	5th period 7 cascade	
$3.9601\ldots$	Period 4 cascade	
	6th period 7 cascade	
$3.9777\ldots$	Period 6 cascade	
	7th period 7 cascade	
$3.9902\ldots$	3rd and last period 5 cascade	
	8th period 7 cascade	
	period 6 cascade	
	9th and last period 7 cascade	
4	Fully chaotic	

Computational Properties

Crutchfield (1994) investigates the *statistical complexity* of the logistic map as a function of λ. The statistical complexity is essentially a measure of the size of a stochastic finite state machine that can predict the statistical properties of a system. He finds that the map has *low* statistical complexity both when the map is periodic and when it is chaotic (essentially random), and has *highest* statistical complexity at the onset of chaos (via the period doubling route, or the intermittent route). At this point, there is a *phase transition>* in the complexity of the

□ Fig. 10
Different classes of behavior of three different period 6 cycles. Top shows the time series of one period; bottom shows the geometry of the behavior in a cobweb diagram. (a) λ=3.6266; (b) λ= 3.8418; (c) λ=3.93755.

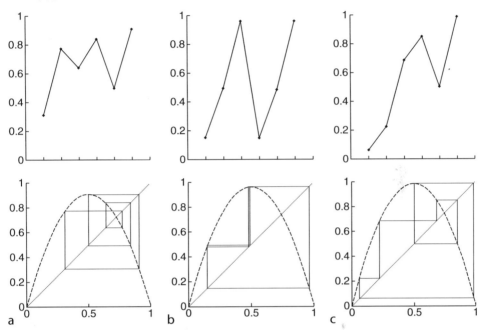

machine needed to predict the behavior, and these parameter values indicate the highest computational capacity.

Despite these observations, the majority of computational applications of the logistic map exploit its completely chaotic behavior, with λ at or near 4, and use this behavior to implement random number generators (Kanso and Smaoui 2009; Phatak and Rao 1995; Ulam and von Neumann 1947), encryption (Kocarev and Jakimoski 2001; Pareeka et al. 2006), etc.

2.2.3 Coupled Map Lattices

Kaneko (1983, 1984, 1985, 1986) introduces one-dimensional coupled map lattices (CMLs), where an array of N iterated maps are coupled together locally, with the following local dynamics ϕ:

$$r_{i,t+1} = \phi_0(r_{i,t}) + \frac{\varepsilon}{2}\left(\phi_c(r_{i-1,t}) - 2\phi_c(r_{i,t}) + \phi_c(r_{i+1,t})\right) \tag{4}$$

where ε is the coupling strength. Each element in the lattice evolves under the dynamics of the local process ϕ_0, with an additional interaction contribution $\varepsilon\phi_c$ from its neighbors, while passing on a similar amount of its own to its neighbors (under periodic boundary conditions). The local state $r_i \in \mathcal{R}$ and the local dynamics $\phi : \mathcal{R}^3 \to \mathcal{R}$ define the global state $\mathbf{r} \in \mathcal{R}^N$ and global dynamics $f : \mathcal{R}^N \to \mathcal{R}^N$. The system is parameterized by the coupling strength ε, as well as by any parameters in ϕ.

An initial value of $r_{i,0}=\kappa$, where all the maps have the same initial value, is trivial, since all maps evolve in lock-step. Kaneko (1985) investigates several cases, one of which is where ϕ_0 is the logistic map with λ in the period 3 window with cycle (r_1^*, r_2^*, r_3^*), with $r_{i \leq N/2,0}=r_1^*$, $r_{N/2 < i,0}=r_2^*$, or with random $r_{i,0}$, and with $\phi_c=\phi_0$. The dynamics exhibits the period doublings, intermittencies, and chaos of the logistic map, and in addition exhibits spatial patterns and structures similar to 1D CAs. Crutchfield and Kaneko (1987) examine the properties of this class of system in some detail.

Subsequent work generalizes the approach to topologies other than 1D nearest-neighbor (2D, tree-structured, irregular, larger neighborhoods), and allows the coupling to be asymmetric. In particular, Holden et al. (1992) provide a generic formalism, and investigate coupled map lattices in terms of their computational properties. A CML approach to the density classification problem has been evolved (Andersson and Nordahl 1998). Open CMLs are also being exploited computationally (see ❷ Sect. 3.3.1).

2.2.4 Note on Dimensionality and Topology

Papers on CMLs tend to describe them as having "discrete time, discrete space, and continuous state" (Kaneko 1986). Here we describe them as "continuous space," because here we are talking purely about the *state space*.

A CML of N maps laid out in 1D line in (physical) space has an (abstract) state space of $\mathcal{R}^N$. The "dimensionality" of the layout in physical space (a 1D line of maps) is unrelated to the dimensionality of the state space (of ND, because there are N maps); if the same N maps were instead laid out in a 2D grid, or even a 27D grid, say, it would not affect the dimensionality of the state space.

Instead, this "dimensionality" is related to the *topology* of the connections between the maps, and hence the potential information flow between the maps (❷ *Fig. 11*).

Compare the information flow visible in a 1D (physical space) CA (❷ *Fig. 4*) and that not visible in an RBN with a similar number of cells (❷ *Fig. 7*). For this reason, it makes sense to talk of the (physical spatial) dimension of a CA (it has a regular local topology), but not of an RBN (it has an irregular graph topology).

◻ **Fig. 11**

Dimensionality versus topology. Examples for three $N=5, k=2$ networks with different topologies. (a) Regular $N=15$ CA-like structure arranged in a 1D spatial lattice (periodic boundary conditions); (b) three $N=5, k=2$ RBNs linked in a 1D spatial lattice (that is, random connections with the groups of 5 nodes, lattice links between the groups); (c) general $N=15, k=2$ RBN.

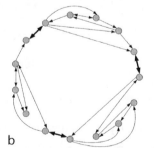
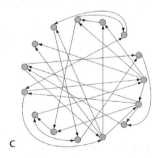

a   b   c

When the physical spatial layout moves from discrete to continuous, the state space moves from being finite (discrete physical space, finite number of cells) or countably infinite (discrete physical space, countably infinite number of cells) to uncountably infinite (continuous physical space). See ❷ Sect. 2.3.4.

2.2.5   Numerical Errors and the Shadowing Lemma

Chaotic systems, such as the logistic map with $\lambda=4$, display sensitive dependence on initial conditions: trajectories with nearby initial conditions diverge exponentially (compare ❷ *Fig. 9g* and ❷ *9h*).

Such systems are usually studied by numerical simulations, which generate *pseudo-orbits*, because of numerical noise. Additionally, the real physical systems that these maps (and below, differential equations) model are themselves subject to noise during their execution, and during measurement, and so potentially execute pseudo-orbits (or pseudo-trajectories) with respect to the model systems. The question naturally arises: What is the relation of these pseudo-orbits to the model orbits? Are they representative of the modeled dynamics?

Fortunately the answer is (a qualified) "yes." The *Shadowing Lemma* states that, for certain classes of system, the pseudo-orbit *shadows* (stays close to) some true orbit of the (modeled) system for all time; this true orbit has a slightly different initial condition from the pseudo-orbit. This result has been extended to a wider class of systems, including those studied here, that the pseudo-orbit shadows some true orbit of the system for "a long time" (Hammel et al. 1988).

However, are the true orbits of the model that are shadowed by these pseudo-orbits themselves representative of the underlying dynamics; that is, are they typical true orbits? This is harder to answer, and is clearly false for some particular cases. For example, consider the binary shift map (❷ Eq. 3). For any limited precision binary arithmetic implementation, *all* pseudo-orbits converge to 0 when the number of iterations (shifts) exceeds the binary numerical precision. But this case appears to be an exception, because most numerical trajectories do not behave like this, and Hayes and Jackson (2005) state:

▶ If otherwise reliable-looking pseudo-trajectories *are* atypical, they must be atypical in an extremely subtle way, because researchers have been making apparently reliable, self-consistent, peer-reviewed conclusions based on numerical simulations for decades.

So we continue here in assuming that the simulated pseudo-orbits, and the pseudo-trajectories of actual physical systems, are in general representative of the true dynamics defined by the equations.

2.3   Continuous Space, Continuous Time

We now consider continuous spaces, with $\mathcal{X} = \mathcal{R}$, with continuous time dynamics $t \in \mathcal{R}$. Let the state vector be $\mathbf{r} \in \mathcal{R}^N$. The dynamics of a particular system is determined by its particular transition function $f : \mathcal{R}^N \to \mathcal{R}^N$, with $\dot{\mathbf{r}} = f(\mathbf{r})$. Hence the system is defined by a set of N coupled first-order ordinary differential equations (ODEs).

We can recast higher order equations into this normal form by adding new variables. For example, consider the 1D equation for damped simple harmonic motion

$$\ddot{r} + \kappa \dot{r} + \omega^2 r = 0 \tag{5}$$

Let $r_1 = r, r_2 = \dot{r}$. Then, rearranging, we can get the 2D normal form version:

$$\dot{r}_1 = r_2 \quad ; \quad \dot{r}_2 = \omega^2 r_1 - \kappa r_2 \tag{6}$$

Note how this normalization takes the single-state variable r, and results in two-state variables r_1 (r, position) and r_2 ($\dot{r}$, velocity). In a continuous system, and particularly when the state variables are position **r** and momentum $m\dot{r}$, the state space is also called the *phase space*.

Physical systems embody their own specific dynamics. If that dynamics can be controlled and exploited in a computational manner, it can be used to reduce the load on, or even replace, conventional classical digital control in embedded systems (Stepney 2007). Understanding the dynamical behavior of complex material systems from a computational perspective is also a necessary step along the way to understanding biological systems as information processing systems (Stepney 2008).

For a good overview of continuous dynamical systems from an embodied computational perspective, see Beer (1995). Abraham and Shaw (1987) provide an excellent visual description of various concepts such as attractors and bifurcations. For more background, see a textbook such as that by Strogatz (1994).

2.3.1 Kinds of Attractors

In these continuous time systems, trajectories are continuous paths through a continuous state space. There are four distinct kinds of attractor that can occur.

Point Attractors
A point attractor is a single point in state space that attracts the trajectories in its basin. An example is the equilibrium position and zero velocity that is the unique end state of damped simple harmonic motion (● *Fig. 12a*).

□ **Fig. 12**

Attractors and their transient trajectories: (a) point attractor: damped SHM, $\ddot{r} + \lambda\dot{r} + r = 0$, with $\lambda = 0.5$; (b) limit cycle attractor: van der Pol oscillator, $\ddot{r} + \lambda(r^2 - r)\dot{r} + r = 0$, with $\lambda = 2$.

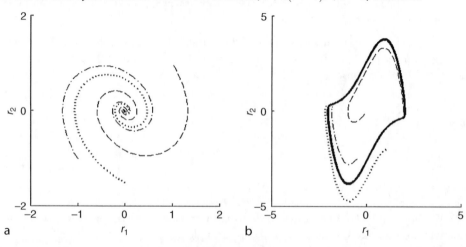

Limit Cycle Attractors
A limit cycle is a closed loop trajectory that attracts nearby trajectories to it. See, for example, ❯ *Fig. 12b*.

Toroidal Attractors
A toroidal attractor is a 2D surface in state space with periodic boundary conditions: shaped like a torus. Trajectories are confined to the surface of the torus. If the *winding number* (the number of times the trajectory loops around in one dimension while it performs one loop in the other dimension) is rational, the trajectory is periodic, otherwise it is *quasiperiodic*, and eventually covers essentially the entire surface of the torus.

Strange Attractors
A strange attractor attracts trajectories to its region of state space, but within this region, nearby trajectories diverge exponentially: it exhibits sensitive dependence on initial conditions, and thus chaotic behavior. This combination of attraction and divergence requires at least three dimensions in which to occur. The detailed structure of a strange attractor is usually fractal.

Example: Rössler Strange Attractor
The Rössler system is defined by

$$\dot{r}_1 = -r_2 - r_3$$
$$\dot{r}_2 = r_1 + ar_2 \tag{7}$$
$$\dot{r}_3 = b + r_3(r_1 - c)$$

It is a family of dynamical systems that displays a range of kinds of dynamics, some with strange attractor behavior, some without. It exhibits the period doubling cascade route to chaos (❯ *Fig. 13a*).

■ **Fig. 13**
Rössler system: (a) period doubling cascade to chaos, with $a=b=0.2$, $c=2.5, 3.5, 4.0, 5.0$; (b) Rössler strange attractor.

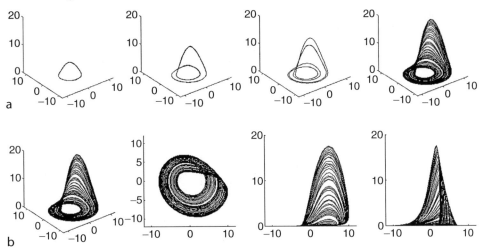

The Rössler strange attractor occurs when $a=0.2, b=0.2, c=5$ (❯ *Fig. 13b*). It is the simplest strange attractor, with only one nonlinear term.

Example: Lorenz Strange Attractor

The Lorenz strange attractor is defined by

$$
\begin{aligned}
\dot{r}_1 &= 10(r_2 - r_1) \\
\dot{r}_2 &= 28r_1 - r_2 - r_1 r_3 \\
\dot{r}_3 &= r_1 r_2 - 8r_3/3
\end{aligned} \tag{8}
$$

(See ❯ *Fig. 14*.) It is a member of a family of dynamical systems that displays a range of kinds of dynamics, some with strange attractor behavior, some without.

Sensitive dependence on initial conditions is popularly known as *the butterfly effect*. Lorenz suggests (Lorenz 1993, p. 14) that the name may have arisen from the title of a talk he gave in 1972, "Does the Flap of a Butterfly's Wings in Brazil Set Off a Tornado in Texas?," coupled with the butterfly-like shape of Lorenz attractor seen from some directions (❯ *Fig. 14*).

Reconstructing the Attractor

Given a physical continuous dynamical system with a high-dimensional state space (state vector $\mathbf{r}_t$), one can determine properties of its dynamics, given only scalar discrete time series observations (time series data $r_\tau, r_{2\tau}, \ldots, r_{n\tau}, \ldots$, where r_t is some scalar projection of the state vector $\mathbf{r}_t$). In particular, one can distinguish chaos (motion on a strange attractor) from noise.

The process of reconstructing the attractor from this data involves constructing a d-dimensional state vector $\hat{\mathbf{r}}_t$ from a sequence of time-lagged observations:

$$
\hat{\mathbf{r}}_{n\tau} = \left(r_{n\tau}, r_{(n+k)\tau}, r_{(n+2k)\tau}, \ldots, r_{(n+(d-1)k)\tau}\right) \tag{9}
$$

d should be $>2d_a$, where d_a is the dimension of the system's attractor; d should also be as small as possible, to avoid fitting noise; k should be large enough that the attractor is sufficiently sampled, but not so large that the correlations are lost. Taken's *embedding theorem* then relates the invariants of motion, including attractor structure, of $\hat{\mathbf{r}}$ to those of $\mathbf{r}$. For more on this process, and other techniques for analyzing chaotic systems, see Ott et al. (1994).

Relationship to Discrete State Space Attractors

Wolfram (1984b) draws a rough analogy between his class 1, 2, and 3 CAs (see ❯ Sect. 2.1.5) and point, limit cycle, and strange attractors, respectively. He mentions that the class 4 CAs (the ones conjectured universal) "behave in a more complicated manner." This might be thought to imply that there is no continuous analog of the discrete class 4 systems, and hence

■ Fig. 14
Lorenz strange attractor.

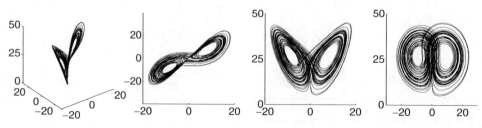

no universal computational properties in continuous matter. This is not so, as we see below (❯ Sects. 2.3.4 and ❯ 2.3.5).

Kauffman (1990) calls RBNs whose attractor cycle length increases exponentially with *n*, "chaotic." He emphasizes that this does not mean that flow on the attractor is divergent (it cannot be, in a discrete deterministic system); the state cycle is the analog of a 1D limit cycle. However, there is an analogy: Exponentially long cycles cover a lot of the state space before repeating (chaotic strange attractors never repeat), and "nearby" states (1 bit different) potentially *do* diverge (even possibly onto another attractor). However, in the discrete system, there is no direct analog of "nearby states diverging exponentially, but staying on the same attractor," since there is usually no concept of distance between states in discrete dynamical systems, and if there were, successive hops through state space can be of any size: there is no simple "continuity" from which to diverge.

2.3.2 Computation in Terms of Attractors

We can interpret computation as finding which attractor basin the system is in, by following its trajectory to the relevant attractor. The output could be (some projection of) the computed attractor, including a subspace of the state space. Most instances of analog computing fall in this domain.

The programming problem is in finding the relevant dynamics, now restricted to natural (albeit engineered) material system properties, which are not arbitrary. The aim is to minimize the engineering required to implement the desired dynamics, by exploiting the natural dynamics.

Continuous systems computing in this way can exhibit *robustness*. A small perturbation to the system might shift it a small distance from its attractor, but its subsequent trajectory will converge back to the attractor. There can be a degree of continuity in the attractor basins, such that a small perturbation tends to remain in the same basin, unlike the discrete case.

It is not necessary for a dynamical system to have a complex (chaotic, strange) dynamics in order to be interesting or useful. Kelso (1995, p. 53) makes this point eloquently:

▶ Some people say that point attractors are boring and nonbiological; others say that the only biological systems that contain point attractors are dead ones. That is sheer nonsense from a theoretic modeling point of view, as it ignores the crucial issue of what fixed points refer to. When I talk about fixed points here it will be in the context of collective variable dynamics of some biological system, not some analogy to mechanical springs or pendula.

That is, the dynamics, including the underlying attractor structure, is part of the specific model, in particular, what state variables are used to capture the real-world system. State variables can capture more sophisticated concepts than simple particle positions and momenta.

2.3.3 Continuous Time Logistic Equation

The logistic growth equation is one of the simplest biologically based nonlinear ODEs. It is a simple model of population growth where there is exponential growth for small populations, but an upper limit, or *carrying capacity*, that prevents unbounded growth. The equation was suggested by Verhulst in 1836 (see, e.g., Murray 1993, p. 2):

$$\dot{r} = \rho r(1 - r/\kappa) \tag{10}$$

It is rare among nonlinear ODEs in that it has an analytic solution

$$r(t) = \frac{\kappa\, r_0\, e^{\rho t}}{\kappa + r_0(e^{\rho t} - 1)} \tag{11}$$

It has a single point attractor at $r_\infty = \kappa$: whatever the initial population r_0, it always converges to the carrying capacity κ.

Contrast this smooth behavior with the very different complex periodic and chaotic behavior of its discrete time analog: the logistic map, ❏ Sect. 2.2.2. This is a general feature: The discrete time analog of simple ODEs can exhibit similarly complex behavior as the logistic map. This does *not* mean, however, that ODEs themselves are unable to display computationally interesting dynamics.

2.3.4 Infinite Dimensions: PDEs

Consider the reaction–diffusion (RD) equation, which models chemical species reacting (nonlinearly) locally with each other, and diffusing (linearly) through space. The relevant state variables are the concentrations of the reacting diffusing chemicals, and are functions of time, and also of space: $r_i(t, \mathbf{x})$. For a two-chemical species, the RD equation is

$$\begin{aligned}
\frac{\partial r_1}{\partial t} &= f_1(r_1, r_2) + k_1 \nabla^2 r_1 \\
\frac{\partial r_2}{\partial t} &= f_2(r_1, r_2) + k_2 \nabla^2 r_2
\end{aligned} \tag{12}$$

Each r_i, since it is a function of continuous space $\mathbf{x}$, can be thought of as an (uncountably) infinite-dimensional state variable. Rather than having a state vector with a finite number of indices r_i, we can consider an infinite-dimensional state vector indexed by position, $r(\mathbf{x})$. The state variable at each position can itself have multiple components (such as the two chemical concentrations, above), leading to $\mathbf{r}(\mathbf{x})$. The space derivative is used to define the dynamics in terms of a local (infinitesimal) neighborhood.

There is a natural link between partial differential equation (PDE) systems and cellular automata. CAs are one natural way to simulate PDEs in a discrete domain (Adamatzky et al. 2005; Wolfram 1985, prob. 9). Care is needed in this process to ensure that the CA models the correct PDE dynamics, and does not introduce artefacts due to its own discrete dynamics (Weimar and Boon 1994; Wolfram 1986a). Despite this caveat, there are some exact correspondences: *ultradiscretization* (Tokihiro et al. 1996) can be used to derive CA-like rules that preserve the properties of a given continuous system, for a class of integrable PDEs; *inverse ultradiscretization* (Kunishima et al. 2004) transforms a CA into a PDE, preserving its properties. A different approach represents CA configurations using continuous *bump functions* (Omohundro 1984), and derives a PDE that evolves the bumps to follow the given CA rule.

The dynamical theory of these infinite-dimensional spaces is not as well developed as in the finite case. Much of the work concentrates on PDEs whose dynamics can be rigorously reduced to a finite subspace, so that the existing dynamical systems theory is applicable. See, for example, Robinson (2001) and Teman (1997).

2.3.5 Reaction–Diffusion Computers

Reaction–diffusion computers (Adamatzky et al. 2005) use chemical dynamics. The relevant state variables are the concentrations of the reacting diffusing chemicals, which are functions of time and space: $r_i(t,\mathbf{x})$. For a two-chemical species, the relevant RD equation is given by ❯ Eq. 12.

Reaction–diffusion systems have a rich set of behaviors, exhibiting spatial-temporal patterns including oscillations and propagating waves. The computation is performed by the interacting wave fronts; the output can be measured from the concentrations of the reagents. RD systems have been used to tackle a wide variety of computation problems (e.g., image processing (Kuhnert et al. 1989), robot navigation (Adamatzky et al. 2004)); here we look at two that demonstrate computation exploiting the natural dynamics, and one that demonstrates the potential for universal computation.

Voronoi Diagrams

An RD computer can solve a 2D Voronoi problem: given a set S of points in the plane, divide the plane into $|S|$ regions R such that every point in a given region $R_i(s_i)$ is closer to s_i than to any other $s_j \in S$.

This problem can be solved directly by a 2D RD computer (Tolmachiev and Adamatzky 1996) (❯ Fig. 15). One reagent forms a substrate; the second reagent marks the position of the data set of points S. The data-reagent diffuses and reacts with the substrate-reagent, forming waves propagating from each data point, and leaving a colored precipitate. Waves meet at the borders of the Voronoi regions, since their constant speed of propagation implies that they have traveled equal distances from their starting points. When waves meet, they interact and form no precipitate. So the lack of precipitate indicates the computed boundaries of the Voronoi regions.

◻ Fig. 15
A reaction–diffusion (RD) computer solving a Voronoi problem. (Figure from Tolmachiev and Adamatzky 1996, Fig. 3.)

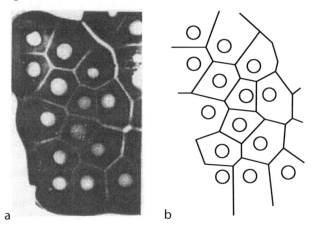

a   b

Shortest Path Searching

A propagating wave technique can be used to find the shortest path through a maze or around obstacles (Steinbock et al. 1995; Agladze et al. 1997; Ito et al. 2004). The maze can be encoded in the chemical substrate, or by using a light mask. A wave is initiated at the start of the maze, and a series of time lapse pictures are taken as the wave propagates at a uniform speed, which provide a series of equidistant locations from the starting point; these are used in a post-processing phase to construct the shortest path (❷ *Fig. 16*).

Logic Gates

Propagating waves can be confined to channels ("wires") and interact at junctions ("gates") so arranged such that the interactions perform logical operations. See, for example, Motoike and Adamatzky (2004), Sielewiesiuk and Gorecki (2001), and Toth and Showalter (1995). Hence the continuous RD system dynamics can be arranged by careful choice of initial conditions to simulate a digital circuit.

2.3.6 Generic Analog Computers

In general, analog computers gain their efficiency by directly exploiting the physical dynamics of the implementation medium. There is a wide range of problem-specific analog computers, such as the reaction–diffusion computers described above, but there are also general purpose analog computers.

For example, Mills has built implementations of Rubel's general purpose extended analog computer (Rubel 1993). The computational substrate is simply a conductive sheet (Mills 2008a; Mills et al. 2006), which directly solves differential equations; the system is "programmed" by applying specific potentials and logic functions at particular points in the sheet. Mills has developed a computational metaphor to aid programming by analogy (Mills 2008b), and is currently developing a compiler to enable straightforward solution of given differential equations (Mills 2008, personal communication).

◻ **Fig. 16**

Computer simulation of a shortest path computation in mazes with obstacles. (From Ito et al. 2004.)

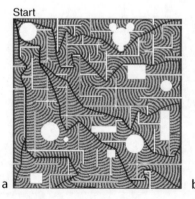
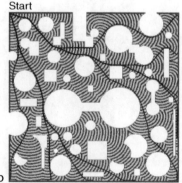

a   b

2.4   Hybrid Dynamical Systems

So far, all the systems considered have homogeneous dimensions, for example, all $\mathcal{S}$ or all $\mathcal{R}$. More complicated dynamical systems have heterogeneous dimensions. For example, coupling a classical finite state machine with a continuous system would yield a hybrid system with a dimensionality like $\mathcal{S}^M \times \mathcal{R}^N$.

It is likely that the topology of such a hybrid system would consist of relatively weakly coupled sub-components (❷ *Fig. 11b*), which should help in their analysis from a dynamical systems point of view.

2.5   Summary

The classes of autonomous dynamical systems that have been discussed are summarized in ❷ *Table 3*.

From a computational perspective, one important classification dimension is the implementation: whether the computational dynamics is implemented "naturally," that is, directly by the physical dynamics of the underlying medium, or whether it is implemented in terms of a virtual machine (VM) itself implemented on that underlying dynamics.

Discrete systems tend to be implemented in terms of VMs. This has the advantage that the computational dynamics is essentially independent of the underlying medium (witness the diversity of systems that implement Boolean logic, for example), and so can be analyzed in isolation. It has the disadvantage of the computational overhead imposed by the VM layer.

One goal of continuous dynamical systems is to provide a computational dynamics closely matched to the physical dynamics, with corresponding gains in efficiency. The downside is that such systems are more likely to be constrained by their physical dynamics, and so are less likely to be Turing-universal computational systems.

3   Open Dynamical Systems

3.1   Openness as Environmental Inputs

The systems described so far are *autonomous* or *closed*. They have an *initial condition* (identifying one state $\mathbf{x}_0$ from the $\mathcal{X}^N$ possible), and then the fixed (non-time-dependent) dynamics

■ Table 3
Classification of the different kinds of autonomous dynamical systems, in terms of their state space and time evolution

	Discrete: $t \in \mathcal{N}$	Continuous: $t \in \mathcal{R}$
$s \in \mathcal{S}^N$	Finite CAs ; RBNs	
$s \in \mathcal{S}^\infty$	Infinite CAs ; TMs	
$r \in \mathcal{R}$	Iterated maps	
$r \in \mathcal{R}^N$	CMLs	ODEs
$r \in \mathcal{R}^\infty$		PDEs
$\mathbf{x} \in \mathcal{S}^M \times \mathcal{R}^N$	Hybrid	

proceeds with no input or interference from the outside world. They move to an attractor, the result of the computation, or they may not discover an attractor, in which case the computation has no result. This is the classical, "ballistic" style of computation exemplified by the Turing machine, or a closed dissipative system relaxing to equilibrium.

Open, or nonautonomous, systems, on the other hand, have dynamics that are governed by parameters that change over time. These parameters are inputs from the environment.

Consider an open dynamical system with N degrees of freedom: its state can be defined by an ND state vector $\mathbf{x}(t) \in \mathcal{X}^N$. The state space is $\mathcal{X}^N$. Now there is also an input space $\mathcal{P}$, and an output space $\mathcal{Q}$. The dynamics f maps the current state and input to the next state and output; $f : \mathcal{X}^N \times \mathcal{P} \to \mathcal{X}^N \times \mathcal{Q}$.

There is a similarity here to a parameterized *family* of dynamics (❷ Sect. 2.2.1). But here the parameter p is a function of t, and is considered an *input* to the dynamics, a way of modulating or controlling the dynamics, for example, moving it between periodic and chaotic attractor behaviors.

3.1.1 Timescales

Understanding open systems is significantly more challenging than understanding closed systems, and depends in part on the relationship between the timescale on which the input is changing and the timescale on which the dynamics is acting. Dynamical systems have a "natural" timescale: the time needed to discover the attractor. As Beer (1995) says

▶ Because ... the flow is a function of the parameters, in a nonautonomous dynamical system the system state is governed by a flow which is changing in time (perhaps drastically if the parameter values cross bifurcation points in parameter space). Nonautonomous systems are much more difficult to characterize than autonomous ones unless the input has a particularly simple (e.g., periodic) structure. In the nonautonomous case, most of the concepts that we have described above (e.g., attractors, basins of attraction, etc.) apply only on timescales small relative to the timescale of the parameter variations. However, one can sometimes piece together a qualitative understanding of the behaviour of a nonautonomous system from an understanding of its autonomous dynamics at constant inputs and the way in which its input varies in time.

That is, an input changes the dynamics of the system, by changing to a different member of the parameterized family. This new member might have moved attractors, or be on the other side of a bifurcation point with different kinds of attractors. A system immediately after an input will be in the same position in its state space, but the underlying attractor structure of that space may have changed.

So, if the input is changing slowly with respect to the dynamics, the system is able to complete any transient behavior and reach the changed attractor before the input changes the dynamics yet again, even if it has passed through a discontinuous, catastrophic bifurcation point. On these timescales, the system is able to "track" the changing dynamics, and so its behavior can be analyzed piecewise, as a sequence of essentially unchanging systems. Even so, such systems can exhibit *hysteresis*: Restoring a parameter to a previous value may not necessarily restore the system to its corresponding previous state, if this path through parameter space crosses catastrophic bifurcation points.

If the input is changing quickly with respect to the dynamics, then the system is unable to respond to changed dynamics before it has changed again. It will mostly be exhibiting transient behavior.

Most interesting and complex is the case where the input is changing on a timescale similar to that of the dynamics: then the system is influenced by its dynamics, but it may never quite, or only just, reach any attractor before the next change occurs.

The situation can get even more complicated, when the input parameter p is a function of space as well as time, $p(\mathbf{x}, t)$. For example, it might be a temperature gradient, or magnetic field gradient, which can also drive the system.

3.1.2 Computation in Terms of Trajectories

Since such open systems need not reach a "halting state" of being on an attractor, the computational perspective is necessarily broader. The computation being performed can be viewed as the *trajectory* the system takes through the changing attractor space: which attractor basins are visited, in which order.

The discussion of timescales implies that, for useful computation, the dynamical timescale of the system should not be significantly slower than the input timescales.

3.1.3 Environmental Constraints

In the simplest open case, it is assumed that the system can be provided with any input, regardless of what it is doing, or has done (❷ *Fig. 17a*). The input may as well be considered random. This is not particularly interesting, except in the cases where the system can somehow exploit noise, for example, using some kind of informational analog of a ratchet mechanism (e.g., Dasmahapatra et al. (2006)).

This case is formally equivalent to a closed system, as the sequence of arbitrary inputs could conceptually be provided at the start, embedded in an (expanded) state $\mathcal{X}^N \times seq\ \mathcal{P}$ as part of the initial condition, along with some pointer to the "current time" value, and the dynamics updated to allow access only to the current value (e.g., see Cooper et al. (2002),

◻ **Fig. 17**
Inputs. (a) arbitrary input stream; (b) constrained inputs from a structured environment; (c) constrained inputs from an interacting environment in a feedback loop.

where such an approach is taken to embed inputs into the initial state of the model of a formal language). So there is no new computational capability, except as provided by the (potentially, much) larger state space.

Type B: Environmentally Constrained Input Stream

More interesting is the case where the inputs come from an environment that has some rich dynamical structure that the system can couple to and exploit (❷ *Fig. 17b*). Here, the environment is an autonomous dynamical system, unaffected by any inputs of its own.

Since the environment is autonomous, its sequence of inputs could again conceptually be provided at the start. However, this case is qualitatively different from the previous one. We are now assuming that there is some *structure* in the environment, and hence in the sequence of inputs. This implies that there are regions of the system state space that are never explored, parts of its underlying dynamics that are never exercised. As before when talking of virtual machines (❷ Sect. 2.1.3), the computation is restricted to a subspace: the subspace and its trajectories here correspond to the computation in the context of the structured environment; that is, the inputs provide *information* that the system need not itself compute. And since the environment may be unboundedly large, the sequence of inputs may represent an unboundedly large amount of computation provided to the system.

Type C: Feedback Constrained Input Stream

Most interesting is the case where the environment and system are both open dynamical systems in a rich feedback loop. Then outputs from the system will alter the environment, and affect its subsequent inputs (❷ *Fig. 17c*). So the actual sequence of inputs cannot even conceptually be provided at the start.

Again, the environment's inputs will be constrained to a region of state space, but here this region is (partly) determined by the system: the environment and systems are coupled, the dynamics of each perturbing the trajectory of the other, in a feedback loop.

Such an environment may well contain other open systems similar to the system being considered. And again, since the environment may be unboundedly large, it may represent an unboundedly large amount of computation provided to the system, but here, the specific computation provided is affected by what the system does, because of the feedback coupling. This opens up the possibility for a system to offload some of its computational burden onto the environment (see ❷ Sect. 3.4.4).

Beer (1995) points out that such a coupled environment and system together form a higher-dimensional autonomous dynamical system, with its own attractors, and, because of its larger state space, that this combined system can "generate a richer range of dynamical behavior than either system could do individually."

3.2 Open Discrete Space, Discrete Time Dynamical Systems

The autonomous discrete systems have an *initial condition* (identifying one state $\mathbf{s}_0$ from the $|\mathbf{S}|^N$ possible). In the finite case, they always "halt" on an attractor cycle, this halting state (cycle) being the result of the computation; in the infinite case, they either halt on such a cycle, or they may not discover an attractor, in which case the computation has no result.

In the open discrete space, discrete time case, where $\mathcal{X} = \mathbf{S}$, the time evolution of the state, and the output function, are given by $(\mathbf{s}_{t+1}, q_{t+1}) = f(\mathbf{s}_t, p_t)$.

3.2.1 Modulating the Dynamics

The input can provide a "kick" or perturbation, changing (a few bits of) the state at a particular timestep. This may move the system into a different attractor basin.

The input might also "clamp" the system into a particular substate (by fixing the value of some bits for many timesteps). This not only perturbs the system at the point where the bits are clamped, but can also change the global dynamics (if the natural dynamics would change the value of the clamped bits, for example), producing new attractors, and changing or removing existing attractors. For example, clamping some bits in a CA can result in "walls" across which information cannot flow, isolating regions, and hence changing the dynamical structure of the system. (See ❷ *Fig. 18*.) Such a simple partitioning is harder to achieve in a more irregularly connected structure such as an RBN.

3.2.2 Perturbing Random Boolean Network State

Kauffman (1990) defines a *minimal perturbation* to the state of an RBN to be flipping the state of a single node at one timestep. Flipping the state of node i at time t is equivalent to changing its update rule at time $t-1$ to be $c_{i,t} = \neg \phi_i(\chi_{i,t-1})$. Such a perturbation leaves the underlying dynamics, and hence the attractor basin structure, the same; it merely moves the current state to a different position in the state space, from where it continues to evolve under the original dynamics: it is a transient perturbation to the state.

◨ **Fig. 18**
Dependence of CA dynamics on a "clamped" bit. The upper plot shows ordinary periodic boundary conditions; the lower plot shows the same initial conditions, but with the central bit "clamped" to 0. (a) ECA rule 26; (b) ECA rule 110.

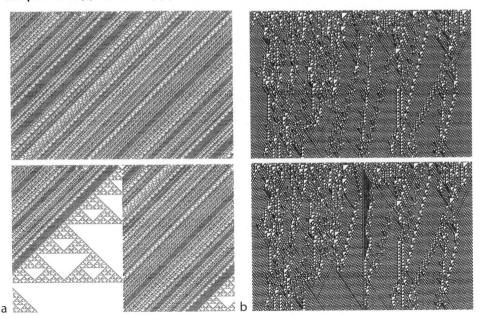

a   b

Kauffman (1990) describes the *stability* of RBN attractors to minimal perturbations: If the system is on an attractor and suffers a minimal perturbation, does it return to the same attractor, or move to a different one? Is the system *homeostatic*? (Homeostasis is the tendency to maintain a constant state, and to restore its state if perturbed.)

Kauffman (1993) describes the *reachability* of other attractors after a minimal perturbation: If the system moves to a different attractor, is it likely to move to any other attractor, or just a subset of them? If the current attractor is considered the analog of "cell type," how many other types can it *differentiate* into under minimal perturbation?

Kauffman's results are summarized in ❯ *Table 4*, which picks out the $k=2$ networks as having "interesting" (non-chaotic) dynamics (a small number of attractors, with small cycle lengths) and interesting behavior under minimal perturbation (high stability so a perturbation usually has no effect; low reachability so when a perturbation moves the system to another attractor, it moves it to one of only a small subset of possible attractors).

3.2.3 Perturbing Random Boolean Network Connectivity

Kauffman (1990) also defines a *structural perturbation* to an RBN as being a permanent mutation in the connectivity or in the Boolean function. So a structural perturbation at time t_0 could change the update rule of cell i at all time $t>t_0$ to be $\phi_i'(\chi_{i,t})$ or change the neighborhood of cell i at all time $t>t_0$ to be $\chi'_{i,t}$. Since the dynamics is defined by all the ϕ_i and χ_i, such a perturbation changes the underlying dynamics, and hence the attractor basin structure: It is a permanent perturbation to the dynamics, yielding a new RBN.

Such a perturbation could have several consequences: a state previously on an attractor cycle might become a transient state; a state previously on a cycle might move to a cycle of different length, comprising different states; a state might move from an attractor with a small basin of attraction to one with a large basin; a state might move from a stable (homeostatic) attractor to an unstable attractor; and so on.

Kauffman (1993) relates structural perturbation to the *mutation* of a cell; if there is only a small change to the dynamics, this represents mutation to a "similar" kind of cell.

3.3 Open Continuous Space, Discrete Time Dynamical Systems

In the open continuous space, discrete time case, where $\mathcal{X} = \mathcal{R}$, the time evolution of the state, and the output function, are given by $(\mathbf{r}_{t+1}, q_{t+1}) = f(\mathbf{r}_t, p_t)$.

◻ Table 4

Dynamics of RBNs for different k (adapted from Kauffman (1993, Table 5.1))

k	Stability	Reachability
1	Low	High
2	High	Low
>5	Low	High

3.3.1 Open Coupled Map Lattices, and Chaos Computing

Sinha and Ditto (1998, 1999) investigate the computational properties of coupled chaotic logistic maps (ϕ_0=the logistic map with λ=4, see ❷ Sect. 2.2.3) with open boundaries. The coupling function ϕ_c is a threshold function: ϕ_c=**if** $r<\theta$ **then** 0 **else** $r-\theta$, and is unidirectional. This unidirectional relaxation propagates along the lattice until all elements have a value below threshold (all the $r_i \leq \theta$), at which point the next timestep iteration of the logistic dynamics occurs (so the timescale of the relaxation is much shorter than the timescale of the logistic dynamics). The boundary of the lattice is open, so the final element can relax below threshold by removing its excess value from the system. Depending on the particular threshold value, and the current state of the system, this process can result in nonlinear avalanches of relaxation along the lattice. Transients are short (typically one dynamical timestep), and the system displays a variety of attractor dynamics, stable to small amounts of noise, and determined by the input threshold parameter.

Sinha and Ditto (1998, 1999) discuss how to use this system to perform computation. A single lattice element with two inputs (either external, or from the coupled output of other lattice elements) and appropriate input value of its threshold can implement a universal NOR logic gate in a single iteration of the logistic dynamics, where the amount of output encodes the result. Hence, networks of such elements can be coupled together to implement more complicated logic circuits. Sinha and Ditto (1999) note that it is the chaotic properties in general, not the logistic map in particular, that give these coupled systems their computational abilities, and suggest a possible implementation based on nonlinear lasers forming a coupled chaotic Lorenz system. They dub their approach *chaos computing*.

Additionally, Sinha and Ditto (1998, 1999) discuss how to implement other functions, such as addition, multiplication, and least common multiple, by suitable choice of numerical encoding, coupling, and input thresholds. These choices program the underlying chaotic dynamics to perform computation directly, rather than emulating a virtual machine of compositions of logic gates (Sinha and Ditto 1999):

▶ dynamics can perform computation not just by emulating logic gates or simple arithmetic operations, but by performing more sophisticated operations through self-organization rather than composites of simpler operations.

Despite this observation, their subsequent work (Ditto et al. 2008; Sinha and Ditto 2006) concentrates on using more complicated (but realizable) chaotic dynamics to implement robust logic circuits, that can be readily reconfigured merely by altering thresholds. They emphasize the openness of their approach (Sinha and Ditto 2006):

▶ it can yield a gate architecture that can dynamically switch between different gates, without rewiring the circuit. Such configuration changes can be implemented either by a predetermined schedule or by the outcome of computation. Therefore, the flexibility of obtaining different logic operations using varying thresholds on the same physical element may lead to new dynamic architecture concepts

3.4 Open Continuous Space, Continuous Time Dynamical Systems

In the open continuous space, continuous time case, $\mathcal{X} = \mathcal{R}$, the time evolution of the state, and the sequence of outputs, are given by $(\dot{\mathbf{r}}, q(t)) = f(\mathbf{r}, p(t))$.

Note again the similarity to a parameterized *family* of dynamics (❷ Sect. 2.2.1), with the (input) parameter p being a function of t. Examples of such input parameters might include the temperature T of the system, or the value of an externally imposed magnetic field **B** permeating the system.

3.4.1   Ott, Grebogi, Yorke (OGY) Control Laws

There are unstable periodic orbits in strange attractors. Small perturbations in the control parameter can be used to keep the system in one of these; the required perturbations are calculated using the OGY control laws (Ott et al. 1990). (It is not necessary to know the underlying dynamics to do this; application of the control laws involves calculating the required parameter from observations of the system.) There can be long transient behavior before the system gets "close" to the desired periodic orbit, and noise can result in bursts of chaotic behavior. The approach can also be used to switch between different periodic orbits with different characteristics (with some transient chaotic behavior).

Ott et al. (1990) note that a chaotic system is potentially more flexible, because this approach can be used to hold it in a variety of different periodic orbits, whereas this range of behaviors would require a range of separate systems with non-chaotic dynamics. Sinha and Ditto (❷ Sect. 3.3.1) make similar observations about their coupled nonlinear maps, although there they claim a lower computational burden (the thresholds can be simply calculated, and stored in a lookup table) and shorter (essentially zero) transient behaviors.

3.4.2   Liquid Crystal Systems

Liquid crystals are a form of matter that lies on the boundary between solids and liquids (sometimes called "the fourth phase of matter"). A liquid crystal has both complex dynamics (the molecules can flow and rotate) and complex structure (the molecules are ordered on length scales much bigger than their individual sizes).

Such materials can perform computation. Harding and Miller (Harding and Miller 2004, 2005; Harding et al. 2006) have demonstrated that a liquid crystal chip can be programmed to act as a tone discriminator (a simple arbitrary input system, where the inputs are tones of two different frequencies that are to be discriminated) and as a robot controller (a constrained feedback system, where the inputs from the environment depend on the robot's position, and the robot's outputs change its position).

It is currently unclear how the liquid crystal performs its computations: In the referenced cases, the material was programmed using an evolutionary algorithm. It would be interesting to analyze these results from a dynamical systems perspective.

3.4.3   NMR Logic Gates

Nuclear magnetic resonance (NMR) uses radio frequency pulses to manipulate nuclear spins in a magnetic field. Depending on the particular values chosen from a rich potential set of parameters (frequency, phase, duration, delay, and more), pulses can be combined in various ways to produce different outputs. These pulses and the outputs can be interpreted as encoding binary values. Under these interpretations, the system can be used to implement a single universal logic gate, a circuit of these gates combined in parallel and in sequence that implement other logic gates, and a half-adder circuit (Roselló-Merino et al. 2010).

Under these interpretations, this is a Type A open system: Any arbitrary set of Boolean inputs is permitted, and the system implements the corresponding logic gate computation on these inputs. Under the wider view of the full parameter space, this is a Type B open system, with the inputs being restricted by the environment (the experimenter) to values that can be interpreted as binary bits. Hence, we see how setup corresponds to a logic gate virtual machine (❯ Sect. 2.1.3) implemented on the underlying dynamics of the nuclear spin system.

3.4.4 Embodiment

Beer (1995) takes a dynamical systems approach to adaptive agents in a changing environment. For example, a robot agent adapts its walking behavior depending on the kind of terrain it is moving across. So an agent receives sensory input from a structured environment (it sees its current surroundings), which affects its internal dynamics (its state changes), which affects its outputs (its leg movements), which in turn affects the environment (at a very minimum, the agent moves to a new location in the environment, and hence sees new surroundings). So "a significant fraction of behavior must be seen as emerging from the ongoing interaction between an agent and its environment." Beer is talking here of Type C open dynamical systems (❯ Sect. 3.1.3), coupled to their environment in a feedback loop.

Beer's aim is to use the language and concepts of dynamical systems theory to develop a theoretical framework for designing adaptive agents. Certain computational tasks, such as planning, can be greatly simplified by exploiting input from the structured environment (e.g., the agent seeing where it is). This relates to Brooks' design principle: "use the world as its own model" (Brooks 1991).

Beer emphasizes that, because of the coupling with the environment, "an agent's behavior properly resides only in the dynamics of the coupled system," and hence cannot be understood or analyzed in isolation. Indeed, an agent is adapted to some environments and not others. Beer analyzes this idea of adaptive fit, and determines that it requires that an agent maintains its trajectory within a certain volume of its state space (which volume may change with time) under perturbations from the environment. This is related to an *autopoietic* (Maturana and Varela 1980) (self-creating and self-maintaining) view of adaptive fitness, and helps define what is needed for *homeostasis*.

This area also links with a whole burgeoning subdiscipline of using dynamical systems theory to model the brain and its cognitive processes. However, that is deemed to be outside the scope of this particular discussion. The interested reader is referred to Kelso (1995).

3.5 External Control Versus Internal Dynamics

Here we have considered the parameter $p(t)$ to be an externally provided input, moving a dynamical system between members of its parameterized family. Alternatively, we could have a case where the parameter p was another degree of freedom, or dimension, of the dynamical system, where the dynamics of the system itself affects the value of p (❯ *Fig. 19*).

This potentially gives a model of a *self-organizing* system. For example, Melby et al. (2000) describe a logistic map (❯ Sect. 2.2.2) where the value of the dynamical variable r_t is fed through a low-pass filter to (slowly) affect the parameter λ: The system self-adjusts to values

□ **Fig. 19**
Two sources of parameter variation: external control versus internal dynamics.

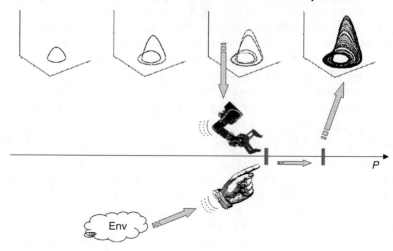

of λ at the edge of chaos, and does so even when subject to an external force attempting to drive it back into the chaotic region (Melby et al. 2002). The situation is a little more complicated in the presence of noise: The system still self-adjusts to suppress chaos, but a power-law distribution of chaotic outbreaks occurs (Melby et al. 2005).

An alternative view is of a hierarchy of coupled dynamical systems, where the outputs at one level couple to the control parameter(s) at another level. Abraham and Shaw (Abraham 1987; Abraham and Shaw 1987) explore this idea in more detail.

4 Constructive Dynamical Systems

So far, we have considered predetermined fixed state spaces of a given dimensionality. However, this is not the case even for classical computational systems. Their data structures define their abstract state space: every time new memory is allocated in the course of the computation, the state space grows in dimension; memory deallocation shrinks the state space. A dynamical systems perspective on computation needs to consider cases where the state space itself dynamically changes dimensionality, $X^{N(t)}$.

In such cases, the dynamics *constructs* the state space, and the state space constrains the dynamics. Closed autonomous systems can exhibit such growth, classical computation being one such example. In open systems, this process can be thought of as the flow of information, matter, energy being recruited to construct new dimensions of the system: the system inputs "food" (constructs dimensions), and excretes "waste" (collapses dimensions). This process of modifying its own dimensionality affects the possible dynamics (recall, e.g., that strange attractors require at least a 3D continuous space ❷ Sect. 2.3.1).

This is an aspect of dynamical systems theory that has received very little attention: the current theory assumes a predefined, fixed state space. Because of the paucity of theory in this area, this section is intended to illustrate the kinds of processes that need to be incorporated in such a theory.

Metadynamics

One approach to incorporating growth might be to model the system with an infinite number of dimensions, confining the dynamics to the finite-dimensional subspace corresponding to the current state space. As new dimensions are needed, the dynamics expands into the preexisting dimensions. However, this approach appears to be a mathematical "trick" that hides the essential properties of the underlying system: precisely what confines the dynamics to a subspace, what causes it to grow into new dimensions, and what determines the topology (the way the information flows between dimensions)? (This approach also has a small technical issue: how to distinguish a system that is not using a particular dimension, from one that is using it, but is currently confined to the zero value of that dimension: how to distinguish "absence" from the "presence of nothing." This can in turn be solved by a further mathematical trick, of introducing some special value, $\perp$, to make this distinction. But the tricks are piling up.)

A different approach might be to have these new dimensions start in some small "curled up" form; a dimension could "uncurl" or "unfold" sufficiently to support the current state value along that dimension, and curl back up when no longer needed. Alternatively, the dimensions might be fractal in nature, with the fractal dimension of the state space increasing, gradually "fattening up" the new dimension.

This suggests that the state space itself might have a *metadynamics*. The metadynamics (the high-level dynamics of the state space) can be studied in isolation from the low-level dynamics (trajectories of the system through the state space).

Another form of metadynamics is to allow the topology (see ❷ Sect. 2.2.4) to change dynamically, even if the dimensionality is constant. Such "network dynamics" allows the neighborhood function (❷ Sect. 2.1.5) to be a function of time, $v(t)$.

Baguelin et al. (Baguelin et al. 2003; Moulay and Baguelin 2005) consider a meta-dynamical transition rule, which changes the state space (and the corresponding dynamical state transition rule) on a slower timescale than the system's dynamics: the system spends most of its time evolving under its dynamics, punctuated by relatively few metadynamical state space changes. The overall system dynamics is defined by "concatenating" the various lower-level dynamical behaviors at the metadynamical change points. Baguelin et al. (Baguelin et al. 2003; Moulay and Baguelin 2005) apply this approach to interacting populations of bacteria and phages (the dynamics) that are also evolving (a slower timescale metadynamics).

Whatever approach is used, it needs to be able to cope with different types of dimensions, as the dynamics results in new types of state variables being constructed (e.g., in evolution, new species can be considered to occupy new types of dimensions). If such types cannot be statically predetermined, but computed only by the unfolding dynamics, it is harder to see how to incorporate this into an analysis based on a known, preexisting set of state spaces.

Timescales

Important timescales in a constructive dynamical system are the speed at which the state space grows compared to dynamics on that space. Where growth timescales are much less than the dynamical timescales (a change in the number of dimensions tends to happen after the system has relaxed to an attractor), piecewise approximations may be made in a way similar to slow open systems (❷ Sect. 3.1.1), as in the metadynamics approach outlined above. But in faster-changing systems (e.g., L-Systems ❷ Sect. 4.1.2, where the dimensionality can change every iteration timestep) this is not possible, and new analysis techniques must be sought.

Computation in Terms of Construction

In an autonomous (closed) constructive system, the result of the computation (if the computation halts) could still be considered as the attractor (final macrostate), and additionally include the structure (dimensionality and topology) of the final grown state space.

In an open, non-halting constructive system, the result of the computation could still be considered as the trajectory through a growing state space, and additionally include the structure (dimensionality and topology) of the state space along that trajectory.

4.1 Constructive Discrete Dynamical Systems

4.1.1 Classical Computation

Turing Machine

A Turing machine can be thought of as a growing system: The tape is of finite but unbounded length, and can be considered to grow (lazily) whenever a new tape position is required. It can also be considered to shrink if the last symbol on the tape is erased.

So the question of growth can be linked to undecidability: Whether the dimensionality of the state space stays bounded (if the computation halts or loops), or grows without limit, is formally undecidable.

Object Orientation

Object orientation systems are particularly constructive in this sense: new objects are created, increasing the dimensionality of the state space; when they go out of scope, the state space shrinks again.

Such systems are usually designed on the software engineering principle of ensuring "weak coupling" by small interfaces, which keeps components separate and modular by restricting the information flow. That is, the topology of the dynamics is deliberately restricted.

Contrast the connectivity of RBNs, which have a "small-world" topology, and require a large number of connections to be cut to partition the graph (❷ *Fig. 11*). This enables rapid information flow, potentially enabling different classes of dynamics. The consequences of these design decisions are rarely addressed from a dynamical systems perspective.

Closed or Open

These classical computational systems can be closed (e.g., the TM model), or open (interactive systems). The open systems are usually analyzed as Type A (❷ Sect. 3.1.3), where the external environment can potentially provide any input.

4.1.2 Rewriting Systems

Imagine trying to "grow" an RBN where, if it got to a certain state, a new node would appear, or an existing node would disappear (with consequent rewiring). The appearance or disappearance of a node would change the dimensionality of the state space (to $N \pm 1$), and also the underlying attractor structure, and hence change the computation being performed.

Similarly, imagine trying to "grow" a 1D CA. One could add a new cell, growing the dimensionality of the state space from S^N to S^{N+1}. Where should the cell be added in the line? The CA dynamics is symmetric under renumbering shifts, and so it should be possible to add the cell anywhere with equal ease. But unless it is added at special position $i=N$, subsequent cells have to be renumbered. This illustrates the tyranny of a global coordinate system: "The introduction of numbers as coordinates ... is an act of violence" (Weyl 1949). Instead we would like a coordinate-free, or purely topological, approach to growth.

L-Systems

Lindenmayer's L-Systems (Prusinkiewicz and Lindenmayer 1990) are such a coordinate-free approach to growth: They are generative grammars that define how symbols in a string are rewritten ("grow") depending on their local context (their topology, their neighborhood symbols), not on any global coordinate system.

The simplest D0L-Systems (deterministic, context free) can be described in the notation of this chapter as follows. The state $\mathbf{s}$ is a string of elements (symbols) drawn from the finite alphabet S, so $\mathbf{s} \in S^*$. The local dynamics ϕ (rewriting rules, or productions) is given by $\phi : S \to S^*$. At each timestep, each element in the state string is updated (rewritten) in parallel: $s_{i,t+1} = \phi(s_{i,t})$, and the resulting substrings concatenated to form the new state string. (Note here that indices are used merely to identify the string elements in order to help define the local dynamics. They are not a fundamental part of the string data type, which supports operations such as insertion and deletion, unlike the fundamentally indexed array data type.)

In conventional use, one starts from some initial state (axiom) $\mathbf{s}_0 \in S^+$, and follows the dynamics for several iterations to reveal the "grown" structure. However, the dynamics does not necessarily result in a different number of dimensions (a changed length string). For example, Danks et al. (2007, 2008) use L-Systems to model protein folding: there the rewriting models the change in conformation of the (rendered) string, and does not result in any change in the length of the string itself.

In Parametric L-Systems (Prusinkiewicz and Lindenmayer 1990), each symbol has associated parameters, and so the state space is $(S \times \mathcal{P})^*$.

In the context-free case, there is no information flow between the dimensions: Each individual dimension grows into new dimensions in a manner based on its own substate alone. Context-sensitive L-Systems (Prusinkiewicz and Lindenmayer 1990) couple the dimensions together, and define a tree-structured topology on the growing space. In standard L-Systems, the topology is explicitly encoded in the state string, by use of reserved symbols to define branching.

L-Systems also include a *rendering* step, where the symbols in the string are interpreted as commands in some language, typically a turtle graphics language, to construct a representation of the string (Prusinkiewicz and Lindenmayer 1990). In standard L-Systems this rendering produces a geometrical structure that follows the tree-structured topology of the state space, typically to produce rendered images of plants. However, other renderings are possible, and can produce non-geometric outputs and outputs with different topologies. For example, Hornby and Pollack (2001) interpret the symbols as instructions for generating topological descriptions of neural networks.

These simplest L-Systems are autonomous dynamical systems, and any computation performed in the rendering step is separate from the dynamics of the growth process.

Parametric L-Systems, where the values of (some) parameters are inputs, are Type A or Type B open dynamical systems (❷ Sect. 3.1.3). In typical L-System applications, they are used as Type B, because the input is assumed to be coming from some constrained environmental source (e.g., day length, or other weather conditions).

Environmentally sensitive L-Systems (Prusinkiewicz et al. 1994) are also Type B open dynamical systems (❯ Sect. 3.1.3) of a different form. The dynamics again takes into account external inputs (e.g., the amount and direction of sunlight, or collision with static obstacles, when modeling plant growth), but here the particular rendering process is coupled into the dynamics, as the input is a function of the specific geometry of the rendered result (rendered leaves shading other leaves, etc.), not just the topology of the abstract state.

Open L-Systems (Měch and Prusinkiewicz 1996) are Type C open dynamical systems (❯ Sect. 3.1.3). They include a feedback process, so the environment is affected by the L-System (e.g., the environment might be the water supply, and the L-System the growing roots that are affecting the amount of water). Again, this approach couples the particular rendering process into the dynamics.

P-Systems

Păun's P-systems (Păun 2000), and other membrane computing formalisms, are another form of rewriting systems with an underlying tree-structured topology. Here the tree structure models not the branching of a plant, but the nesting of membranes, or containers. Membranes contain symbols, and rules that act on the symbols, and on membranes (creating, connecting, and dissolving them). The computational model successively applies the rules until no more are applicable; the computation has halted, and the symbols in a given membrane are the result of the computation. Much theoretical effort is expended on determining the computational power of different classes of membrane systems: Turing-complete systems exist. So this prototypical system has a rather classical view of computation, but there is a plethora of variants with different computational models.

Topological Approach

Michel and coworkers (Giavitto and Michel 2002) contrast the "absolute space" coordinate-based philosophy of Newton, with the relative space, coordinate-free philosophy of Leibniz, from a computational perspective. Their approach to (discrete) "dynamical systems with a dynamical structure $(DS)^2$" is topological, focussing on the neighborhood relation and induced subcollections (essentially a set of related elements in a neighborhood). The neighborhood relation can specify structures ranging from the regular spatial neighborhood of a CA to the logical neighborhood of a data structure such as a list, graph, or array (Giavitto et al. 2005). The approach then specifies global dynamics in terms of parallel application of local transformations that rewrite subcollections (Spicher et al. 2004; Giavitto et al. 2005). This approach subsumes dynamic rewriting systems such as L-Systems and P-systems, as well as more static-structured CA-like rewritings (Giavitto et al. 2005).

4.2 Constructive Continuous Dynamical Systems

Gordon Pask was a member of the Cybernetics movement. In the 1950s, he built a system that could "grow" into an "ear" (a sensor capable of detecting sound waves, or magnetic fields) by adaptively laying down suitable conducting filaments (❯ *Fig. 20*) in response to environmental inputs and a structured "reward" input.

◻ Fig. 20

Pask's schematic indicating the relationship between the electrode array and the ferrous sulfate medium (adapted from Pask (1960), as cited by Cariani (1993, Fig. 1)).

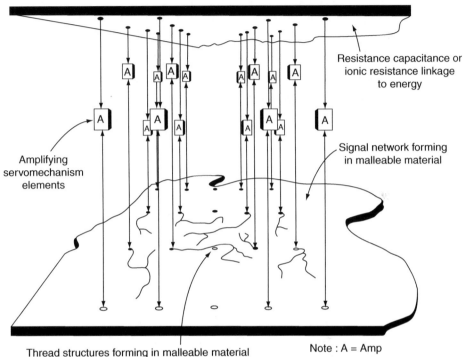

Cariani (1993) provides an excellent review of this work, and some implication in terms of "organizationally closed" (i.e., able to construct their own input filters), "informationally open" systems. In an open, adaptive, self-constructing system

▶ the dimensionality of the signal space can increase over time as new informational channels evolve. Hill-climbing is thus accomplished not only by following gradients upwards, but also by changing the dimensionality of the problem landscape when one can go no further using those dimensions already available.

4.3 Constructive Hybrid Dynamical Systems

Winfree and coworkers (Winfree 2003; Rothemund et al. 2004) are implementing "algorithmic self assembly." They use carefully designed DNA fragments to implement nanoscale "sticky tiles," which self-assemble into crystalline structures. The self-assembly can be interpreted as a computation: Which computation is performed depends on the tiles' design that determines which tiles they will stick to. Their running example is the implementation of a rule 90 ECA (which from a single "on" state grows a Sierpinski gasket). Other tile designs can implement other CAs, or even a representation of the tape of a TM.

The model of this system is a hybrid system. There is an abstract tile assembly model that captures the discrete dynamics of tiles (the particular CA or TM behavior). This is augmented with a kinetic tile assembly model, which captures continuous properties that account for errors in the assembly process.

They speculate that a similar process could be used to construct molecular electronic circuits and other nanoscale devices. That is, design a constructive dynamics to grow a structure that then has an autonomous computational dynamics.

5   Discussion and Conclusions

Classical computational systems can be analyzed in terms of autonomous discrete dynamical systems. The dynamical systems approach can be extended to continuous systems, to give a computational perspective on embodied computational devices. It can be extended again, to include open dynamical systems coupled to a structured dynamical environment, and again to include constructive, or growing systems. The dynamical systems theory is, however, less well developed in these latter cases.

In addition to the straightforward computational perspective, this dynamical systems viewpoint gives insight into important properties of complex computational systems: robustness and emergence.

5.1   Robustness

Homeostasis

Robustness is the ability to maintain function in the presence of perturbations, stresses, or errors. The system exhibits *homeostasis*. Classical computation, on the other hand, is notoriously *fragile*: A single bit change can completely alter the behavior of the system.

From a dynamical systems computational point of view, robustness is the ability to find the attractor even when perturbed from the current trajectory. A system will find the attractor from anywhere within its basin of attraction: If a perturbation nevertheless maintains the basin, the system will be robust to that perturbation.

Here we see the source of classical computation's fragility: extremely small basins of attraction, such that any perturbation is likely to move the system to a different basin, hence to a different attractor, resulting in a different computation.

Compare this situation to the dynamical structure of $k=2$ RBNs (❷ Sect. 2.1.6). There a perturbation is likely to leave the system in the same basin. "Small attractors located inside a volume of states constituting their basins of attraction are the natural image of stable systems exhibiting homeostasis." (Kauffman 1993, p. 467) So robustness requires large basins of attraction, which implies relatively few macrostates compared to the number of microstates.

Continuous systems have greater potential for robustness. They have a notion of locality: A small perturbation moves the system a small distance in its state space, and is therefore likely to be in the same basin. (Although it is true that as well as fractal strange attractors, there can be fractal-structured basins; in such cases, a small perturbation might end up in one of many other basins.) Alternatively, the small perturbation could be a small change to a parameter. Provided the parameter does not cross a bifurcation point, this will result in only a small change to the dynamics.

Kitano (2004) explicitly casts a discussion of various kinds of biological robustness in terms of dynamical systems attractors.

Precision

Robustness comes at the price of precision, however. Many microstates per macrostate means a loss of precision: we no longer distinguish these microstates. Also, real-valued output from a continuous system cannot be measured with arbitrary precision. Similar to the case when considering the timescales, the precision of the computation and output should be matched to the precision of the environment and inputs.

Homeorhesis

Waddington (1957) notes that homeostasis is too restrictive a term when considering growing systems, since they do not have a "steady" state to which to return: They are embarked on some trajectory through their ever-changing state space. He introduces the term *homeorhesis* (Waddington 1957, p. 32], meaning "constant flow," in that systems actually maintain a constant trajectory through developmental space. He deprecates the alternative term, *canalization*, because it may (Waddington 1957, p. 43–44]

▶ suggest too concrete an image to be suitable as a name for the abstract quality to which it refers; but this seems a less important failing than those involved in the alternative term homeostasis.

Note that this implies that the developmental trajectory is in some sense an attractor of the growth process: A perturbation away from the growth trajectory is attracted back to that *trajectory*, not merely to some final attractor state. In biological terms, this is robust morphodynamics.

5.2 Emergence

The attractors and their basins, and the bifurcation points of parameterized systems, are *emergent properties* of their dynamics (Stepney et al. 2006).

Although the microstates are changing on an attractor cycle, this can lead to a stable macrostate if observed on a timescale longer than the attractor period. In the dynamical systems view, everything is process (motion on an attractor), but when it is viewed on a suitable timescale, it can behave like a thing.

▶ An attractor functions as a symbol when it is viewed . . . by a *slow observer*. If the dynamic along the attractor is too fast to be recorded by the slow-reading observer, he then may recognize the attractor only by its averaged attributes . . . , but fail to recognize the trajectory along the attractor as a deterministic system. (Abraham 1987)

The slow observer does not see the intricate dynamics on the attractor, just some averaged behavior, and this dynamics becomes an atomic component in its own right. What is lost is the microstructure; what remains is a stable pattern that becomes an entity in its own right. This connects directly with the concepts of relative timescales, where there is a sufficient difference between a fast timescale (of the underlying dynamics of the system), and a slower timescale (of the inputs perturbing the system ❷ Sect. 3.1.1; of the metadynamic timescales of state space change ❷ Sect. 4; of the slow observer of these processes).

These new high-level emergent entities can have their own state space and (meta)dynamics, particularly when coupled to other dynamical systems such as the environment, or in some hierarchical structure. So their (higher level) dynamics will exhibit attractors; motion on these attractors will appear as emergent entities to (even slower) observers; and so on.

5.3   The Future

These discussions imply that we need a dynamical systems theory of open, constructive computational systems, where new dimensions of state space, and new types of dimensions, are constructed by the computation as it progresses. This will provide one of the tools we need to enable us to study, understand, design, and reason about robust emergent computational systems, covering the range from classically discrete, to nonclassical embodied systems.

Acknowledgments

My thanks to Ed Powley for providing ❷ *Figs. 1,* ❷ *3,* ❷ *4,* and ❷ *18,* and for drawing attention to the work on exact correspondences between CAs and PDEs in ❷ Sect. 2.3.4. My thanks also to Andy Adamatzky, Leo Caves, Ed Clark, Simon Hickinbotham, Mic Lones, Adam Nellis, Ed Powley, and Jon Timmis for comments on a previous draft.

References

Abraham RH (1987) Dynamics and self-organization. In: Yates FE (ed) Self-organizing systems: the emergence of order. Plenum, New York, pp 599–613

Abraham RH, Shaw CD (1987) Dynamics: a visual introduction. In: Yates FE (ed) Self-organizing systems: the emergence of order. Plenum, New York, pp 543–597

Adamatzky A (ed) (2002) Collision-based computing. Springer, London

Adamatzky A, Arena P, Basile A, Carmona-Galan R, Costello B, Fortuna L, Frasca M, Rodriguez-Vazquez A (2004) Reaction-diffusion navigation robot control: from chemical to VLSI analogic processors. IEEE Trans Circuits Syst 51(5):926–938

Adamatzky A, Costello BDL, Asai T (2005) Reaction-diffusion computers. Elsevier, Boston, MA

Agladze K, Magome N, Aliev R, Yamaguchi T, Yoshikawa K (1997) Finding the optimal path with the aid of chemical wave. Physica D 106:247–254

Andersson C, Nordahl MG (1998) Evolving coupled map lattices for computation. In: EuroGP'98, Paris, April 1998. LNCS, vol 1391. Springer, New York, pp 151–162

Baguelin M, LeFèvre J, Richard JP (2003) A formalism for models with a metadynamically varying structure. In: Proceedings of the European Control Conference, Cambridge, UK

Beer RD (1995) A dynamical systems perspective on agent-environment interaction. Artif Intell 72:173–215

Berlekamp ER, Conway JH, Guy RK (1982) Winning ways for your mathematical plays, vol 2. Academic, New York

Bossomaier T, Sibley-Punnett L, Cranny T (2000) Basins of attraction and the density classification problem for cellular automata. In: Virtual Worlds 2000, Paris, July 2000. LNAI, vol 1834. Springer, Heidelberg, pp 245–255

Brooks RA (1991) Intelligence without representation. Artif Intell 47:139–159

Cariani P (1993) To evolve an ear: epistemological implications of Gordon Pask's electrochemical devices. Syst Res 10(3):19–33

Cook M (2004) Universality in elementary cellular automata. Complex Syst 15(1):1–40

Cooper D, Stepney S, Woodcock J (2002) Derivation of Z refinement proof rules: forwards and backwards rules incorporating input/output refinement. Tech. Rep. YCS-2002-347, Department of Computer Science, University of York

Crutchfield JP (1994) The calculi of emergence. Physica D 75:1154

Crutchfield JP, Kaneko K (1987) Phenomenology of spatio-temporal chaos. In: Bin-Lin H (ed) Direction in Chaos, World Scientific, Singapore, pp 272–353

Culik K II, Yu S (1988) Undecidability of CA classification schemes. Complex Syst 2(2):177–190

Danks G, Stepney S, Caves L (2007) Folding protein-like structures with open L-systems. In: ECAL 2007,

Lisbon, Portugal, September 2007. LNAI, vol 4648. Springer, Heidelberg, pp 1100–1109

Danks G, Stepney S, Caves L (2008) Protein folding with stochastic L-systems. In: ALife XI, Winchester, UK, MIT Press, Boston, MA, pp 150–157

Dasmahapatra S, Werner J, Zauner KP (2006) Noise as a computational resource. Int J Unconventional Comput 2(4):305–319

Ditto WL, Murali K, Sinha S (2008) Chaos computing: ideas and implementations. Phil Trans R Soc A 366:653–664

Drossel B (2008) Random Boolean Networks. In: Schuster HG (ed) Reviews of nonlinear dynamics and complexity, vol 1, Wiley, Weinheim, arXiv:0706.3351v2 [cond-mat.stat-mech]

Durand B, Formenti E, Varouchas G (2003) On undecidability of equicontinuity classification for cellular automata. In: Morvan M, Rémila E (eds) Discrete models for complex systems, DMCS'03, Lyon, France, June 2003. DMTCS, vol AB, pp 117–128

Gács P, Kurdyumov GL, Levin LA (1978) One dimensional uniform arrays that wash out finite islands. Probl Peredachi Inf 12:92–98

Giavitto JL, Michel O (2002) Data structure as topological spaces. In: Unconventional models of computation, Kobe, Japan, October 2002. LNCS, vol 2509. Springer, Heidelberg, pp 137–150

Giavitto JL, Michel O, Cohen J, Spicher A (2005) Computation in space and space in computations. In: UPP 2004, France, September 2004. LNCS, vol 3566. Springer, Berlin, pp 137–152

Hammel SM, Yorke JA, Grebogi C (1988) Numerical orbits of chaotic processes represent true orbits. Bull Am Math Soc 19(2):465–469

Hao BL, Zheng WM (1998) Applied symbolic dynamics and chaos. World Scientific, Singapore

Harding SL, Miller JF (2004) A tone discriminator in liquid crystal. In: CEC 2004, Portland, Oregon. IEEE Press, pp 1800–1807

Harding SL, Miller JF (2005) Evolution in materio: A real-time robot controller in liquid crystal. In: Proc. NASA/DoD Conference on Evolvable Hardware, Washington, DC. IEEE Press, pp 229–238

Harding SL, Miller JF, Rietman EA (2006) Evolution in materio: exploiting the physics of materials for computation. arXiv:cond-mat/0611462

Hayes W, Jackson KR (2005) A survey of shadowing methods for numerical solutions of ordinary differential equations. Appl Numer Math 53:299–321

Holden AV, Tucker JV, Zhang H, Poole MJ (1992) Coupled map lattices as computational systems. Chaos 2 (3):367–376

Hornby GS, Pollack JB (2001) Body-brain coevolution using L-systems as a generative encoding. In:

GECCO 2001, Morgan Kaufmann, San Francisco, CA, pp 868–875

Ito K, Aoki T, Higuchi T (2004) Design of an excitable digital reaction-diffusion system for shortest path search. In: ITC-CSCC 2004, Japan

Kaneko K (1983) Transition from torus to chaos accompanied by frequency lockings with symmetry breaking. Prog Theor Phys 69(5):1427–1442

Kaneko K (1984) Period-doubling of kink-antikink patterns, quasiperiodicity in antiferro-like structures and spatial intermittency in coupled logistic lattice. Prog Theor Phys 72(3):480–486

Kaneko K (1985) Spatiotemporal intermittency in coupled map lattices. Prog Theor Phys 74(5):1033–1044

Kaneko K (1986) Lyapunov analysis and information flow in coupled map lattices. Physica D 23:436–447

Kanso A, Smaoui N (2009) Logistic chaotic maps for binary numbers generations. Chaos, Solitons & Fractals 40(5):2557–2568

Kauffman SA (1990) Requirements for evolvability in complex systems. Physica D 42:135–152

Kauffman SA (1993) The origins of order. Oxford University Press, New York

Kelso JAS (1995) Dynamic Patterns. MIT Press, Cambridge, MA

Kitano H (2004) Biological robustness. Nat Rev Genet 5:826–837

Kocarev L, Jakimoski G (2001) Logistic map as a block encryption algorithm. Phys Lett A 289 (4–5):199–206

Kuhnert L, Agladze K, Krinsky V (1989) Image processing using light-sensitive chemical waves. Nature 337:244–247

Kunishima W, Nishiyama A, Tanaka H, Tokihiro T (2004) Differential equations for creating complex cellular automaton patterns. J Phys Soc Jpn 73:2033–2036

Land M, Belew RK (1995) No perfect two-state cellular automata for density classification exists. Phys Rev Lett 74(25):5148–5150

Li TY, Yorke JA (1975) Period three implies chaos. Am Math Monthly 82(10):985–992

Lorenz EN (1993) The essence of chaos. UCL Press, London

Maturana HR, Varela FJ (1980) Autopoiesis and Cognition. D. Reidel, Boston, MA

May RM (1976) Simple mathematical models with very complicated dynamics. Nature 261:459–467

Melby P, Kaidel J, Weber N, Hübler A (2000) Adaptation to the edge of chaos in a self-adjusting logistic map. Phys Rev Lett 84(26):5991–5993

Melby P, Weber N, Hübler A (2002) Robustness of adaptation in controlled self-adjusting chaotic systems. Fluctuation Noise Lett 2(4):L285–L292

Melby P, Weber N, Hübler A (2005) Dynamics of self-adjusting systems with noise. Chaos 15:033902

Metropolis N, Stein M, Stein P (1973) On finite limit sets for transformations on the unit interval. J Comb Theory 15(1):25–43

Mills JW (2008a) The architecture of an extended analog computer core. In: UCAS-4, Austin, TX, USA

Mills JW (2008b) The nature of the extended analog computer. Physica D 237(9):1235–1256

Mills JW, Parker M, Himebaugh B, Shue C, Kopecky B, Weilemann C (2006) "Empty Space" computes: The evolution of an unconventional supercomputer. In: Proceedings of the 3rd ACM computing frontiers conference, New York, pp. 115–126

Mitchell M, Crutchfield JP, Das R (1996) Evolving cellular automata with genetic algorithms: a review of recent work. In: Goodman ED, Uskov VL, Punch WF (eds) Evolutionary computation and its applications: EvCA'96, Moscow

Motoike IN, Adamatzky A (2004) Three-valued logic gates in reaction-diffusion excitable media. Chaos Solitons Fractals 24:107–114

Moulay E, Baguelin M (2005) Meta-dynamical adaptive systems and their application to a fractal algorithm and a biological model. Physica D 207:79–90

Murray JD (1993) Mathematical biology, 2nd edn. Springer, New York

Měch R, Prusinkiewicz P (1996) Visual models of plants interacting with their environment. In: SIGGRAPH '96, ACM, New Orleans, LA pp 397–410

Omohundro S (1984) Modelling cellular automata with partial differential equations. Physica D 10:128–134

Ott E, Grebogi C, Yorke JA (1990) Controlling chaos. Phys Rev Lett 64(11):1196–1199

Ott E, Sauer T, Yorke JA (eds) (1994) Coping with chaos. Wiley, New York

Pareeka NK, Patidara V, Sud KK (2006) Image encryption using chaotic logistic map. Image Vis Comput 24(9):926–934

Pask G (1960) The natural history of networks. In: Yovits MC, Cameron S (eds) Self-organizing systems. Pergamon, New York

Păun G (2000) Computing with membranes. J Comput Syst Sci 61(1):108–143

Phatak SC, Rao SS (1995) Logistic map: a possible random-number generator. Phys Rev E 51(4):3670–3678

Powley EJ, Stepney S (2009a) Automorphisms of transition graphs for elementary cellular automata. J Cell Autom 4(2):125–136

Powley EJ, Stepney S (2009b) Automorphisms of transition graphs for linear cellular automata. J Cell Autom 4(4):293–310

Prusinkiewicz P, Lindenmayer A (1990) The algorithmic beauty of plants. Springer, New York

Prusinkiewicz P, James M, Měch R (1994) Synthetic topiary. In: SIGGRAPH '94, ACM, Orlando, FL, pp 351–358

Rendell P (2002) Turing universality of the game of life. In: Adamatzky A (ed) Collision-based computing. Springer, London, chap 18

Robinson JC (2001) Infinite-dimensional dynamical systems: an introduction to dissipative parabolic PDEs and the theory of global attractors. Cambridge University Press, Cambridge, MA

Roselló-Merino M, Bechmann M, Sebald A, Stepney S (2010) Classical computing in nuclear magnetic resonance. Int J Unconventional Comput 6 (3–4):163–195

Rothemund PWK, Papadakis N, Winfree E (2004) Algorithmic self-assembly of DNA Sierpinski triangles. PLoS Biol 2(12):e424

Rubel LA (1993) The extended analog computer. Adv Appl Math 14:39–50

Sielewiesiuk J, Gorecki J (2001) Logical functions of a cross-junction of excitable chemical media. J Phys Chem A 105(35):8189–8195

Sinha S, Ditto WL (1998) Dynamics based computation. Phys Rev Lett 81(10):2156–2159

Sinha S, Ditto WL (1999) Computing with distributed chaos. Phys Rev E 60(1):363–377

Sinha S, Ditto WL (2006) Exploiting the controlled responses of chaotic elements to design configurable hardware. Phil Trans R Soc A 364:2483–2494

Sloane NJA (2008) The on-line encyclopedia of integer sequences. http://www.research.att.com/~njas/sequences/ (accessed 10 November 2008)

Spicher A, Michel O, Giavitto JL (2004) A topological framework for the specification and the simulation of discrete dynamical systems. In: ACRI 2004, Amsterdam, October 2004. LNCS, vol 3305. Springer, Heidelberg, pp 238–247

Steinbock O, Tóth A, Showalter K (1995) Navigating complex labyrinths: optimal paths from chemical waves. Science 267:868–871

Stepney S (2007) Embodiment. In: Flower D, Timmis J (eds) In silico immunology, Springer, New York, chap 12, pp 265–288

Stepney S (2008) The neglected pillar of material computation. Physica D 237(9):1157–1164

Stepney S (2009) Visualising random Boolean network dynamics. In: GECCO 2009, ACM, New York

Stepney S, Polack F, Turner H (2006) Engineering emergence. In: ICECCS 2006, IEEE, Stanford, CA pp 89–97

Strogatz SH (1994) Nonlinear dynamics and chaos. Westview, Boulder, CO

Sutner K (2005) Universality and cellular automata. In: Machines, computations, and universality 2004, Saint Petersburg, September 2004. LNCS, vol 3354. Springer, Heidelberg, pp 50–59

Teman R (1997) Infinite-dimensional dynamical systems in mechanics and physics, 2nd edn. Springer, New York

Tokihiro T, Takahashi D, Matsukidaira J, Satsuma J (1996) From soliton equations to integrable cellular automata through a limiting procedure. Phys Rev Lett 76(18):3247–3250

Tolmachiev D, Adamatzky A (1996) Chemical processor for computation of Voronoi diagram. Adv Mater Opt Electron 6(4):191–196

Toth A, Showalter K (1995) Logic gates in excitable media. J Chem Phys 103:2058–2066

Ulam SM, von Neumann J (1947) On combination of stochastic and deterministic processes. Bull Am Math Soc 53(11):1120, (abstract 403)

Waddington CH (1957) The strategy of the genes. Allen and Unwin, London

Weimar JR, Boon JP (1994) Class of cellular automata for reaction-diffusion systems. Phys Rev E 49 (2):1749–1752

Weyl H (1949) Philosophy of mathematics and natural science. Princeton University Press, Princeton, NJ

Winfree E (2003) DNA computing by self-assembly. The Bridge 33(4):31–38

Wolfram S (1984a) Computation theory of cellular automata. Commun Math Phys 96:15–57

Wolfram S (1984b) Universality and complexity in cellular automata. Physica D 10:1–35

Wolfram S (1985) Twenty problems in the theory of cellular automata. Phys Scr T9:170–183

Wolfram S (1986a) Cellular automata fluids: basic theory. J Stat Phys 45:471–526

Wolfram S (1986b) Random sequence generation by cellular automata. Adv Appl Math 7:123–169

Wolz D, de Oliveira PPB (2008) Very effective evolutionary techniques for searching cellular automata rule spaces. J Cell Autom 3(4):289–312

Wuensche A (2002) Finding gliders in cellular automata. In: Adamatzky A (ed) Collision-based computing. Springer, London, chap 13

Wuensche A, Lesser M (1992) The global dynamics of cellular automata. Addison-Wesley, Reading

Yates FE (ed) (1987) Self-organizing systems: the emergence of order. Plenum, New York

Index

Printed by Publishers' Graphics LLC